Data Science, Machine Intelligence, and Law

Volume 3

Data-driven technologies, such as AI systems and robotics, have the potential to improve our lives, our society and our economy. At the same time, the development and use of these technologies raise unresolved legal and ethical issues. Our new peer reviewed book series brings together works from different legal fields, dealing with the legal, social and economic impact of these technologies, exploring on the one hand the opportunities for law and legal practice, and on the other hand, the ethical and legal frameworks they require. The series especially welcomes book proposals dealing with the legal and/or ethical aspects of data science and the emerging discipline of machine intelligence, which encompasses perception (sensing), AI (planning) and robotics (acting). In this way, our book series aims to contribute to an informed dialogue across the various disciplines about the chances and risks of data-driven systems.

Hans Steege • Ilaria Amelia Caggiano •
Maria Cristina Gaeta • Benjamin von Bodungen
Editors

Autonomous Vehicles and Civil Liability in a Global Perspective

Liability Law Study across the World in relation to SAE J3016 Standard for Driving Automation

Springer

Editors
Hans Steege
Wolfsburg, Germany

Ilaria Amelia Caggiano
Department of Legal and Economic Sciences
Università degli Studi Suor Orsola Benincasa
Naples, Italy

Maria Cristina Gaeta
Department of Legal and Economic Sciences
Università degli Studi Suor Orsola Benincasa
Naples, Italy

Benjamin von Bodungen
Bird & Bird LLP
Frankfurt/Main, Germany

ISSN 2730-5899 ISSN 2730-5902 (electronic)
Data Science, Machine Intelligence, and Law
ISBN 978-3-031-41991-1 ISBN 978-3-031-41992-8 (eBook)
https://doi.org/10.1007/978-3-031-41992-8

This Springer imprint is published by the registered company Springer Nature Switzerland AG
The registered company address is: Gewerbestrasse 11, 6330 Cham, Switzerland

Paper in this product is recyclable.

Preface

Autonomous driving inspires the imagination. Digitalization and connectivity as well as automation of vehicles are continuously increasing and promise to be a disruptive technology. Numerous opportunities are associated with this new form of mobility. The most prominent aspect is Vision Zero, in which no more traffic fatalities occur. But, at the same time, autonomous driving also brings with it entirely new concepts of urban planning, the reclaiming of rural areas and their connectedness to metropolitan mobility infrastructures, developing smart cities, and the reduction of emissions. Finally, this new form of driving harbors the potential for greater sustainability and makes automotive mobility accessible to the elderly and people with disabilities. Of course, all of this also has an impact on the traditional use and perception of vehicles, away from the status symbol and a strong understanding of ownership toward a shared economy.

The development of autonomous driving is far from complete. Along the way, five levels of automated driving are typically identified, based on the SAE J3016 classification. With increasing automation, the human driver is replaced by the vehicle. Eventually, the human driver will be completely removed as a fallback level and tomorrow's vehicle will only carry passengers. It is precisely this increasing automation that raises numerous questions of liability. Who is liable in case of an accident involving an autonomous vehicle? The passengers, the owner, or the operator of such vehicle or fleet of vehicles? In any case, can the manufacturer be held liable? Furthermore, insurance laws vary from one legal system to another, and insurance in favor of the victims of road accidents is not always able to cover cases of accidents caused by autonomous vehicles. If such an insurance scheme exists, it will often—fully or at least partially—compensate the injured parties for their losses. In other cases, the risk of an insurance regulation gap with detrimental consequences for road accident victims is evident, and then a regulatory solution seems to be necessary, in order to protect victims of accidents where an automated vehicle is involved.

This book, born from the network and the collaboration between two research institutions strongly focused on the law of automation in an interdisciplinary perspective (Research Centre of European Private Law—ReCEPL—at Università degli Studi Suor Orsola Benincasa, Naples, Italy, and the Interdisziplinäres Institut für Automatisierte Systeme e.V.—RifaS—Hannover, Germany), aims to respond to the questions posed above, addressing the topic of civil liability for autonomous vehicles, thereby starting from the core issue: who is the tortfeasor in the event of an accident involving an automated or autonomous vehicle?

Civil liability represents a crucial area for the regulation of automation, especially considering its role as incentive for the spreading of innovation through the protection of the public, and the identification of definite rules for all the actors within this ecosystem.

To this extent, we intend to offer a comprehensive overview of the existing models of civil liability for automated and autonomous driving with examples from the five continents, not only to cover the most relevant markets of the automotive industry but also bearing in mind that automated and autonomous vehicles will in all likelihood cross borders and thus may be involved in road accidents abroad.

The comparative analysis provided is complemented by other illustrative and auxiliary material (such as tables and the analysis of questionnaires presented to the authors) to provide a consultation and analytical tool that we believe will be helpful for practitioners, judges, academic scholars, and regulators.

With regard to the analysis of the respective national liability regimes undertaken in this book, two principal scenarios arise.

The first one concerns the case where States rely on general liability concepts (which vary from semi-objective to objective/strict liability) and road traffic rules, or there are some autonomous car-related rules in place. Consequently, there is typically no conclusive legislation available that allows us to designate the tortfeasor in all cases in which an automated car is involved.

In the second scenario, the respective States have enacted specific legislation that generally displaces the traditional liability scheme with the principle of strict liability, whereby the owner or operator of the vehicle and the manufacturer are mainly considered to be the liable parties.

In both scenarios, the need to adapt the system of civil liability in the context of road traffic becomes more than clear, also considering the adequate allocation of risks.

In this light, a country-specific analysis of the allocation of risk between the different stakeholders is carried out in this book and, in many cases, also an initial evaluation of those insurance law principles that are crucial in relation to "autonomous" traffic accidents is provided.

After a comprehensive introduction to the aforementioned issues at the beginning of this book, the technical aspects of the automated and autonomous vehicle are explored first, to better understand the functionality of the specific type of driving technology involved as well as its potential risks. The technology-focused chapter also pays particular attention to the functioning of the human–machine interface (HMI) in order to set the stage for the subsequent analysis of the distribution of liability between man and machine in the jurisdictions examined.

The subsequent chapters are then arranged by continent and country and follow a common structural pattern. Starting from a critical analysis of the state of the art, the chapters strive to achieve two aims:

1. First, to assess whether there is a civil liability regulation in place in each country that serves the purpose of protecting the injured person, in the event of a traffic accident due to a self-driving car. In this context, the allocation of costs of insurance and compensation among the different actors involved will also be considered.
2. Second, to verify whether the legislation in force on civil liability is sufficient to ensure a comprehensive and systematically consistent regulation of autonomous vehicles or, alternatively, it is still necessary to modify the existing regulation or to introduce an *ad hoc* framework of rules for autonomous cars.

More precisely, following the illustration of the liability rules in relation to more conventional vehicles (up to SAE level 2) as well as the determination of the relevant liability addressees and legal requirements to establish their liability, the issue of civil liability for conditionally and highly automated vehicles (SAE levels 3 and 4) is examined. Are there already any existing regulations in this respect? Have new liability subjects been added to the legal regime? Finally, these questions are also explored for fully automated—autonomous—driving at SAE level 5.

If there are special legal regulations for autonomous driving, the individual chapters closely examine how this has changed the previously applicable liability law. In the context of conditional and high automation, special attention is thereby paid to the rights and (remaining) obligations of the vehicle driver. In contrast, in the case of fully automated vehicles, the question of alternative liability addressees to the driver, such as the owner and/or the manufacturer, would seem to be far more urgent.

With this approach, we would like to enable the reader to gain a comprehensive overview of the state of the art on civil liability issues that arise in road traffic across all SAE J3016 levels.

The state of the art is also graphically represented in a map regarding legal regulations around the world, at the beginning of this book. In this political map, countries are divided between those without regulations for (semi-)autonomous vehicles and those which have already modified their existing legal regime or introduced a new one, thereby representing different liability schemes and identifying different liable subjects according to the rules of a specific legal system.

An analysis of practical cases relating to the allocation of responsibility for autonomous vehicle damages in the selected legal systems is included in the respective chapters, along with an illustration of the current state of the legislation, judicial decisions, and literature relating to road accidents caused by (semi-) autonomous vehicles.

One final note to the methodology of this volume is worth making here: We have tried to focus not only on the legal analysis of existing laws and their operation in practice and thus to describe the current state of the art. Instead, especially in view of the technical insights provided at the beginning of the book, we attempt to provide an interdisciplinary view, which can help the reader to better understand the technical, social, and economic implications of self-driving vehicles and consequently address them through legal rules and their application. What is more, the insights gained from the evaluation of the questionnaire provided to all authors are based on an empirical legal studies methodology.

This exciting project would not have been possible without the support of Springer and our dedicated authors. We would like to express our sincere gratitude for their ongoing commitment.

We very much hope that you will enjoy reading this book and look forward to receiving your feedback, critique, and suggestions for improvement.

Wolfsburg, Germany — Hans Steege
Naples, Italy — Ilaria Amelia Caggiano
Naples, Italy — Maria Cristina Gaeta
Frankfurt/Main, Germany — Benjamin von Bodungen
April 2023

Contents

Introduction

Hans Steege, Ilaria Amelia Caggiano, Maria Cristina Gaeta, and Benjamin von Bodungen

1 General Overview on Autonomous and Automated Vehicles: Main Issues for a Global Perspective and the Scope of *the* Book[1]

Since the 1990s, when the first adaptive control system was introduced, there has been an increase in automatization of vehicles that progressively utilize more advanced driver assistance technologies and require less and less human

The author of Sects. 1 and 5 is Prof. Ilaria Amelia Caggiano, Full professor of Private law and Vice-director of the Research Centre of European Private Law (ReCEPL), Università degli Studi Suor Orsola Benincasa. The author of Sects. 2, 3, 4.2 is Dr. Maria Cristina Gaeta, Appointed Lecturer in Private Law and Scientific Secretary of the Research Centre of European Private Law (ReCEPL), Università degli Studi Suor Orsola Benincasa. The authors of Sect. 4.1 are Prof. Dr. Benjamin von Bodungen and Dr. Dr. Hans Steege.

[1] The notion of autonomy in relation to vehicles has not been uniformly defined by law. Autonomy means the ability to make decisions (the quality or state of being self-governing, according to the Merriam Webster Dictionary) and does not correspond to automation, which is a technical concept and according to the classification adopted in this book can be of different degrees (*i.e.* partial, high, full). Therefore, on the basis of some legislative choices (see, *e.g.* the chapter

H. Steege
Wolfsburg, Germany

I. A. Caggiano (✉) · M. C. Gaeta
Department of Legal and Economic Sciences, Università degli Studi Suor Orsola Benincasa, Naples, Italy
e-mail: ilaria.caggiano@unisob.na.it; mariacristina.gaeta@unisob.na.it

B. von Bodungen
Bird and Bird LLP, Frankfurt/Main, Germany
e-mail: benjamin.von.bodungen@twobirds.com

H. Steege et al. (eds.), *Autonomous Vehicles and Civil Liability in a Global Perspective*, Data Science, Machine Intelligence, and Law 3,
https://doi.org/10.1007/978-3-031-41992-8_1

intervention. This process will be completed by putting into circulation fully autonomous vehicles (*i.e.* self-driving cars), which operate with no human input, but rather by means of automated driving functions which have been trained by machine learning methods that are often regarded as Artificial Intelligence (AI).[2] This final step of vehicle automation is expected to generate even higher levels of safety and a reduction of accidents, as well as economic growth (*e.g.* for the construction of infrastructures and more sophisticated vehicles), rationalization of traffic flows, and a new model of mobility that can be *per se* beneficial also to social and environmental sustainability (*e.g.* by ridding historic city centers of unused vehicles).[3]

The externalities identified seem to far exceed the concurrent social, ethical, and legal costs of automatization in terms of wealth re-distribution also in the labour market, ethical choices taken by automated systems, risks for the rights of individuals due to the pervasive processing of their data by a high number of controllers,[4] and, finally, administrative costs for changing existing laws.

Therefore, there is an amplified and integrated attention by supranational institutions, contracting States,[5] and individual States on the regulation of this domain.

---

on Germany) and some other evaluation (see, *e.g.* the chapter on Austria we have referred to autonomy as a condition stemming from full automation, enabling the self-driving function. However, focusing on the sole concept of autonomy, there can be different degrees of vehicle's autonomy, referring to functions different from driving. To this extent, we also use the binomial (fully) autonomous / semi-autonomous in this introduction and the questionnaire set out therein.

[2]For a complete overview on the evolution of autonomous/self-driving cars with particular regard to liability rules, see Gaeta (2019b). For the level of automation used in the book, *see* the final part of this section.

[3]Calabresi and Al Mureden (2021). See in this book, C. Pek, S. van Waveren, Autonomous Vehicles – A Technical Introduction.

[4]For an extensive analysis of the issues concerning automated cars, as connected vehicles transferring data, see the European Data Protection Board, Guidelines 01/2020 on processing personal data in the context of connected vehicles and mobility related applications, version 2.0, 2021; European Data Protection Board, Guidelines 02/2021 on virtual voice assistants. Version 2.0, 2021. *See* also, European Data Protection Supervisor, 'TechDispatch #3: Connected Cars', 2019. On the European common space on mobility, see the Communication from the Commission to the European Parliament, the Council, the European Economic and social Committee and the Committee of the Regions, 'A European strategy for data', COM(2020) 66 final. On data protection and self-driving cars, *see* also Gaeta (2019a), pp. 15–23; Gaeta (2018), pp. 147–179.

[5]*E.g. see* the recent revised EU General Safety Regulation, in force from July 2022, which gradually introduces advanced safety systems in new motor vehicles, as intelligent speed assistance, alcohol interlock installation facilitation, driver drowsiness and attention warning, advanced driver distraction warning, emergency stop signal, reversing detection, and an Event Data Recorder, with the objectives of ensuring the proper functioning of the internal market and in order to offer a high level of safety, the protection of vulnerable road users, and environmental performance. *See* Regulation (EU) 2019/2144 of the European Parliament and of the Council of 27 November 2019. See, also the amendment to the UN 1968 Convention on Road Traffic (signed in Vienna), at article 1 and 34 *bis* introducing the definitions of 'automated driving system' and 'dynamic control', and that of moving vehicle and combination of vehicles, encompassing vehicles using an automated driving system, which enable Contracting States to enact legislation of automated vehicles without violating the Vienna Convention.

Automated and fully autonomous vehicles are products, devices or combinations of devices processing personal data, and subject-matter of a number of transactions, including consumer transactions. The most relevant feature is that they embed algorithms trained with Machine Learning (ML) and Deep Learning (DL), which are subsets of AI. On the basis of these functions, automated and autonomous vehicles are able to autonomously interact with their environment and decide how to act/function, on the basis of their ML and DL training.[6] On a general level, these characteristics have posed new ethical questions (concerning both the testing phase of autonomous vehicles and the phase of placing them on the market), as well as copyright issues,[7] and evoke the urgency to regulate the market, in view of the risks associated with different sectors, by imposing requirements and obligations on AI developers, providers and operators (as in the EU forefront proposals).[8]

[6]For a definition of AI, see the High-Level Expert Group on Artificial Intelligence, Ethics Guidelines for trustworthy AI, 8 April 2019, according to which (p. 36): '(AI) systems are software (and possibly also hardware) systems designed by humans that, given a complex goal, act in the physical or digital dimension by perceiving their environment through data acquisition, interpreting the collected structured or unstructured data, reasoning on the knowledge, or processing the information, derived from this data and deciding the best action(s) to take to achieve the given goal. AI systems can either use symbolic rules or learn a numeric model, and they can also adapt their behaviour by analysing how the environment is affected by their previous actions'. According to the Artificial Intelligence Act proposal, art. 3 a(1) of the AI Act, AI is defined as: a "software that is developed with one or more of the techniques and approaches listed in Annex I and can, for a given set of human-defined objectives, generate outputs such as content, predictions, recommendations, or decisions influencing the environments they interact with". However, according to the Amendments of the European Parliament, adopted on June 14, 2023, to the proposal for a regulation of the European Parliament and of the Council laying down harmonized rules on artificial intelligence (Artificial Intelligence Act), and especially art. 3 (1), the definition of Artificial Intelligence System has been proposed more generally as it follows: *"a machine-based system that is designed to operate with varying levels of autonomy and that can, for explicit or implicit objectives, generate outputs such as predictions, recommendations, or decisions, that influence physical or virtual environments"*. It must be pointed out that in the field of automated vehicles, there is no self-learning function of each automated vehicle. Instead, algorithms incorporated in vehicles are those trained with ML and DL methods, before receiving type approval.

[7]*See, e.g.*, European Parliament resolution of 20 October 2020 on a framework of ethical aspects of artificial intelligence, robotics and related technologies, 2020/2012(INL); Horizon 2020 Commission Expert Group to Advise on Specific Ethical issues raised by driverless mobility Report, Ethics of Connected and Automated Vehicles: recommendations on road safety, privacy, fairness, explainability and responsibility, 2020; European Parliament resolution of 20 October 2020 on intellectual property rights for the development of artificial intelligence technologies, 2020/2015 (INI).

[8]*See, e.g.*, European Parliament and Council, Proposal for a regulation on laying down harmonised rules on artificial intelligence (artificial intelligence act) and amending certain union legislative acts, 21.04.2021, COM (2021) 206 final.

AI impacts the regime of liability in case of damages caused by products embedding it, as emerging from the European regulatory proposals,[9] even leading to the idea of recognizing them as legal persons, not products anymore.[10]

As far as autonomous vehicles (partially or fully) are concerned, the intensifying application of AI to the driving-related functioning of such vehicles means that they represent "*a significant potential to cause harm to one or more persons, in a manner that is random and goes beyond what can reasonably be expected*", where "*the significance of the potential depends on the interplay between the severity of possible harm, the likelihood that the risk causes harm or damage and the manner in which the AI-system is being used*".[11]

AI operates differently in autonomous and automated vehicles, according to the degree of automation applied (partial, high, full), and can affect the entire driving function or merely one or a limited number of driving-related functions.

Moreover, with respect to the use of data, a data breach or a cybersecurity breach of the Advanced Driver Assistance Systems (ADAS)[12] can occur, as well as a malfunctioning of the driving software.[13]

The crucial question addressed in this book concerns the identification of the tortfeasor (the subject responsible) and the liability rules for damages resulting from "vehicles crashes" when at least one autonomous or partially autonomous (*i.e.* automated or semi-autonomous) vehicle is involved in a harmful event connected

---

[9] Along with a European legal framework for AI to address fundamental rights and safety risks specific to the AI systems (see, above footnote n. 7), the EU Commission has proposed other inter-related legal initiatives to contribute to building trustworthy AI; this includes the proposed Directive on adapting non-contractual civil liability rules to artificial intelligence (AI Liability Directive) and that on regarding liability for defective products. For references, *see* below footn. 20 and footn. 21.

[10] The idea of an "Electronic Personhood" for robots has been advanced, in addition to a number of scholars, by the European Parliament resolution of 16 February 2017 with recommendations to the Commission on Civil Law Rules on Robotics (2015/2103(INL)) on Civil Law Rules on Robotics, where the Parliament: 'Calls on the Commission, when carrying out an impact assessment of its future legislative instrument, to explore, analyse and consider the implications of all possible legal solutions, such as: [. . .] f) creating a specific legal status for robots in the long run, so that at least the most sophisticated autonomous robots could be established as having the status of electronic persons responsible for making good any damage they may cause, and possibly applying electronic personality to cases where robots make autonomous decisions or otherwise interact with third parties independently'. However, this legislative choice cannot be considered as a necessity for answering issues of liability and damages caused by robots, which can be addressed by other tools (alternatively or cumulatively, insurance and separate funds for indemnification) referred to things (products).

[11] See the definition of high-risk artificial intelligence under sec. 15 of the European Parliament resolution of 20 October 2020 with recommendations to the Commission on a civil liability regime for artificial intelligence (2020/2014 (INL)).

[12] See, European Union Agency for Cybersecurity (ENISA), Cybersecurity challenges in the uptake of artificial intelligence in autonomous driving, EU30568EN, 2021.

[13] They are those technologies that collect data on the performance of the car and on the space-time context of its circulation, informing the driver and reaching up to make suggestions to the driver or, even, to take partial or total control of the vehicle.

with its driving function. Who should be considered the tortfeasor in these cases is the subject of debate across the world and one of the most intriguing questions in the field of civil and traffic liability.

As stated above, while, on the one hand, automation in vehicles is expected to reduce the number of accidents on average, on the other hand, AI functioning may cause unexpected harm or, the existing liability regimes may undercompensate or not compensate the injured party at all.[14]

Uncertainties in compensating or indemnifying the loss may stem from various reasons: difficulty in identifying the cause of the harm because of the complexity and opacity of AI systems, flawed data (either collected by the vehicles itself or other providers) that result in an incorrect decision of the vehicle, error of the device both in case of the decision made by a human or the vehicle,[15] and unexpected technical defects.

In the face of any of these occurrences, the legal response may vary according to the tort law system adopted by the relevant State, which may have opted for a fault-based liability or a strict liability regime.

Yet, if one considers that automatization reduces the level of human control over singular acts, it would seem appropriate then to apply a strict liability model in this context.

Having said this, one must still bear in mind that even if the relevant legal system adopts a no fault and rather risk-based liability approach, in which the distribution of risk facilitates compensation of the claimant (the victim of the road accident), it might be difficult:

1. to identify the subject of liability (the owner of the car, the producer, and in turn the producer of the application/AI system involved), and
2. to trace and detect the cause/causes of the accident in an environment as complex as that of autonomous and semi-autonomous vehicles, where also human error may intervene.

Therefore, legal systems need to consider these aspects arising in practice, when they regulate civil liability regarding semi-autonomous or autonomous vehicles. Obviously, regulation of civil liability can be addressed in general civil law provisions, special acts, or traffic law provisions where road traffic offenses are regulated as well.

Finally, automatization requires adjustments in insurance regimes, and it appears difficult to evaluate how the insurance policies should be developed.[16]

---

[14] See, for a complete reference to these aspects, Karner et al. (2021). European Union, where a specific use case is dedicated to Autonomous vehicles (p. 64 ff.).

[15] Karner et al. (2021), p. 9.

[16] See European Parliament in-depth analysis, Artificial Intelligence in road transport, PE 654.212, January 2021; European Commission Report, The safety and liability implications of Artificial Intelligence, the Internet of Things and robotics, 19 February 2020, COM 2020/64 final; European Parliament Study, A common EU approach to liability rules and insurance for connected and

At the same time, it is well understood that as liability models differ, so do standards and burden of proof in each jurisdiction, and, again, also these differences have an impact on the gaining of compensation by the claimant.[17]

Therefore, it is well clear that providing an overview, from a liability law perspective, of the current regulation of autonomous and automated driving is only a part of a manifold phenomenon.[18] However, the analysis, with respect to certain selected jurisdictions, of the relevant rules governing liability for accidents involving autonomous vehicles and those with varying levels of automation, is the indispensable basis for a comparative view of the legal framework at national and international level, which can contribute to the legal discussion on this topic.

In order to clearly distinguish the different levels of vehicle automation, the SAE International standard J3016 (firstly published in 2014 and then lastly amended in 2021) has been selected as reference parameter. It is the classification typically applied in this context and defines six different automation levels. The automation levels identified by the SAE international standard are based on HMI analysis. In more detail, they relate to the degree of the driver's involvement in driving activities, which decreases proportionally as vehicle automation increases. The SAE automation levels are defined as follows: (L0) no automation, (L1) driver assistance, (L2) partial automation, (L3) conditional automation, (L4) high automation, and (L5) full automation. An in-depth analysis of the various stages of vehicle automation, as well as the related SAE standard is presented in the technical chapter which precede the legal analysis.

This introductory chapter is structured as follows: First, a summary on the state of the art of the legislation at the international level and in the European Union, to which a large number of jurisdictions examined in this book belong, is provided. This is followed by an overview of the regulation of civil liability for accidents caused by autonomous vehicles in each country considered, based on the questionnaire compiled by the authors. Next, two maps are provided that give a global overview of these results. Finally, the authors offer a number of conclusions.

---

autonomous vehicles, PE 615.635, February 2018; European Parliament Report, Autonomous driving in European transport (2018/2089(INI)), 5 December 2018.

[17] Karner et al. (2021), p. 10, where "*the claimant has to prove that an AI System caused her loss, and evidence only supports a 51% likehood that this was indeed the case, she will win the case in full* [...] in those jurisdictions that are satisfied with a mere preponderance of evidence, whilst will get nothing in those that ask for a higher standard of proof (casual link).

[18] The global regulatory framework for automated vehicles reaches much further and includes traffic law, criminal law and insurance law. A more complete scenario including these aspects has been depicted in the book as complementary, for some legal systems (*e.g.* Chapter on French law), in addition to the liability aspects dealt with in all chapters.

2 State of the Art of the Legal Regulation on Autonomous Vehicles at International and European Level

2.1 International Soft Law Regulation on Autonomous Vehicles as Smart Robots

Self-driving cars can be conceptualized as robotic vehicles of varying levels of automation (from SAE Level 1 to 5). At least at the higher levels of automation, self-driving cars are equipped with AI. As a consequence, the regulation of AI is applicable in the present context.

So far, no binding regulation for AI exists but, as will be seen, a proposal for the regulation of AI (AI Act or AIA),[19] as well as a proposal for the AI Liability Directive (AILD),[20] has been put forward after several interventions by international and European institutions. Furthermore, the proposal for the revision of the PLD directive[21] is very relevant with regard to smart products (including autonomous vehicles).

However, even if an initial regulation on artificial intelligence is likely to be approved in the near future, this may not be sufficient to ensure an adequate regulatory framework for autonomous vehicles. Indeed, a general law on AI would not necessarily be sufficient for the regulation of a specific type of AI embedded in robotics, such as self-driving cars. Traditionally, there has always been special traffic and insurance regulation in place concerning the operation of conventional vehicles, to complement the legislation on the driver's liability, which suggests that this is likely to be advisable in the case of self-driving vehicles as well. More precisely, concerning road traffic, the Vienna Convention[22] is a very relevant international treaty that, since 1968, establishes principles to govern and harmonize traffic laws at international level. One of the fundamental principles of this Convention has traditionally been the requirement that every vehicle must have a driver who is fully in control and responsible for the behaviour of the vehicle at all times. In 2016, a reform of the Convention opened up the possibility of satisfying the driver-related requirement of permanent vehicle control also in the context of automated driving, for ratifying countries.[23] Then, in January 2021, amendments to Article 1 and a new

[19] European Parliament and Council, Proposal for a regulation on laying down harmonised rules on artificial intelligence (artificial intelligence act) and amending certain union legislative acts, 21 April 2021, COM(2021)206 final (revised 3 November 2022).

[20] European Parliament and Council, Proposal for a Directive on adapting non-contractual civil liability rules to artificial intelligence (AI Liability Directive), 28. September 2022, COM(2022) 496 final.

[21] European Parliament and Council, Proposal for a Directive on liability for defective products, 28 September 2022, COM(2022) 495 final.

[22] Vienna Convention on Road Traffic, 9 November 1968.

[23] On 23.03.2016, the amendments to the Vienna Convention on Road Traffic of 1968 entered in force (in particular, to articles 8 and 39) that authorized the use of automated driving technologies

Article 34 *bis* to the Vienna Convention were proposed[24] that, specifically for the operation of vehicles using automated driving systems, waived the requirement that every moving vehicle must have a driver. These amendments entered into force on 14 July 2022.[25]

At international level, as well as at the European one, Institutions are called to intervene to guarantee the protection of the human being in relation to potential threats flowing from new technologies. Indeed, the technology is not neutral and, in addition to entailing a series of advantages, it also implies continuous risks for individuals who have a structural condition of vulnerability in relation to them. Therefore, new technology, especially considered in its more current manifestation of AI, must be properly regulated.

The International Institutions, the European Union and its Member States have ever since been founded on the values of human dignity, equality, solidarity, freedoms, and respect for human rights, but today, specific ethical and legal principles on automation and AI are indispensable in order to put these new technologies at the service of their human users.

At all levels of regulation, the awareness of the need for a concrete strategy and regulation of AI has matured.

In particular, a fundamental study on this topic is represented by the UNESCO - COMEST, Preliminary Study on the Ethics of Artificial Intelligence, 26.02.2019, published two years after the Report on the Ethics of Robotics of 2017. Furthermore, in May 2019, the Council of Europe adopted some recommendations aimed at maximizing the potential of AI systems and preventing a negative impact on human rights in the public and private sector (Unboxing artificial intelligence: 10 steps to protect human rights, 2019).

In the same direction, the Council of Europe appointed an *ad hoc* committee for Artificial Intelligence (CAHAI), in September 2019, that aims to assess the impact of AI on the individual and the society, as well as on existing soft law and hard law instruments that deal with AI. Subsequently, in December 2020 the CAHAI published a feasibility study that examines the potential elements of a legal framework for the development, design and application of AI, based on Council of Europe standards in the field of human rights, democracy and the rule of law.

transferring driving tasks to the vehicle, provided that these technologies are in compliance with the United Nations vehicle regulations (so-called UNECE-Regulations) or can be overridden or switched off by the driver. The amendment is available here: https://unece.org/DAM/trans/doc/2014/wp1/ECE-TRANS-WP1-145e.pdf

[24] UNECE, Proposal of amendment to Article 1 and new Article 34 bis to the convention,14 January2021 (Retrieved 6 April 2022) available here: https://treaties.un.org/doc/Publication/CN/2021/CN.5.2021.Reissued.15012021-Eng.pdf

[25] UNECE, Acceptance of amendment to Article 1 and new Article 34 bis to the convention", 14 January 2022 (Retrieved 6 April 2022), available here: https://treaties.un.org/doc/Publication/CN/2022/CN.26.2022-Eng.pdf

2.2 *Autonomous Vehicles Regulation in Europe*

At the European level, the main hard law regulation that can be applicable is the so-called Product liability directive or PLD (Directive 85/374/EEC,[26] whose need for revision has been intensively evaluated and which can now be expected to be soon replaced with a revised PLD.[27] The current draft provides for a new definition of 'products' in the modern digital and circular economy, and also addresses the problems faced by claimants in meeting their burden of proof as well as the limits to making compensation claims. Indeed, article 2 of the PLD defines as a product every movable, even if it is incorporated into another movable or immovable, and article 4 of the revised PLD Proposal adds that 'Product' includes digital manufacturing

[26] Council Directive 85/374/EEC of 25 July 1985 on the approximation of the laws, regulations and administrative provisions of the Member States concerning liability for defective products [1985] OJ L 210/29, that will be referred to as PLD.

[27] In relation to the PLD, it is important to underline that before the proposed PLD revision of 2022, in 2018 the European Commission intervened with the last quinquennial report on the PLD (European Commission Report to the European parliament, the Council and the European economic and social committee on the application of the Council Directive on the approximation of the laws, regulations, and administrative provisions of the Member States concerning liability for defective products (85/374/EEC), 7 May 2018, COM(2018)246 final), which was complemented by a formal evaluation of the directive (European Commission Staff Working Document on the evaluation of Council Directive 85/374/EEC of 25 July 1985 on the approximation of the laws, regulations and administrative provisions of the Member States concerning liability for defective products SWD (2018), 7 May 2018, 157 final). The quinquennial report covered the period from 2011 to 2015, while the formal evaluation aimed to assess the «functioning and the performance» of the PLD for the period 2000–2016. In the quinquennial report, the EU Commission affirmed that many products available on the market today have characteristics that were considered science fiction in the past century and our society has to face new challenges related to digitisation, the Internet of Things (IoT), Artificial Intelligence (AI) (included Robotics), and cybersecurity. With specific reference to AI, the report of the European Commission defined it as one of the most important technologies of the twenty-first century and focused on maximising the benefits of AI. In this context, product safety and liability are fundamental aspects in agreeing on a policy response that enables European societies, businesses, and consumers to benefit from AI. In compliance with the EU Commission formal evaluation of the PLD, an evaluation of the PLD has been carried out to assess its performance, as the directive has never been evaluated since its entry into force and, during these more than thirty years, completely new products have been designed and fabricated due to the development of new technologies. Hence, it has become necessary to verify whether the PLD is really effective, efficient, coherent and relevant and has EU added value, being able to regulate also "smart products", i.e. automated products or even products equipped with AI. Moreover, as part of the evaluation, the Commission launched a public consultation on the evaluation of Directive 85/374/EEC in January 2017. The objective of this consultation was to collect stakeholders' feedback on the application and performance of the PLD, including how this relates to challenges raised by new technological developments. In the Commission's view, the results of the formal evaluation and of the public consultation showed that the PLD (text in force) continued to be an adequate tool and to meet its objectives today, also in the light of new technological developments, but the literature and the following studies of the EU Authorities seem to demonstrate the opposite and the proposal for the revision of the PLD underlines this need for amendment.

files and software; therefore, both motor vehicles and the automatic devices installed on them fall within this category of product.

A product is to be considered defective under article 6 of the PLD if it does not offer the security that a person is entitled to expect taking into account all circumstances, including the presentation of the product, the reasonable use of the product, and the time when the product was put into circulation. The revised PLD adds to this listing the effect on the product of any ability to continue to learn after deployment, as well as the effect on the product of other products that can reasonably be expected to be used together with the product (as added by art. 6, PLD revision). In future, the assessment of product defectiveness, which is to be carried out on a case-by-case basis, will also have to take into account the latter parameters which will most likely be relevant for self-driving vehicles.

As can be seen from the definition of a defective product, the concept of defect is strictly connected to the concepts of safety of the product user and third parties affected by the product. Therefore, the definition of a safe product contained in the general product safety regulation (art. 2, lit. b) Directive 2001/95/EC)[28] can be considered as an additional reference parameter, in order to assess the existence of a defect. On the basis of this provision, the product is safe if, under normal or reasonably foreseeable conditions, it presents no risk or only the minimum risks compatible with the use of the product and considered acceptable and consistent with a high level of protection for health and safety of persons.[29] Therefore the safety of the product is defined as the general absence of risks even if the existence of some minimum risks may be acceptable.[30] On the basis of this general statement, the risks the consumer is exposed to must be quantitatively determined (which in any case may be minimal risks only).[31]

Having defined the self-driving car as a product, Directive 771/2019/EU[32] has to be at least mentioned here. It regulates the sales contracts of 'goods with digital elements' concluded between sellers and consumers and is applicable also to any

[28] Directive of the European Parliament and of the Council 2001/95/EC of 3 December 2001 on general product safety [2001] OJ L11/4.

[29] The requirements that a safe product must satisfy are listed in art. 3, Directive 2001/95/EC.

[30] As the liability of the producer is closely connected to the concept of safety, it is important to ascertain whether safety standards represent a minimum or a maximum threshold of producer's liability. In general, safety standards are intended as minimal requirements, which merely allow the distribution of the product on the market. Indeed, as settled in the article 7, lit. d), PLD, liability is excluded only if the damage occurred because of the specific feature set by a legal rule. Furthermore, the existence of a higher safety standard, which has actually been achieved by other producers (so-called alternative design), may be adequate to trigger the producer's liability even though he had respected the safety standards. On safety standards, see Al Mureden (2019), p. 895 ff.; Al Mureden and Ruffolo (2019), p. 1704 ff...

[31] For further information on the consumer expectation, see Palmerini and Bertolini (2014), p. 167 ff.

[32] Directive 771/2019/EU of 20 May 2019 of the European Parliament and of the Council on certain aspects concerning contracts for the sale of goods, amending Regulation (EU) 2017/2394 and Directive 2009/22/EC, and repealing Directive 1999/44/EC [2019] OJ L 136/28.

tangible movable items that incorporate or are inter-connected with digital content or digital service in a functional way (art.1, para 5, lit. b)). On the basis of this definition, it would seem that the directive could also be applied to single ADAS or the entire self-driving car.

After identifying the relevant hard law on the topic, it appears very relevant to shed some light on the recent European initiatives in the field of Robotics and AI, with particular regard to self-driving cars, as they are robots that according to the respective level of automation can be equipped with AI. Starting in 2017, various initiatives have been launched at the European level in the field of Robotics, AI and, more specifically, self-driving cars.[33]

Our analysis starts with the European Parliament Resolution, that already in 2017[34] called on the European Commission to submit a directive proposal for civil law rules on Robotics and AI (which also includes self-driving cars), as well as non-legislative acts (such as guidelines and codes of ethical conduct). The purpose of the EU resolution was to address the main issues foreseeable in the next 10–15 years, taking into account the Charter on Robotics attached to the Resolution. In addition, the European Parliament considers that the automotive sector is in most urgent need of efficient European Union and global rules, in order to ensure the cross-border development of self-driving cars, the exploitation of their economic potential and the benefits from the technology. In the same vein, the Declaration of Amsterdam of 14 and 15 April 2016[35] on Cooperation in the field of connected and automated driving highlighted the need to develop and maintain a joint program with other European countries, in order to support these goals and to remedy the problems arising from the development of this new type of driving.

Few years later, the EU Parliament revisited the topic of regulation, highlighting the need to resolve legal issues related to Robotics and AI in general, which, again, also include AI-enabled vehicles (European Parliament resolution on a comprehensive European industrial policy on AI and Robotics of 2019),[36] as well as ethical

[33] For the analysis of the recent developments of the European Institutions, please refer to Bertolini A (2020) Artificial Intelligence and Civil Liability. Study for the Policy Department for Citizens' Rights and Constitutional Affairs Directorate-General for Internal Policies. European Union, PE 621.926. The study was commissioned by the Policy Department C at the request of the Committee on Legal Affairs and is focused on the definition and liability issues related to AI. For an analysis of the European state of the art on product liability in relation to new technologies see also: Machnikowski (2016); Amidei (2019), p. 1715 ff.

[34] European Parliament resolution of 16 February 2017 with recommendations to the Commission on Civil Law Rules on Robotics (2015/2103(INL)).

[35] Declaration of Amsterdam of 14 and 15 April 2016, Cooperation in the field of connected and automated driving.

[36] European Parliament resolution of 12 February 2019 on a comprehensive European industrial policy on artificial intelligence and robotics (2018/2088(INI)). See also European Parliament Report of 5 December 2018 on autonomous driving in European transport (2018/2089(INI)). On the topic of intelligent transport systems, first of all see Directive 2010/40/EU of the European Parliament and of the Council of 7 July 2010 on the framework for the deployment of Intelligent Transport Systems in the field of road transport and for interfaces with other modes of transport [2010] OJ L

ones (European Parliament resolution on a framework of ethical aspects of AI, Robotics and related technologies of 2020).[37]

Following a request of the European Parliament, the European Commission published a Communication on AI for Europe[38] in April 2018 and, subsequently in December 2018 the Coordinated Plan on Artificial Intelligence.[39] In these two documents, the EU Commission highlights the importance of AI, stating that AI is undoubtedly one of the most strategic technologies of our century and that the way we approach AI will define the world we live in. In the same vein, in April 2019 the European Commission published the Communication on Building Trust in Human-Centric AI,[40] that puts the human being at the centre of studies on AI, assuming that AI should be at the service of its human user who should trust it. After less than one year, in February 2020, the EU Commission published the White Paper on AI[41] and the 'Report on the safety and liability implications of AI, the IoT and Robotics',[42] that sets the broader implications for a liability and safety framework in the field of AI, IoT and Robotics.

Furthermore, in 2018 the Commission appointed the High-Level Expert Group on Artificial Intelligence (AI HLEG), which was tasked with supporting the implementation of the European Strategy on Artificial Intelligence. Since its appointment, the AI HLEG has published the 'Ethics Guidelines for trustworthy AI',[43] a document on the definition of AI,[44] the 'Policy and investment recommendation for

207, welcoming the communication from the Commission and taking into account the legislative initiatives relating to intelligent transport systems (ITS and C-ITS). We also underline that the Parliament urges the Commission to present a data and cybersecurity strategy, as set out in European Parliament resolution of 13 March 2018 on a European strategy on Cooperative Intelligent Transport Systems (2017/2067(INI)).

[37] European Parliament resolution of 20 October 2020 on a Framework of ethical aspects of artificial intelligence, robotics and related technologies, (2020/2012(INL)).

[38] Communication from the Commission to the European Parliament, the European Council, the Council, the European Economic and social committee and the Committee of the regions, on Artificial Intelligence for Europe, 25 April 2018, COM(2018) 237 final.

[39] Communication from the commission to the European Parliament, the European Council, the Council, the European Economic and Social Committee and the Committee of the Regions, Coordinated plan on Artificial Intelligence, 7 December 2018, COM(2018) 795 final.

[40] Communication from the commission to the European Parliament, the Council, the European Economic and Social Committee and the Committee of the Regions, Building Trust in Human-Centric Artificial Intelligence, 8 April 2019, COM(2019) 168 final.

[41] White Paper of the European Commission, on Artificial Intelligence - A European approach to excellence and trust, 19 February 2020, COM(2020) 65 final.

[42] Report from the Commission to the European Parliament, the Council and the European Economic and Social Committee, Report on the safety and liability implications of Artificial Intelligence, the Internet of Things and robotics, 19 February 2020, COM(2020) 64 final.

[43] High-Level Expert Group on Artificial Intelligence, Ethics Guidelines for trustworthy AI, 8 April 2019.

[44] High-Level Expert Group on Artificial Intelligence, A definition on AI: Main capabilities and Disciplines, 8 April 2019.

trustworthy AI',[45] and the 'Assessment List for Trustworthy AI (ALTAI)',[46] all mainly focused on how technologies, specifically AI, can be trustworthy and can be put at the service of their human user.

As a matter of fact, in the field of connected and automated driving (CAD), the European Union trend in the recent past has clearly been to invest in the autonomous driving sector.[47] Indeed, in May 2018, the European Commission presented a European strategy for connected and automated mobility, which outlines a common approach between the European Union, the Member States and the stakeholders, in order to make Europe a world leader in the use of connected and automated mobility, bringing about a huge change in terms of reducing the number of victims of road accidents, harmful emissions from transport and traffic congestion.[48]

In the same period, the EU Parliament addressed the topic of civil liability with specific reference to autonomous vehicles. As a matter of fact, in February 2018, the EU Parliament published a study on a common EU approach to liability rules and insurance for connected and autonomous vehicles,[49] that underlines the necessity to revise the current legislative EU framework for liability rules and insurance for connected and autonomous vehicles. The reason for this study was to provide an EU policy response to current gaps and limitations of the existing regulations, in order to ensure legal coherence and better protection of the consumers' rights, as well as to generate economic added value, considering the positive economic impact of self-driving cars.

Then, the same Parliament published a second study on Civil liability, but for artificial intelligence in general, in September 2020.[50]

Specifically with regard to product liability, in 2019 the EU Commission also set up an expert group on liability and new technologies, which has two formations: (1) The 'product liability formation', that was established to assist the Commission in drawing up guidance on the directive, and (2) the 'new technologies formation', that was tasked with assessing the implications of emerging digital technologies for the wider liability frameworks at EU and national level.

---

[45] High-Level Expert Group on Artificial Intelligence, Policy and Investment Recommendation for Trustworthy AI, 26 June 2019.

[46] High-Level Expert Group on Artificial Intelligence, the Assessment List for Trustworthy Artificial Intelligence (ALTAI), 17 July 2020.

[47] Gatt (2020), p. 1 et seq. Professor Gatt highlights, indeed, that we can see how both European bodies are deeply aware of the need to create a framework of rules for the development of an anthropocentric AI in the sense of protecting and empowering the human being in relation to new technologies.

[48] Communication from the commission to the European Parliament, the Council, the European Economic and Social Committee, the Committee of the Regions, On the road to automated mobility: An EU strategy for mobility of the future, 17 May 2018, COM(2018) 283 final.

[49] Study of the European Parliament, A common EU approach to liability rules and insurance for connected and autonomous vehicles, PE 615.635, February 2018.

[50] Study of the European Parliament, Civil liability regime for artificial intelligence, PE 654.178, September 2020.

Soon after, the expert group on liability and new technologies published its 'Report on liability for artificial intelligence.'[51] Starting from the application of the existing European laws to emerging digital technologies and in line with what had already been stated by the EU Parliament, the report points out the necessity to amend the existing liability regimes, with the aim of providing specific regulations for emerging digital technologies rather than a general law for new technologies. This is thought to be the most suitable way of reflecting the diversity of such technologies that entail a diverse range of potential risks.

In June 2020, an independent expert group published a report on 'Ethics of connected and automated vehicles (CAVs)',[52] which aims to support the stakeholders in addressing ethical issues in the development and regulation of CAVs. The ethical aspects explored in the report also include that of the electronic personhood of robots.

Among the twenty recommendations, the fourth suggestion provides that it is important to consider the revision of traffic rules in order to promote safety of CAVs, and recommendation twenty pushes for the creation of fair and effective rules for granting compensation to victims of CAVs' crashes.

Finally, after years of focusing on soft law intervention, the European Commission started to work on hard law draft regulations: first of all the mentioned Proposal for the AI Act (AIA) in 2021 and subsequently the AILD proposal, as well as the PLD revision proposal in 2022. The AI Act proposal divides AI technologies into high-risk, low-risk and prohibited AI systems. Based on the level of risk, the AIA provides for more or less strict rules that producers must respect. The proposed AILD, on the other hand, aims to introduce uniform rules for access to information and alleviate the burden of proof in relation to damage caused by AI systems, establishing broader protections for victims (whether consumers or businesses) and promoting the AI sector through increased guarantees.

In general terms, autonomous vehicles seem to be considered as high-risk AI systems, but will not be directly regulated by the AI Act proposal. Indeed, Article 2, para. 2 of the AI Act also in the version approved by the EU Parliament in June 2023, provides that high-risk AI systems covered by EU harmonisation legislation listed in Annex II, section B, are generally not subject to the AI Act. In particular, Annex II, section B, AIA, lists Regulations 2018/858 and 2019/2144, which, in turn, also address automated driving systems. Hence, as it stands at present, the AI Act will not be directly applicable to autonomous vehicles. At the same time, however, the AIA remains highly relevant also for self-driving-cars regulations in view of its Recital 29, which stipulates that the requirements under the AI Act must be taken

---

[51] Report from the Expert Group on Liability and New Technologies, New Technologies Formation, Liability for Artificial Intelligence and other emerging digital technologies, (Publication Office of the European Union, 2019).

[52] Report from the Horizon 2020 Commission Expert Group to advise on specific ethical issues raised by driverless mobility (E03659), Ethics of Connected and Automated Vehicles: recommendations on road safety, privacy, fairness, explainability and responsibility (Publication Office of the European Union, 2020).

into account when adopting future delegated or implementing acts in relation to Regulation 2018/858 and 2019/2144.

The PLD revision proposal and the AILD proposal are complementary and aim to provide a comprehensive framework of protection in the event of harm caused by AI (including autonomous driving systems). In fact, the former confirms the principle of strict producer liability (also in relation to intelligent products, such as autonomous vehicles or their components) regarding to B2C relationship, while the latter concerns non-contractual fault-based civil law claims at national law level for damages caused by AI and has the objective of ensuring compensation for both consumers and businesses.

To sum up, the recent years (in particular 2017–2021) have been characterized by a propensity of the EU to self-question about the regulation of Robotics and AI, among which self-driving cars play a very important role and require regulation. The protagonists of the debate have been the European Parliament and the European Commission. While the former seemed certain from the outset of the need to intervene to provide specific rules for robotics (including autonomous vehicles), the European Commission initially seemed doubtful and attempted to protect the effectiveness of the PLD. However, the tendency of the EU Commission to favor regulation of new technologies is now also quite evident.

The introduction of *ad hoc* binding rules that regulate new technologies is now foreseeable and they will most surely clarify, first of all, the rules governing liability in the event of damage caused by AI systems.

Concerning autonomous vehicles, it appears to be general understanding now that the current EU regulations in force have not been designed specifically with a view to solving the challenges posed by the use of autonomous vehicles and that the current regulatory framework of rules will no longer be adequate in the face of the new risks emanating from the increase in automation, connectivity and vehicle complexity.

In order to push in this direction, experimentation is fundamental, as we will see. Fully automated vehicles are not yet ready to be driven on public roads and many technical problems still need to be solved to make these vehicles fully capable of perceiving the surrounding environment. On the other hand, testing will help to improve the awareness of the risks and to understand how to properly regulate the new phenomena triggered by the application of advanced technologies.

3 Questionnaire Analysis

Having outlined the overall context, this section analyzes the answars to a questionnaire submitted by the authors of the individual chapters of this volume related to the jurisdiction examined by them in each case.[53] In particular, the authors were asked,

[53] The questionnaire is reproduced in full in Annex 1 at the end of the book.

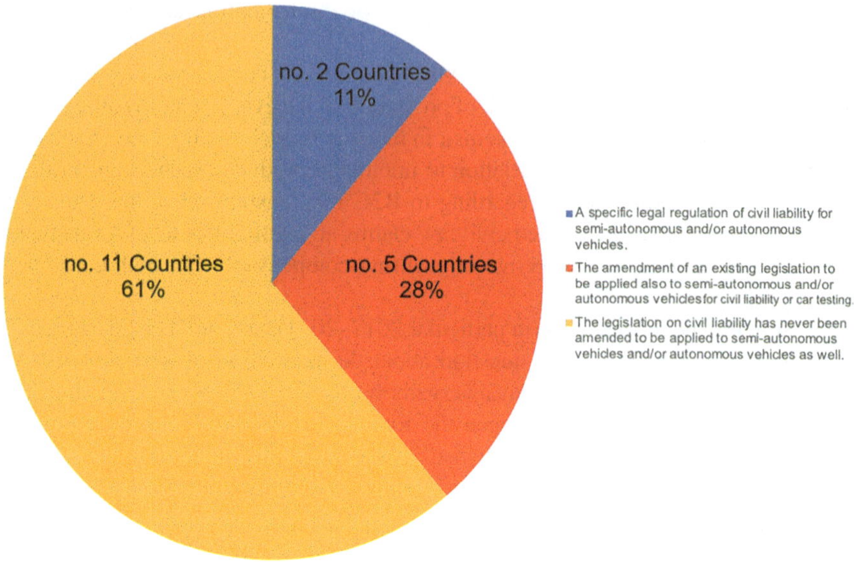

Fig. 1 Self-driving cars regulation in the selected Countries (Q2)

first, to indicate whether legislation on autonomous vehicles has been introduced in their legal systems and, if so, to indicate to what extent. Secondly, they were asked to answer several questions on the civil liability regime for damage caused by autonomous vehicles, identifying the potentially responsible subjects according to the different levels of automation according to the SAE standard, in compliance with the relevant national legislation in force.

The graph below (Fig. 1) displays the state of the art on the regulation of autonomous vehicles in the selected countries (no. 18 countries in total). More precisely, civil liability regulation has been taken into account, but also road traffic law and driving test regulations have been considered in order to verify whether and in what terms regulatory changes have already been effected in relation to (semi-) autonomous vehicles, with regard to each selected legal system. Indeed, before being able to concretely assess the liability regime for an accident caused by a self-driving car in each selected legal system, it was necessary to verify how these legal systems have addressed the development of these types of vehicles.

Within the selected countries, eleven (61%) of them indicated that national civil liability legislation has never been amended with the purpose of specifically addressing semi-autonomous or fully autonomous vehicles. Five countries (28%) confirmed that there has been an amendment of the existing national legislative framework in order to make it applicable also to (semi-)autonomous vehicles, with regards to the civil liability profiles, or only in relation to the possibility of testing autonomous vehicles on the road. Finally, two countries (11%) indicated that an *ad hoc* regulation has been introduced for semi-autonomous or fully autonomous vehicles, or that it is going to be approved in due course.

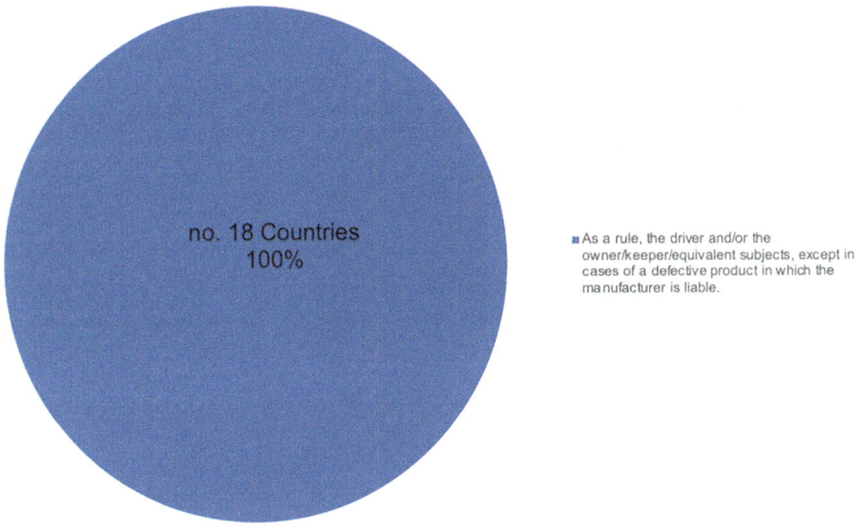

Fig. 2 Liability rules for conventional vehicle (Level 0–2). Q5: 'In the event of an accident caused by a conventional vehicle (up to SAE level 2), who is usually held responsible in your legal system?'

With respect to the specific case of civil liability arising from a road accident caused by a vehicle, it was verified the subjects of liability vary depending on the increase in the vehicle's level of automation.

In greater detail, for traditional vehicles or vehicles equipped with basic advanced driver assistance systems (level 0 to 2 of the SAE standard), the possible tortfeasors are the same in all the countries analysed (*i.e.* eigtheen countries—100%), as indicated in the graph below (Fig. 2). The principal tortfeasor is the driver, followed by the owner and by subjects having an equivalent legal status, with the exception of the manufacturer's liability in the event of a defective product, considering the vehicle and its components as products.

For SAE Levels 3–4, the liability scheme is the same, whereby liability generally rests with the driver and the equivalent subjects. At the same time, however, the cases in which the manufacturer is liable increase in consideration of the fact that product defects can be expected to be far more frequent.

In contrast, with the achievement of total automation, and thus in the case of fully autonomous vehicles, the situations in which the manufacturer is liable increase considerably (this is the case in five of the selected countries). This is a consequence of the fact that road traffic accidents are far more likely to be the consequence of a design, production, or information defect, but obviously the case in which the liability rests on the owner, keeper, or equivalent subject is not fully excluded. As shown in the graph below (Fig. 3), the driver is omitted as a liability addressee because the vehicle drives itself and all those on board are merely passengers. In this regard, it appears significant to point out that among the five countries that consider

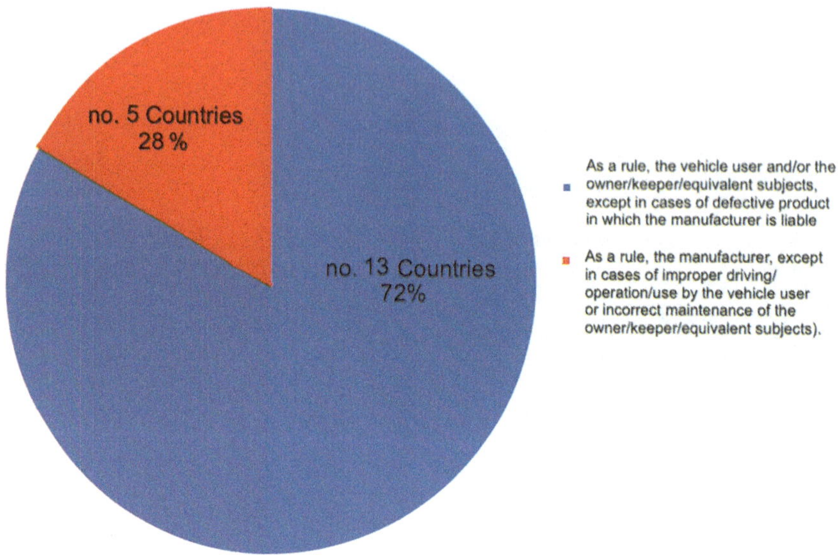

Fig. 3 Liability rules for fully autonomous vehicle (Level 5). Q7. 'In the event of an accident caused by a fully automated vehicle (SAE level 5), who is usually held responsible in your legal system?'

the manufacturer as the main tortfeasor or as an alternative liability addressee to the vehicle owner, there are two countries that have introduced specific regulations for autonomous vehicles. This data can be viewed as indicative of the fact that future regulations of the phenomenon of self-driving vehicles will presumably develop in the direction of producer and/or owner liability.

4 Chapter Analysis

4.1 Analysis of Non-EU Countries

The legal framework of many countries is characterized by the fact that it rests on case law whereas, in comparison, the legal framework of the continental European countries is based on codified law. Having said that, it is of most interest how the different legal systems deal with new technologies such as autonomous driving. One way of looking at this could be whether the existing laws are deemed applicable and sufficient in the case of autonomous driving, or new laws have been adopted instead. Another major issue related to the question of who is to be held liable in case of an accident involving vehicles of the various SAE levels is whether the respective legal system suffers from gaps in the protection of injured parties or exhibits a shift of liability towards certain liability addressees.

4.1.1 Singapore

Singapore has implemented a regulatory sandbox for testing and serial operation of autonomous vehicles by amending its Road Traffic Act. These amendments are explicitly intended to tackle the levels 3, 4 and 5 of SAE J3016. In addition, an industrial standard "Technical Reference 68" (TR 68) has come into force. Admittedly, these amendments do not regulate the liability for accidents involving vehicles at SAE Levels 3 to 5. They merely provide a regulatory framework and, in the case of TR 68, do not even have binding effect.

Hence, in the case of SAE Level 0 to 2 cars, liability addresses can be the driver, the owner and the manufacturer. The driver can be held liable for negligent conduct that breached a duty of care if this led to a third party suffering damage. Unlike in countries with strict owner or keeper liability, in Singapore vehicle owners are not liable for damage caused by the driver. In addition, the manufacturers will be held liable for vehicle malfunctions. Such product liability is regulated by common law and requires the breach of a duty of care on the part of the producer.

With regard to SAE Levels 3 and 4, the Land Transport Authority has implemented a regulatory framework for the testing and use of autonomous vehicles which makes the purchase of liability insurance obligatory. With regard to liability in relation to the use of a SAE Level 3 system, drivers can be held liable for a negligent omission if they violated their duty to respond to a system request and, in particular, to take over vehicle control. But unlike the driver liability for a vehicle at SAE Levels 0 to 2, liability is limited to a smaller range of duties, such as for instance taking over the control after the car requested this. Keeper liability is again not available at SAE Levels 3 and 4. In contrast, as far as the manufacturer is concerned, he is held liable for a malfunction or a defect under the same common law principles on negligence as for SAE Levels 0 to 2. Software programmers do not fall under a law, nor does case law exist to this effect. Their liability will therefore depend on whether the courts determine a duty of care on the part of the programmer.

As for the user of a SAE Level 5 vehicle, liability depends on whether the automation functions must be activated or checked before activation. If so, the human user may be held liable under the same preconditions as the driver at the preceding automation levels. Likewise, the liability of vehicle owners, manufacturers and software programmers for autonomous vehicles at SAE Level 5 will have to be determined on the basis of the same principles as those applied at the previous automation levels.

4.1.2 South Africa

The legal system of South Africa is flexible enough to accommodate for motor vehicle accidents involving vehicles at SAE Levels 0–3. More specifically, the Road Accident Fund has been set up for the purpose of compensating those who have suffered certain loss or damage from the negligence or other wrongful act of the driver or owner of a motor vehicle. A claim against the wrongdoer himself is only available in limited cases. In addition, the Consumer Protection Act places producers under strict liability for any harm caused by the supply of defective goods or inadequate instructions. Finally, damage claims for harm caused negligently or intentionally may also exist under common law principles of tort liability.

In contrast, in the case of SAE Levels 4 and 5, the conventional road traffic liability regime reaches its limits in view of the fact that there is no longer a human driver in the vehicle. What is more, the South African Supreme Court of Appeal has held very recently that strict supplier liability under the Consumer Protection Act will only apply where there is a consumer-supplier relationship, thus excluding third-party liability claims raised, e.g., by drivers of other vehicles, cyclists, or pedestrians. One way of addressing this legal lacuna might be to extend the *actio de pauperie* to self-driving vehicles. This common law action places strict liability on the owner of a domestic animal for injury and damage caused if the animal acted contrary to its nature when causing the damage.

4.1.3 China

In China, local rules are in place which regulate the testing of autonomous vehicles in local areas. For cars of SAE Levels 0 to 2, the existing law allocates the liability among the involved parties. Therefore, Art. 76 of the Chinese Law on Road Traffic Safety states that the insurance company shall compensate damages, resulting from an accident. Having insurance is compulsory. In case of fault, the motor vehicle party is liable for the damages. This includes the vehicle holder (also the owner, lessee or borrower, etc.) and the driver of the car. The fifth chapter of China's new Civil Code then differentiates in Art. 1209 Civil Code and states that if the owner and user are not identical in person, the user shall be liable, whereas the owner is liable if he is at fault for the damage.

In case of a SAE Level 3 or 4 vehicle, it is possible that the automated driving function or the driver are responsible for an accident and the resulting damage. Liability of the driving function is not provided for *de lege lata*. If the damage is caused by a malfunction or defect of the system, liability addressees can be the manufacturer or the seller. Nevertheless, the driver or the owner must compensate the injured person first and can then take recourse to the manufacturer or seller. Their liability is subordinated in comparison to the owner or driver. Liability and its addressees are determined by existing laws and regulations. Main subjects for compensation are the driver and owner of the vehicle, which becomes virulent if the mandatory insurance amount is not sufficient to fully cover the damage.

Finally, there is no law or regulation for SAE Level 5 cars either. *De lege ferenda* liability addressees could be the driving function or the manufacturer. However, liability of the system itself is ruled out *de lege lata*. It also remains uncertain whether the owner will be held liable.

4.1.4 Japan

Japan has no law regulating the civil liability for autonomous vehicles. Nevertheless, accidents will be covered by existing law such as the Civil Code, the Act on Securing Compensation for Automobile Accidents and the Product Liability Act. These claims are complemented by an insurance policy. This being said, there is an amendment of the Road Traffic Act and the Road Transport Vehicle Act to provide a legal framework for the testing and use of SAE J3016 Level 3 and 4 vehicles.

The focus of the Civil Code is on driver liability for SAE Level 1 to 2 cars. The driver must observe several duties, such as to react to external incidents in the vehicle environment. With respect to SAE Level 3 and 4, it is not so much the issue of whether driver liability applies at all, but instead how far-reaching it is. This raises the question whether the driver can be held liable even before a takeover request was issued by a SAE Level 3 driving function. In contrast, driver liability is obsolete in the context of operating a SAE Level 5 driving function that performs the driving task without any human intervention whatsoever. In practice, Japanese tort law is of minor importance in comparison to the Act on Securing Compensation for Automobile Accidents and the Product Liability Act. Even though there are a lot of unclear issues that require further discussion, the existing legal framework is able to deal with autonomous vehicles up to SAE Level 5.

4.1.5 Korea

In Korea, the Motor Vehicle Accident Compensation Guarantee Act provides for strict vehicle owner liability with very limited exceptions, as well as an obligation to purchase a liability insurance policy. Claims against the driver for damages attributable to his negligent conduct can be made under the Civil Code. If an accident is due to the defect of a vehicle, its manufacturer is subject to claims under the Product Liability Act. In view of a recent amendment to the Motor Vehicle Accident Compensation Guarantee Act, all of this equally applies to vehicles at SAE Level 3.

In contrast, the Ministry of Land, Infrastructure, and Transport has announced that the amended Act does not extend to SAE Levels 4 and 5. Admittedly, this is not fully clear from the statutory text, which generically refers to "autonomous vehicles". It is an ongoing discussion whether future legislation should preserve driver liability at SAE Level 4 to the extent there is a requirement for the human being to intervene in abnormal situations. At SAE level 5, on the other hand, driver liability is not an option. Rather, the apportionment and sharing of liability among vehicle owners, manufacturers and software developers takes center stage at this most advanced automation level.

4.1.6 Columbia

There is no law addressing autonomous vehicles in Colombia either. The liability regime in Columbia for road traffic accidents flows from the Civil Code, which was shaped and differentiated by judgements and is based on established case law. The main liability addressee for accidents is the keeper of a vehicle. This is first and foremost the driver, because he has control over the vehicle and its direction. Keepers are also the owner of the car or a transport company which it is affiliated with. Product liability allows taking recourse against the manufacturer if a defect can be proven.

Due to the fact that under Columbian law driving a vehicle is a dangerous activity, strict liability applies, which means that fault is generally of no relevance. An exculpation is only possible in restricted cases, such as force majeure or an act of God. Passengers of a car are no liability addresses because they do not carry out a dangerous activity. The latter applies to SAE Level 0 to 2 cars, and also to all other levels of automation under the SAE nomenclature. The car or the driving function itself cannot be held liable. This does not change for SAE Level 3 to 5 cars because a thing such as a car does not have legal capacity. In addition, this would discharge manufacturers, owners and holders from their liability.

The general liability regime for SAE Level 0 to 2 cars also applies for SAE Levels 3 and 4. The only difference is that the driver is not liable if he cannot intervene in the performance and control of the driving function. Not surprisingly, this equally applies with respect to liability for SAE Level 5 vehicles. As there is no human driver present at this automation level, there is no such liability addressee available. But nevertheless, the owner or transport company are liable as they are still considered the vehicle keeper. In addition, the software programmer can be held liable. All in all, it can be said that even if there is no existing law for SAE Level 0 to 5 cars, the existing and well traditioned laws do apply and that there is no liability gap.

4.1.7 U.S.A

In the U.S., federal legislation and safety standards promulgated by the National Highway Traffic Safety Administration (NHTSA) are expected to displace the state regulation of motor vehicles in due course. In the interim, state legislation remains authoritative, with state tort law providing the default rules to determine liability in cases of accidents involving automated vehicles. Traditionally, driver liability is thereby determined by the tort law of negligence, whereas a crash caused by a vehicle defect exposes the manufacturer to strict product liability in most states.

The characteristic of SAE Levels 2 and 3 is the interface between human and machine driving, which brings about safety issues if drivers are lulled into complacency or inattention despite the necessity to react promptly to take-over requests at the human-machine interface (HMI). Against this background, manufacturers must adopt fault-tolerant product designs if they want to protect themselves effectively against strict product liability.

At SAE Levels 4 and 5, manufacturers' liability for properly deployed automated vehicles is the core issue. It is available for crashes caused directly by self-driving vehicles. In the event of programming bugs, the manufacturer will generally be subject to strict liability. If, on the other hand, a fully functioning autonomous vehicle crashes, then the manufacturer is not automatically liable. It is arguable in this case that consumers cannot justifiably expect perfect safety, which is why the manufacturer will incur liability for such crashes only if the design of the automation system causes the vehicle to perform in an unreasonably dangerous manner or if the manufacturer failed to adequately warn consumers about the inherent risk that even a fully functioning autonomous vehicle may crash.

4.1.8 Australia

In Australia, so far, only SAE Level 0–2 vehicles are permissible on public roads. Where a human driver causes an accident, the injured person may avail itself of a civil liability claim under common law negligence. Liability of the registered vehicle owner for actions of the driver is conceivable in accordance with the principles of vicarious liability. Manufacturers and suppliers face liability for defective products both at common law and under current legislation. While the former establishes fault-based liability, the Australian Consumer Law provides that losses or damage due to safety defects of goods may be recovered without the need to prove fault on the part of the manufacturer. With a view to higher levels of automation, the Australian Road Rules form the basis of the road traffic rules that have been enacted by the Australian States and Territories. They currently rest on the notion of a human driver as the dominant and controlling party during vehicle operation. Hence, a substantial overhaul of the legislative framework addressing in-service vehicle safety is required to accommodate vehicles at SAE Levels 3–5, which to varying degrees can operate without a human driver.

Legislative reform is well under way in this regard. The National Transport Commission has worked closely with government and industry and published its final recommendations for a national Automated Vehicle Safety Law in 2021, which introduce various new categories of stakeholders and potential liability addressees. For instance, the entity producing or importing automated driving systems to Australia is placed under an obligation to maintain and update the automated driving system throughout its product life cycle. However, as no statutory right of action is currently envisaged, injured parties must rely on the tort of negligence and other statutory actions such as product liability under the Australian Consumer Law referred to above. Similarly, it is to be assumed that the duties of remote drivers will be fleshed out in the new law. In contrast, the fallback ready user at SAE Level 3 will most likely continue to be subject to existing State and Territory legislation for SAE Level 0–2 vehicles with necessary modifications (e.g., the insertion of an obligation to resume vehicle control without undue delay when required). This also applies with respect to other persons with influence on in-service vehicle safety, such as repairers and modifiers. The responsibilities of telecommunication service providers regarding vehicle connectivity, on the other hand, will presumably be dealt with at national level.

4.1.9 United Kingdom

In Great Britain, the Automated and Electric Vehicles Act (AEVA) came into force in 2018. The law was passed due to the necessity to close a gap in third party motor insurance in the context of damages caused by automated vehicles. Since then, the Law Commissions has conducted several public consultations. The AEVA provides a definition of automated vehicles. Insurers are held liable under strict liability if an accident involving an automated vehicle occurs. Only in certain cases where no compulsory insurance is available, the vehicle owner will be held liable. The focus of the liability concept for automated vehicles is placed on the insurer, who compensates the drivers and owners and regulates the damages.

4.2 *EU Member States*

Within the international and EU framework of hard law and soft law, the legislators of the selected EU countries have adopted differing legislative approaches.

4.2.1 Belgium

More precisely, in 2018 Belgium introduced rules to allow the testing of autonomous cars but a specific regulation for the event of an accident involving self-driving vehicles is not provided.

Indeed, existing liability mechanisms are applicable to more traditional vehicles (0–2 of SAE standard of automation level). For vehicles of SAE Level 3, the sharing of driving functions between the human user and the on-board driving software triggers the question as to whether fault liability solutions can be applied to a partially autonomous vehicle. On the one hand, it is possible to suppose that road victims should continue to benefit from the no-fault compensation regime of art. 29bis of the Act of 21 November 1989 on Mandatory Liability Insurance for Auto Motor Vehicles, while drivers will have to resort to general fault liability (pursuant to art. 1382 BCC) in order to obtain the compensation of damages. On the other hand, there are views supporting the possibility for the victim to invoke product liability as a basis for bringing a claim. For the regimes that may be applicable to vehicles at SAE Level 5, the possible scheme should be manufacturer liability as vehicle control shifts from the human operator to the machine.

4.2.2 Germany

In Germany, the intention of the Federal Government of the time was to demonstrate the country's leadership in the development of self-driving vehicles. To this end, regulations concerning conditional vehicle automation were added to the Federal Road Traffic Act (StVG) in 2017, and subsequently legal provisions for highly automated vehicles were introduced in 2021.

Concerning SAE Level 0–2 vehicles, the statutory framework makes a careful distinction between the driver, the vehicle keeper and the manufacturer's potential liability. Sect. 7 StVG mandates strict liability of the vehicle keeper for damages to persons or items of property caused during the operation of a vehicle. In addition, Sect. 18 StVG imposes fault-based liability on the driver. The driver is also liable under general tort law standards set out in Sect. 823 of the German Civil Code (BGB), which provides for fault-based liability. Finally, the manufacturer of a defective product is liable for damage inflicted on product users and third parties due to the deficiency of his product. Such liability follows from the strict liability regime set out in the Product Liability Act, as well as the general tort law regime referred to above, which takes its starting point in the fault-based liability provided for in Sect. 823 BGB.

SAE Levels 3 and 4 have already been taken up by the 2017 and 2021 amendments to the Federal Road Traffic Act (Sect. 1a and following StVG), respectively. However, the legislative amendments do not affect the traditional liability structure, which imposes liability for personal injury and property damage on the keeper of a motor vehicle regardless of fault; in addition, the German Compulsory Insurance Act (PflVG) requires the keeper to take out liability insurance. The recently enacted provisions on the operation of vehicles with highly automated driving functions impose specific obligations on the parties involved but do not change the keeper's liability.

Finally, no specific regulation exists for SAE Level 5 driving. Although it seems possible to apply the existing rules also to fully automated vehicles, intervention of the legislator seems desirable in the medium term.

4.2.3 France

The law applicable for testing autonomous vehicles in France was introduced in 2016 (Experimental Ordinance), then implemented in 2018 with a regulation amended in 2020. It was initially developed for the sole purpose of allowing testing of these new vehicles but, in 2018, a report was published on the French autonomous vehicle development strategy (Mobility Orientation). In 2019, the Mobility Orientation Law entered into force, proposing an in-depth reform of the general framework of mobility policies; however, this has not yet been implemented.

A few years later, two Ordinances were adopted by the Government in 2021: the Data Ordinance and the Criminal Liability Ordinance. The former sets out in detail the persons authorised to access vehicle data, while the latter regulates the criminal exemption of the user of the autonomous vehicle and also provides for an information obligation towards consumers who buy or hire an autonomous vehicle from a professional, according to Consumer Code provisions. Furthermore, the Criminal Liability Ordinance was completed by a Decree of the same year (Decree on criminal liability) that includes especially a set of definitions (delegated driving vehicle).

Also, in France, the traditional liability scheme provides for driver liability or owner liability with the exception of producer liability. As the automation level increases, the justification for assuming producer liability increases proportionally.

4.2.4 Spain

In Spain, the case of self-driving car damages has not been regulated but it appears practicable that the compensation for damages shall follow the general system of valuation for pecuniary damage or personal injury. Currently, the Spanish law in force is the *Codigo Civil* (with specific reference to art. 1902 and following) and the *Ley de responsabilidad civil y seguro en la circulación de vehículos de motor* and the *Ley de Tráfico*. At the same time, however, the concept of autonomous vehicle is defined in the Instruction 15/V-113 of 2015, of the Directorate General of Traffic, on the authorisation of tests or research trials carried out with automated driving vehicles (from SAE Level 3 to 5) on roads open to traffic in general.

Vehicle categories from SAE Level 0 to 4 comply with the scheme of the civil liability regulated in the Law on Civil Liability and Insurance in the Circulation of Motor Vehicles.

For fully autonomous vehicles (SAE Level 5), the liability, in the absence of a driver, by virtue of the *pro damanato* principle, should fall on the owner of the vehicle and its insurance company, according to the principles of strict liability. Furthermore, if the loss was due to the defective state of a vehicle, the owner could seek compensation from the manufacturer, who will always be liable under the strict liability scheme.

The obligation to take out compulsory insurance applies to all driving automation levels.

4.2.5 Italy

The Italian legal system does not have a specific regulation on civil liability for autonomous vehicles. Currently, the applicable regulation is composed of art. 2054 of Italian Civil Code as well as to the Italian Consumer Code and Italian Private Insurance Code.

For levels 0–2 of the SAE standard, the rule is the driver liability, and the possible joint and several liability of the owner of the vehicle and equivalent subject. The regulation also provides for liability of the manufacturer in case of a vehicle defect, considering it as a product according to the Consumer Code provisions. Car insurance is mandatory, and the insurer has to compensate for the damages that occurred according to the third-party liability scheme, while first-party liability scheme is the exception applicable in defined hypothesis related to the direct compensation, according to Private Insurance Code. It provides that, in specific cases, the injured person can directly approach to his/her insurance company to claim compensation for damages in case of an accident caused by another party, without having to request such payment from the insurance company of the damaging party.

In the case of (semi-)autonomous vehicles, i.e. SAE Level 3 and 4 of driving automation, it is possible to extensively apply the existing regulation.

Concerning level 5 of the SAE standard, the introduction of specific regulation for self-driving cars is needed, as the traditional roles of civil liability will completely change. Fully autonomous vehicles will overcome the traditional central figure of liability in road traffic: the driver. Indeed, the autonomous vehicle is able to drive itself, through the driving function, and the main subjects involved in the driving relationship are the owner of the vehicle for having bought the vehicle that caused the damage and the producer of the vehicle for having produced a defective vehicle.

In the meantime, in 2017 a regulation for testing autonomous vehicles on selected roads (so-called smart roads) was authorised and, then, implemented in 2018.

4.2.6 Netherlands

In Dutch law, there are no specific regulations for autonomous vehicles, which have not been admitted yet by the Road Traffic Government Service; so the existing rules can be applied extensively. At the same time, however, in 2018 the testing of semi-autonomous cars (up to level 4) was permitted on public roads (Law 2018/347) if a license has been granted by the Ministry of Traffic. Different parties may be liable in the case of accidents with a self-driving car, but the main parties involved are the driver, the owner and the keeper, and the manufacturer. In practice, in compliance with the art. 6:162 and ff. of the Dutch Civil Code (BW), the driver may be held liable on the basis of negligence if he could and should have intervened to prevent an accident not only for traditional vehicles (levels 0–2), but also for semi-autonomous ones (levels 3–4).

On the other hand, the owner or keeper will primarily be held liable for self-driving car damages (level 5 of automation), as art. 185 Road Traffic Act (WVW) imposes strict liability for all accidents. Besides owner or keeper liability, the possessor may be considered liable on the basis of art. 6:173 BW.

Finally, the liability of the manufacturer, regulated under the BW, would be the same for SAE Level 3–4 cars, while, arguably the concrete obligations of the manufacturer would be more stringent, as more safety tests and precautions would be required of a SAE Level 5 car.

4.2.7 Austria

In the Austrian legal system, according to § 102, para 3a, Motor Vehicle Act (KFG), the Automated Driving Regulation (AutomatFahrV) was adopted in 2016 for driving tests in specific cases, that have been implemented in 2022: autonomous minibuses (§ 7 AutomatFahrV), provided in 2016, and then automated vehicles for passenger and freight transportation (§ 7a AutomatFahrV, § 7b AutomatFahrV), in 2022; vehicles with motorway pilots for automatic lane changing (§ 8 AutomatFahrV), in 2016, and automated driving on and off highways (§ 8a AutomatFahrV), in 2022; self-driving army vehicles (§ 9 AutomatFahrV), in 2016, and automated parking services (§ 9a AutomatFahrV) as well as automated work machines (§ 9b AutomatFahrV) in 2022. Moreover, in March 2019, a basis in section 3 of the AutomatFahrV was created for approved driver assistance systems (such as parking assistance and motorway assistance).

Concerning civil liability, there is no shift in liability and the general civil liability regulations provided for vehicles with SAE Level 0 to 2 automation under the Austrian Civil Code (ABGB) and the Railway and Motor Vehicle Liability Act (EKHG) remain applicable. In relation to semi-autonomous vehicles (SAE Levels 3–4), the driver can transfer certain driving tasks to the assisting systems, but he/she remains responsible for taking over the driving tasks again at any time, although the duties of care may be set lower.

For SAE Level 5 automation, the requirements for liability are, in principle, the same as for all the other levels. In particular, the holder of a vehicle could generally be held liable under EKHG, but liability under ABGB may be difficult to argue and especially to prove fault-based liability. In addition, the manufacturer of the autonomous vehicle is generally liable under the Austrian Product Liability Act (PHG) for damages caused by the malfunction of a vehicle.

4.2.8 Romania

In Romania, the liability for traffic accidents caused by traditional vehicles (levels 0–2) is regulated only under the conditions of the liability for one's own actions (fault liability) according to artt. 1349, 1357–1371, 1381–1395 of Civil Code. For vehicles equipped with driver assistance systems (SAE Levels 3–4), liability, as set out by the express provisions of art. 1376 in the Civil code, is objective, independent of any culpable conduct and implies the legal guardian's strict liability. Also concerning autonomous vehicles (SAE Level 5), the legal guardian is held liable (strict liability), as there are no legal regulations in Romania for autonomous vehicles. In case of a car defect, the Romanian law on strict producer liability is applicable.

4.2.9 Sweden

Swedish liability legislation regarding vehicles focuses mainly on three liability subjects: the manufacturer, the owner, and the driver. The liability rules in force (Tort Liability Act - Skadeståndslagen) are adapted for driving at SAE Levels 0–3, while the legislation regarding traffic insurance is such that it could be applied to all levels of automated vehicles. For vehicles of SAE Levels 4 and 5, Swedish liability law requires adaptations, especially with regard to the driver's responsibilities during automated travel as, currently, Swedish law depends on the courts to develop the concept of 'driver', that also include the driver of automated vehicles, and to apply the driver's liability also in a context of automated vehicles. Therefore, an intervention by the legislator is needed.

Concerning product liability, also in Sweden vehicles are considered products. For this reason, according to the Product Liability Act, which complements the Product Safety Act, the manufacturer bears strict liability also for autonomous vehicle defects. However, the public reports SOU of 2018 and 2021 proposed a tightening of the owner's liability, who is envisaged to be held liable for breaches of traffic rules by the autonomous vehicle, unless the breaches were due to a defect in the driving system that does not fall within the owner's control.

5 Conclusions

The comparative analysis of the various jurisdictions dealt with in this book demonstrates that, as the level of automation increases, the applicable liability rules may change. In this regard, the respective legal systems have so far responded differently, thereby providing for the liability of the driver, of the owner, of the manufacturer and/or directly of the insurer.

Even if in a number of jurisdictions there are no new national laws or regulations in place that specifically address the responsibility for damage to persons or property brought about by autonomous vehicles, the existing liability rules seem to be able to respond, on a large scale, to the current state of automation and to the interests of the parties. Indeed, in the context of road traffic liability, many jurisdictions traditionally provide for strict liability of the vehicle owner or keeper, which can be beneficial to the victim and can equally be applied to vehicles at SAE Levels 3–5.

Moreover, as already mentioned above, qualifying autonomous and automated vehicles as products leads to the application of strict manufacturer liability towards any injured person, as generally laid down across all national jurisdictions. As highlighted in the following chapters of this book, the manufacturer's liability is, then, set out in a separate set of rules to be considered alongside the liability rules specifically addressing road traffic.

In sum, damage caused by or in connection with the operation of autonomous (or automated) vehicles is generally regulated in multiple layers of liability that include extra-contractual liability rules for road traffic accidents and insurance regulation, rules on manufacturer liability if a product defect can be identified, as well as any other special rules laid down by legislation for autonomous (and automated) vehicles.

With regard to the manufacturer's liability regulation for product defectiveness, important issues under consideration are those arising from automated driving functions embedding AI.

As already mentioned above,[54] the use of AI in products has started to be considered, for example, by the European Union in the two complementary sets of rules proposed in the context of the ongoing revision of the product liability directive (PLD) and the suggested AI liability directive (AILD).

Although an in-depth analysis of these proposals is outside the scope of this book, a few initial considerations will be presented here in order to address the question of the future eligible tortfeasors in accidents caused by autonomous or automated vehicles in the European Union, and to start delineating a possible legislative model on a transnational level.

Under the currently envisaged revision of the PLD, it is still not fully clear to what extent AI functioning will have a specific impact on the defectiveness of a product. The effects of certain AI functions, namely the ability to continue to learn after deployment (art. 6, para. 1, c)), must be taken into account when considering a product as defective; however, in these cases liability can be averted by the development risk defense, if the defectiveness of a product was only discovered after placing it on the market and at the time of its deployment the manufacturer could not discover such defectiveness on the basis of the then existing state of scientific and technical knowledge (art. 10, para. 1, e)). In practice, it is conceivable that flawed decisions of the learning functions of an AI device cause damage, whereby it remains uncertain whether such damage is to be considered the result of a product defect or instead could not have been detected in advance. Similarly, the lack of safety resulting from the interoperability with other products, *e.g.* cars or infrastructures, may be decisive for the defectiveness of an automated vehicle (art. 6, para. 1, d)).

These prospective issues in identifying defectiveness seem to be driven by the attempt to ensure a balance between the interests of manufacturers (*e.g.* avoiding chilling effects) and those of the injured persons, but could lead to uncertainties.

Other choices of the proposed new PLD seem to facilitate the application of the manufacturer's liability regime to damages caused by AI products: This pertains to the rules establishing presumptions of defectiveness, presumptions of a causal link between the defectiveness of the product and the damage, and the presumptions of both, to be admitted by the judge, in cases of technical and scientific complexity (art. 9, parr. 2, 3, 4). Other rules that facilitate the application of the envisaged manufacturer liability are those regarding the number of economic operators potentially liable (depending on the specific case, the final manufacturer, the manufacturer of a

[54] See above, Sect. 2.2. in this chapter.

defective component, importers within the EU, as well as each distributor of the product) and the availability of joint liability where two or more operators are involved in the manufacturing of the defective product (artt. 7, 11).

In addition, under the envisaged European legal framework for Artificial Intelligence, the victim of an accident caused by an autonomous/automated vehicle will be allowed to fairly claim compensation for damages not due to product defectiveness (see the AI Liability Directive Proposal). The two Directives proposed (on Product Liability and AI Liability) are intended to be complementary, although there are a number of misalignments, for example with regard to the envisaged claimants and defendants (natural persons or persons; manufacturers or providers), as well as the classification of AI into high-risk, low-risk and prohibited AI systems, according to the AI Act (taken into consideration only in the AILD, while not expressly in the PLD).

Under the AILD proposal, in case of a fault-based claim for damages under national law, the legal tools foreseen are those of a rebuttable presumption of the causal link between the fault of the defendant and the output produced by the AI system (or the failure of the AI system to produce an output), and the presumption of non-compliance with the relevant duty of care in case of the provider's failure to disclose evidence (artt. 4 and 3).

In conclusion, both proposals, in relation to different types of wrongdoings, aim to enable EU Member States' legal systems to adequately deal with liability claims for damage caused by AI-enabled products, by holding the manufacturers and providers of such products accountable as tortfeasors. This approach deserves attention if one considers the pioneering role around the world that the European Union has played in many other sectors (*e.g.* data protection), which it can equally assume in the field of AI, and, consequently, in that of autonomous vehicles.

Returning to the special laws on road traffic liability examined in this book, it is worth noting that many jurisdictions have meanwhile embarked on enacting new rules specifically addressing self-driving vehicles. These regulations regularly concern SAE Level 3 and/or Level 4 vehicles. In contrast, rules specifically for SAE Level 5 vehicles have yet to be established. Figures 4 and 5 provide a geographic illustration of the outcome of the comparative study undertaken in this book with respect to the global state of liability law with regard to self-driving vehicles. It demonstrates at first glance which of the jurisdictions selected for the purposes of this comparative analysis have already put in place legal regulations in order to address vehicle automation. More precisely, in the maps below countries are divided between: (i) those without any regulations for (semi)autonomous vehicles in force or still under evaluation (highlighted in pink), (ii) those which have already modified their existing legal regime on liability or only on driving testing (highlighted in light blue), (iii) and those that have introduced a specific regulation for self-driving cars (highlighted in green). Considering the evolution and peculiarities of the United States (in yellow), a separate map is provided.

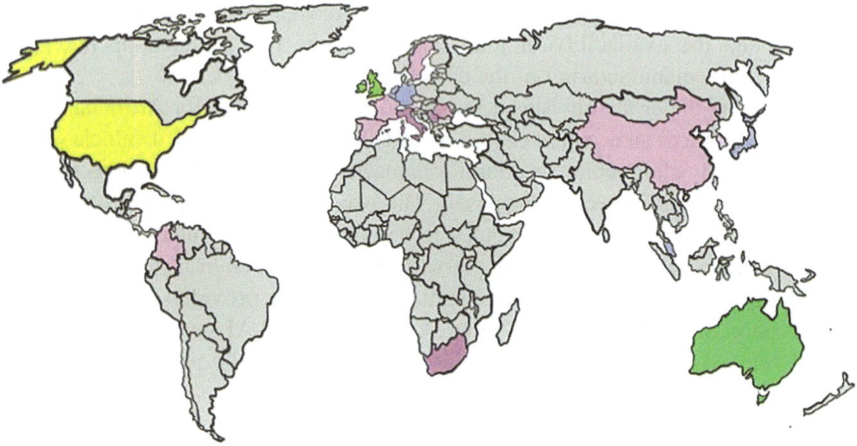

Fig. 4 Map on autonomous vehicles regulation across the world in the selected Countries (The Author of the map is Ms. Simona Latte, Law Degree and II level Master degree in DPO and Privacy Law at Università degli Studi Suor Orsola Benincasa, professional graphic designer)

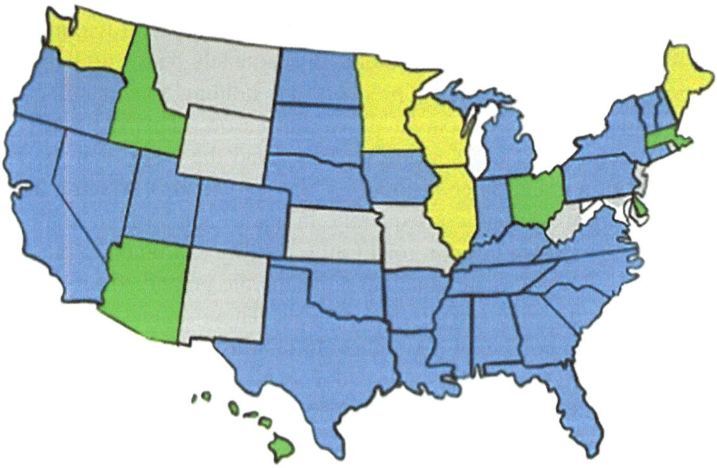

Fig. 5 Map on autonomous vehicles regulation in the United States of America (The Author of the map is Ms. Simona Latte, Law Degree and II level Master degree in DPO and Privacy Law at Università degli Studi Suor Orsola Benincasa, professional graphic designer.)

References

Al Mureden E (2019) Autonomous cars e responsabilità civile tra disciplina vigente e prospettive de iure condendo. Contr impr 3: 895 ff

Al Mureden E, Ruffolo U (2019) Autonomous vehicles e responsabilità nel nostro sistema ed in quello statunitense. Giur it 7:1704 ff

Amidei A (2019) Intelligenza Artificiale e product liability: sviluppi del diritto dell'Unione Europea. Giur it 7:1715 ff

Calabresi G, Al Mureden E (2021) Driveless cars. Il Mulino, Bologna

Gaeta MC (2018) La protezione dei dati personali nell'Internet of Things: l'esempio dei veicoli autonomi. Dir inf e informatica 1:147–179

Gaeta MC (2019a) Data protection and self-driving cars: the consent to the processing of personal data in compliance with GDPR. Commun Law 24(1):15–23

Gaeta MC (2019b) Liability rules and self-driving cars: the evolution of tort law in the light of new technologies. ES, Naples

Gatt L (2020) 'Per un'intelligenza artificiale antropocentrica. Diritto Mercato Tecnologie, p 1 et seq

Karner E, Koch BA, Geistfeld MA (2021) Comparative law study on civil liability for artificial intelligence, p 9

Machnikowski P (ed) (2016) European product liability. An analysis of the state of the art in the era of new technologies. Intersentia, Cambridge

Palmerini E, Bertolini A (2014) Regulating robotics: a challenge for Europe. Upcoming issues of EU law for the IURI Committee, EU Parliament Workshop 2:167 ff

Autonomous Vehicles: A Technical Introduction

Christian Pek and Sanne van Waveren

Abstract Automated and fully autonomous vehicles have the potential to revolutionize mobility in various areas, mostly notably are the opportunities to improve traffic safety and mobility access. This chapter provides a technical overview of the basic working principles of such vehicles: from processing of sensor data to planning safe and comfortable motions for the vehicle to execute. The chapter concludes by summarizing major challenges that need to be solved on the way to releasing commercial vehicles that are always reliable and safe.

1 Introduction

Automated driving is widely transforming the classical mobility sector as we know it. Over the past decades, technological advances have accelerated the automation of driving functions, including, but not limited to, cruise control, emergency braking, and automated traffic sign recognition. At the highest level of automation, there are fully autonomous vehicles (AVs) and driverless cars.

A key motivation for driving automation is safety. Every year, millions of people are involved in severe or even fatal traffic accidents,[1] and almost half of the traffic fatalities are among vulnerable road users such as pedestrians and cyclists. Advanced driver-assistance software intervenes when the system detects an unsafe situation. The software takes over control from the driver and employs preventive countermeasures to keep the vehicle safe. For example, a system can alert a driver about

[1] WHO (2021).

C. Pek (✉)
Department of Cognitive Robotics, Delft University of Technology, Delft, The Netherlands
e-mail: c.pek@tudelft.nl

S. van Waveren
School of Interactive Computing, Georgia Institute of Technology, Atlanta, GA, USA
e-mail: sanne@gatech.edu

H. Steege et al. (eds.), *Autonomous Vehicles and Civil Liability in a Global Perspective*, Data Science, Machine Intelligence, and Law 3,
https://doi.org/10.1007/978-3-031-41992-8_2

unsafe lane changes, or it can avoid collision by automatically activating emergency braking when the vehicle is at risk of driving into obstacles in front of it. The availability of such systems in modern vehicles has already shown to drastically decrease the number of fatalities in traffic.[2]

In addition, AVs can offer several benefits to drivers in terms of comfort and mobility.[3,4] They can relieve drivers from the complex and potentially stressful task of driving, improve mobility for people who otherwise could not drive independently, and support drivers in difficult driving conditions, e.g., on slippery roads or shorter sightline due to heavy rain or snow.

Automated driving also has an impending role in transport and logistics companies, allowing for easier and autonomous transportation of passengers and goods, and increasing energy efficiency, e.g., through truck platooning.[5] Increased comfort and efficiency in traffic are key driving forces behind the development of AVs.

However, fully automated driving remains challenging in real-world traffic scenarios,[6,7] in which the AV needs to deal with uncertainties, changes in lighting conditions, and other traffic participants, among others. German and American researchers demonstrated pioneering AV developments between the 1980s and 2000s.[8,9] The Grand DARPA Challenge[10] in 2005 and DARPA Urban Challenge[11,12] in 2007 are examples of large-scale competitions to foster autonomous driving developments and research. These competitions enabled breakthroughs in AV research and showcased the challenges of fully autonomous driving in both rural and urban settings. As a result of these challenges, companies like Google and Daimler started to heavily invest in AV research.

The automation level of a vehicle can be categorized in six automated driving levels according to the Society of Automotive Engineers[13] (SAE), as illustrated in Fig. 1. The levels range from no automation (level 0) to full automation (level 5). In level 5, no driver needs to be present in the vehicle anymore and the vehicle can perform the driving task under all conditions, e.g., weather or traffic. Level 4 only supports full driving support under certain conditions, e.g., no bad weather. The levels 0 to 3 require a driver to always be present and to take over the driving task in case of uncertainties or malfunctions of the system. The remainder of this chapter

[2] Yue et al. (2018).

[3] Martínez-Díaz and Soriguera (2018).

[4] Bagloee et al. (2016).

[5] Tsugawa et al. (2011).

[6] Mallozzi et al. (2019).

[7] Koopman and Wagner (2016).

[8] Bagloee et al. (2016).

[9] Bengler et al. (2014).

[10] Thrun et al. (2006).

[11] Urmson et al. (2007).

[12] Buehler et al. (2009).

[13] Mallozzi et al. (2019).

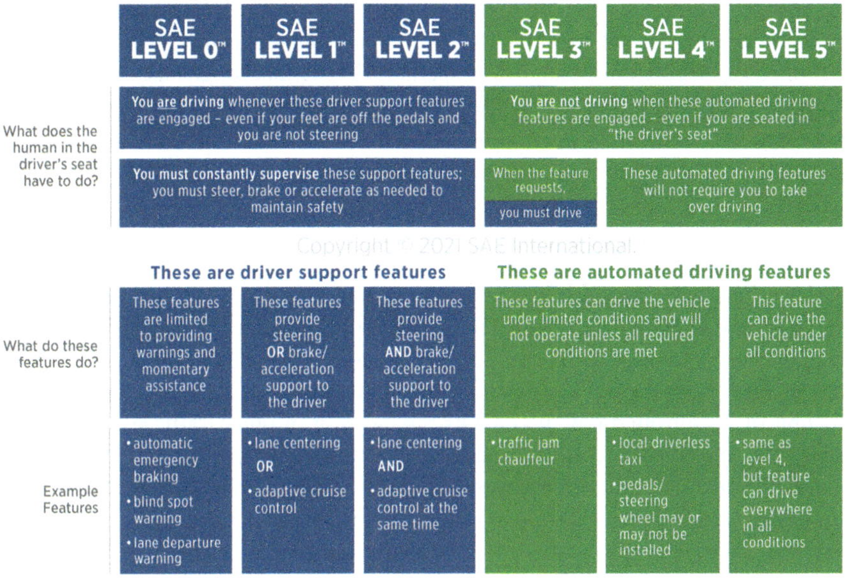

Fig. 1 Six automation levels of vehicles according to SAE. Figure taken with permission from SAE International

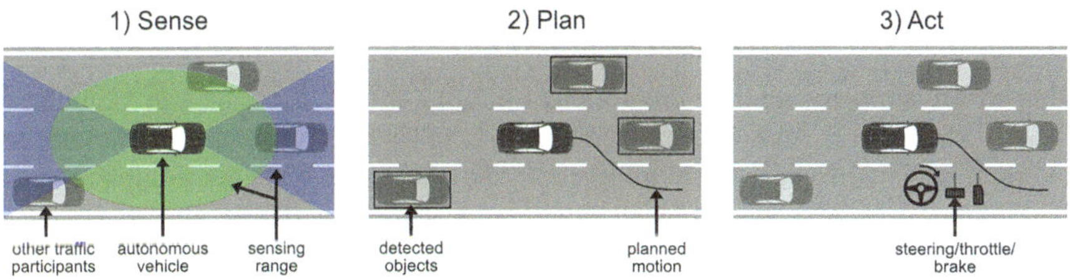

Fig. 2 Overview of the sense-plan-act architecture of robotic systems

outlines the underlying techniques of fully autonomous driving (level 5), and the technical challenges that come with it.

In general, AVs are advanced robotic systems and roughly function according to the sense-plan-act architecture of many robotic systems,[14,15] see Fig. 2. During their operation, AVs create a model of their environment, including the road network and other traffic participants. The environment model is created by processing various sensor data, which is referred to as the sense layer of the sense-plan-act architecture. Based on this environment model, AVs subsequently plan a safe and comfortable motion towards their destination, i.e., the plan layer. Finally, this motion is translated

[14] Mallozzi et al. (2019).

[15] Brooks (1986).

into precise steering, throttle, and brake commands to control the vehicle actuators, corresponding to the act layer in the sense-plan-act architecture. This process is performed repeatedly during the AV's operation to account for changes in the traffic scenario. The following sections summarize the main working principles of each of the layers.

2 Environment Model

During their operation, AVs create an internal representation[16] of their environment based on available information from sensors, maps, and remote data (e.g., information on traffic accidents). This environment model provides the AV with an understanding of its surroundings. It contains the location of other traffic participants, a map of the road network, and information about traffic signs and traffic jams. AVs use the information in the environment model to decide their next action and to plan safe motions (see Sect. 4). In this section, we summarize what sensors AVs are equipped with, what information the environmental model typically provides, and what algorithms are used to create the environment model from those sensor data.

2.1 Sensors

Human drivers perceive their environment through their senses, such as vision and hearing. Similarly, AVs require dedicated senses to perceive and algorithms to interpret their surrounding environment.[17] To this end, AVs are equipped with multiple sensors that are arranged around the vehicle with the aim to create a 360-degree view. Figure 3 illustrates the placement of various sensors around the vehicle and their corresponding sensing range. In general, these sensors can be categorized into proprioceptive and exteroceptive sensors.[18] Proprioceptive sensors measure internal variables that capture the state of the vehicle itself (e.g., its position and velocity). Exteroceptive sensors measure external variables that capture the state of the vehicle's surroundings (e.g., the positions of other traffic participants). These internal and external variables are typically measured through a combination of data from multiple sensors. The fusion of multiple sensor measurements[19] ensures that information is gathered with high accuracy and ensures that extra sensors are available in case of faults or uncertainties of other sensors, i.e., a high degree of

[16] Van Brummelen et al. (2018).

[17] Yeong et al. (2021).

[18] Van Brummelen et al. (2018).

[19] Fayyad et al. (2020).

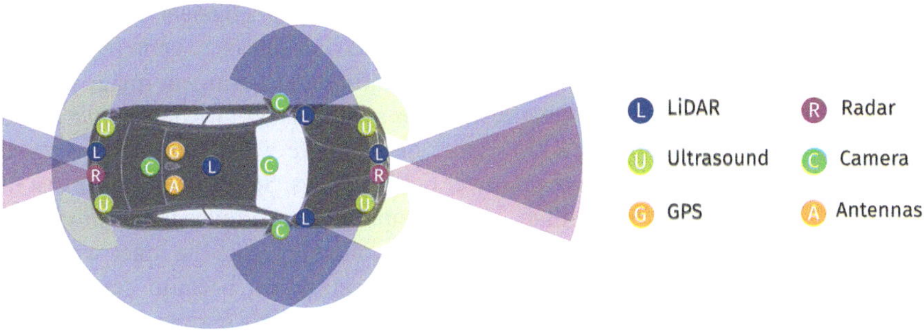

Fig. 3 Example sensor setup of an autonomous vehicle with illustrated sensing ranges

redundancy. In the following paragraphs, we elaborate on the most common sensors in AVs.

2.1.1 Proprioceptive Sensors

Proprioceptive sensors measure the internal state of the AV, which is necessary to estimate the position of the vehicle in the world and what its dynamics are, e.g., its velocity and acceleration. For instance, encoders in the vehicle's actuators, such as the steering wheel or the drivetrain, provide the AV with the current steering wheel angle, the tires' rotational velocities, or the current gear. Inertial Measurement Units (IMUs) use gyroscopes and accelerometers to estimate the vehicle's orientation in the world (roll, pitch, and yaw) and accelerations along the vehicle axes.

For instance, the classical assistance system *dynamical stability control* is responsible to keep a vehicle controllable when certain wheels loose traction, e.g., on slippery roads. In this system, IMUs are used to detect when the vehicle becomes unstable by checking whether the motion of vehicle diverges from the intended motion of the driver, also known as under- and over-steering. Global positioning systems (GPS) maintain the exact position of the vehicle with respect to given reference systems. Using GPS, the vehicle can locate in which city or in which highway lane it is driving. Differential GPS systems provide even higher accuracies up to centimeter resolutions by making use of reference base stations that are precisely placed at known locations.[20]

2.1.2 Exteroceptive Sensors

Exteroceptive sensors provide information about external variables, such as the position, velocity, and shape of surrounding traffic participants. The exteroceptive

[20] Rezaei and Sengupta (2007).

sensors can be categorized according to the underlying physical measurement principle, e.g., whether they use light or ultrasound. For instance, cameras provide visual information by gathering light and color information using an image sensor. The image sensors can have different resolutions and field of views depending on the application area, e.g., finding parking spots may require wider field of views and detecting traffic signs may require narrower field of views. Stereo cameras consist of multiple, separated sensors that can compute three-dimensional images[21] (two-dimensional image with additional depth information) like the inner workings of the human eye. Using camera images is advantageous, because cameras are relatively cheap sensors and there exist various robust processing algorithms to process the images. Yet, cameras have difficulties in harsh lighting conditions, such as direct sunlight or darkness.

Light detection and ranging (LiDAR) sensors emit light into the environment and measure the time it takes for the light to be reflected by an object and to return to the sensor.[22] By emitting multiple beams in various layers and rotating the sensor, LiDARs can create a 360-degree point cloud in which each point describes the distance to an object. Dense point clouds allow one to detect objects and to estimate their shapes. Modern LiDARs provide a high accuracy of the depth information and are not prone to different lighting conditions, i.e., they work even in full darkness. However, LiDARs are still expensive sensors and require high computing power to process the large amount of data points.

Radio detection and ranging (Radar) sensors work in a similar way. Instead of light, Radars emit radio signals and detect the reflected signal. In contrast to LiDARs, Radars are often better at estimating velocities of objects by leveraging the Doppler effect.[23] This effect results in a frequency shift when the radio signals are reflected from objects that are moving. The Doppler effect allows one to easily compute the velocity difference by analyzing the frequency shift, which allows radars to compute velocities with high accuracy. Radars require antennas that need to be designed for the desired detection range, which is why companies usually distinguish between short-range and long-range radars. These sensors are cheaper than LiDARs and are already in use in automotive applications, such as cruise control systems. Nonetheless, the drawbacks of Radars are that they provide less resolution, require accurate dynamic information of the AV to compensate for the AV's own motion, and may interfere with other electromagnetic signals in the environment which degrades the sensor performance.

Lastly, ultrasound sensors emit ultrasound to measure distances to objects. However, ultrasound sensors usually have a limited range so that they are often only used in applications that require sensing of objects in closer proximity to the AV, e.g., automated parking.

---

[21] Köhler et al. (2015).

[22] Wang et al. (2017).

[23] Patole et al. (2017).

Determining the optimal sensor setup for an AV is a difficult process. Though, more sensors usually offer a higher accuracy of the data and provide redundancy in case certain sensors fail. To process all the data that come from different sensor modalities (e.g., camera images and LiDAR point clouds), the data need to be fused into a single data stream, which requires exact sensor calibration and fusion algorithms that are robust against different sources of sensor noise. The resulting data streams can become quite large so that the AV might collect multiple gigabytes of data per minute. This large amount of data needs to be processed in real-time to ensure that AVs always consider the latest available information about their environment. Thus, AV sensor processing usually requires high-speed data transfer, major computing power, and efficient algorithms.

2.2 *Object Detection and Tracking*

AVs use their sensor data to detect objects in the environment and estimate their properties,[24] e.g., the shape and the type of the object. In general, objects can be classified into static objects, which are stationary, and dynamic objects, which are moving. For instance, traffic lights or trash cans are static objects, and pedestrians or other vehicles can be dynamic objects. Identifying if an object is static or dynamic is important for the AV to plan safe motions. Moreover, presumably static obstacles, such as parked cars, are dynamic obstacles that might start moving, which means that the AV needs to identify them as a dynamic obstacle instead. The most common required information about objects involves their position, orientation, shape (e.g., a box with dimensions), type, and velocity. This information is extracted by several classes of algorithms, such as object detection and tracking algorithms.

Object detection algorithms aim to determine object information from the gathered sensor data. The difficulty, however, is that these algorithms need to extract this information from noisy or partial data. For instance, in the case of point clouds (see Fig. 4), the object detection algorithm needs to extract such object information from only a few data points, e.g., the point clouds only describe one or two sides of the obstacles only. In particular, the shape estimation becomes difficult when the full shape is not visible in the sensors' sensing range. Partial views can be obtained when the cameras can only see an object from certain sides or when the object is being partly occluded by other objects, e.g., a pedestrian may be blocked by parked vehicles and hence, is only partially visible to the camera.

Whereas detection algorithms usually consider the obstacle detection at a specific time instance, tracking algorithms keep track of this object information over time to provide higher accuracies and less uncertainty in the estimation. For instance, keeping track of a vehicle's velocity over time allows one to reason about the validity of new velocity estimates. Updating the information over time can

[24] Steyer et al. (2018b).

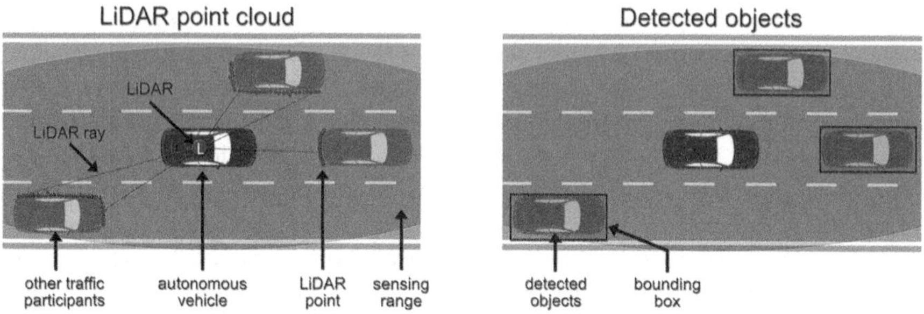

Fig. 4 LiDAR rays create a point cloud when reflected from obstacles in the environment. Object detection algorithms use this information to estimate the shape of the obstacles and provide bounding boxes

counteract errors in object detection. In another example, an object detection algorithm may incorrectly detect two adjacent bicycles that are driving close to each other, as a single car. The tracking algorithm keeps track of the objects over time and can use this stored information to determine that the two bicycles were previously detected as two separate objects.

One of the challenges in object tracking is to assign new sensor information to the correct object. For instance, the algorithm needs to determine that the new information belongs to the two previously seen bicycles. In case the information belongs to a new object, a new track needs to be created while removing old unused tracks.[25] To assign correct properties to an object or the desired parameters in algorithms, it is often helpful to also identify the type of an object, e.g., whether it is a car, a car with a trailer, a truck, a bicyclist, or a pedestrian. Identifying an object's type often requires analyzing its shape and its appearance,[26] e.g., passenger cars usually have rear windows compared to trucks and cyclists might wear helmets. Therefore, object classification algorithms also incorporate the camera images of the scene. State-of-the-art methods apply machine learning methods to mark the type of the object within the images.[27]

Object detection, tracking, and classification algorithms allow AVs to create a precise model of other objects in their environment by fusing and processing sensor data. Each object is usually assigned a bounding box that captures its shape, a state containing the position, orientation, and velocity, as well as its type and other characteristics, e.g., if a pedestrian is inattentive. Moreover, these algorithms also estimate the uncertainties of their results, e.g., the detected object is of type truck with 90% probability. These uncertainties are helpful, e.g., to decide when the AV needs to be more cautious.

[25] Steyer et al. (2018a).

[26] Ziegler et al. (2014b).

[27] Gao et al. (2018).

2.3 High-Definition Maps and Additional Information

Besides information on the AV's internal state and objects in the environment, AVs require additional information on their environment to safely perform driving tasks in real traffic.[28] Most importantly, they require a map that provides detailed information on lanes, drivable areas, and parking spaces among others.[29] These maps are used to plan safe motions that follow lanes without leaving the drivable areas. In contrast to maps used in classical navigation systems, the maps in AVs are high-definition (HD) maps, providing much finer resolution, up to centimeter level, and more information about how lanes are connected.[30] In general, only HD maps are sufficiently detailed to allow AVs to localize themselves precisely in their surroundings and to plan safe motions. The creation of HD maps is usually done offline, before the AVs operation on the road, and requires dedicated sensor equipment and algorithms to attain high accuracy. For these reasons, HD maps are often recorded by specialized mapping companies. It is especially difficult to detect lane markings, particularly when they are worn down, and to filter out objects that are not persistent in the map, e.g., temporary traffic cones. Ideally, AVs are also equipped with online mapping algorithms that refine the existing map during operation with new information obtained from sensors, e.g., construction site markings.

In addition to the information about drivable and non-drivable areas, maps often encode additional elements in different layers that are necessary for automated driving. For instance, the map's base layer may describe all the roads and lanes, a second layer encodes how the lanes are connected to each other, such as through intersections, and other layers include the position of other traffic elements, such as traffic lights and traffic signs. These elements are necessary, e.g., to predict the motion of other traffic participants (see Sect. 3) and to plan the AV's own motion (see Sect. 4) in accordance with traffic laws. Moreover, maps often encode traffic rules and mark in which areas those traffic rules are valid. For instance, different speed limits are valid for different parts of a highway. Additional layers may describe the location of construction sites, traffic jams, weather information, or height information, e.g., of bridges or tunnels. Figure 5 illustrates an example map with additional information such as the location of traffic lights and traffic signs.

Due to the large amount of information within a map, it is challenging to design a map format that allows easy access of information while being data efficient in storage.[31] Popular map formats include OpenDrive,[32] NDS,[33] and Lanelets.[34]

[28] Ziegler et al. (2014b).

[29] Seif and Hu (2016).

[30] Van Brummelen et al. (2018).

[31] Liu et al. (2020).

[32] Dupuis et al. (2010).

[33] Kwoczek et al. (2015).

[34] Poggenhans et al. (2018).

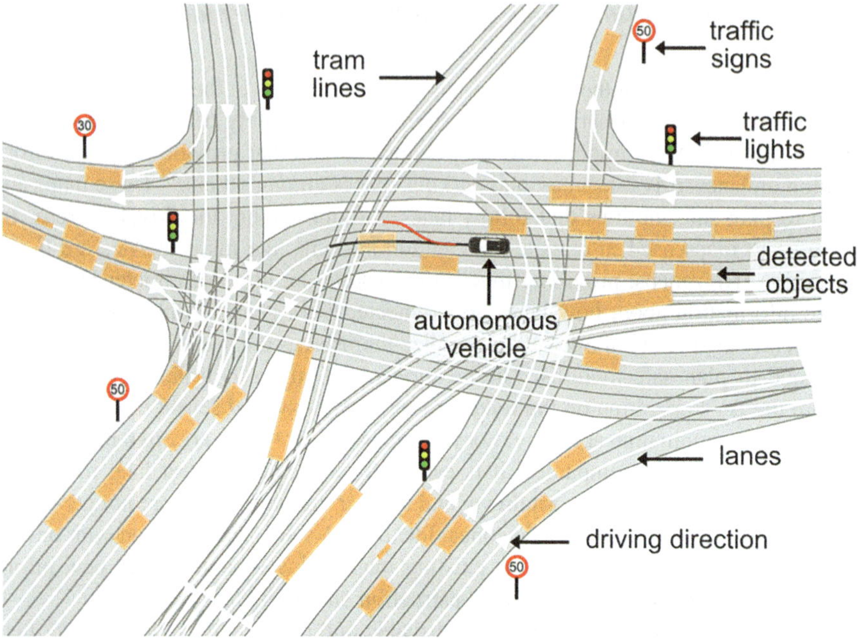

Fig. 5 Example illustration of a map representation in the Lanelet format with additional traffic information for the Stachus area in Munich, Germany

Another challenge is to update specific information of the map online,[35] during the AV's operation on the road, e.g., because the map has changed, or the current state of traffic lights is required.

3 Predicting Other Traffic Participants

The nature of traffic scenarios is inherently dynamic, with other traffic participants around the AV constantly performing various maneuvers.[36] AVs need to consider these behaviors of co-traffic to plan safe motions for themselves that do not interfere with other traffic participants, e.g., do not result in collisions. In general, other traffic participants may be other AVs, human-driven vehicles, or pedestrians. Human drivers are often able to intuitively predict the intentions of other traffic participants. For AVs, however, it is rather difficult to accurately predict the intentions and future motions of other traffic participants.[37] The difficulties lie in the uncertain nature of

[35] Van Brummelen et al. (2018).

[36] Lefèvre et al. (2014).

[37] Gulzar et al. (2021).

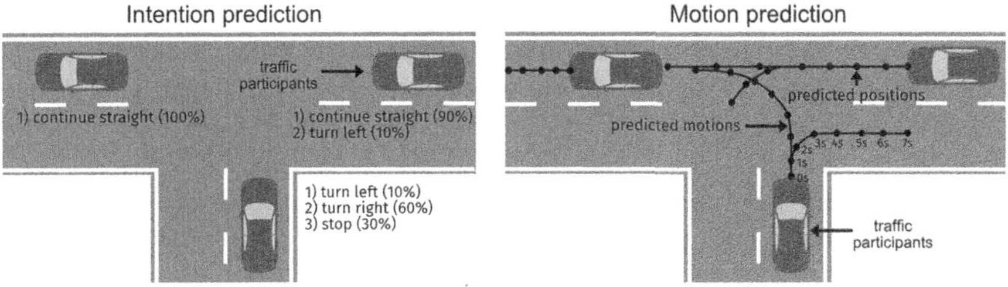

Fig. 6 In intention prediction, algorithms predict the probabilities of possible intentions/maneuvers of other traffic participants. In motion prediction, algorithms predict the actual motion of traffic participants over time

human decisions, that might be irrational at times, and the complexity of encoding the cognitive processes that human drivers use to make predictions. Prediction algorithms aim at predicting the future intentions and motions of other traffic participants by creating models and incorporating the available information in the AVs environment model.

3.1 High-Level Intentions

In many situations, the AV needs to base its decisions on the high-level intentions of other traffic participants.[38] Such high-level intentions may comprise lane changes, turns, braking maneuvers, reversing or parking maneuvers for vehicles, or stopping and crossing roads for pedestrians.[39] To predict high-level intentions of a traffic participant, algorithms usually first determine the feasible intentions, i.e., which intentions the traffic participants can have given their current state, including but not limited to their position, orientation, and velocity, and location in the road network. For instance, the vehicle in Fig. 6 is driving towards a T-junction and can only perform a stop maneuver, a left turn, or a right turn maneuver. When also considering traffic rules, e.g., a right-of-way rule, and interaction with other traffic participants, this vehicle might not turn left when a vehicle is coming from the right-hand side. Considering the set of feasible intentions, prediction algorithms assign probabilities to each of the possible intentions, allowing the AV to analyze which of the intentions is most likely.

The computation of intention probabilities often requires estimating the desired goals of traffic participants. Intentions that help traffic participants to get closer to their goal are usually more likely than intentions that lead further away from goals. In addition, the interaction with other traffic participants plays an important role in

[38] Lefèvre et al. (2014).

[39] Deo et al. (2018).

estimating probabilities.[40] For instance, a car that is driving behind a slower truck might overtake the truck to be able to drive at a higher velocity, or a pedestrian might suddenly cross the road to get to the bus on the other side. Modeling and evaluating possible interactions between traffic participants is a difficult task, since the behavior of one traffic participant might, directly or indirectly, influence the behaviors of all other traffic participants in a traffic scene. For this reason, prediction algorithms often consider a window of past data of traffic participants (e.g., the last 10 seconds) to provide better estimates.

3.2 *Motion Prediction*

Although predicting the intentions of other traffic participants supports AVs in their decision-making process, high-level intentions are not enough for safe motion planning. For safe motion planning, it is necessary to estimate the exact positions and orientations of traffic participants over time.[41] For example, when a vehicle performs a left turn, as illustrated in Fig. 6, where exactly will this vehicle be located 1, 2, or 3 seconds ahead in time and with which orientation? With this motion prediction information at hand, AVs can determine which space on the road is possibly occupied with another traffic participant at a certain point in time. When planning its motion, the AV can avoid these occupied regions to not collide with any other traffic participant. For this safety reason, it is necessary to provide motion predictions as accurately as possible and to provide uncertainty bounds on it, e.g., the prediction might hold with 95% probability.

Motion predictions are often done by considering the current states of traffic participants as well as their possible intentions.[42] This information is, for example, fed into a motion model that simulates the possible motion of the traffic participants in the environment for a given time horizon, e.g., 5 seconds. The motion model returns the states of the traffic participant for the queried points in time considering the provided time horizon. These motion models are often similar in nature to the motion planner of the AV, since they produce motions for a given traffic scenario. Yet, motion prediction algorithms do not aim to execute such motions on a real vehicle and focus more on determining possible alternative motions given the intentions of traffic participants. In general, these algorithms can output one motion prediction or multiple, even for a single intention. They assign probabilities to the resulting motions to denote which motion prediction is more likely than another. The algorithms and models can be designed by engineers or developed in a data-driven fashion by extracting models from large datasets of example motions, e.g., recorded in real traffic.

[40] Schulz et al. (2017).

[41] Petti and Fraichard (2005).

[42] Schulz et al. (2018).

4 Motion Planning

Motion planning comprises the steps that are necessary to move the AV, from finding a route to the destination to controlling the throttle and steering wheel actuators.[43,44,45] The goal for the motion planning modules is to generate comfortable and safe motions. Motions are safe when they are collision-free with respect to the environment model and the motion prediction of other traffic participants in the current traffic scenario.[46] To reduce the computational complexity, the motion planning of AVs is often decomposed into four levels,[47] as illustrated in Fig. 7. First, algorithms determine a rough route to the destination. Second, the appropriate high-level decisions, such as turning left or right, are decided to reach that goal destination. These decisions are mapped to a fine-granular trajectory and finally, converted to the actuator commands, i.e., throttle and steering wheel positions.

4.1 Route Planning

The first step in motion planning is to determine a route to the desired destination that the AV should follow. Route planning is done similarly as in present day navigations systems in commercial vehicles. The route planning module uses the AVs map and current position to plan a route to the desired destination. This route is usually optimized for certain efficiency criteria such as being the shortest or fastest route.[48] Besides the map information, AVs can use their communication system to obtain real-time traffic information to adjust the route during the AV's operation. The route spans the entire way from the current position of the AV to the desired destination. It describes coarse information which roads the AV should follow. Route planning algorithms are often based on path planning approaches from the classical artificial intelligence field. Yet, more intelligent routings are nowadays achieved by integrating advanced machine learning components that have been trained on large scale traffic data, e.g., to provide traffic congestion estimates. In Fig. 7, the AV is tasked to drive to its destination city "Stockholm". The route planner determines a route covering the roads the AV needs to drive along towards the destination.

[43] Gonzalez et al. (2016).

[44] Paden et al. (2016).

[45] Claussmann et al. (2019).

[46] Petti and Fraichard (2005).

[47] Paden et al. (2016).

[48] Nha et al. (2012).

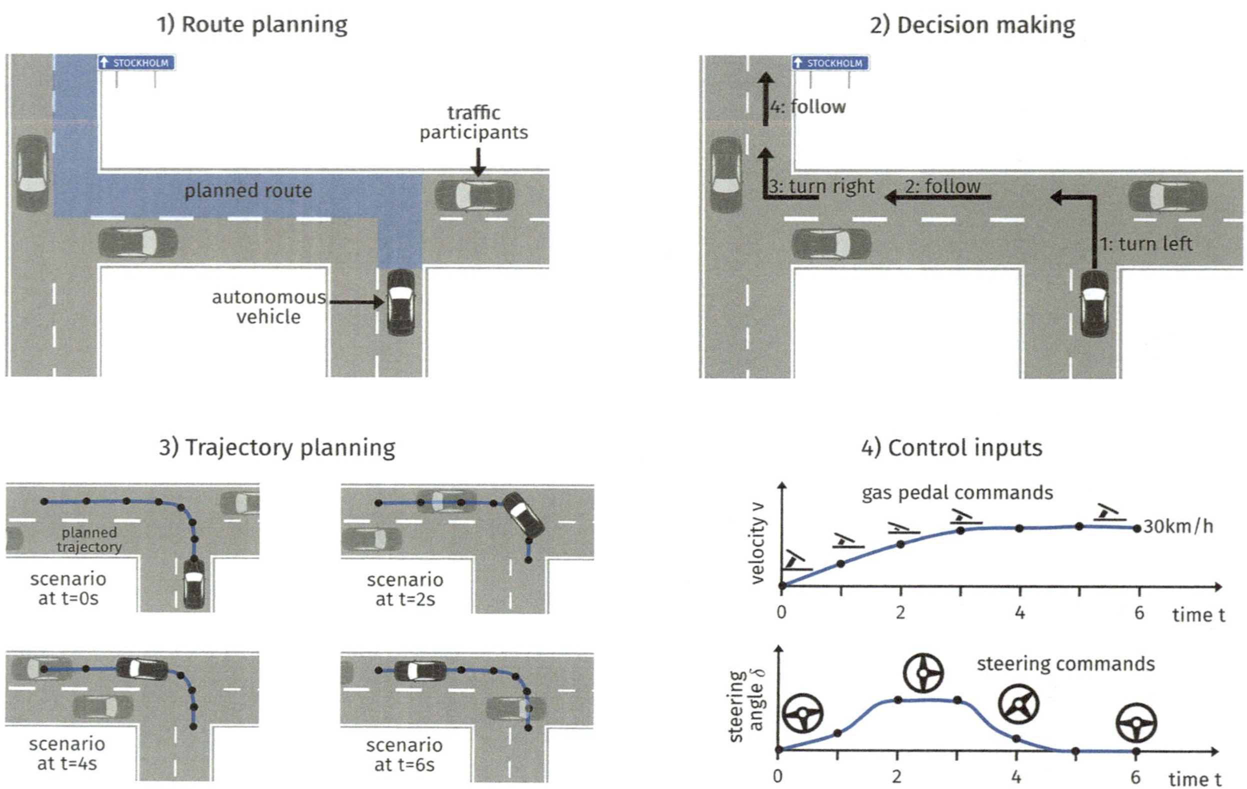

Fig. 7 Motion planning for autonomous vehicles is often decomposed into four layers. First, the AV plans a route to its destination. Subsequently, it plans a sequence of maneuvers to follow the route. These maneuvers are translated into a precise trajectory using a trajectory planner. Lastly, this trajectory is tracked by a controller that provides commands to the throttle, brake, and steering wheel actuators

4.2 *Decision Making*

Routes only describe rough information on how to reach the destination. They do not provide the AV with information on possible road changes and how to get there. AVs need to plan a series of high-level decisions to follow the planned route.[49] These high-level decisions usually cover distinct maneuvers that the AV can perform to follow the planned route.[50] For instance, high-level decisions may be turning left, turning right, changing lanes, following a lane, or stopping. The series of high-level decisions represent a more fine-grained way to reach the destination while accounting for the current traffic situation. The considered time horizon of decision-making algorithms ranges from multiple seconds to multiple minutes. Often, the decisions are recomputed multiple times per second (e.g., every 200 milliseconds) or on demand when the traffic situation changes.

To generate safe decisions sequences, the algorithms take the traffic situation and traffic rules into account. For instance, the decision-making algorithm may decide to perform a lane change earlier to get the highway exit when there is a traffic jam ahead of the AV or to yield to another traffic participant at a T-junction. Decision-making algorithms may be as simple as a rule-based system ("if X happens, do Y"), a complex state machine with conditions, or even be a deep learning-based system that has been trained on millions of driving examples. The challenge of decision-making is to anticipate the future intentions or motions of other traffic participants and to perform proactive decisions to closely follow the determined route. Moreover, the decisions need to be comfortable for passengers and should not disrupt traffic flow. In Fig. 7, the AV's decision-making algorithm may determine a sequence of four maneuvers to follow the route to its destination "Stockholm". The AV needs to first turn left, follow the lane and at the end of the lane turn right and follow the lane to the destination.

4.3 *Trajectory Planning*

Although decisions provide more detailed instructions on how to follow the path, these decisions may be executed in different variations, e.g., lane changes can be executed by merging quickly or slowly into the adjacent lane. Trajectory planning aims at generating the exact motion, or trajectory, of the AV from the given sequence of decisions or single decisions. Trajectories usually describe the planned future positions, velocities, and orientations of the AV at given future time steps over a

[49] Schwarting et al. (2018).

[50] Hubmann et al. (2017).

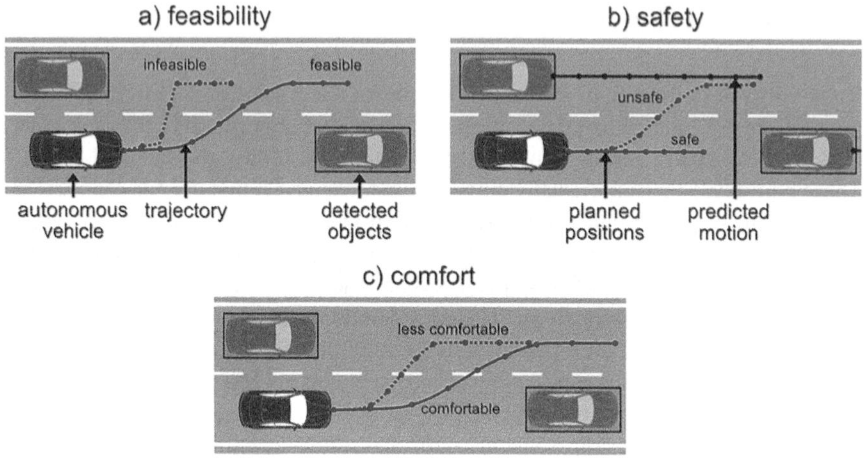

Fig. 8 Planned motions of the AV need to be feasible with respect to its vehicle dynamics, be safe considering the predictions of other traffic participants, and be comfortable for passengers

predetermined planning horizon,[51,52,53] usually in the range of 5 seconds to 12 seconds. Trajectories are computed at higher frequency than decisions. Usually, AVs recompute trajectories 10 to 50 times per second.

Trajectory planning algorithms solve the two-point boundary value problem.[54] This problem describes finding a trajectory from a given initial state (i.e., the current position, orientation, and velocity of the AV) to a target state (i.e., closely following the route) over time. To be drivable by the AV and accepted by passengers of the AV, this trajectory further needs to be:

(a) dynamically feasible: trajectories are not allowed to violate the vehicle's driving dynamics, e.g., have abrupt steering angle changes as illustrated in Fig. 8. Otherwise, the AV cannot follow the trajectory or might even become uncontrollable.
(b) safe: trajectories need to be collision-free with respect to the predicted future motions of other traffic participants. Only if the planned positions of the AV (and its shape) are never intersecting with any other position (and shapes) of other traffic participants and objects over time, the AV can safely execute the trajectory, as illustrated with the unsafe lane change in Fig. 8.
(c) comfortable: trajectories can lead the AV to harsh accelerations or quick lane changes. These motions are often perceived as unsafe by passengers. For these reasons, trajectories usually optimize a comfort assessment criterion, e.g., small accelerations when performing lane changes as illustrated in Fig. 8.

---

[51] Ziegler et al. (2014a).
[52] Werling et al. (2010).
[53] Gutjahr et al. (2016).
[54] LaValle (2006).

More details on perceived safety and comfort of trajectories can be found in Sect. 5.4. Figure 7 illustrates a planned trajectory of the AV for our example of driving to the destination Stockholm. This figure also illustrates the positions of the AV and other traffic participants over time (considering the motion predictions) to validate that the planned trajectory is collision-free over its planning horizon.

4.4 Control Inputs

The last layer in the motion planning hierarchy is the control layer. This layer translates the planned trajectory into precise control inputs for the AV's throttle, brake, and steering wheel. The aim of the controllers is to follow the planned trajectory as close as possible over time. Therefore, controllers operate at a high frequency (of up to 1000 computations per second) to account for possible deviations during/after the execution of the previous control input. These deviations occur, since the used vehicle models are never perfect, and the AV is a dynamical system with physical limitations. There are often residual errors that cannot be modeled, are caused by uncertain sensor measurements, or cannot be directly measured with sensors, e.g., road surface changes or mechanical wear of tires.

Control errors should not accumulate over time since they cause deviations from the planned trajectory and thus might cause unsafe situations. These unsafe situations are a result of the planning hierarchy: only the trajectory planner is usually accounting for collision-free trajectories (to reduce computational complexity) and the control layer is responsible for the precise execution of such trajectories. Therefore, in each control loop, the controllers compute the errors between the current motion of the AV and the planned trajectory. Considering these errors, they generate new control inputs to steer the AV along its trajectory while minimizing the errors. The generated control inputs need to be feasible given the actuator limits (e.g., not requesting a steering angle larger than what is considered feasible by the AV's steering system) and not lead to undesired temporal behaviors of the AV such as instabilities or oscillations. The challenges for controller design are to identify vehicle models that are as accurate as possible, proactively accounting for possible errors, and providing performance guarantees of the resulting controller.

5 Important Technical Challenges

The research and development of AVs has made significant progress in various areas since the presentations of first proof-of-concept vehicles. Yet, the promised market introduction years of various companies have been postponed and the world only sees test vehicles or vehicles with very limited operation up until now on roads. The most crucial factor in the development is to build AVs with high reliability and it emerges more and more that human drivers are highly reliable drivers contrary to media reports.[55] Therefore, building an AV that is as reliable as a human driver or even better than a human driver requires sophisticated research techniques as well as large scale validation of the systems on real roads. The following paragraphs summarize a selection of four important technical challenges that are seen as crucial building blocks towards reliable AVs, namely accounting for uncertainties,[56] ensuring safe motions,[57] improving machine learning models,[58] and increasing social autonomy.[59]

5.1 Accounting for Uncertain and Partially Observable Traffic Scenarios

To enable fully autonomous driving with high reliability, AVs need to account for the various uncertainties in traffic scenarios during their operation.[60] These uncertainties range from uncertain sensor measurements and control uncertainties to uncertain motion predictions of other traffic participants and possibly occluded traffic participants. Humans can often intuitively reason about situations with high uncertainties, however, teaching an AV to properly address and react to uncertainties is considerably harder. Not properly considering these uncertainties might lead to unsafe situations from which the AV will not be able to recover.

Uncertainties in measurements and controllers can be addressed with appropriate models and providing estimates of the remaining uncertainties. These estimates can be used to plan more conversative but safer trajectories for the AV. For instance, measurement uncertainties in the current state of other traffic participants (e.g., position and velocity) can be incorporated in prediction modules by enlarging the possibly occupied space of other traffic participants, to create a safe buffer. Figure 9 illustrates that the uncertain position of a traffic participant can be approximated with a geometric object, such as a circle, that covers all the possible positions within a given uncertainty bound (e.g., the position is in within this circle with a probability of 95%). Yet, it is difficult to appropriately model uncertainty from the sensor level, since it is difficult to propagate all small errors from the physical measurement to the algorithmic analysis of the data. State-of-the-art approaches aim at creating better models and more fine-grained uncertainty notions across different processing steps.

Uncertainties that arise from other traffic participants and the traffic scenario itself are often more challenging, since they are usually partially observable by the

[55] Kalra and Paddock (2016).

[56] Schwarting et al. (2018).

[57] Koopman and Wagner (2016).

[58] Roy et al. (2021).

[59] Maurer et al. (2016).

[60] Schwarting et al. (2018).

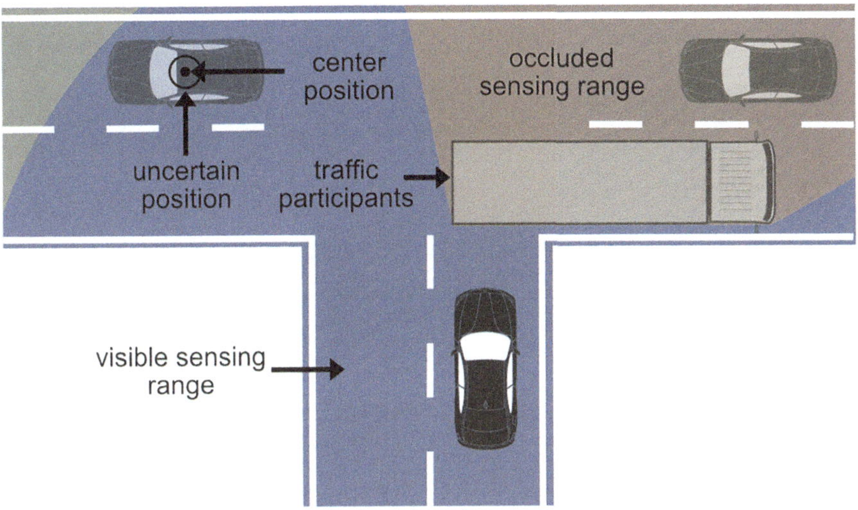

Fig. 9 Measured positions of other traffic participants may be uncertain due to sensor uncertainties. Other traffic participants may be occluded through another traffic participant, making it difficult for the AV to observe the whole environment

AV. For instance, the AV can observe the current position of another traffic participant but not its intention and future motion. These variables need to be estimated through prediction algorithms. The longer these algorithms need to predict into the future, the more uncertain the predictions become. In addition, human drivers might behave irrationally or are hard to accurately predict due to possible and sudden changes in their driving behavior, leading to potentially wrong predictions. Yet, in other situations in which other traffic participants are outside of the sensor range of the AV or occluded by another traffic participant (as illustrated in Fig. 9), the AV cannot even observe all traffic participants.

Such traffic situations require new reasoning algorithms that aim to estimate missing variables while providing possible estimation errors. However, all the uncertainties can never be completely reduced, resulting in residual risks for the AV during operation.

5.2 *Ensuring Safe Motions*

The highest priority in the development of AVs is to ensure that planned motions of the AV are always safe for passengers and surrounding traffic participants.[61] Planned motions of the AV are only safe if (1) all safety-relevant objects have been detected, (2) the utilized motion prediction of other traffic participants is correct, and (3) the

[61] Schwarting et al. (2018).

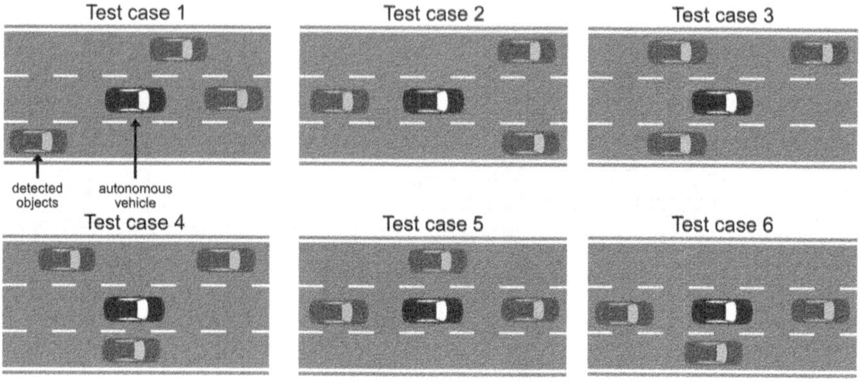

Fig. 10 Illustration of six possible test cases in a simple three lane highway scenario with three other traffic participants. The scenarios are created by moving the traffic participants and changing their velocities

trajectory can be followed by the AV without errors.[62] The automotive industry has often solely relied on testing their systems (e.g., adaptive cruise control systems) on a large variety of test cases.[63] Figure 10 illustrates six possible test cases for a three-lane highway scenario with three other traffic participants. These test cases can be created in simulation by varying the positions and velocities of all traffic participants. However, just testing alone is not enough for autonomous systems that should act without human intervention and corrections. Due to the complexity of AVs and the uncertainties of traffic scenarios, developers need to create billions of test cases to conduct statistical analyses. Even for the allegedly simple scenario in Fig. 10 with three additional traffic participants, one can generate more test cases than the number of atoms existing in the whole universe[64] (estimated to be between 10^{78} to 10^{82}). Testing becomes infeasible for large numbers of test cases and selecting a smaller number of representative test cases instead is challenging and not straightforward.[65] Finally, testing can only reveal the presence of failures (when a test case fails) but never that there are no failures in the system, since the failure might occur in a yet untested scenario.

Formal verification approaches can support the safety certification process of AVs in addition to testing by rigorously proving that the AV system adheres to a given design specification.[66] These approaches can verify whether a given model of a system (e.g., a motion planner) adheres to a given formal specification (e.g., a safety specification). The term formal refers to the fact that the specification as well as the proof itself are coded in a mathematically rigorous framework that allows the

[62] Petti and Fraichard (2005).

[63] Koopman and Wagner (2016).

[64] Pek et al. (2019).

[65] Koopman and Wagner (2017).

[66] Wongpiromsarn and Murray (2008).

deduction of properties. For instance, a designer can provide a mathematical way to compute a safe distance that the AV needs to always maintain to a preceding vehicle.[67,68] A formal verification approach can then prove that the model of the motion planner always satisfies this safe distance specification or when it violates it, it may provide the designer with an example of where the system violates the safe distance.

Classical verification approaches perform the verification process offline during the design process, meaning that all necessary information for the proof needs to be present. However, for uncertain variables, such as the positions or future motions of traffic participants, offline verification may not be applicable. For this reason, new online verification approaches are developed that perform the verification process online during the operation of the AV. For instance, every time the AV plans a new trajectory, this trajectory is verified against the safety specification.[69,70] If successful, the AV is allowed to execute the verified trajectory, otherwise, the AV may execute a fail-safe trajectory, or back-up plan, that serves as a safe alternative trajectory, e.g., stopping the vehicle.[71] Online verification approaches have shown great success, but more research is needed to provide rigorous guarantees. Moreover, regulators need to standardize a suitable safety specification that AVs need to fulfill[72] and discuss the social dilemma in which an AV cannot avoid an accident anymore.[73,74]

5.3 Improving Machine Learning Models

It is expected that commercial AVs will rely on complex machine learning (ML) components for their operation, ranging from environment modeling[75] to decision making.[76] Yet, analyzing the performance and safety of many ML components, such as neural networks, is a challenging and often even an infeasible task.[77] This problem conflicts with the goal of building reliable and safe AVs. ML components use example data to learn tasks. For instance, developers may provide the ML system with camera images of a traffic scenario in which each type of a traffic participant (e.g., vehicle, bus, pedestrian, or bicyclist) is marked with a label

[67] Pek and Althoff (2018).
[68] Shalev-Shwartz et al. (2017).
[69] Pek et al. (2020).
[70] Vaskov et al. (2019).
[71] Pek and Althoff (2020).
[72] Anderson et al. (2014).
[73] Bonnefon et al. (2016).
[74] Awad et al. (2018).
[75] Fayyad et al. (2020).
[76] You et al. (2018).
[77] Burton et al. (2017).

indicating the type. The ML system can then learn to identify types of traffic participants and locate them in new images.

The accuracy of ML systems heavily depends on the amount and quality of the data that is used for training.[78] Complex tasks may require hundreds of millions of examples to train a system that can solve a desired task with high accuracy. Obtaining such a large amount of data is often difficult. Recording millions of camera images across the world takes time, especially when developers also need to provide the labels for each data point. Many companies employ simulations to synthesize artificial data for training. Yet, these simulators cannot produce realistic data (also called sim-to-real gap[79]). Furthermore, the training data needs to be diverse enough so that the system can generalize well to new, yet unseen data samples. For instance, if the training dataset only contains camera images with a classical city background, the system may struggle with new images that are captured outside of cities with forest backgrounds. This distribution shift in data may result in failures and unsafe actions of ML systems since the data is biased towards certain properties of the training data. Identifying these distribution shifts and biases in data is difficult and requires high expertise by the developers.

In addition to acquiring data, ML systems are often essentially black-box systems, in which it is difficult to understand how the system has come to a certain output and how much residual uncertainty about the output remains. These properties make it challenging to use them in highly safety-critical systems such as AVs, where unsafe decisions can lead to catastrophic outcomes. Current research aims at making the decision process of ML systems interpretable for users. Moreover, these systems are equipped with techniques to output a residual uncertainty estimate about their decisions. These changes ideally help to identify the cause of incorrect decisions and to not rely on decisions with high uncertainty. Yet, these approaches are not yet applicable to many large ML systems due to the sheer size and complexity of them (e.g., a neural network with millions of artificial neurons). Lastly, researchers are aiming at improving the data-efficiency of ML systems to require less training data and use a mixture of synthetic and real data for generalized decision.

5.4 Increasing Social Autonomy

Until all vehicles are fully autonomous, AVs need to consider the drivers of other vehicles. Even when all motor vehicles on the road operate fully autonomously, AVs will always interact with other road users, like pedestrians and cyclists.[80] Many factors are considered in the interactions between traffic participants, for example, vehicle speed, eye contact, head or arm gestures to signal intent. Section 3 discussed

[78] Jordan and Mitchell (2015).

[79] Zhao et al. (2020).

[80] Klein et al. (2004), p. 200.

how AVs can predict and understand other traffic participants' behaviors and intentions using prediction algorithms. In a similar fashion, AVs need to communicate their intent and planned behavior to their passengers and traffic participants around the vehicle. This section discusses how AVs should communicate their state, intent, and planned motion through Human-Machine Interfaces (HMIs), how pedestrians perceive AVs, and the role of driving styles and perceived safety of AVs.

Human-machine interfaces can be divided into internal, which communicate with the passengers in the vehicle, and external, which communicate with traffic participants around the vehicle. The communication can be through messages, audio or visual, using lights, or sounds. The next paragraphs discuss the communication within the vehicle first, followed by the communication to traffic participants around the vehicle.

Internal human-machine interfaces (iHMIs) are designed to help remain passengers aware of the internal status of the vehicle, including intention, planned actions, levels of uncertainty about those planned actions.[81] Such iHMIs need to inform the passenger of the AV's plan, changes to the initial plan, and it needs to enable the passenger to give instructions to the AV on what to do.[82] These interfaces are important, because higher levels of automation shift the role of a human driver to a more supervisory role. Yet, uncertainty is detected by the system, the vehicle can still request the driver to take over control. When automated driving functions take over, people are no longer actively driving, and this can lead to driver disengagement.[83] Prolonged periods of uninterrupted, reliable autonomous driving may also lead to overreliance or too much trust in the system, and lead to misuse and disuse.[84] This can be a problem, when the passenger needs to stay alert and aware of the vehicle and its surroundings to act when the vehicle requests intervention. To facilitate such shared driving autonomy, iHMIs need to consider important human factors such as acceptance, trust, and driver's reactions to system failures.[85] In addition to determining which decisions to leave to the driver and what information to display to facilitate their situation awareness, it is important for driving comfort to allow the driver to intervene when the AVs intentions do not align with their own. For example, when the driver prefers a route different than that chosen by the AV, they should be able to overwrite this route.

External human-machine interfaces (eHMIs) are used to communicate with traffic participants around the vehicle. While automated driving reduces the risks for accidents caused by human factors, such as tiredness, drowsiness, or distraction, the removal of a human driver in fully autonomous vehicles means that other traffic participants cannot rely on non-verbal communication, such as eye contact or gestures, to signal intentions and receive acknowledgments. This change in how

[81] Beggiato et al. (2015).

[82] Hancock et al. (2019).

[83] Saxby et al. (2007).

[84] Lee and See (2004), p. 20.

[85] Hancock et al. (2019).

pedestrians and cars communicate affects pedestrians' perceived safety. For example, eye contact between a pedestrian and a driver before the pedestrian crosses a road confirms that a driver notices the crossing pedestrian. The driver could wave their hand to signal that they noticed the pedestrian and let them cross.

As fully autonomous driving is not yet fully developed, on-road interactions between fully AVs and pedestrians remain less studied. In one study with a mock-up driverless car, the driver wore a costume to make it appear as if there was no one behind the wheel to make a conventional car appear autonomous.[86] This study showed that pedestrians tend to stick with familiar interaction patterns with the vehicle. For example, they would look for eye contact with the driver. However, the results also showed people's ability to adapt when eye contact was not possible, due to the absence of a human driver. While the current laws of many countries require AVs to have a human operator in the driver seat, this study provides insights into how people interact with a fully autonomous, driverless car. Once vehicles are fully autonomous driving, passengers may be in the front seats performing secondary tasks that are not related to the driving itself, which can be stressful to pedestrians and bicyclists around the vehicle who might not know if the vehicle is autonomous and might assume AVs will always yield. AVs can adopt interaction patterns with pedestrians like those of conventional cars. They can implicitly communicate their intents, by adapting the vehicle motion itself. For example, a car can slow down to leave a larger gap when entering an intersection to signal their intention to let someone else go first. Similarly, by adopting a certain driving style,[87] AVs might implicitly communicate different intents. For example, aggressive driving styles are likely to impact pedestrian trust of AVs negatively.[88] Yet, adopting a defensive driving style might invite people to take advantage of the autonomous system, e.g., by unrightfully taking the right of way because the autonomous car will make a forced stop. People know that autonomous vehicles will force stop when something is in front of them and could take advantage of this. Alternatively, via external HMIs, such as the use of displays or lights can be used to convey the AVs intent by means of explicit communication. Such external HMIs to communicate the AVs intent and status have shown to improve pedestrians' sense of safety and contribute to a positive experience with AVs. However, lights are difficult to see in bright daylight and additional communication channels may overload traffic participants.

AVs can adopt interaction patterns with pedestrians like those of conventional cars. They can implicitly communicate their intents, by adapting the vehicle motion itself. For example, a car can slow down to leave a larger gap when entering an intersection to signal their intention to let someone else go first. Similarly, AVs can adopt a more or less aggressive driving style to incorporate people's preferences[89] or to implicitly communicate different intents. For example, aggressive driving styles

[86] Rothenbücher et al. (2016).

[87] Karlsson et al. (2021).

[88] Jayaraman et al. (2019).

[89] Karlsson et al. (2021).

are likely to impact pedestrian trust of AVs negatively.[90] Yet, adopting a defensive driving style might invite people to take advantage of the autonomous system, e.g., by unrightfully taking the right of way because the autonomous car will make a forced stop. People know that autonomous vehicles will force stop when something is in front of them and could take advantage of this. Alternatively, via external HMIs, such as the use of displays or lights can be used to convey the AVs intent by means of explicit communication. Such external HMIs to communicate the AVs intent and status have shown to improve pedestrians' sense of safety and contribute to a positive experience with AVs. However, lights are difficult to see in bright daylight and additional communication channels may overload traffic participants.

On-road safety and perceived safety are both crucial in traffic. Fully autonomous driving brings new challenges when it comes to responsibility, does the AV have any form of responsibility? If the driver is not the one making the driving decisions, and even performing secondary tasks while being driven, are they responsible for the AV's actions and decisions? Understanding how the level of autonomy relates to user acceptance is crucial for the acceptance of AVs.[91] Brown[92] points out the potential Uncanny Valley problem, which describes the phenomenon that people feel a sense of unease when seeing or interacting with highly human-like robots, that strongly resemble but are not exactly human. If AVs behave highly similarly to human driving, but not quite yet exactly like humans, people might experience negative attitudes or feelings towards those vehicles.

6 Conclusions

Autonomous vehicles have the potential to revolutionize mobility for many. They can increase traffic safety, improve traffic flow, relieve the driver from the driving task, which may be difficult or tiring, and enable mobility for people who have no or limited access to mobility. The research and development of AVs has drastically increased since the mass introduction of classical driver assistance systems and various companies around the world work on releasing commercial vehicles to the market for each automation level. AVs operate by creating a precise model of their environment through sensors, planning safe and comfortable motions, and controlling the vehicle's actuators to accurately follow these motions.

Although the basic principles of complex robotic systems such as AVs have not changed much over the years, the algorithms and techniques that are being used in AVs are advancing at rapid speeds. These advancements are necessary to provide high safety and reliability levels for AVs. Unsafe decisions may result in economic losses but also endanger the lives of passengers and surrounding traffic participants.

---

[90] Jayaraman et al. (2019).

[91] Hewitt et al. (2019).

[92] Brown (2017).

These high requirements and adverse and challenging conditions, such as densely populated areas or harsh weather conditions, make it difficult to deploy AVs in real-world traffic on large scale.

Many challenges still need to be solved to release commercial AVs into the market that are reliably, safe and accepted by users and other traffic participants. Uncertainties in various components of AVs pose dangers to derive incorrect and unsafe decisions. Algorithms need to be more robust against uncertainties and it requires more understanding of uncertainties and how to measure residual risks. Safety verification techniques support the development process of AVs by rigorously proving that certain components always satisfy a given safety specification. Yet, the application of such techniques requires law makers to decide on common safety standards for AVs. These standards are particularly important since AVs will be equipped with complex machine learning systems. These systems are difficult to verify, require large amounts of training data, and their decisions often cannot be interpreted by developers. Lastly, it is important to remember that AVs need to operate with and round humans. This circumstance requires more research on the interaction between AVs and humans to provide comfortable rides that are perceived as safe but also new ways of interacting to make decisions and intentions of AVs understandable. To conclude, recent advances in algorithms and techniques for safe and reliable AVs have shown initial success but challenges of robustness against uncertainties, human-AV interaction, and legal safety standards for AVs need to be addressed before we can realize the full potential of commercial AVs to revolutionize mobility.

References

Anderson JM, Kalra N, Stanley KD, Sorensen P, Samaras C, Oluwatola OA (2014) Autonomous vehicle technology: a guide for policymakers. Rand Corporation, Santa Monica, CA

Awad E, Dsouza S, Kim R, Schulz J, Henrich J, Shariff A, Bonnefon JF, Rahwan I (2018) The moral machine experiment. Nature 563(7729):59–64

Bagloee SA, Tavana M, Asadi M, Oliver T (2016) Autonomous vehicles: challenges, opportunities, and future implications for transportation policies. J Mod Transp 24(4):284–303

Beggiato M, Hartwich F, Schleinitz K, Krems J, Othersen I, Petermann-Stock I (2015) What would drivers like to know during automated driving? information needs at different levels of automation. In 7. Tagung Fahrerassistenzsysteme

Bengler K, Dietmayer K, Farber B, Maurer M, Stiller C, Winner H (2014) Three decades of driver assistance systems: Review and future perspectives. IEEE Intell Transp Syst Magaz 6(4):6–22

Bonnefon JF, Shariff A, Rahwan I (2016) The social dilemma of autonomous vehicles. Science 352(6293):1573–1576

Brooks R (1986) A Robust layered control system for a mobile Robot. IEEE J Robot Automat 2(1): 14–23

Brown B (2017) The social life of autonomous cars. Computer 50(2):92–96

Buehler M, Iagnemma K, Singh S (2009) The DARPA Urban challenge: autonomous vehicles in city traffic. In: Springer tracts in advanced Robotics, vol 56. Springer

Burton S, Gauerhof L, Heinzemann C (2017) Making the case for safety of machine learning in highly automated driving. In: International Conference on Computer Safety, Reliability, and Security

Claussmann L, Revilloud M, Gruyer D, Glaser S (2019) A review of motion planning for highway autonomous driving. Transact Intell Transp Syst 21(5):1826–1848

Deo N, Rangesh A, Trivedi MM (2018) How would surround vehicles move? A unified framework for maneuver classification and motion prediction. IEEE Transact Intell Vehicles 3(2):129–140

Dupuis M, Strobl M, Grezlikowski H (2010) Opendrive 2010 and beyond–status and future of the de facto standard for the description of road networks. In: Driving Simulation Conference Europe

Fayyad J, Jaradat MA, Gruyer D, Najjaran H (2020) Deep learning sensor fusion for autonomous vehicle perception and localization: a review. Sensors 20(15):4220

Gao H, Cheng B, Wang J, Li K, Zhao J, Li D (2018) Object classification using CNN-based fusion of vision and LIDAR in autonomous vehicle environment. IEEE Transact Ind Inf 14(9): 4224–4231

Gonzalez D, Perez J, Milanes V, Nashashibi F (2016) A review of motion planning techniques for automated vehicles. IEEE Transact Intell Transport Syst 17(4):1135–1145. https://doi.org/10.1109/TITS.2015.2498841

Gulzar M, Muhammad Y, Muhammad N (2021) A Survey on motion prediction of pedestrians and vehicles for autonomous driving. IEEE Access 9:137957–137969

Gutjahr B, Pek C, Gröll L, Werling M (2016) Efficient trajectory optimization for vehicles using quadratic programming. Automatisierungstechnik 64(10):786–794

Hancock PA, Nourbakhsh I, Stewart J (2019) On the future of transportation in an era of automated and autonomous vehicles. Proc Natl Acad Sci 116(16):7684–7691

Hewitt C, Politis I, Amanatidis T, Sarkar A (2019) Assessing public perception of self-driving cars: the autonomous vehicle acceptance model. In: International Conference on Intelligent User Interfaces

Hubmann C, Becker M, Althoff D, Lenz D, Stiller C (2017) Decision making for autonomous driving considering interaction and uncertain prediction of surrounding vehicles. In: Intelligent Vehicles Symposium (IV), 2017 IEEE, 1671–1678. IEEE http://ieeexplore.ieee.org/abstract/document/7995949/

Jayaraman SK, Creech C, Tilbury DM, Yang XJ, Pradhan AK, Tsui KM, Robert LP Jr (2019) Pedestrian trust in automated vehicles: role of traffic signal and AV driving behavior. Front Robot AI 6:–117

Jordan MI, Mitchell TM (2015) Machine learning: trends, perspectives, and prospects. Science 349(6245):255–260

Kalra, N, Paddock SM (2016) Driving to safety: how many miles of driving would it take to demonstrate autonomous vehicle reliability? RAND Corporation, Santa Monica http://www.rand.org/pubs/research_reports/RR1478.html

Karlsson J, van Waveren S, Pek C, Torre I, Leite I, Tumova J (2021) Encoding human driving styles in motion planning for autonomous vehicles. In: IEEE International Conference on Robotics and Automation

Klein G, Woods DD, Bradshaw JM, Hoffman RR, Feltovich PJ (2004) Ten challenges for making automation a 'team player' in joint human-agent activity. IEEE Intell Syst 19(6):91–95

Köhler S, Goldhammer M, Zindler K, Doll K, Dietmeyer K (2015) Stereo-vision-based Pedestrian's intention detection in a moving vehicle. In: IEEE International Conference on Intelligent Transportation Systems

Koopman P, Wagner M (2016) Challenges in autonomous vehicle testing and validation. SAE Int J Transport Saf 4(1):15–24

Koopman P, Wagner M (2017) Autonomous vehicle safety: an interdisciplinary challenge. IEEE Intell Transport Syst Mag 9(1):90–96

Kwoczek S, Di Martino S, Rustemeyer T, Nejdl W (2015) An architecture to process massive vehicular traffic data. In: International Conference on P2P, Parallel, Grid, Cloud and Internet Computing

LaValle SM (2006) Planning algorithms. Cambridge University Press

Lee JD, See KA (2004) Trust in automation: designing for appropriate reliance. Human Fact 46(1): 50–80

Lefèvre S, Vasquez D, Laugier C (2014) A survey on motion prediction and risk assessment for intelligent vehicles. ROBOMECH 1(1):1–14

Liu R, Wang J, Zhang B (2020) High definition map for automated driving: overview and analysis. J Navig 73(2):324–341

Mallozzi P, Pelliccione P, Knauss A, Berger C, Mohammadiha N (2019) Autonomous vehicles: state of the art, future trends, and challenges. Automotive Syst Softw Eng:347–367

Martínez-Díaz M, Soriguera F (2018) Autonomous vehicles: theoretical and practical challenges. Transport Res Proc 33:275–282

Maurer M, Christian Gerdes J, Lenz B, Winner H (2016) Autonomous driving — technical, legal and social aspects. Springer, Berlin

Nha VTN, Djahel S, Murphy J (2012) A comparative study of vehicles' routing algorithms for route planning in smart cities. In: International Workshop on Vehicular Traffic Management for Smart Cities

Paden B, Cap M, Yong SZ, Yershov D, Frazzoli E (2016) A survey of motion planning and control techniques for self-driving urban vehicles. IEEE Transact Intell Veh 1(1):33–55. https://doi.org/10.1109/TIV.2016.2578706

Patole SM, Torlak M, Wang D, Ali M (2017) Automotive radars: a review of signal processing techniques. IEEE Signal Proc Mag 34(2):22–35

Pek C, Althoff M (2020) Fail-safe motion planning for online verification of autonomous vehicles using convex optimization. IEEE Transact Robot 37(3):798–814

Pek C, Koschi M, Althoff M (2019) An online verification framework for motion planning of self-driving vehicles with safety guarantees. AAET-Automatisiertes und vernetztes Fahren

Pek C, Manzinger S, Koschi M, Althoff M (2020) Using online verification to prevent autonomous vehicles from causing accidents. Nat Mach Intell 2(9):518–528

Pek C, Althoff M (2018) Efficient computation of invariably safe states for motion planning of self-driving vehicles. In: 2018 IEEE/RSJ International Conference on Intelligent Robots and Systems (IROS). IEEE, Madrid, Spain, pp 3523–3530. https://doi.org/10.1109/IROS.2018.8593597

Petti S, Fraichard T (2005) Safe motion planning in dynamic environments. In: Proceedings of the IEEE International Conference on Intelligent Robots and Systems, 2210–2215

Poggenhans F, Pauls JH, Janosovits J, Orf S, Naumann M, Kuhnt F, Mayr M (2018) Lanelet2: a high-definition map framework for the future of automated driving. In: IEEE International Conference on Intelligent Transportation

Rezaei S, Sengupta R (2007) Kalman filter-based integration of DGPS and vehicle sensors for localization. IEEE Transact Contr Syst Technol 15(6):1080–1088

Rothenbücher D, Li J, Sirkin D, Mok B, Ju W (2016) Ghost driver: a field study investigating the interaction between pedestrians and driverless vehicles. In: IEEE international symposium on robot and human interactive communication

Roy N, Posner I, Barfoot T, Beaudoin P, Bengio Y, Bohg J, Brock O (2021) From machine learning to Robotics: challenges and opportunities for embodied intelligence

Saxby DJ, Matthews G, Hitchcock EM, Warm JS (2007) Development of active and passive fatigue manipulations using a driving simulator. Proc Human Fact Ergon Soc Annual Meet 51(18): 1237–1241

Schulz J, Hirsenkorn K, Löchner J, Werling M, Burschka D (2017) Estimation of collective maneuvers through cooperative multi-agent planning. In: Proceeding of the IEEE Intelligent Vehicles Symposium, pp 624–631

Schulz J, Hubmann C, Löchner J, Burschka D (2018) Interaction-aware probabilistic behavior prediction in urban environments. In: IEEE/RSJ International Conference on Intelligent Robots and Systems

Schwarting W, Alonso-Mora J, Rus D (2018) Planning and decision-making for autonomous vehicles. Ann Rev Control Robot Auton Syst 1:187–210

Seif HG, Hu X (2016) Autonomous driving in the ICity—HD Maps as a key challenge of the automotive industry. Engineering 2(2):159–162

Shalev-Shwartz S, Shammah S, Shashua A (2017) On a formal model of safe and scalable self-driving cars

Steyer S, Tanzmeister G, Lenk C, Dallabetta V, Wollherr D (2018a) Data association for grid-based object tracking using particle labeling. In: IEEE International Conference on Intelligent Transportation Systems

Steyer S, Tanzmeister G, Wollherr D (2018b) Grid-based environment estimation using evidential mapping and particle tracking. IEEE Transact Intell Veh 3(3):384–396

Thrun S, Montemerlo M, Dahlkamp H, Stavens D, Aron A, Diebel J, Mahoney P (2006) Stanley: the Robot that won the DARPA grand challenge. J Field Robotic 23(9):661–692

Tsugawa S, Kato S, Aoki K (2011) An automated truck platoon for energy saving. In: IEEE/RSJ International Conference on Intelligent Robots and Systems

Urmson C, Andrew Bagnell J, Baker CR, Hebert M, Kelly A, Rajkumar R, Rybski PE et al (2007) Tartan Racing: A Multi-Modal Approach to the Darpa Urban Challenge. http://repository.cmu.edu/robotics/967/

Van Brummelen J, O'Brien M, Gruyer D, Najjaran H (2018) Autonomous vehicle perception: the technology of today and tomorrow. Transport Res Part C Emerg Technol 89:384–406

Vaskov S, Larson H, Kousik S, Johnson-Roberson M, Vasudevan R (2019) Not-at-fault driving in traffic: a reachability-based approach. In: 2019 IEEE Intelligent Transportation Systems Conference (ITSC), 2785–2790. IEEE, Auckland, New Zealand, https://doi.org/10.1109/ITSC.2019.8917052

Wang H, Wang B, Liu B, Meng X, Yang G (2017) Pedestrian recognition and tracking using 3D LiDAR for autonomous vehicle. Robot Auton Syst 88:71–78

Werling M, Ziegler J, Kammel S, Thrun S (2010) Optimal trajectory generation for dynamic street scenarios in a Frenet frame. In: Proceedings of the IEEE International Conference on Robotics and Automation, 987–993

WHO (2021) Global status report on road safety, 2021. https://www.who.int/publications/i/item/9789241565684

Wongpiromsarn T, Murray RM (2008) "Formal verification of an autonomous vehicle system. In: Conference on Decision and Control

Yeong DJ, Velasco-Hernandez G, Barry J, Walsh J (2021) Sensor and sensor fusion technology in autonomous vehicles: a review. Sensors 21(6):2140

You C, Lu J, Filev D, Tsiotras P (2018) "Highway Traffic Modeling and Decision Making for Autonomous Vehicle Using Reinforcement Learning. IEEE Intelligent Vehicles Symposium

Yue L, Abdel-Aty M, Wu Y, Wang L (2018) Assessment of the Safety Benefits of Vehicles' advanced driver assistance, connectivity and low level automation systems. Accid Anal Prev 117:55–64

Zhao W, Queralta JP, Westerlund T (2020) Sim-to-Real transfer in deep reinforcement learning for robotics: a survey. In: IEEE Symposium Series on Computational Intelligence

Ziegler J, Bender P, Dang T, Stiller C (2014a) Trajectory Planning for Bertha—a Local, Continuous Method. In: Intelligent Vehicles Symposium Proceedings, 2014 IEEE, IEEE, pp 450–457. http://ieeexplore.ieee.org/xpls/abs_all.jsp?arnumber=6856581.

Ziegler J, Bender P, Schreiber M, Lategahn H, Strauss T, Stiller C, Dang T et al (2014b) Making Bertha Drive—an autonomous journey on a historic route. Intell Transp Syst Mag IEEE 6(2): 8–20

Part I
Africa

South Africa: Motor Vehicle Collisions Liability Regime Applied to Automated Vehicles

Jessica Anne Steele and Rakhee Dullabh

Abstract South Africa is home to some of the world's most dangerous roads, with 12,921 people killed by motor vehicle accidents in 2018, the majority of which were caused by human error (Road Traffic Management Corporation (2019) *State of Road Safety Report: Calendar* available at https://www.arrivealive.co.za/documents/State%20of%20Road%20Safety%20Report%202019.pdf Accessed 19 October 2020.). According to the eNatis system there were 12,283,777 licensed drivers on the roads (This was as at 28 February 2017. See http://www.enatis.com/index.php/statistics Accessed 19 October 2020.) and 12,932,873 registered vehicles (This was as at 31 August 2020. See http://www.enatis.com/index.php/statistics Accessed 19 October 2020.). These statistics demonstrate that there is an increased need to regulate and improve the safety of South African roads. The introduction of driver assistance technologies over the last decade has been accompanied by the suggestion that these technologies will improve road safety and reduce the occurrence of accidents. Legal regimes and principles often play catch-up to ever-changing and advancing technology—this is no different in South Africa. In this chapter we consider how automated vehicles ("**AVs**"), at all levels of automation as per the SAE J3016 standard will be regulated in South Africa and discuss the existing laws and regulations that govern liability arising from motor vehicle collisions in South Africa.

An application of the relevant law reveals that the South African legal regime is currently flexible enough to continue to apply to motor vehicle collisions involving SAE Level 0 to 3 AVs. However, when one begins to consider how the relevant law may be applied to SAE Level 4 to 5 AVs it becomes clear that the delictual principles are no longer capable of being satisfied. South Africa's legislative enactments which

J. A. Steele (✉)
Technology, Media and Telecommunications, ENS, Johannesburg, South Africa

Baker McKenzie LLP, London, UK

R. Dullabh
Technology, Media and Telecommunications, ENS, Johannesburg, South Africa

Baker McKenzie, Melbourne, VIC, Australia

H. Steege et al. (eds.), *Autonomous Vehicles and Civil Liability in a Global Perspective*, Data Science, Machine Intelligence, and Law 3,
https://doi.org/10.1007/978-3-031-41992-8_3

seek to compensate individuals for harms suffered on the roads or as a result of defective products still rely on the common law delictual principles, leaving a lacuna in the law when it comes to SAE Level 4 to 5 AVs. To address this lacuna, an argument to extend the *actio de pauperie*, however public policy considerations may impede the success of such arguments.

1 Introduction

South Africa has an uncodified legal system. South African law consists of a number of sources, being the *Constitution of the Republic of South Africa, 1996* ("**Constitution**") which is the supreme law of the country, legislation (acts of the national and provincial legislatures, and governmental regulations), judicial precedent, the common law (rules developed by previous decisions of superior courts, and rules and principles discussed in the 'old Roman-Dutch authorities'), custom (or conventions), customary law, international law and the writings of authoritative publicists of the law. For the purposes of this chapter, and in the context of liability arising from motor vehicles, legislation, common law principles and case law are the most relevant sources of law and will be considered in order to explain how liability will be apportioned in respect of level 0–2 AVs and to explore how liability may be apportioned in respect of level 3–5 AVs.

2 Relevant Law

2.1 Legislation

2.1.1 National Road Traffic Act

The *National Road Traffic Act*[1] was promulgated to manage and control the use of South African roads and to ensure that vehicles and the drivers thereof use the roads in a lawful and safe manner. The *National Road Traffic Act* regulates a number of matters such as, the licensing of vehicles and drivers, to creating the offences and penalties in respect of negligent, reckless[2] or inconsiderate driving.[3] It will become clear in the discussion below regarding SAE level 4 and 5 AVs, that the *National Road Traffic Act* will need to be amended to cater for these categories of automation.

---

[1] Act 93 of 1996.

[2] Section 63 of Act 93 of 1996.

[3] Section 64 Act 93 of 1996.

2.1.2 Road Accident Fund Act

The *Road Accident Fund Act*[4] is, for the purposes of liability in respect of accidents involving AVs, one of the most important sources of law. The *Road Accident Fund Act* applies only to motor vehicle accidents which occur within the borders of South Africa.[5] The *Road Accident Fund Act* established the Road Accident Fund ("**RAF**"), a juristic person,[6] which is funded by a compulsory levy on fuel purchases per litre.[7]

The purpose of the RAF is to compensate those who have suffered personal loss or damage as a result of a road accident caused by the wrongful driving of a motor vehicle.[8] The RAF indemnifies the driver or owner of the motor vehicle against liability incurred as the result of certain loss or damage caused to another person in road accidents, and in this regard the RAF steps into the shoes of the common law wrongdoer. The liability of the common-law wrongdoer in respect of loss or damage resulting from bodily injury to or the death of any person will only be permitted in certain circumstances, for example, where the RAF is unable to pay or where the claim for damages is for emotional shock sustained by a victim not actually involved in the accident.[9] By substituting the common law wrongdoer with the RAF, the victim of a road accident caused by the wrongful driving of a motor vehicle is assured compensation provided, of course, that all the requirements of the *Road Accident Fund Act* are met.[10]

The liability of the RAF is set out in section 17(1) of the *Road Accident Fund Act,* which provides that the RAF will:

> (a) subject to this Act, in the case of a claim for compensation under this section arising from the driving of a motor vehicle where the identity of the owner or the driver thereof has been established;
>
> (b) subject to any regulation made under section 26, in the case of a claim for compensation under this section arising from the driving of a motor vehicle where the identity of neither the owner nor the driver thereof has been established,
>
> be obliged to compensate any person (the third party) for any loss or damage which the third party has suffered as a result of any bodily injury to himself or herself or the death of or any

---

[4] Act 56 of 1996.

[5] Claims arising from motor vehicles accidents in Namibia, Lesotho, Botswana and Swaziland are dealt with in accordance with the legislation promulgated in those countries along with certain reciprocal agreements which regulate the enforcement of third party claims.

[6] Section 2(1) of Act 56 of 1996.

[7] The RAF fuel levy is currently at 207 cents per litre for the 2019/20 financial year. To access these details see. https://www.freightnews.co.za/article/fuel-and-road-accident-fund-raf-levy accessed on 21 October 2020.

[8] Section 2 of Act 93 of 1996.

[9] Klopper (2020).

[10] The victim is not guaranteed *full* payment of compensation in all cases, rather the *Road Accident Fund Act* places a statutory limit on certain types of claims which can be made. For further reading on the statutory limit of the RAF's liability refer to section 17 of Act 56 of 1996.

bodily injury to any other person, caused by or arising from the driving of a motor vehicle by any person at any place within the Republic, if the injury or death is due to the negligence or other wrongful act of the driver or of the owner of the motor vehicle or of his or her employee in the performance of the employee's duties as employee: Provided that the obligation of the Fund to compensate a third party f or non-pecuniary loss shall be limited to compensation for a serious injury as contemplated in subsection (1A) and shall be paid by way of a lump sum.

It is clear that the *Road Accident Fund Act*, distinguishes between those motor vehicle accidents where the victim has been able to identify the owner or driver of the motor vehicle concerned, and those motor vehicle accidents where the victim has not been able to identify the owner or driver of the motor vehicle concerned. The former category is commonly referred to as an "identified claim" whilst the latter category is commonly referred to as a "hit-and-run claim".[11]

The basis of the RAF's liability is delictual, as such the common law principles discussed below are relevant to understanding when and how the RAF will be liable.[12] In order to assess where liability will arise on the part of the RAF in respect of the different levels of AVs it is necessary to consider the essential components of the RAF's liability.[13]

The RAF's liability is, first and foremost, dependant on whether or not the victim is able to prove negligence or another unlawful act on the part of the driver of another motor vehicle, or the owner of the motor vehicle. Whether or not negligence can be proved will depend on the particular set of facts and circumstances. From various case law general principles have been established over a number of years, to the effect that, the assessment of negligence is routed in certain rules which are founded on an "...expression of the measure of attentiveness and care expected by society from all persons who are actors in vehicular traffic and are stated in the form of certain general and specific legal duties owed by drivers, pedestrians and train drivers".[14]

The RAF will only pay compensation for serious bodily injury or financial loss resulting from an injury or the death of a breadwinner. The RAF will therefore not compensate a victim for damage to property.[15]

The loss suffered must have been caused by or arising out of the driving of a motor vehicle. There is therefore a requirement to prove a causal connection between

[11] The most important distinction when it comes to a hit-and-run claims and identified claims is that the *Road Accident Fund* applies different rules regarding the prescription of a victim's claim in these instances. Section 23 of Act 56 of 1996 deals with the prescription of claims where the identity of the driver or owner of the motor vehicle is known, while hit-and-run claims are dealt with in terms of section 17(1)(b) of Act 56 of 1996 read with regulation 2 of the *Road Accident Fund Regulations, 2008* (Published under Government Notice R770 in Government Gazette 31,249 of 21 July 2008).

[12] Klopper (2020), p. 270.

[13] It should be noted that there are certain other heads of damages or categories of claims which are excluded by the *Road Accident Fund Act*, this is not relevant for the purposes of determining liability however.

[14] Klopper (2012).

[15] See for example *Oberholzer v Santam* 1970 (1) SA 337 (N) at p181.

the loss suffered and the motor vehicle accident.[16] There are two additional and important components, namely, the meaning of the words "driving" and "motor vehicle". To understand the meaning and scope of the statutory definitions one must refer to numerous case law and the particular facts and circumstances of those cases.

In terms of the *Road Accident Fund Act*, a "motor vehicle" is defined to mean "any vehicle designed or adapted for propulsion or haulage on a road by means of fuel, gas or electricity, including a trailer, a caravan, an agricultural or any other implement designed or adapted to be drawn by such motor vehicle" and a "motor car" is defined as "a motor vehicle designed or adapted for the conveyance of not more than 10 persons, including the driver".[17] The facts of every case must be considered in order to determine whether the vehicle involved is a motor vehicle in terms of the *Road Accident Fund Act.*[18] In *Chauke v Santam LTD*[19] the court held that just because a vehicle can be used on a public road does not mean that it was 'designed' for propulsion on a road, finding that the correct approach to the interpretation of whether a motor vehicle in question was designed for propulsion on a road is to give it a common sense and objective meaning.[20]

The requirement of "driving" is the conduct requirement[21] and "...is crucial to the enquiry of whether the Road Accident Fund is liable or not and is the criterion which distinguishes a claim against the Road Accident Fund from any other ordinary common law delictual claim".[22] According to Klopper,

> Driving has a circumscribed technical meaning and is constituted by the intentional starting of a vehicle in order to drive it, setting a vehicle in motion, exercising control over the vehicle while it is in motion by using its controls including direction indicators and hooter and then bringing such vehicle to a standstill and all related and required conduct to achieve this.[23] [Our emphasis]

The *Road Accident Fund Act* also extends the meaning of "driving" with the presumptions contained in section 20, namely:

1. For the purposes of this Act a motor vehicle which is being propelled by any mechanical, animal or human power or by gravity or momentum shall be deemed to be driven by the person in control of the vehicle.

---

[16] Klopper (2020).

[17] Section 1 of Act 56 of 1996.

[18] The particular facts of the case are important, if the particular motor vehicle in the case law and the motor vehicle in the relevant matter are not identical, the outcome may be different. For example in *Mutual & Federal Insurance Co. Ltd v Day* 2001 (3) SA 775 (SCA) the court *a quo* and full bench held that a particular make of forklift was a motor vehicle for the purpose of the *Road Accident Fund* Act, while in *Chauke v Santam LTD* 1997 (1) SA 178 (A) the Appelate Division held that a different make of forklift was not a motor vehicle.

[19] See citation at note 22 above.

[20] 1997 (1) SA 178 (A) 181–182.

[21] See the discussion of the requirements for a delictual action below.

[22] Klopper (2009), p. 2.

[23] Klopper (2009), p. 2.

2. For the purposes of this Act a person who has placed or left a motor vehicle at any place shall be deemed to be driving that motor vehicle while it moves from that place as a result of gravity, or while it is stationary at that place or at a place to which it moved from the first mentioned place as a result of gravity.
3. Whenever any motor vehicle has been placed or left at any place, it shall, for the purposes of this Act, be presumed, until the contrary is proved, that such vehicle was placed or left at such place by the owner of such vehicle.

The interpretation given to "driving" and these presumptions will be an important consideration in the discussion below where these principles must be applied to SAE level 4 and 5 AVs.

2.1.3 Consumer Protection Act

The *Consumer Protection Act*[24] was enacted in order to, amongst other things, promote and protect the economic interests of consumers and to protect consumers from hazards to their well-being and safety.

The relevance of the *Consumer Protection Act* under this topic is that of product liability. Under the common law, product liability is dealt with in terms of the law of delict and the law of sale.[25] As such, in order to claim damages for a defective product, all the elements of delict, discussed briefly below, must be present. The difficulty which arises and which the *Consumer Protection Act* seeks to remedy is the complexity of proving fault in these types of cases, most especially in light of the fact that technologies have become so advanced that even expert determination is almost rendered impossible.[26] Over and above this, insofar as the law of sale is concerned, a consumer can only seek damages from the seller or manufacture of goods in respect of a defective product.[27]

The applicability of the *Consumer Protection Act* is determined by a number of considerations. Some of the notable considerations include whether there is an exemption[28] or whether the consumer is a juristic person whose asset value exceeds the threshold value determined by the Minister of Trade and Industry.[29] The discussion regarding the *Consumer Protection Act* therefore assumes that the *Consumer Protection Act* is applicable, however, this will need to be determined in each case.

[24] Act 68 of 2008.

[25] It is beyond the scope of this chapter to consider the potential common law product liability implications further than what is described in the body of this chapter.

[26] Kaashifa (2017).

[27] Basson (2011).

[28] See section 5(2)(c) of Act 68 of 2008.

[29] Section 5(2)(b) of Act 68 of 2008; at the time of writing this threshold was set at R2,000,000.00 (Two Million Rands) in terms of the *Determination of Threshold in Terms of the Consumer Protection Act, 2008 (act no. 68 of 2008)* published under Government Notice 294 in Government Gazette 34,181 of 1 April 2011.

Chapter 2 of the *Consumer Protection Act* sets out a number of fundamental consumer rights, among these and relevant to this chapter is the consumer's right to safe, good quality goods.[30] Stemming from this right is section 61 of the *Consumer Protection Act* which provides for liability for damage caused by goods. Section 61 of the *Consumer Protection Act* introduced strict liability in certain defined circumstances.[31] Thus, unlike when a claim is brought under the *Aquilian* action, fault need not be proved.[32] The section provides that a producer, importer, distributor or retailer (collectively referred to below as "supplier") will be held liable for any harm that is caused by, amongst other things, the supply of unsafe or defective goods or inadequate instructions or warnings in certain circumstances.[33] The liability of the supplier is assumed notwithstanding whether the harm resulted from the supplier's negligence or not.[34] If, in the circumstances, more than one person is liable, then their liability will be joint and several.[35] In order to bring a claim under the *Consumer Protection Act*, the consumer must establish a causal link between the unsafe product characteristic, failure, defect or hazard and the harm suffered.[36]

The *Consumer Protection Act* does make certain defences available to suppliers.[37] These defences are limited and will apply only to certain categories supplier.

The *Consumer Protection Act* further defines the types of harm for which a supplier may be held liable, the relevant categories for this chapter, amongst others, include (i) the death of, or injury to, any natural person[38]; (ii) any loss of, or physical damage to any property, irrespective of whether it is movable or immovable property[39]; and (iii) any economic loss that results from the aforegoing.[40] When one compares the heads of damages that may be claimed under the *Consumer Protection Act* and those described under the *Road Accident Fund Act* above it becomes important to consider whether the strict liability regime created by the *Consumer Protection Act* may be applicable if the facts are such as to found a claim based on product liability.

---

[30] Section 55 of Act 68 of 2008.

[31] Neethling and Potgieter (2015).

[32] Botha and Joubert (2011) pp. 305–319.

[33] s61(1)(a)-(c) of Act 68 of 2008.

[34] Section 61(1) of Act 68 of 2008.

[35] Section 61(3) of Act 68 of 2008.

[36] Mahabeer (2012).

[37] Section 61(4) of Act 68 of 2008.

[38] Section 5(a) of Act 68 of 2008.

[39] Section 5(c) of Act 68 of 2008.

[40] Section 5(d) of Act 68 of 2008.

2.2 Common Law

The common law principles of delictual liability are relevant to this chapter. A delict is a civil wrong, and can be more narrowly defined as a wrongful and blameworthy conduct which causes harm to a person.[41] The classic remedy for a delict is compensation in the form of damages for the harm caused.

The essential requirements for delictual liability are (i) harm sustained by the plaintiff; (ii) conduct on the part of the defendant which is (iii) wrongful; (iv) a causal connection between the conduct and the plaintiff's harm; and (v) fault or blameworthiness in the form of negligence or intention on the part of the defendant.[42] In order to be successful in a delictual claim, all of these elements must be present.

The harm elements speaks to the requirement that the plaintiff must have suffered some form of loss or injury in order to found a cause of action. This element is therefore the starting point, however, it is not necessary for the purposes of this chapter to delve into this requirement any further.

The conduct requirement may take the form of an act or omission. Importantly, conduct must be voluntary, in that the defendant in question should not have been compelled to act or not act. Furthermore, it is required that the conduct must be that of a *human*, barring some exceptions.[43]

At common law, finding a causal link consists of a two-stage enquiry into factual and legal causation.[44] This was recognised in the leading case of *Minister of Police v Skosana*[45]:

> Causation in the law of delict gives rise to two rather distinct problems. The first is a factual one and relates to the question as to whether the negligent act or omission in question caused or materially contributed to . . . the harm giving rise to the claim. If it did not then no legal liability can arise and *cadit quastio*. If it did, then the second problem becomes relevant, *viz.* whether the negligent act or omission is linked to the harm sufficiently closely or directly for legal liability to ensue or whether, as it is said, the harm is too remote. This is basically a juridical problem in which considerations of legal policy may play a part.[46]

The enquiry into factual causation focuses primarily on the facts, while the enquiry into legal causation considers normative issues.[47] The test that is most often used in respect of the enquiry into factual causation is referred to as the 'but-for' or *conditio*

---

[41] Van der Walt and Midgley (2016).

[42] *Evins v Shield Insurance Co Ltd* 1980 2 All SA 40 (A) 58; *Edouard v Administrator, Natal* 1989 4 All SA 309 (D); 1989 2 SA 368 (D) 391; *HL & H Timber Products (Pty) Ltd v Sappi Manufacturing (Pty) Ltd* 2001 (4) SA 814 (SCA) para 13; *McCarthy Ltd t/a Budget Rent a Car v Sunset Beach Trading 300 CC t/a Harvey World Travel* 2012 (6) SA 551 (GNP) para 12.

[43] Namely in respect of juristic persons. However, the law does provide for compensation in instances where harm is caused by animals.

[44] Van der Walt and Midgley (2016).

[45] 1977 (1) SA 31 (A).

[46] 1977 (1) SA 31 (A) 34E-G

[47] Loubser et al. (2012).

sine qua non test.[48] The application of the but-for test involves "... a hypothetical enquiry as to what probably would have happened but for the wrongful conduct of the defendant."[49] The court hypothetically inserts (in the case of an omission) or removes (in the case of a commission) the conduct of the defendant and asks whether the harm would nevertheless have resulted.[50] If the harm would not have resulted then no liability can arise.

In *First National Bank of South Africa Ltd v Duvenhage*[51] the Supreme Court of Appeal opined that factual causation should be limited "... so as to contain liability within reasonable bounds".[52] The enquiry into legal causation on the other hand involves policy considerations to determine whether the loss suffered is too remote to warrant the imposition of liability.[53]

The element of wrongfulness is generally assumed in cases where the harm arose from physical damage to the plaintiff's person or property. [54] The requirement of wrongfulness calls for proof of a harmful result caused by the defendant in a legally reprehensible or unreasonable manner.[55]

Fault or blameworthiness is the subjective element of a delict, it focuses on the blameworthy state of mind or conduct of the defendant.[56] Once the wrongful character of the defendant's conduct has been established, the question of his or her fault arises, and the question as to whether there is either intent or negligence on the part of the defendant forms a basis for imposing liability for the defendant's wrongful conduct. In order to be at fault, the defendant must be accountable, this means the ability to distinguish between what is right and wrong and also to have the capability of acting in accordance with that distinction.[57] For example, the defendant may lack accountability—he or she was intoxicated to the point of not being in control of his or her action or otherwise unconscious.[58] Fault is determined by reference to the foreseeability and preventability of the harm caused by the defendant in light of the facts and circumstance of the case.[59] There are also certain instances where fault need not be proved, these are referred to as cases which give rise to strict

[48] 1990 (1) SA 680 (A) 700E-G.

[49] 1990 (1) SA 680 (A) 700F-H.

[50] 1990 (1) SA 680 (A) 700F-H.

[51] 2006 (5) SA 319 (SCA).

[52] 2006 (5) SA 319 (SCA) para 2.

[53] *Fourway Haulage SA Pty (Ltd) v SA National Roads Agency Ltd* 2009 (2) SA 150 (SCA) para 31.

[54] *Betko Products CC v Grasso (Pty) Ltd* [2010] JOL 25246 (WCC) para 10.

[55] Neethling J, Potgieter JM (2015) Law of Delict. 7ed LexisNexis South Africa.

[56] *Nyaguse v Skinners Auto Body Specialists & another* [2008] JOL 21160 (ZH) 2.

[57] Certain factors such as youth, mental incapacity, intoxication as a result of alcohol consumption or drugs and provocation may exclude accountability. Klopper (2012).

[58] While the elements to found criminal liability should not be conflated with the delictual elements, the following cases contain important principles regarding the element of conduct which do apply in the delictual sense: *R v Dhlamini* 1955 (T); *S v Eadie* 2002 (1) SACR 663 (SCA); *S v Chretien* 1981 (1) SA 1097 (A).

[59] [2008] JOL 21160 (ZH) 2.

liability. Strict liability is either established through common law principles or legislation.[60] There is also an argument to be made that the specific delictual remedy, the *actio de pauperie,*[61] may be relied upon.

For the purposes of this chapter as it relates to motor vehicle accidents in the context of the different levels of AVs, it is submitted that negligence, will most likely be the form of fault that will need to be proven. Furthermore, as is elaborated on below, fault, in the absence of any strict liability regime, will also likely be the most difficult requirement to be proved by a plaintiff in respect of SAE level 4 and 5 AVs.

The standard used to judge the conduct of a person is that of the reasonable person[62] or *bonus paterfamilias.*[63] In *Kruger v Coetzee*, Holmes JA stated that negligence will be established where (a) a reasonable person in the position of the defendant would (i) have foreseen the reasonable possibility of his or her conduct causing harm to another's person or property resulting in patrimonial loss, and (ii) have taken reasonable steps to guard against that possibility; and (b) the defendant failed to take those steps.[64] As with all of the preceding principles discussed in this chapter, whether a defendant should have foreseen the possibility of the harm arising or whether the defendant should have taken steps to avoid such harm will always depend on the facts and circumstances of each case, in this regard there can be no hard and fast rule.

When the principles above are applied to the context of a motor vehicle accident, the driver is required to behave and respond as a reasonable person would in the circumstances and must "... be capable of reasonably foreseeing the damage flowing from his/her negligent act and must also take reasonable steps to prevent damage from occurring".[65] Accordingly, a reasonable person would foresee the possibility that pedestrians, cyclists and animals may be on the roads, and must drive in accordance with that foresight.[66] Failing which, the driver will have been negligent.

A breach of a statutory provision, such as the *National Traffic Act* discussed above does not automatically give rise to a finding of negligence. For this to have been the case, the *National Traffic Act* would have to have included civil remedies. However, it would be difficult for a court to ignore the duties which are created by the *National Traffic Act* when deciding on negligence and when applying the

---

[60] Refer to the section above regarding the Consumer Protection Act.

[61] Refer to Sect. 3.4 below.

[62] The reasonable person as opposed to man is now the accepted terminology. It refers to the personification of society's perception of the acceptable standard of conduct within society under certain circumstances. It was pointed out in *Santam Insurance Co Ltd v Nkosi* 1978 (2) SA 784 (A) at 791-792A, that the reasonable man is not a "timorous faintheart always in trepidation lest he or others suffer from injury", nor is the reasonable man a person "given to anxious conjecture and morbid speculation".

[63] *Kruger v Coetzee* 1966 (2) SA 428 (A) 430.

[64] 1966 (2) SA 428 (A) 430.

[65] Klopper (2012).

[66] Klopper (2012).

reasonable person test.[67] In *Bekker v Du Toit*, for example, it was held that conduct contrary to a traffic ordinance in force during that time, "... could only be relied upon as evidence of negligence if it included a possibility of danger for other traffic which a reasonable man would avoid or minimise by precautionary measures."[68] On the contrary, the *dicta* in *Fourie v Road Accident Fund*[69] the court applied the *res ipsa liquitur* doctrine in the circumstances where the plaintiff, in seeking to avoid a collision, drove on the wrong side of the road. Thus, it was stated that there is an inference, rather than a presumption, of negligence.[70] While this doctrine exists, the court is not compelled to use it and must still focus its enquiry on all the evidence to determine where the balance of probabilities lies.[71]

3 Application of the Relevant Law

The above brief synopsis of the relevant law has now laid the foundations upon which to apply these principles to the different levels of AVs. Each level of AV discussed below is discussed in the context of the Society of Automotive Engineering's ("**SAE**") J3016 Standard.[72] The J3016 Standard "... defines six levels of driving automation, from SAE Level Zero (no automation) to SAE Level 5 (full vehicle autonomy). It serves as the industry's most-cited reference for automated-vehicle (AV) capabilities".[73]

For the purposes of applying these principles below, the focus is predominantly on whether fault can be proved. Sight, should however, not be lost of the fact that all the essential elements of the delict must be present, whether in a true delictual claim or a claim against the RAF.

An important and relevant term which we will rely on in this chapter and which coined and defined by the SAE J3016 Standard is that of "dynamic driving tasks"[74] which means:

> ... all the real-time operational and tactical functions required to operate a vehicle in on-road traffic, excluding the strategic functions such as trip scheduling and selection of destinations and waypoints, and including without limitation:
>
> lateral vehicle motion control via steering (optional)

---

[67] Klopper (2012); for examples see: *East London Municipality v Van Zyl* 1959 (2) SA 514 (E); *Pretorius v South British Insurance* 1963 (3) SA 8 (W); *Bekker v Du Toit* 1974 (3) SA 248.

[68] 1974 (3) SA 248 (O) 249.

[69] [1999] 3 All SA 661 (O).

[70] [1999] 3 All SA 661 (O) 667.

[71] [1999] 3 All SA 661 (O) 667.

[72] In this chapter we have used many of the phrases contained in the SAE J3016 Standard itself for the purposes of clarity.

[73] Shuttleworth (2019).

[74] SAE International (2018) Surface Vehicle Recommended Practice – Taxonomy and definitions for terms related to dirving automation systems for on-road motor vehicles.

longitudinal vehicle motion control via acceleration and deceleration (optional)
monitoring the driving environment via object and event detection, recognition, classification, and response preparation (operational and tactical)
object and event response execution (operational and tactical)
manoeuvre planning (tactical)
enhancing conspicuity [*sic*] via lighting, signalling and gesturing etc… (tactical).[75]

3.1 Liability for SAE Level 0–2 AVs

Under this section the following levels of AVs are considered: (i) SAE Level 0 – No Driving Automation, the SAE describes this level of AV as that all dynamic driving tasks are performed by the human driver, despite possibly being enhanced by certain safety mechanisms[76]; (ii) SAE Level 1 – Driver Assistance, the SAE describes this level as that the human driver performing the dynamic driving tasks, but that there is a limited set of tasks which can be performed by the driving automation[77] system[78]; and (iii) SAE Level 2 – Partial Driving Automation, similarly the SAE describes this level as that the human driver still performs the dynamic driving tasks, but that there is a limited set of tasks which can be performed by the driving automation system.[79] The distinction between SAE Level 1 and 2 is that with the former, there will be only one advanced driver assistance feature, while the latter will have two or more advanced driver assistance features.[80]

The SAE J3016 Standard explains that SAE Level 1 and SAE Level 2 has functions which can perform only parts of the dynamic driving tasks, which means that a human driver must perform all the remaining dynamic driving tasks. At the same time, the human driver must oversee the performance of the function while being used. As a result, these functions are for convenience and "…support, but do not replace…"[81] a human driver.

In the event of an accident involving a SAE Level 0–2 AV, because a human driver is still in control and is not replaced by the automated driving system, the current liability regime will apply. Thus, if the driver of the SAE Level 0–2 AV is found to have been negligent (or otherwise at fault), the RAF will be liable for the resultant serious bodily injury or financial loss resulting from an injury or the death

[75] SAE (2018) International Surface Vehicle Recommended Practice.

[76] SAE (2018) International Surface Vehicle Recommended Practice.

[77] "Driving automation" is defined as "the performance by hardware/software systems of part or all of the [dynamic driving tasks] on a sustained basis", SAE International Surface (2018) Vehicle Recommended Practice.

[78] SAE International (2018) Surface Vehicle Recommended Practice.

[79] SAE International (2018) Surface Vehicle Recommended Practice.

[80] For a practical and easy to understand description of the SAE Levels with examples, see: Hyatt and Paukert (2018).

[81] SAE International (2018) Surface Vehicle Recommended Practice.

of a breadwinner where such loss was caused by or arising out of the driving of a motor vehicle.

Similarly, the driver or owner of the SAE Level 0–2 AV may be held liable where all the other requirements of a delict have been met.

Interestingly, as one moves from an SAE Level 0 AV to the SAE Level 1 AV or SAE Level 2 AV, the possibility of a driver being found to have been negligent may arguably increase in likelihood. This inference is drawn from the *dicta* in *Bridgman NO v Road Accident Fund,*[82] in this case the driver of a motor vehicle had collided with a cyclist. In determining whether the driver had been negligent, the court emphasised that prior to the collision the driver took no evasive action, whether by braking or by counter-steering the vehicle.[83] In considering the driver's failure to take evasive action to avoid colliding with the cyclist, Nel J held as follows:

> I am not persuaded that evasive action could not have been taken by Mr Brown. Undue weight should not be attached to the views expressed by courts twenty years, thirty years, or even longer ago about the options open to motorists in comparable circumstances. Due to improved braking and steering mechanisms, the manoeuvrability of a modern motor-car is vastly superior to that of the motor-car of a few decades ago.[84] [our emphasis]

There are two important elements that are brought to the fore by this quote. In the first instance, Nel J acknowledges that as motor vehicles have improved over time, the judgment of a driver's response to a situation must accordingly change and in the second instance, the improved safety mechanisms of vehicles mean that drivers are required to understand and make use of these improvements to ensure that the appropriate response is carried out in a particular situation. In this regard, Nel J held that "[t]he duty of a motorist to maintain a proper lookout involves not only the physical act of looking, but also a reasonably prudent reaction to whatever might be seen"[85] and perhaps this quote could have been concluded by adding, "and utilising the enhanced safety mechanisms and functions of the vehicle available to him".

In a very different scenario, if an AV collided with another vehicle (but as a result of a failure or defect in respect of the functioning of the AV) and the driver in the circumstances was not negligent by failing to take proper evasive action, then in such scenario the driver of the AV could have a product liability claim against the manufacturer.

Whether or not the driver will be found to be negligent in this example will depend on the prevailing facts and circumstances of the case and the principles which have previously been applied by the courts in respect of what has been referred to as a 'sudden emergency'. In a sudden or unexpected emergency situation, the conduct of the driver will be judged in light of the nature of the emergency and the time in which the driver had to respond.[86] The guiding principle, however, being

[82] [2002] 1 All SA 1 (C).

[83] [2002] 1 All SA 1 (C) 5.

[84] [2002] 1 All SA 1 (C) 9.

[85] [2002] 1 All SA 1 (C) 9.

[86] *Samson v Winn* [1977] 1 All SA 442 (C) 447.

that a driver is expected "... to comply with a high standard of judgment and courage when in control of motor cars on public roads and they must be able to cope with the unexpected without losing control of their vehicles".[87] That said, it has been held that where the emergency situation is one which places the life of the driver in peril, then his or her conduct will not be judged at such a high standard and the driver will not likely be found to have been negligent.[88]

As a result, in the example above, it may be that the driver will have a product liability claim against the manufacturer. In addition, the driver may also be liable to any person that was injured or who suffered a loss, if the driver is found to have been negligent in the circumstances.

The above example, must also be distinguished from the failure to maintain a vehicle. In circumstances where the owner of a vehicle has failed to properly maintain the vehicle, then the RAF or the owner of the vehicle, as the case may be, may be found to have been negligent on this basis.[89]

3.2 Liability for SAE Level 3 AVs

Under this section we consider SAE Level 3 – Conditions Driving Automation. It is from this SAE Level that true driving automation becomes a reality. At SAE Level 3 the majority of the dynamic driving tasks are performed by the automated driving system within a certain set of previously defined parameters (these parameters are referred to by the SAE as the "operation design domain").[90] Under SAE Level 3, a human driver must be at the ready to take over the dynamic driving tasks in the event of any automated driving system failures, for example.

In respect of a SAE Level 3 AV, it is assumed that the human driver is by necessity the dynamic driving task "fallback".[91] As such, while the SAE Level 3 AV may be performing all of the dynamic driving tasks, the human driver is required to remain aware and to take over the dynamic driving tasks where required. Under the *National Traffic Act*, "driver" is defined as "any person who drives or attempts to drive any vehicle or who rides or attempts to ride any pedal cycle or who leads any

[87] [1977] 1 All SA 442 (C) 448.

[88] This was the reasoning in *Samson v Winn*, in this case the driver was being attacked by a man wielding a panga and accidently pressed the accelerator while swerving onto the incorrect side of the road and in another case, *Mfihlo v Port Elizabeth Municipal Council* 1976 (3) S.A. 183 (E) where a driver was held not be negligent after swerving onto the incorrect side of the road because he (perhaps) mistakenly believed that a man wielding a brick was going to throw it at him.

[89] See for example, *Santam v Kemp* 1971 (3) SA 305 (A); *Maatla v Road Accident Fund* (11,690/11) [2015] ZAGPPHC 129 (6 March 2015); *Abrahams v Road Accident Fund* (15,863/2013) [2016] ZAWCHC 99 (12 August 2016).

[90] SAE International (2018) Surface Vehicle Recommended Practice.

[91] SAE International (2018) Surface Vehicle Recommended Practice.

draught, pack or saddle animal or herd or clock of animals, and "drive" or any like word has a corresponding meaning".[92] It is argued that the *National Traffic Act* will still be applicable to a driver of a SAE Level 3 AV because of the requirement that the driver remain the dynamic driving task fall back.

Further to the above, the conventional duties such as, the duty to keep a proper lookout still applies. In the event of a collision caused by or arising out of the driving of a SAE Level 3 AV, where the driver negligently failed to take over the dynamic driving tasks, either when prompted to do so by the AV itself or because of a potentially hazardous situation, for example, a dog running into the road, then the same principles discussed above in respect of SAE Level 0–2 AVs will apply.

With regard to SAE Level 3 AVs, it has been seen that certain safety features installed by manufacturers are being actively overridden, for example, an after-market device which was invented to trick the Tesla Autopilot Systems into thinking that the driver's hands were on the steering wheel.[93] In other instances, the explicit warning and requirement that a driver remain vigilant and alert when using a SAE Level 3 AV has simply been ignored.[94] It is argued that if a collision is caused and the facts show that the driver has failed to properly follow the safety precautions or has overridden the safety mechanisms, then the form of fault may likely be intention, in the form of *dolus eventualis*[95] or what has been referred to as 'conscious negligence'. Conscious negligence would arise where the driver foresees the remote possibility of harm occurring but takes an unjustifiable risk in continuing with his or her course of action.[96] In this instance, the driver would "foresee the possibility of harm and yet be negligent in respect of that harm ensuing, e.g. by unreasonably underestimating the degree of possibility or unreasonably failing to take steps to avoid that possibility".[97] It is also most likely that the driver in question would be criminally charged in this example.[98]

---

[92] Section a of Act 93 of 1996.

[93] Hawkins (2018).

[94] See for example the following news article where a driver was charged with speeding and sleeping: Anonymous (2020) Alberta RCMP charge Tesla driver for speeding and sleeping. Available via Royal Canadian Mounted Police. https://www.rcmp-grc.gc.ca/en/news/2020/alberta-rcmp-charge-tesla-driver-speeding-and-sleeping. Accessed 22 October 2020.

[95] This is where resultant harm was subjectively foreseen, but the person in question continued with their course of conduct despite such foresight; Burchell (2013).

[96] Burchell (2013); see also *S v Humphreys* 2013 (2) SACR 1 (SCA) where a driver of a mini bus was found guilty of culpable homicide for crossing a railway line despite warnings to the contrary, killing a number of children.

[97] *S v Ngubane* 1985 (3) SA 677 (A) 685A – H.

[98] This is not to say that criminal liability is not possible in the previous examples, negligent driving that results in the death of another person will most often lead to a charge and likely conviction of culpable homicide and in extreme cases murder.

3.3 *Liability for SAE Level 4–5 AVs*

Under this section the following levels of AVs are considered: (i) SAE Level 4 – high driving automation, the SAE describes this level as that all of the dynamic driving tasks are performed by the automated driving system, including where there is a requirement to avoid a risk situation without the requirement that a human driver takeover[99]; and (ii) SAE Level 5 – full driving automation the SAE describes this level as that all of the dynamic driving tasks are performed by the automated driving system, including where there is a requirement to avoid a risk situation without the requirement that a human driver takeover.[100] The distinction between SAE Level 4 and SAE Level 5 are the parameters in which these AVs are able to function, SAE Level 4 may have limited parameters, for example, the use of a SAE Level 4 AV may be confined to a specific geographical area or operating speed.[101] On the other hand, an SAE Level 5 AV would not have such parameter constraints and would be able to function as a motor vehicle does today, but without a human driver being necessary.

As a result of the automated driving system performing all dynamic driving tasks and that the person using the AV in question is a passenger, the definition of 'driver' in the *National Road Traffic Act* will no longer be applicable. The offences set out in the *National Road Traffic Act* will also likely not apply as they require a person to drive the vehicle in question.[102] The *National Road Traffic Act*, however, requires that all motor vehicles be registered prior to being utilised on the roads, as such it most likely that as SAE Level 4 AVs and SAE Level 5 AVs are introduced, the *National Road Traffic Act* will be amended to cater for this.[103]

When considering where liability will rest in respect of collisions involving SAE Level 4 AVs and SAE Level 5 AVs it is argued that it must be looked at from two temporal perspectives. The first is that, according to the SAE J3016 Standard, the driver or dispatcher will verify the operational readiness of the AV.[104] In this first category the owner of the AV may perform this task, and if any negligence occurs in this verification exercise which can be attributed to the owner, then the usual principles above insofar as proving the requirements of a delict will most likely apply. This is assuming that it is the owner, and not a supplier of the AV, that performs this assessment. Furthermore, the presumptions regarding "driving" as set

[99] This is referred to as dynamic driving task fallback which is a separate function to that performing the dynamic driving task itself. SAE International (2018) Surface Vehicle Recommended Practice.

[100] SAE International (2018) Surface Vehicle Recommended Practice.

[101] Hyatt K, Paukert C (2018) Self-driving cars: A level-by-level explainer of autonomous vehicles. Available via CNet. https://www.cnet.com/roadshow/news/self-driving-car-guide-autonomous-explanation/. Accessed 20 October 2020.

[102] See for example the offences of inconsiderate driving or driving while under the influence of intoxicating liquor in section 64 and 65 of Act 93 of 1996.

[103] Section 3 of Act 93 of 1996.

[104] SAE International (2018) Surface Vehicle Recommended Practice.

out in the *Road Accident Fund* would be capable of being met in these circumstances.

The second category in time, is when the AV is in motion and driving. In this instance there are challenges with applying the normal liability regime as a result of the fact that there is no human driver, the driver in fact being classified as a passenger.[105] As such, both in respect of trying to prove a claim against the RAF, the driver or the owner all the requirements of a delict will not likely be present, particularly the requirement of conduct, wrongfulness and fault. There may even be an argument in respect of the enquiry into legal causation that the conduct of the owner is too remote to impose liability.

It is in this regard that the principles relating to strict liability imposed on suppliers, in particular section 61 of the *Consumer Protection Act*, becomes relevant. It is most likely that the manufactures or other suppliers of SAE Level 4 AVs and SAE Level 5 AVs will be held liable in circumstances where a person suffers harm as a result of a collision caused by a SAE Level 4 AV or SAE Level 5 AV, where the automated driving system is performing all dynamic driving tasks.

In the event that an owner of a SAE Level 4 AV or SAE Level 5 AV suffers harm as a result of a collision caused by or arising out of the driving of the SAE Level 4 AV or SAE Level 5 AV, then he or she would likely have a claim against the supplier in terms of section 61 of the *Consumer Protection Act*, provided that the harm arose wholly or partly as a consequence of, the AV being unsafe, a failure, defect or hazard of the AV or inadequate instructions or warnings having been provided.[106] The difficulty will arise in instances where the person who has suffered harm is not being carried by the SAE Level 4 AV or SAE Level 5 AV which caused the collision, for example, he or she may be in a different vehicle, may be a cyclist or pedestrian.

In *Eskom Holdings Limited v Halstead-Cleak*[107] a cyclist had come into contact with a low hanging live power line spanning a footpath that he was cycling on.[108] The cyclist had sought to rely on section 61 of the *Consumer Protection Act* based on Eskom's role as the sole supplier or producer of electricity on the national grid and its control of all power lines not falling under the control of any local authority or municipality.[109] The Supreme Court of Appeal held that by virtue of the absence of a consumer-supplier relationship on the facts section 61 of the *Consumer Protection Act* was not applicable.[110] As such, where there is no consumer-supplier relationship the person suffering harm will not have a cause of action under the *Consumer Protection Act*.

[105] SAE International (2018) Surface Vehicle Recommended Practice.

[106] Section 61(1)(a)-(c) Act 68 of 2008.

[107] [2017] JOL 37573 (SCA).

[108] [2017] JOL 37573 (SCA) para 2.

[109] [2017] JOL 37573 (SCA) para 3.

[110] [2017] JOL 37573 (SCA) para 25.

In this scenario the victim of the collision may pursue a claim based on the common law principles of public liability which is based on delictual principles. Thus, all the requirements of a delict must be met in order to hold the supplier liable. In this regard, the same difficulty of proving fault on the part of a supplier which led, in part, to the introduction of a strict liability regime would confront a claimant in these instances. In *Wagener v Pharmacare Ltd, Cuttings v Pharmacare Ltd*[111] the Supreme Court of Appeal was asked to develop the common law of delict to impose strict liability on manufacturers of defective products in instances where no contractual privity exists between the claimant and the manufacturer, the Supreme Court of Appeal declined to develop the law, stating that "[i]f strict liability is to be imposed it is the legislature that must do it".[112] This decision was made prior to the introduction of the *Consumer Protection Act* and as such the legislative response has been to limit section 61 to where there is a consumer relationship.

It is therefore unlikely that, in a future where SAE Level 4 AVs and SAE Level 5 AVs are present, the court will be willing to develop the common law principles to the extent of developing almost all the requirements of delict in order to impose liability. As a result, the legislature will need to consider the introduction of appropriate laws or changes to the current laws to cater for this scenario.

3.4 An Argument to Extend the Actio de Pauperie

Unfortunately, the legislature is often more reactive than proactive and as is with the *Wagener* case above, we may have to wait for the court to direct that the legislature fill this lacuna. Does this mean that a person who suffers harm as a result of, caused by or arising from the driving of a SAE Level 4 AV or SAE Level 5 AV and who is not an owner or passenger, but a pedestrian, driver in another vehicle or cyclist is without remedy until then? It is argued that the public policy considerations would require that the harmed person be capable of claiming compensation as a result of any harm suffered.[113] As in the *Wagener* case, there is also the constitutional right to bodily integrity which must be considered when determining whether the common law should be developed in accordance with the Constitution's spirit, purport and objects.[114] A controversial argument to be made, is that the law should be developed in order to extend the *actio de pauperie*. This common law action gives rise to another form of strict liability and is an exception to the law of delict which ordinarily requires that fault be proved.[115]

[111][2003] 2 All SA 167 (SCA).

[112][2003] 2 All SA 167 (SCA) para 38.

[113][2003] 2 All SA 167 (SCA) para 17.

[114][2003] 2 All SA 167 (SCA) para 9; section 12(2) of the Constitution.

[115]Wagner (2017).

The *actio de pauperie* was first recognised as being a part of South African law in *O'Callaghan v Chaplin*[116] where the following was stated "[b]y our law, therefore, the owner of a dog that attacks a person who was lawfully at the place where he was injured, and who neither provoked the attack nor by his negligence contributed to his own injury, is liable, as owner, to make good the resulting damage".[117]

The *actio de pauperie* is therefore an action in terms of which the owner of a domestic animal can incur delictual liability for injury and damage caused by an animal. The requirements of the *actio de pauperie* are that (i) the wrongdoer must have been the owner of the animal in question; (ii) the animal must be domesticated; (iii) it must be shown that the animal acted contrary to its nature when causing the damage; and (iv) the victim must have been lawfully present at the location in which the damage was caused.[118]

It is argued, that to extend the *actio de pauperie* would not require the legal gymnastics which would be required if the *Aquilian* action were to be developed, other than to replace the animal requirement with a SAE Level 4 AV and SAE Level 5 AV. This may sound peculiar, but when one considers the following statement in *Veiera v Van Rensburg*[119] "The owner of a dog that bites is liable, not because of any negligence on his part, even though he has been careful, and even though he has never known the dog to bite before. His liability is based simply upon his ownership." [*Our emphasis*][120] The concept brings about a practical manner in which to hold the owner of the SAE Level 4 AV and SAE Level 5 AV, who is responsible for the AV being on the road in the first instance, despite no fault, control or otherwise on his part, simply by virtue of being an owner.

That said, there may be an argument that public policy would militate against this. For example, if AVs prove to reduce accidents and increase the safety of the roads, public policy considerations may require that the law encourage the use of these AVs by limiting the scope of liability that may be imposed.[121] In this regard, as has already been tabled in parliament but since rejected, the RAF's liability may become a no-fault regime. However, there are already arguments and examples that the introduction of AVs brings about the moral hazard that humans will become complacent about their safety as automation performs these tasks for us.[122]

---

[116] 1927 AD 310.

[117] 1927 AD 310 at 320.

[118] Wagner K (2017) Dog owners beware: strict liability for dog attacks. Available via De Rebus. http://www.derebus.org.za/dog-owners-beware-strict-liability-dog-attacks/. Accessed 23 October 2020.

[119] [1953] 3 All SA 170 (T).

[120] [1953] 3 All SA 170 (T) 174.

[121] Greenemeier (2018).

[122] For example, Google, having achieved level 3 in 2012, found that human drivers were too trusting and slow to retake control from the system in the event of trouble see Greenemeier L (2018) "Intelligent to a Fault: When AI Screws Up, You Might Still be to Blame" Available avia Scientific American: https://www.scientificamerican.com/article/intelligent-to-a-fault-when-ai-screws-up-you-might-still-be-to-blame1/. Accessed 22 October 2020.

Legislation

Constitution of the Republic of South Africa, 1996.
Consumer Protection Act No, 68 of 2008.
Determination of Threshold in Terms of the Consumer Protection Act, 2008 (act no. 68 of 2008) published under Government Notice 294 in Government Gazette 34,181 of 1 April 2011.
National Road Traffic Act No, 53 of 1996
Road Accident Fund Act Act No. 56 of 1996.
Road Accident Fund Regulations, 2008 (Published under Government Notice R770 in Government Gazette 31,249 of 21 July 2008).

Case Law

Abrahams v Road Accident Fund (15,863/2013) [2016].
Bekker v Du Toit 1974 (3) SA 248.
Betko Products CC v Grasso (Pty) Ltd [2010] JOL 25246 (WCC).
Bridgman NO v Road Accident Fund [2002] 1 All SA 1 (C).
Chauke v Santam LTD 1997 (1) SA 178 (A)
Da Silva v Coutinho 1971 (3) SA 123 (A).
East London Municipality v Van Zyl 1959 (2) SA 514 (E).
Edouard v Administrator, Natal 1989 4 All SA 309 (D).
Eskom Holdings Limited v Halstead-Cleak [2017] JOL 37573 (SCA).
Evins v Shield Insurance Co Ltd 1980 2 All SA 40 (A).
First National Bank of South Africa Ltd v Duvenhage 2006 (5) SA 319 (SCA).
Fourie v Road Accident Fund [1999] 3 All SA 661 (O).
Fourway Haulage SA Pty (Ltd) v SA National Roads Agency Ltd 2009 (2) SA 150 (SCA).
HL & H Timber Products (Pty) Ltd v Sappi Manufacturing (Pty) Ltd 2001 (4) SA 814 (SCA);
*International Shipping Co (Pty) Ltd v Bentley*1990 (1) SA 680 (A).
Kruger v Coetzee 1966 (2) SA 428 (A).
Maatla v Road Accident Fund (11,690/11) [2015] ZAGPPHC.
McCarthy Ltd t/a Budget Rent a Car v Sunset Beach Trading 300 CC t/a Harvey World Travel 2012 (6) SA 551 (GNP).
Mfihlo v Port Elizabeth Municipal Council 1976 (3) S.A. 183 (E)
Minister of Police v Skosana 1977 (1) SA 31 (A).
Mutual & Federal Insurance Co. Ltd v Day 2001 (3) SA 775 (SCA).
Nyaguse v Skinners Auto Body Specialists & another [2008] JOL 21160 (ZH).
O'Callaghan v Chaplin 1927 AD 310.
Oberholzer v Santam 1970 (1) SA 337 (N).
Pretorius v South British Insurance 1963 (3) SA 8 (W).

R v Dhlamini 1955 (T).
S v Chretien 1981 (1) SA 1097 (A).
S v Eadie 2002 (1) SACR 663 (SCA).
S v Humphreys 2013 (2) SACR 1 (SCA).
S v Ngubane 1985 (3) SA 677 (A).
Samson v Winn [1977] 1 All SA 442 (C).
Santam Insurance Co Ltd v Nkosi 1978 (2) SA 784 (A).
Santam v Kemp 1971 (3) SA 305 (A).
Veiera v Van Rensburg [1953] 3 All SA 170 (T).
Wagener v Pharmacare Ltd, Cuttings v Pharmacare Ltd [2003] 2 All SA 167 (SCA).

References

Anonymous (2020) Alberta RCMP charge Tesla driver for speeding and sleeping. Available via Royal Canadian Mounted Police. https://www.rcmp-grc.gc.ca/en/news/2020/alberta-rcmp-charge-tesla-driver-speeding-and-sleeping. Accessed 22 Oct 2020

Basson J (2011) The south African law on products liability- Quo Vadis? South Afr J Ind Eng 12(1)

Botha MM, Joubert EP (2011) Does the consumer protection act 68 of 2008 provide for strict product liability? - a comparative analysis. Tydskrif vir Hedendaagse Romeins-Hollandse Reg 74:305–319

Burchell J (2013) Principles of Criminal Law, 4th edn. Juta South Africa

eNatis (2019) Statistics. http://www.enatis.com/index.php/statistics. Accessed 19 Oct 2020

Greenemeier L (2018) Intelligent to a Fault: When AI Screws Up, You Might Still be to Blame. Available via Scientific American: https://www.scientificamerican.com/article/intelligent-to-a-fault-when-ai-screws-up-you-might-still-be-to-blame1/. Accessed 22 Oct 2020

Hawkins AJ (2018) Autopilot Buddy' that tricks Tesla vehicles declared 'unsafe' by US. Available via The Verge. https://www.theverge.com/2018/6/19/17479316/tesla-autopilot-buddy-aftermarket-nhtsa. Accessed 22 Oct 2020

Hyatt K, Paukert C (2018) Self-driving cars: a level-by-level explainer of autonomous vehicles. Available at: https://www.cnet.com/roadshow/news/self-driving-car-guide-autonomous-explanation/. Accessed 23 Oct 2020

Kaashifa DA (2017) A critical examination of section 61 of the consumer protection act 68 of 2008 and its implications for those in the supply chain. University of Kwazulu Natal

Klopper HB (2009) Accidental starting of a motor vehicle and section 20(1) of the Road Accident Fund Act of 1996. Tydskrif vir Hedendaagse Romeins-Hollandse Reg

Klopper HB (2012) The law of collisions in South Africa. 8th edn. Product Developer: Tiger Chetty

Klopper HB (2020) RAF Practitioners Guide. Product Developer: Sophia Kahn

Loubser M et al (eds) (2012) The law of delict in South Africa, 2nd edn. Oxford University Press, South Africa

Mahabeer L (2012) Strictly Speaking. Available via Mondaq.: http://www.mondaq.com/southafrica/x/172162/Product+Liability/Strictly+speaking. Accessed 20 Oct 2020

Neethling J, Potgieter JM (2015) Law of Delict, 7th edn, LexisNexis, South Africa

Road Traffic Management Corporation (2019) State of Road Safety Report: Calendar. https://www.arrivealive.co.za/documents/State%20of%20Road%20Safety%20Report%202019.pdf. Accessed 19 Oct 2020

SAE International (2018) Surface Vehicle Recommended Practice – Taxonomy and definitions for terms related to driving automation systems for on-road motor vehicles

Shuttleworth J (2019) SAE Standards News: J3016 automated-driving graphic update. Available via SAE International. https://www.sae.org/news/2019/01/sae-updates-j3016-automated-driving-graphic#:~:text=The%20J3016%20standard%20defines%20six,%2Dvehicle%20(AV)%20capabilities.&text=The%20latest%20J3016%20graphic%20is%20a%20living%20document.,-It%20will%20continue. Accessed 23 Oct 2020

Van der Walt JC, Midgley R (2016) Principles of delict, 3rd edn. LexisNexis/Butterworths Morningside, Durban

Wagner K (2017) Dog owners beware: strict liability for dog attacks. 1 February 2017 Available at De Rebus: http://www.derebus.org.za/dog-owners-beware-strict-liability-dog-attacks/. Accessed 23 Oct 2020

Part II
America

Civil Liability for Motor Vehicle Crashes in the United States: From Conventional Vehicles to Autonomous Vehicles

Mark A. Geistfeld

Abstract As of June 1, 2020, thirty-five states and the District of Columbia have enacted statutes expressly addressing autonomous vehicles. These regulations largely address the preconditions for operating autonomous vehicles, including rules that govern operators and technical features of the automated driving systems. Only a few states have established independent frameworks for determining liability in the event of a crash. Federal legislation governing autonomous vehicles has yet to be enacted, but developments thus far strongly suggest that federal regulations will ultimately determine the important safety performance standards required of autonomous vehicles. Federal law is supreme over state law, and so federal regulations will preempt any conflicting state law requirements and immunize a compliant autonomous vehicle from tort liability. However, if a vehicle does not comply with an applicable federal safety standard and crashes as a result, the responsible party will be subject to state tort liability for the injuries. But until federal legislation or regulation has been enacted, state law will wholly govern the space. In the absence of state legislation, state tort law will determine liability for the crash of an autonomous vehicle. In the vast majority of states, manufacturers will be subject to strict products liability for any defects that cause their vehicles to crash.

Financial support was provided by the Filomen D'Agostino and Max E. Greenberg Research Fund of the New York University School of Law.

M. A. Geistfeld (✉)
New York University School of Law, New York, NY, USA
e-mail: geistfeld@mercury.law.nyu.edu

H. Steege et al. (eds.), *Autonomous Vehicles and Civil Liability in a Global Perspective*, Data Science, Machine Intelligence, and Law 3,
https://doi.org/10.1007/978-3-031-41992-8_4

1 Introduction

In the U.S., tort law is the most common source of civil liability for motor vehicle crashes. The majority of tort cases in the state courts involve automobile accidents allegedly caused by a driver's negligence.[1] By eliminating the human driver, autonomous vehicles will eliminate these tort claims. The manufacturer will instead be responsible for the driving performance of the autonomous vehicle, potentially subjecting it to tort liability for a crash under the rules of strict products liability.[2]

The liability rules governing the crashes of autonomous vehicles, however, will depend on legislative developments. Tort law is a matter of state law in the U.S., and typically is based on the common law rather than legislation. At the state level, the common law of torts can be modified by state legislation (subject to any state constitutional limitations). The state regulation of motor vehicles, whether by statute or under the common law of torts, can then be preempted (displaced) by federal law.[3] The preemptive effect of a federal statute extends to regulations promulgated by federal agencies pursuant to such statutory authority.

In 2017, the U.S. House of Representatives passed a bill—without opposition—that would establish a federal regulatory framework for accelerating the development and safe use of automated driving technologies. A Senate committee then unanimously recommended passage of a similar bill, which ultimately died on the Senate floor. Committees from both the House and Senate have since been working together, and have issued thirteen sections of a newly proposed bill that so far largely track the provisions of the previously proposed legislation. There have not been any further developments, however, since the onset of the Covid-19 pandemic in spring 2020.[4]

Although federal legislation governing autonomous vehicles has yet to be enacted, both prior and ongoing legislative efforts provide strong indications of

---

[1] According to court data collected from seventeen states, auto tort cases in 2015 ranged from a low of 32 percent of all tort cases (Missouri) to 75 percent (Texas), with auto tort cases exceeding 50 percent of all tort cases in the majority of these states. See 2015 Statewide Auto Tort Cases per 100,000 Population, National Center for State Courts, Court Statistics Project, <http://www.courtstatistics.org/Civil.aspx> (follow "Launch DataViewer" hyperlink; then select "Statewide Auto Tort Rates (Chart)"). "Auto tort cases arise from auto accidents and typically allege personal injury, property damage, or wrongful death resulting from negligent operation of a motor vehicle." *Id.*

[2] For comprehensive discussion of how state tort law is likely to apply to the crash of an autonomous vehicle, see Geistfeld (2017), p. 1611 (hereinafter "Roadmap"). The ensuing discussion is derived from this article in ways not fully communicated by the varied references.

[3] U.S. Constitution art. VI, cl. 2 (commanding that the laws of the United States 'shall be the supreme Law of the Land; . . . any Thing in the Constitution or Laws of any state to the Contrary notwithstanding').

[4] See Maggie Miller, Action on Driverless Cars Hits Speed Bump as Congress Focuses on Pandemic, The Hill (May 20, 2020), <https://thehill.com/policy/technology/498863-action-on-driverless-cars-hits-speed-bump-as-congress-focuses-on-pandemic>.

the final statutory framework. [5] Consistent with prior legislative practices dating back to the 1990s, the statute will probably require the National Highway Traffic Safety Administration (NHTSA) to promulgate safety standards regarding the performance of autonomous vehicles. Recognizing that the rule-making process takes some time to initiate, the proposed bills all require NHTSA to begin that process within a prescribed time period (3-years being the most common requirement). Once NHTSA has initiated the process, it will probably take at least another year before it adopts such standards.

The standards that NHTSA might adopt cannot yet be predicted. For present purposes, however, the issue of greatest relevance involves the impact that these standards will have on the state regulation of autonomous vehicles. According to NHTSA's most recent policy report, "Conflicting State and local laws and regulations surrounding automated vehicles create confusion, introduce barriers, and present compliance challenges. [NHTSA] will promote regulatory consistency so that automated vehicles can operate seamlessly across the Nation.[6] As this policy suggests, federal regulations will ultimately determine many of the safety performance standards required of autonomous vehicles. In the interim, state legislation will govern the space, with state tort law supplying the default rules of liability that courts will apply to determine liability for cases involving the crash of an autonomous vehicle. Requirements for insurance and related matters fall outside the jurisdiction of NHTSA and will remain within the province of state law.

As of June 1, 2020, thirty-five states and the District of Columbia have enacted statutes expressly addressing autonomous vehicles. The statutes largely focus on basic issues related to the introduction and deployment of autonomous vehicles. For example, fourteen have provisions permitting the operation of autonomous vehicles on public roads under specified conditions.[7] Several address the related issues of duty and insurance.[8] Only a few address liability issues, as discussed further below.

[5] For citations to prior legislation and more extensive discussion, see Geistfeld (2018), p. 337 (hereinafter "Regulatory Sweet Spot").

[6] U.S. Dept. of Transp., Automated Vehicles 3.0: Preparing for the Future of Transportation (Oct. 2018), <https://www.transportation.gov/sites/dot.gov/files/docs/policy-initiatives/automated-vehicles/320711/preparing-future-transportation-automated-vehicle-30.pdf>.

[7] See, e.g., Cal. Vehicle Code § 38750 ("Existing law authorizes the operation of an autonomous vehicle on public roads for testing purposes by a driver who possesses the proper class of license for the type of vehicle operated if specified requirements are satisfied."); Iowa Code § 321.515 ("A driverless-capable vehicle may operate on the public highways of this state without a conventional human driver physically present in the vehicle. . . .").

[8] See, e.g., Nev. Rev. Stat. § 482A.060 ("Before a person begins testing an autonomous vehicle on a highway within this State, the person must 1. Submit to the Department proof of insurance or self-insurance acceptable to the Department in the amount of $5,000,000; or 2. Make a cash deposit or post and maintain a surety bond or other acceptable form of security with the Department in the amount of $5,000,000."); Tex. Transp. Code § 545.452 ("An automated motor vehicle may not operate on a highway in this state with the automated driving system engaged unless the vehicle is . . . covered by motor vehicle liability coverage or self-insurance in an amount equal to the amount of coverage that is required under the laws of this state.").

In at least eleven states, governors have issued executive orders related to autonomous vehicles, which typically promote the testing and potential deployment of the technology.[9]

To date, only Tennessee has directly specified the liability framework for autonomous vehicle crashes, requiring that liability "shall be determined in accordance with product liability law, common law, or other applicable federal or state law."[10] The statute also deems the fully engaged and properly deployed autonomous vehicle to be "the driver" for liability purposes, an assignment of responsibility also implicitly adopted by Utah in a statute addressing this technology.[11]

This Chapter surveys the common-law rules and enacted statutes that determine civil liability for both conventional motor vehicles and automated driving technologies, as classified by the SAE J3016 levels of automation.

2 SAE Level 0 (No Driver Automation) and SAE Level 1 (Driver Assistance)

The vast majority of motor vehicles on public roads currently fall into this category. Driver-assistance systems (DAS) are incorporated into conventional vehicles. Level 1 DAS handles a single function of the dynamic driving task under certain operating conditions. This type of technology has been available since the late 1970s with the introduction of anti-lock braking systems, and now includes a host of other features, such as adaptive cruise control and forward-collision prevention systems. "These functions improve the interface between the driver and the vehicle in such a way as to provide better control or more convenient operation but do not fundamentally alter the roles of the driver and vehicle in executing the [dynamic driving task]."[12] Humans are still behind the wheel, so vehicles equipped with Level 1 DAS have not created liability issues fundamentally different from those posed by conventional vehicles not equipped with this technology (Level 0).

The most common form of civil liability for these motor vehicle crashes involves the negligence of one or more drivers. In a detailed study of individual cases, NHTSA found that "the critical reason for the crash was assigned to the driver in 94 percent" of the crashes" and "to the vehicle in an estimated 2 percent of the

[9] For more detailed information, see National Conference of State Legislatures, Autonomous Vehicles State Bill Tracking Database, <https://www.ncsl.org/research/transportation/autonomous-vehicles-legislative-database.aspx>.

[10] Tenn. Code Ann. § 55-30-106.

[11] See Utah Code Ann. § 11 41-26-104(3) (stating that the automated driving system (ADS) "is responsible for compliant operation of an ADS-dedicated vehicle").

[12] Crash Avoidance Metrics P'ship (CAMP) Automated Vehicle Research (AVR) Consortium, Key Considerations in the Development of Driving Automation Systems 2 (2014), http://www-esv.nhtsa.dot.gov/Proceedings/24/files/24ESV-000451.PDF [https://perma.cc/MZ62-TK7C].

crashes."[13] Whether a driver incurs civil liability for a crash is conventionally determined by the tort law of negligence, whereas a crash caused by a defect in the vehicle subjects the manufacturer to strict products liability in the vast majority of states.[14]

All states require drivers to insure against at least some their liabilities in the event of a crash, consisting of bodily injury insurance covering liability for third-party bodily injuries, property insurance covering liability for third-party property damage, and uninsured/underinsured-motorist coverage for losses when the insured is injured by another driver lacking adequate insurance coverage.[15] A few states require drivers to purchase medical payment insurance to cover their own medical costs. These various forms of mandated coverage have created the "world's largest insurance market," and the "widespread insurance use has influenced the actual operation of the traditional system of [tort] liability in several important ways."[16] For example, in many cases insurance adjusters resolve claims based on a simple set of rules instead of a detailed analysis of negligence.[17] But as previously discussed, a significant number of claims result in tort litigation.

In an effort to expand insurance coverage while reducing litigation costs, around two dozen states have some enacted some type of no-fault legislation, with many also restricting the motorist's right to sue.[18] All of these states "ultimately adopted a hybrid system that combined elements of no-fault with traditional tort," typically obligating each vehicle owner to purchase "her own (first-party) insurance against economic loss capped at "fairly modest sums," while "giving 'seriously injured' claimants the right to seek recovery for noneconomic damages" and uncompensated economic losses "through the traditional tort system."[19]

If the crash is caused by a defect in the motor vehicle, those suffering bodily injury or property damage can recover under strict products liability against the manufacturer.[20] If the crash is caused by unsafe conditions of the roadway itself, the injured parties in some cases can recover from the governmental entity responsible for maintaining the safety of the public road.[21]

---

[13] Nat'l Highway Traffic Safety Admin., Traffic Safety Facts DOT HS 812 115 (Feb. 2015), <https://crashstats.nhtsa.dot.gov/Api/Public/ViewPublication/812115>.

[14] Commercial distributors of a defective product are also subject to strict products liability but will not be expressly mentioned in the ensuing discussion.

[15] Anderson et al. (2010), pp. 10–11.

[16] *Id.* 1 & 8.

[17] See generally Engstrom (2011), p. 805.

[18] Engstrom (2012), p. 306 (detailing the individual states).

[19] Id. at 320.

[20] See generally Geistfeld (2020) (hereinafter "Products Liability").

[21] See, e.g., Friedman v. State of New York, 493 N.E.2d 893 (N.Y. 1896) (permitting recovery only when the unsafe road condition stems from "ministerial" rather than "policy" decisions by the governmental authority).

3 SAE Level 2 (Partial Automation) and SAE Level 3 (Conditional Automation)

Levels 2 and 3 involve limited autonomous driving that requires the human operator to monitor conditions and assume control if necessary, thereby creating an interface between automated driving and conventional driving—the point at which the human takes over the dynamic driving task from the automated vehicle. The switch from one driving mode to the other presents a safety problem that does not exist in conventional vehicles lacking this technology. The sustained autonomous operation of these vehicles can make the person behind the wheel complacent or otherwise overly reliant on the DAS. What if road conditions suddenly change and require human intervention, but the driver is not sufficiently attentive to quickly take over the wheel?

Utah has enacted a statute addressing this issue. If a Level 3 vehicle "issues a request to intervene," it remains "responsible for compliant operation of the vehicle until disengagement" of the automated driving features.[22]

Established tort doctrine already obligates manufacturers to adopt fault-tolerant product designs, a doctrine capable of adequately addressing the safety question of how DAS design should address the foreseeable risk that the technology will lull the driver into complacency or inattention. If manufacturers do not design the vehicle's operating system to account for the human errors that will foreseeably occur at the interface of automated driving, they would be subject to strict products liability for the consequent crashes.[23]

4 SAE Level 4 (Full Automation in Limited Conditions) and SAE Level 5 (Full Automation Across All Conditions)

Once the operator has properly deployed an autonomous vehicle, the manufacturer becomes primarily responsible for the vehicle's driving performance. A properly deployed autonomous vehicle can crash for many reasons. For tort purposes, a manufacturer's responsibility is limited to crashes that were proximately caused by the vehicle. The mere fact that a vehicle was involved in a crash is not sufficient for this purpose. For example, the distracted driver of a conventional vehicle could suddenly veer into the path of an autonomous vehicle. The autonomous vehicle would not be a proximate cause of the ensuing crash—its driving behavior did not increase the likelihood that the distracted driver would veer into its path, nor did it

[22] Utah Code Ann. § 11 41-26-104.

[23] See Geistfeld, Roadmap (2017), pp. 1624–1629.

have the opportunity to avoid that outcome.[24] The manufacturer's responsibility, and hence potential liability, is limited to crashes caused by the risks foreseeably created by the autonomous vehicle's driving behavior.

As is true for any motor vehicle, defects in the hardware that proximately cause a crash will subject the manufacturer to strict products liability in almost all states. The distinctive forms of liability for autonomous vehicles depend on the different ways in which a crash can be caused by the operating system of an autonomous vehicle—the hardware and software that executes the dynamic driving task. This problem is most pronounced for Level 4 and Level 5 vehicles, but it also exists for Level 3 vehicles when their operating systems cause the vehicle to crash.

Crashes caused by programming bugs. A coding error could cause the operating system to freeze or crash, in turn causing the vehicle to crash. Because the coding error prevented the operating system from performing its manifestly intended function of executing the dynamic driving task, the manufacturer will be subject to strict liability for physical injuries proximately caused by the "malfunction."[25]

The liability in these cases is truly strict. Due to the complexity of coding, programmers cannot ensure that an operating system will never malfunction. "Unlike mechanical errors, software errors occur in an arbitrary manner that cannot be reasonably prevented via ex ante design or ex post testing. In the commercial-grade software systems being built today, it is impossible to guarantee complete correctness of code."[26] This condition, however, does not justify negligence liability as some commentators maintain.[27] The same condition applies to ordinary products for which the complexity of the manufacturing process makes it impossible to ensure that the final product will always be free from defects.[28] That complexity also considerably limits the effectiveness of negligence liability; it routinely prevents plaintiffs from being able to prove that manufacturers failed to adopt reasonably safe systems of quality control. Consequently, a negligence regime would mean that "as a

[24] *See* Restatement (Third) of Torts: Liability for Physical and Emotional Harm § 30 (Am. Law Inst. 2010) ("An actor is not liable for harm when the tortious aspect of the actor's conduct was of a type that does not generally increase the risk of that harm.").

[25] Restatement (Third) of Torts: Products Liability § 3 cmt. b (Am. Law Inst. 1998) (defining defect in terms of a product that malfunctions by failing "to perform its manifestly intended function").

[26] Choi (2019), p. 44.

[27] See generally id. (arguing that liability for software malfunctions that cause physical products to malfunction be determined by a negligence rule of "crashworthiness" that asks whether the code "adequately incorporates state-of-the-art techniques in software fault tolerance").

[28] "Monitoring the entire manufacturing process in terms of quality ... could prove difficult considering that satisfactory manufacturing quality can only be achieved by rapid screening of materials at the input stage, transformation process stage within the manufacturing hub and finally at the output stage before shipment to the customer." Oduoza (2020). In such a manufacturing environment, it "may not be feasible to evaluate every single item going through the process and also at the output. Rather, acceptance sampling is used for quality assurance." But even this method of quality control is limited. In particular, "[d]uring mass production it is very difficult using conventional quality control methods to find defects of the order of ppm or less at the stage of design and production." Id. at 4.

practical matter, sellers may escape their appropriate share of liability" due to problems of proof.[29] To solve this problem, tort law subjects manufacturers to strict liability for products that malfunction in order to give them an adequate financial incentive to adopt reasonably safe systems of quality control.[30] This rationale for strict liability straightforwardly applies to programming bugs that cause the operating system of an autonomous vehicle to malfunction.

Crashes caused by third-party hacking. A manufacturer's tort obligations encompass the foreseeable risk that a third-party will illegally hack into the operating system, take control of the vehicle, and cause it to crash. The manufacturers liability for these crashes could be governed by the malfunction doctrine. The manufacturer obviously does not intend for the vehicle to be controlled by an unauthorized third party, and the ordinary consumer presumably expects that the operating system, not a hacker, determines the vehicle's driving behavior. The performance of a hacked vehicle satisfies the requirements for a product malfunction that would subject the manufacturer to strict liability in the event of a crash.

If, however, strict liability would produce extensive liabilities that manufacturers cannot adequately avoid—recall the complexity of coding—the rule could disrupt the development and widespread deployment of this life-saving technology. For life-saving products such as vaccines and blood products, courts have immunized sellers from strict liability, instead subjecting them to negligence liability. By analogy, courts could rely on this same policy rationale for immunizing autonomous vehicles from strict liability for malfunctions caused by third-party hacking.[31]

Crashes caused by the fully functioning operating system. If the fully functioning operating system proximately causes an autonomous vehicle to crash, could this performance—the crash itself—constitute a malfunction subject to strict liability? A malfunction is defined by reference to the product's expected performance, and so the liability question depends on how courts formulate the expectation of how a fully functioning operating system should execute the dynamic driving task for an autonomous vehicle.

Under the *Restatement (Third) of Torts,* the relevant expectation involves the product's failure to "perform its manifestly intended function."[32] For the design of an operating system, "[t]he most general objective" is for the autonomous vehicle "to safely reach the specified destination."[33] Indeed, "[m]any manufacturers have ...

---

[29] Restatement (Third) of Torts: Products Liability § 2 cmt. a.

[30] Id. See also Geistfeld, Products Liability (2020), pp. 83–106 (explaining why product malfunctions rely on the same rationale for strict liability applicable to construction or manufacturing defects).

[31] See Geistfeld, Roadmap (2017), pp. 1669–1674.

[32] Restatement (Third) of Torts: Products Liability § 3 cmt. b (Am. Law Inst. 1998).

[33] Furda and Vlacic (2011), p. 10; see also, e.g., Arnay et al. (2016), p. 23 ("The problem we want to solve is safely following a predefined route while avoiding dynamic obstacles."). In general, any "information processing system" embodies a "computational theory [that] corresponds to the goal of computation and an abstract definition of the task." Alpaydin (2016), p. 20. For autonomous

adopted targets and plans for reaching zero injuries and fatalities."[34] Based on this coding objective for an autonomous vehicle's operating system, any crash arguably involves a failure of the vehicle's manifestly intended function, constituting a product malfunction that subjects the manufacturer to strict liability.

Rather than defining a malfunction by reference to the product's manifestly intended function, a substantial majority of states instead evaluate this issue with the consumer expectations test.[35] To satisfy this test, "the product must meet the safety expectations of the general public as represented by the ordinary consumer, not the industry or a government agency."[36] Under this test, the "crucial question in each individual case is whether the circumstances of the product's failure permit an inference that the product's design performed below the legitimate, commonly accepted minimum safety assumptions of its ordinary consumers."[37] At minimum, the ordinary consumer expects that a product will not malfunction. The frustration of that expectation supplies the rationale for subjecting the manufacturer to strict liability for product malfunctions.

To trigger the consumer expectations test, the ordinary consumer must only have well-formed expectations of the product performance in question; he or she does not otherwise have to understand the underlying technology (as with exploding airbags).[38] The consumer, therefore, can have minimum expectations of how safely an autonomous vehicle should perform, such as the expectation that the operating system will not malfunction because of a programming error or bug. The question is whether the ordinary consumer has minimum safety expectations about the minimally safe performance of a fully functioning autonomous vehicle.

To the extent that consumers have expectations about how this new technology will perform, they will not expect perfect safety. After all, automated driving technologies have already caused fatalities.[39] In the event that a fully functioning autonomous vehicle crashes, the unexpected accident will not be a malfunction that

---

vehicles, the "abstract definition of the task" includes arriving safely at the specified destination along with other factors such as minimizing the time or length of trip.

[34] Anders Eugensson et al., Environmental, Safety, Legal and Societal Implications of Autonomous Driving Systems (National Highway Traffic Safety Administration, Research Paper No. 13-0467, 2013), www-nrd.nhtsa.dot.gov/pdf/esv/esv23/23ESV-000467.PDF [https://perma.cc/KC4W-QLVL]; see also Richard Waters, CES 2016: Toyota Poaches Google Exec to Help Lead AI Effort, Financial Times, Jan. 5, 2016, at 15 (reporting that the "ultimate goal" of Toyota's driverless car program is to create "a car that cannot be responsible for a collision").

[35] Branham v. Ford Motor Co., 701 S.E.2d 5, 14 n.12, 15 (finding that only seventeen different states expressly reject consumer expectations in favor of the risk-utility test).

[36] Soule v. General Motors Corp., 882 P.2d 298, 306 (Cal. 1994).

[37] Id. at 309.

[38] See, e.g., Green v. Smith & Nephew AHP, Inc., 629 N.W.2d 727, 742 (Wis. 2001) ("[A] condition not contemplated by the ordinary consumer[] does not inevitably require any degree of scientific understanding about the product itself. Rather, it requires understanding of how safely the ordinary consumer would expect the product to serve its intended purpose.").

[39] See Mark A. Geistfeld, Why the Next Person Hit By a Driverless Car Might Not Be Able to Sue, Time, March 30, 2018, http://time.com/5221393/uber-autonomous-vehicle-death-settlement/

violates the ordinary consumer expectations test. The manufacturer would incur liability for such a crash only if the design of the operating system were not reasonably safe or if the manufacturer otherwise failed to adequately warn consumers about the inherent risk of crash.

5 Reasonably Safe Operating Systems and Adequate Warnings

"[W]ho should be held liable for crashes involving [autonomous vehicles]? Though manufacturers, insurers, news outlets, and academics have all posed this question, they have not found easy answers."[40] As indicated by the discussion so far, the most difficult liability issue involves cases in which the fully functioning operating system causes an autonomous vehicle to crash. The manufacturer would incur liability if the crash were caused by a defect attributable to the design of the operating system that causes the vehicle to perform in an unreasonably dangerous manner.[41] What, then, determines whether the operating system of an autonomous vehicle is defectively designed?

These systems "typically use machine learning algorithms that have been 'trained' to drive by analyzing examples of safe driving, and automatically generalizing about the core patterns that constitute effective driving from these examples."[42] The driving experience of an autonomous vehicle—or more precisely, all vehicles with the same operating system—provides the data for machine learning that enables the operating system to adapt accordingly.[43] As NHTSA explains, while "human driver[s] may repeat the same mistakes as millions before them, an [autonomous vehicle] can benefit from the data and experience drawn from thousands of other vehicles on the road."[44]

This attribute of the technology has important implications for the liability problem. Autonomous vehicles will transform the individualized behavior of human drivers into a collective, systemized form. Each vehicle with the same operating system executes the dynamic driving task in the same manner.

(discussing these cases and proposing that during the period of premarket testing, manufacturers be obligated to pay no-fault compensation to injured bystanders).

[40] Boeglin (2015), p. 174.

[41] See Restatement (Second) of Torts § 402A (applying strict liability to defective products that are "unreasonably dangerous").

[42] Surden and Williams (2016), p. 148.

[43] *See* Lipson and Kurman (2016), p. 17 ("As cars pool their driving 'experience' in the form of data, each car will benefit from the combined experience of all other cars.").

[44] National Highway Traffic Safety Administration, U.S. Department of Transportation, Federal Automated Vehicles Policy: Accelerating the Next Revolution in Roadway Safety 5 (2016), https://one.nhtsa.gov/nhtsa/av/pdf/Federal_Automated_Vehicles_Policy.pdf [https://perma.cc/VSX9-B6J6].

Consequently, the entire fleet is effectively guided by a single driver—the operating system that determines how this class of autonomous vehicles executes the dynamic driving task. Quite unlike the crashes of conventional vehicles that require case-by-case analysis of driver behavior, the crashes of autonomous vehicles with fully functioning operating systems are properly evaluated in relation to the systemized safety performance of the entire fleet.

Whether a fully functioning autonomous vehicle performed in a reasonably safe manner, therefore, does not depend on the circumstances of the crash. The appropriate liability question is whether the operating system has had sufficient learning experience to systematically drive the vehicle in a reasonably safe manner across its intended operating domain.

Premarket testing supplies the driving experience that machine learning uses to improve the safety performance of an autonomous vehicle before it is commercially distributed in the market. The operating system would be defectively designed if it were not adequately tested and unable to make the necessary safety improvements via machine learning.

In the vast majority of states, this question is resolved by the risk-utility test or its substantive equivalent, the modified consumer expectations test.[45] Under this test, a design is defective if there is a reasonable alternative design with a disutility (or increased cost) that is less than the associated reduction of risk (or safety benefit of the design modification).[46]

As applied to the safety issue under consideration, the liability inquiry takes the same form in most states. To identify a reasonable alternative design of the existing operating system, the tort plaintiff must show that the costs of more extensive testing are less than the safety benefit of improving the operating system's systemic safety performance via machine learning. Proof of this type would show that the existing design is unreasonably dangerous or defective as compared to the reasonable alternative design of the operating system that would be generated by more extensive premarket testing (and machine learning).

The costs and benefits of more extensive testing depend on various factors. The requisite amount of premarket testing is an empirical question. Consequently, by adopting a set of factual assumptions that bias the risk-utility analysis in favor of more extensive testing, we can identify the amount of testing that would conclusively satisfy the manufacturer's obligation to supply a reasonably safe operating system.

During the initial period when the market transitions from conventional vehicles to autonomous vehicles, there is a clear benchmark that would conclusively satisfy a

[45] See Geistfeld, Products Liability (2020), pp. 107–143.

[46] See Restatement (Third) of Torts: Products Liability § 2 cmt. d (Am. Law Inst. 1998) (adopting "a reasonableness ('risk-utility balancing') test as the standard for judging the defectiveness of product designs," which "asks whether a reasonable alternative design would, at reasonable cost, have reduced the foreseeable risks of harm posed by the product and, if so, whether the omission of the alternative design by the seller or a predecessor in the distributive chain rendered the product not reasonably safe").

manufacturer's tort obligation to adequately test an autonomous vehicle. Taking into account the safety trade-off between a conventional vehicle and an imperfect but safer autonomous vehicle, a fully functioning autonomous vehicle will systematically drive in a reasonably safe manner if prior driving experience shows that the operating system at least halves the incidence of crashes relative to conventional vehicles.

To see why, consider a set of driving conditions—total miles, proportion spent on expressways, proportion spent in urban areas, and so on—that make it possible to reliably compare the safety performance of the operating system with the safety performance of conventional vehicles. Suppose that under these conditions, on average there are ten fatal crashes involving conventional vehicles driven by human drivers. An autonomous vehicle that halves this rate would be expected to cause only five fatal crashes under these same conditions.

The case for more extensive testing would be strongest if further testing of the autonomous vehicle under these same conditions would virtually eliminate these crashes. At most, then, more extensive testing would eliminate five fatal crashes, and so this risk can be imputed to the current design of the vehicle's operating system.

This risk must then be compared to the cost or disutility of reducing it by altering the current design via more extensive testing (and machine learning for the operating system). The delayed deployment of the autonomous vehicle determines the expected cost of additional testing. At minimum, the cost equals the lost safety benefit of eliminating the five fatal crashes that a conventional vehicle's human driver would otherwise cause across the same set of operating conditions.

Consequently, the *minimum* expected cost of more extensive testing (five fatal conventional vehicle crashes that immediate deployment of the autonomous vehicle would prevent) equals the *maximum* expected safety benefit or reduced risk (five fatal crashes of the autonomous vehicle that would be prevented by further machine learning of the operating system). Expected minimum costs (or disutility) equal expected maximum benefits (or risk reduction), but because the risk-utility analysis so far has been biased in favor of more extensive testing, the balance at this point does not justify such testing. Therefore, more extensive testing is not required by the risk-utility test and by extension, the modified consumer expectations test.

In the vast majority of states, the liability rule yields an identifiable solution to the safety problem of how much testing is required in order to ensure that a fully functioning operating system performs in a reasonably safe manner. This benchmark, however, is only transitional.[47] As the market matures and autonomous vehicles become the norm, a different baseline of comparison will be required in order to account for the continual improvement in the safety performance of autonomous vehicles. For now, however, the baseline for evaluating safety benefits of an autonomous vehicle can be defensibly defined in terms of the relative safety performance of conventional vehicles.

---

[47] Geistfeld, Roadmap (2017), p. 1653.

To prove that the design of a product is defective, the tort plaintiff must identify a reasonable alternative design that is within the "state of the art"—a requirement that could permit the plaintiff to compare the autonomous vehicle's safety performance to other (reasonably designed) autonomous vehicles that have already been commercially distributed by other manufacturers.[48] This baseline is problematic because it would provide a considerable competitive advantage for the "first movers" in the commercial distribution of autonomous vehicles. Due to machine learning, the collective driving experience of a fleet already on the road will significantly enhance the safety performance of each vehicle. Without such experience to draw upon in the design of its operating system, a new entrant would often face a substantial cost disadvantage (of more extensive testing) to match the performance of other commercially available vehicles. Requiring new entrants to equal or exceed the safety performance of autonomous vehicles already on the road, therefore, could easily undermine competition in the market by entrenching the first movers, thereby suppressing the technological innovations otherwise offered by new entrants.

This dynamic presumably will change as the market matures, but those are not the market conditions we now face. To allow new entrants sufficient access to this newly developing market, the liability rule should evaluate the safety performance of a fully functioning autonomous vehicle in relation to conventional vehicles—the type of liability rule described above.

Because autonomous vehicles are projected to reduce fatalities by up to 90 percent relative to conventional motor vehicles, Professors Abraham and Rabin claim that a liability rule defined in relation to the safety performance of conventional vehicles is "insufficiently exacting."[49] But they acknowledge that this formulation of this liability rule is only "transitional,"[50] and so their critique of the rule by reference to the long-term safety potential of autonomous vehicles is entirely beside the point. As they recognize in another portion of their discussion, "any new approach will have to deal with the long and uneven transition to automated technology."[51] The foregoing liability rule accounts for the circumstances that define the transition from conventional motor vehicles to automated vehicles; it will not forever define the required safety performance of fully functioning autonomous vehicles.

Although the transitional performance benchmark can be clearly identified as a matter of state tort law—the fully functioning autonomous vehicle must at least halve the crash rate relative to conventional vehicles—it depends on underlying criteria for evaluating the safety performance of an autonomous vehicle. What are the necessary road conditions? How many miles should be driven on freeways and in

[48] See Restatement (Third) of Torts: Products Liability § 2 cmt. d (Am. Law Inst. 1998) (discussing how the term "state of the art" has been variously defined by a multitude of courts; for some it refers to industry custom or practice; for others it means the safest existing technology that has been adopted for use; for others it means cutting-edge technology).

[49] Abraham and Rabin (2019), p. 145.

[50] Id. 145 n. 40.

[51] Id. at 129.

urban conditions? How many total miles must be logged by an operating system to generate sufficiently reliable crash data? What other metrics are required for adequately measuring safe performance? These issues are not clearly resolved by widely adopted tort rules, making this safety issue particularly appropriate for federal regulation by NHTSA.

As compared to the tort system, NHTSA is better situated to determine the appropriate testing conditions for evaluating the safety performance of an autonomous vehicle. The tort system would resolve these issues through the adversarial presentation of evidence addressing the particular claims before a court. NHTSA, by contrast, would use its specialized expertise to comprehensively address these matters through the administrative rule-making process, giving it a comparative institutional advantage for determining the appropriate testing criteria for evaluating the safety performance of a fully functioning autonomous vehicle.

Once NHTSA has identified the appropriate premarket testing conditions, it can then mandate standards requiring that the testing data show that the autonomous vehicle at least halves the rate of crashes as compared to conventional vehicles. Doing so would fully attain the national interest in uniformity while maximally accounting for the traditional state interest in tort law—the regulatory sweet spot.[52]

One other important source of liability remains. Even when subject to adequate premarket testing and properly deployed, a fully functioning operating system can still cause the autonomous vehicle to crash. The circumstances in which this might occur will be opaque to consumers, and so the underlying tort obligations require the manufacturer to adequately warn about the inherent, foreseeable risk that the fully functioning autonomous vehicle might crash.[53] The failure to do so renders the warning defective and subjects the manufacturer to liability for these crashes.

Because these crashes involve systemized driving performance, aggregate driving data provide a determinate method for the manufacturer to satisfy the tort obligation. For example, when the vehicle is first commercially distributed, the manufacturer can determine the inherent risk of crash from the data showing that the vehicle has been subject to adequate premarket testing. Over time, machine learning will improve the systemic safety capabilities of the operating system, but manufacturers will have ongoing access to the performance data and can communicate the updated risk information to consumers via product warnings.[54]

[52] For more extensive discussion of why this approach best fits the existing federal statutory framework for regulating the safe performance of motor vehicles—a framework likely to be maintained by any new legislation governing autonomous vehicles—see Geistfeld, Regulatory Sweet Spot (2018).

[53] See Restatement (Third) of Torts: Products Liability § 2 cmt. i (Am. Law Inst. 1998) ("Warnings alert users and consumers to the existence and nature of product risks so that they can prevent harm either by appropriate conduct during use or consumption or by choosing not to use or consume.").

[54] Another way to satisfy the tort obligation is by apprising consumers about the full cost of insuring the autonomous vehicle, because the insurance premium is based on the underlying risk of crash. See Geistfeld, Roadmap (2017), pp. 1654–1660.

6 Crashes Involving Bystanders

To "teach" the operating system how to drive, "the programmers feed the software with many traffic situations and specify the correct action for each situation," and the machine-learning algorithm then employs statistical analysis to determine the best way to achieve the desired outcomes.[55] What constitutes the "correct action" for situations in which the autonomous vehicle will inevitably crash and could injure consumers and bystanders? How should the vehicle's operating system be instructed to execute actions that can protect one party at the expense of another? The issue has drawn a great deal of public attention.[56]

The negligence rule provides clear guidance on how a manufacturer must design an autonomous vehicle to protect bystanders. As someone who would be foreseeably threatened by operation of the vehicle, a bystander is encompassed within the manufacturer's duty to exercise reasonable care in designing the vehicle. To satisfy this obligation, the manufacturer must give "impartial consideration" to the interests of bystanders, treating them no differently from its own interest in satisfying consumer demand for the product.[57] The manufacturer, therefore, must initially code or teach the operating system of an autonomous vehicle so that the "correct action" treats consumers and bystanders equally.

For example, an autonomous vehicle with sensors indicating that it is occupied by one person must swerve to avoid hitting a group of pedestrians, even if doing so would threaten grave injury to the lone passenger. If the "correct action" for the operating system were not specified in this manner, its programming or design would unreasonably privilege the interests of the single occupant over the larger number of pedestrians. The manufacturer would then be subject to negligence liability for injuries suffered by the pedestrians. To avoid liability, the manufacturer must design the operating system of the autonomous vehicle to minimize the expected injuries from any given crash, whether the potential victims are occupants or bystanders.[58]

---

[55] Alexander Hars, Inventivio GmbH, Top Misconceptions of Autonomous Cars and Self-Driving Vehicles 4, http://www.driverless-future.com/?page_id=774 [https://perma.cc/UAJ8-EX6N].

[56] See, e.g., Larry Greenemeier, Driverless Cars Will Face Moral Dilemmas, Scientific American (June 23, 2016), http://www.scientificamerican.com/article/driverless-cars-will-face-moral-dilemmas [https://perma.cc/CTV3-5ZGS].

[57] Restatement (Second) of Torts § 283 cmt. e (Am. Law Inst. 1965).

[58] An actor's conduct is "negligent if the magnitude of the risk [foreseeably created by the conduct] outweighs the burden of risk prevention." Restatement (Third) of Torts: Liability for Physical and Emotional Harm § 3 cmt. e (Am. Law Inst. 2010). In problem at hand, the foreseeable risk created by the driving behavior involves the threatened fatal injuries to bystanders, whereas the burden of preventing that risk involves exposing the occupants of the vehicle to the fatal risk. Because the interests of bystanders are given the same weight as the interests of the vehicle's occupants, it would be negligent to design the vehicle so that it chooses to injure a larger number of individuals to prevent the same injury for a smaller number.

In addition to the operating system, the manufacturer must design other aspects of the vehicle to reasonably account for bystander interests. The manufacturer, for example, may have to design the autonomous vehicle so that it can adequately communicate with other vehicles or pedestrians, signaling the driving behavior that they can expect.[59] Once again, the negligence standard of reasonable care (or its substantive equivalent, the risk-utility test) determines the requisite precautions.

7 Conclusion

Although the liability framework for autonomous vehicles will ultimately depend on yet-to-be-enacted federal legislation, state tort law provides ample guidance on the underlying liability issues, at least for the transitional period when autonomous vehicles will share the road with human drivers.

After autonomous vehicles are first commercially distributed, there will be a transitional period when human drivers find it particularly difficult to anticipate the driving behavior of autonomous vehicles, resulting in crashes. A fully functioning autonomous vehicle foreseeably creates the risk of these crashes, making it a proximate cause of the crash. This inherent risk of crash, however, factors into the relative safety performance of the autonomous vehicle and is accordingly addressed by the tort law requirements of adequate premarket testing and disclosure of the inherent risk of crash in the product warning. The vehicle must also be coded and otherwise designed in a manner that equally accounts for the interests of other motorists and users of the roadway (bystanders). If such a fully functioning autonomous vehicle proximately causes a crash with a conventional vehicle, the manufacturer would not be subject to tort liability. The absence of manufacturer liability will channel injury costs from the relatively inefficient tort system, where they are now largely located, into the more cost-effective systems of first-party insurance.[60] The same outcome is ideally attained by no-fault automobile insurance, creating the distinct possibility that the transition from conventional motor vehicles to autonomous vehicles will finally usher in the era of no-fault insurance long championed by those looking to reduce the cost of motor vehicle crashes.[61]

[59] *See* Surden and Williams (2016), pp. 163–174 (discussing how "the activities of autonomous vehicles . . . can be made more predictable through deliberate technological design decisions").

[60] In a tort suit, both parties typically need legal representation, and so by eliminating tort liability, no-fault auto insurance can reduce total costs for consumers by channeling claims for compensation into systems of first-party insurance. See also Geistfeld, Products Liability (2020), pp. 65–71 (providing other reasons why first-party insurance mechanisms reduce costs for consumers relative to the cost of injury compensation supplied through the tort system).

[61] See generally Engstrom (2012) (explaining this rationale for no-fault auto insurance and arguing that the plans adopted by states fatally undermined the objective of reducing insurance costs by placing the insurance obligation on auto insurers rather than first-party insurers).

References

Abraham KS, Rabin RL (2019) Automated vehicles and manufacturer responsibility for accidents: a new legal regime for a new era. Va Law Rev 105:127

Alpaydin E (2016) Machine learning: the new AI, p 20

Anderson JM et al (2010) RAND, The U.S. experience with no-fault automobile insurance: a retrospective, pp 10–11

Arnay R et al (2016) Safe and reliable path planning for the autonomous vehicle Verdino. IEEE Intell Transp Syst Mag 8:22

Boeglin J (2015) The costs of self-driving cars: reconciling freedom and privacy with tort liability in autonomous vehicle regulation. Yale J Law Technol 17:171

Choi B (2019) Crashworthy code. Wash Law Rev 94:39

Engstrom NF (2011) Sunlight and settlement mills. N Y Univ Law Rev 86:805

Engstrom NF (2012) An alternative explanation for no-fault's "demise". DePaul Law Rev 61:303

Furda A, Vlacic L (2011) Enabling safe autonomous driving in real-world city traffic using multiple criteria decision making. IEEE Intell Transp Syst Mag 3:4

Geistfeld MA (2017) A roadmap for autonomous vehicles: state tort liability, automobile insurance, and federal safety regulation. Calif Law Rev 105:1611

Geistfeld MA (2018) The regulatory sweet spot for autonomous vehicles. Wake For Law Rev 53: 337

Geistfeld MA (2020) Principles of products liability, 3rd edn

Lipson H, Kurman M (2016) Driverless: intelligent cars and the road ahead, p 17

Oduoza CF (2020) Quality control perspectives during mass production with a focus on the chemical industry. IntechOpen. https://doi.org/10.5772/intechopen.90203

Surden H, Williams M-A (2016) Technological opacity, predictability, and self-driving cars. Cardozo Law Rev 38:121

Self-Driving Cars Regulation in Colombia

Juan F. Córdoba-Marentes and Obdulio Velásquez-Posada

Abstract Civil liability derived from driving activities has been built, in Colombian case law, based on the doctrine of dangerous activities. In light of this doctrine, the person liable for the damage caused by a vehicle is its keeper. The driver is presumed to be the car's keeper, who has the power of direction and control over the vehicle. Additionally, others who hold the role of keepers of the vehicle, such as its owner or the transport company to which it is affiliated, will also be liable. There is no specific legal regulation for self-driving in Colombia. However, the activity of such vehicles will indeed also be considered a dangerous one, in line with case law, since it is performed with the "use of things or energies that may cause harm to third parties." Furthermore, the one who has programmed the vehicle system, its manufacturer, may share the quality of keeper. For this reason, this liability regime for dangerous activities, in the case of self-driving cars, will be complemented with the defective product liability regime regulated in the Colombian consumer statute and the Electronic Commerce Act guidelines.

1 Introduction

60% of the world's vehicles are in low and middle income countries. Nevertheless, it is in these countries where 93% of road accident fatalities occur (World Health Organization 2021). Colombia, a middle-income country with 17 million vehicles (RUNT 2021), has become one of the countries where vehicle accidents represent the second cause of violent deaths nationwide. Moreover, of the nearly 6000 road deaths between January and October 2021, 41% of them were caused for disobeying traffic signs, 38% for speeding, and 8% for drink-driving (Instituto Nacional de Medicina Legal 2021). In this sense, human factors or errors cause at least 87% of these deaths.

J. F. Córdoba-Marentes (✉) · O. Velásquez-Posada
Universidad de La Sabana, Chía, Colombia
e-mail: juan.cordoba@unisabana.edu.co; obdulio.velasquez@unisabana.edu.co

H. Steege et al. (eds.), *Autonomous Vehicles and Civil Liability in a Global Perspective*, Data Science, Machine Intelligence, and Law 3,
https://doi.org/10.1007/978-3-031-41992-8_5

With the arising of multiple questions and desperate solutions, one highlights above others: Would self-driving cars help prevent most of these accidents? It would be a scenario worth contemplating, notwithstanding the minimum number of self-driving cars on Colombian roads, primarily due to the high price of such vehicles for the lower-middle-income population that constitutes the country. The Colombian economy is incapable of keeping the pace of the constantly emerging technology, registering, in this respect, only 28 new Tesla cars within the frame of 2020 and 2021 in its territory (ANDEMOS 2021).

Thereby, the vast majority of cars in Colombia are conventional—level 0–2—and, in this regard, subject to the accidents' layout just mentioned. Nonetheless, the few self-driving cars circulating on the streets might as well be involved in road accidents, due to failures of those cars, but also due to the poor traffic signaling, the faulty infrastructure, or to the recklessness of other drivers or pedestrians (World Health Organization 2018). In this scenario, traditional legal rules and principles can help determine the limits and scope of its application to self-driving-related accidents.[1] In other words, an essential question about self-driving cars would be that one that considers "who or what should be liable for contractual, criminal or civil misconduct involving artificial intelligence and under what conditions" (Giuffrida et al. 2018).

In order to answer the latter question, it would be necessary to understand the Colombian legal framework that regulates liability for damages caused by cars. In particular, the Colombian legal system belongs to the civil law system. Thereupon, the civil law, as a deductive system, regulates all the hypotheses with its legal texts, pointing out, before the facts, the respective consequences, which in some instances may be able to facilitate a case resolution when this one contemplates facts easily adaptable to the legal description (Rheinstein 1952; Steven 1996).

Against this background, there are two parties in any judicial process of claiming damages in Colombia: the defendant or tortfeasor and the injured party or plaintiff. For a claim against the defendant to proceed, he or she must have legal capacity. Furthermore, when two or more persons have caused damage and possess legal capacity, each of them will be jointly responsible.

Civil responsibility can be direct or indirect. The indirect or vicarious civil responsibility is the so-called civil responsibility for the actions of others. Parents, for example, are liable for the actions of their children while in their care. In the same way, legal entities are directly liable for damages caused by their employees (CSJ 1962).

With this legal framework, the following pages will study Colombian law under the scope of liability for damage caused by cars at different levels of automated driving. The analysis will focus on the dangerous activities doctrine considering the role of the car's keeper and the liability for defective products. Firstly, the legal analysis corresponding to level 0–2 cars will examine the fundamentals of the

[1] For more information concerning the importance and urgency of regulation for autonomous vehicles see: Beale (2018), p. 223; Anderson et al. (2016), p. 134.

liability regime and its current application. In the section referring to level 3–4 cars, the general legal principles will be adapted to the technical characteristics of these vehicles, bearing in mind the matters related to the responsibility of the manufacturer and the programmer of the car software. There will finally be certain conclusions and recommendations for determining liability for the use of self-driving cars.

2 Liability for Level 0–2 Cars

Colombia's automotive fleet consists mainly of Level 0–2 vehicles that require the presence of a human driver. Therefore, this axis will be of utmost importance for legal purposes: the human person should not lose control of the vehicle, and the car should not assume its operation autonomously under any circumstance.

As there is no legal or judicial definition, the new situation of the level 0–2 must be regulated by the traditional principles of contractual and Aquilian liability for damages to passengers and third parties, respectively. The question that arises, then, is whether the existing traditional law may be sufficient to regulate the new technological phenomenon or if, on the other hand, it should require adjustments and, if so, to what extent.

2.1 Civil Liability in Colombia: Background

More than 160 years ago, the Civil Code built a whole legal framework concerning civil liability in Colombia. However, society and law changes have outflowed in new interpretations evident in the growing case law, especially in the Supreme Court of Justice decisions. Regarding civil liability, article 2341 has become one of the most widely interpreted: "The one who has committed a crime or fault, that has inferred damage to another, is obliged to compensate [. . .]". From the understanding of it, it is possible to identify the conditions to declare civil liability:

(i) A conduct, which may consist of an action or omission.
(ii) The presence of negligence or malice (the defendant has acted with fault if a prudent person would not have done the conduct in question under the same external circumstances as the defendant did. In other words, the defendant acted below the standard of behavior expected from a diligent person in the specific circumstances)
(iii) Damages: a plaintiff must always prove pecuniary and non-pecuniary damages. The damage is compensable only if it also meets three requirements: (a) True damage: damage may be previous or future, but there must be sure of its existence, and the judge may achieve this certainty with all the available proof (witnesses, documents, reports of experts or judicial experts, confessions of the parties). (b) Entitlement to claim the damages: the plaintiff must be the

victim, heir, or third party entitled to claim those damages. (c) Unjustified damage: the defendant produced damages without any just cause (this can be the legitimate defense, the order of legitimate authority, the strict fulfillment of a legal duty, or the state of necessity). If the defendant proves any of these causes of justification, the damage is not compensable, and the victim must bear it.

(iv) Causal link between wrongful conduct and harm. In other words, the damage must be imputable to the defendant as the cause of the damage.

These four conditions apply for fault liability, the general rule in Colombia. There is, exceptionally, strict liability in some instances. In contrast to fault liability, strict liability does not require the existence or proof of negligence or malice.

As the Colombian law of torts provides that driving vehicles is a dangerous activity, strict liability applies. Therefore, the only available defense is to present evidence of an extrinsic cause that breaks the causal link. The defendant is not exonerated just by proving the absence of fault or diligence and care, but he or she has to destroy the causal link by proving force majeure or an act of God or by the exclusive action of the victim or a third party. The defendant will be acquitted of paying damages in any of these hypotheses. There is a broader explanation of the dangerous activities doctrine in the following sections.

2.2 Origin of the Concept of Dangerous Activity in Colombia

The irruption of strict liability in France at the end of the nineteenth century caused pressure on the Colombian doctrine and courts. Faced with the political difficulty of reforming the Civil Code's rules that did not provide strict liability, scholars and courts proposed a novel interpretation of an old article to make room for the emerging theories of strict liability.

Article 2356 of the Civil Code states that "[a]s a general rule, any damage that can be attributed to malice or negligence of another person must be repaired by the latter" reiterating the principle of fault liability. Nevertheless, when the same article then introduces the expression "[. . .] they are especially obliged to this reparation [. . .]", it states a series of examples where the defendant should be deemed especially obliged to repair the damage:

> 1. He who recklessly fires a firearm; 2. He who removes the slabs from a ditch or pipe, or discovers them in the street or road, without the necessary precautions to prevent those who pass by there from falling by day or night.

With this explanation, an updated interpretation of the mentioned article made it possible to consider that the said activities were "dangerous" and should be part of a different legal regime that would relieve the victim of the burden of proof in some instances (Zuleta Angel 1974).

Hence, dangerous activity doctrine application began in the 1930s (CSJ 1938a), when the Supreme Court of Justice started to consider the scope of article 2356 and developed a theory about it (Zuleta Angel 1974). Accordingly, through leading cases, the Supreme Court expressly accepted the French doctrine of strict liability (1938b)[2] and interpreted article 2356 as requiring "only that the damage be imputable" (1938a). In this cornerstone case in which a locomotive caused a fire with its sparks, the Court considered that "a warehouse for inflammable substances, an explosives factory, as well as a railroad or an automobile [...] carry or have of their own extraordinary dangerousness, from which individuals generally cannot escape with their sole prudence. Therefore, damages of that kind are presumed to be caused by the respective agent, as would be the case in these examples, the train driver or the Railroad Company" (1938a). In another case, the Court reiterated that "the fault of the author of the damage must be presumed, which imposes on him the obligation to destroy that presumption by demonstrating the existence of one of these three exculpatory factors: fortuitous case, force majeure, the intervention of foreign elements" (1938c).

2.3 *Characterization of the Dangerous Activity Doctrine*

There is a consensus that an activity is dangerous when once deployed, its structure or operation generates more probabilities of damage than an average person is ordinarily able to bear on his own (Constitutional Court 2014). This danger arises because the activity's effects become uncontrollable or unpredictable due to energy and movement multiplication, the uncertainty of the phenomenon's effects, or the capacity to destroy its elements (Tamayo Jaramillo 2007).[3] Therefore, dangerousness is a "matter of fact" as declared by the Court. One that could be applied to both moving and inert things and must be examined with the support of objective criteria in each specific case unless an anticipated legal qualification is to be present.

In this sense, as stated by the Court and different scholars, a fundamental aspect to consider while characterizing all dangerous activity should be that one that determines the required conditions for the liability from the latter's exercise. These are: (a) The realization of activity with the use of things or energies with a high risk of causing harm to others, such as those related to vehicles, railroads, electrical or atomic energy. (b) The damage comes from the danger inherent in the hazardous activity. (c) There is a responsible agent (keeper) for handling the hazardous activity. (d) The victim is not related to the cause of the damage. (e) The damage is effectively caused by an activity qualified as dangerous. In essence, not all damages produced

[2]The Court stated that "the progress of ideas today revalue the ancient dogma that there is no liability without proven fault" (1938b).

[3]A dangerous activity occurs "when a person, in order to carry out a task, adds to his strength a "strange" one, which, by increasing his strength, breaks the balance that previously existed with the associates and places them "in imminent danger of being injured," even though the task "is carried out by observing all the diligence that it demands" (CSJ 1976).

by a business unit give the right to invoke liability grounded on the dangerous activities doctrine (CSJ 1995).

2.4 *The Role of Keepers of Dangerous Activities*

The keepers of a dangerous activity are responsible for it, and therefore, they are obliged to repair the damage caused. However, these activities usually comprise the participation of different stakeholders. For instance, transport activities involve the vehicle owner, a driver, a transport company, and others. All of them have more or less influence on the performance of the activity, which makes it difficult to determine the very one to be responsible or identified as the keeper (Tamayo Jaramillo 2007).

Based on French law, Colombian precedents refer to the keeper as the one who has the intellectual power to use, direct and control the activity (Tamayo Jaramillo 2007), assuming that the keeper is the vehicle owner as he or she can make it work (Santos Ballesteros 1996). The owner can disprove the presumption of being a keeper "if he or she demonstrates that he or she transferred to another person the possession of the thing by a legal title, such as that of a lease, bailment, or that he or she was culpably deprived of it, as in the case of having been robbed or stolen" (CSJ 1972). It is also common to find vehicles under a leasing contract. Although the owner is still the leasing company, there is no doubt that, after delivery, the person who has control over the vehicle and therefore is the keeper, is the leasing beneficiary, who uses the car, and who, without being the owner, should compensate the damages caused with the vehicle (CSJ 2016).

2.5 *Dependent Driver Is Not Liable for a Dangerous Activity*

According to case law, a dependent driver—any person using, driving or operating the vehicle with permission, express or implied, of the owner—is not responsible for the dangerous activity. As long as the worker or the employee is not the owner of the instruments used to carry out the dangerous activity, only the employer is liable, even if the worker materially holds the thing used in such activity (Tamayo Jaramillo 2007).

The driver is not responsible for the dangerous activity since the owner bears the vehicle's use, direction, control, and power. The owner orders the schedules, routes, and maintenance and finally decides to put the vehicle in circulation, creating a risk to others. Therefore, he must bear total liability for the exercise of the dangerous activity (without excluding responsibility for others' actions). The driver will only be liable to the extent that there is proof that he or she committed a fault[4] (Santos Ballesteros 1996).

In sum, from the Supreme Court of Justice's precedents, it is possible to draw, by way of conclusion, who is the one responsible or the keeper of vehicles under Colombian law:

i. The owner, if he or she has not voluntarily given up the possession of it or if, against his will and without any fault on his part, he has lost it.
ii. The legitimate holders of the thing with the power of use and enjoyment (tenants, bailees, administrators, usufructuaries, agents, and depositaries).
iii. Case law has extended the keeper status to all those who take advantage of the vehicle. An example is transportation companies, which are jointly responsible with the vehicle's owner for the dangerous activity while performing such service under their control and responsibility (CSJ 1992).

2.6 Passengers Do Not Carry Out a Dangerous Activity

Colombian law prohibits creating hybrid forms of responsibility, combining contractual and non-contractual civil liability (Tamayo Jaramillo 2007). The Court has held that if a passenger suffers damage, the vehicle's driver cannot argue that the passenger was also performing a dangerous activity by getting into the vehicle, applying the so-called neutralization of dangerous activities.

Invoking the neutralization of dangerous activities (typical institution of non-contractual liability) in a contractual relationship, such as passenger transport, involves combining contractual and liability institutions, creating a prohibited hybrid regime of the two types of liability. Consequently, from a legal point of view, the passenger cannot generate the risk that triggers the accident in which he or she suffers the damage.

In this respect, the Court has stated that "strictly speaking, there can be no question of a collision of dangerous activities and; therefore, the victim can use in its favor -as the doctrine has preached- the presumptions of article 2356 of the Civil

[4] In a 1992 ruling, the Supreme Court made an excerpt from several previous rulings in which the concept of a keeper of dangerous activity had been defined in the following terms: "In summary, the concept of "keeper" of the activity shall be of the individual or legal entity that, at the time of the accident, has effective and independent power of direction, governance, or control over the instrument that generates the damage, whether or not it is the owner, and provided that, by virtue of some factual circumstance, it is not prevented from exercising that power". To the point Santos Ballesteros specifies: "there are also numerous cases of persons who, despite having the thing, are not responsible for it, such as those in charge who have in their possession a thing that their principal has given to them for the exercise of their functions and other persons who cannot behave independently in relation to the thing that has been given to them. An example of the latter situation is that of a person who uses a thing with the owner's authorization when the owner reserves the power of command, direction and control over the thing, that of the 'load' in the same circumstance" (1996).

Code. [The driver's conduct,] concerning the vehicle that transports him or her, is usually passive and, therefore, incapable of creating a risk regarding her physical conduct."

In sum, concerning a level 0–2 car, the responsibility for the damage it may cause will always relapse on its keeper (a human person or a legal entity) and never on the vehicle itself. The fundamental reason relies on the technology itself. A level 0–2 car always demands the presence of a human driver. The human person must be fully aware of the driving environment to assume direct and total vehicle control. Autonomous assistance functions such as braking or acceleration are driving aids. It is the human person who drives the vehicle. Even at level 2, the vehicle only operates some functions without human command, but these are under the driver's responsibility. The keeper is solely responsible as he carries out a dangerous activity.

3 Liability for Level 3–4 Cars

There is no specific legal regulation for level 3–4 cars in Colombia. However, the general principles of responsibility mentioned above can be applied to them. The following is an analysis of level 3–4 car driving as a dangerous activity and an inquiry about who are the keepers of this kind of cars. Finally, there is a comment on liability for defective products and software that make a self-driving car run autonomously.

3.1 Level 3–4 Car Driving as a Dangerous Activity

As happens with level 0–2 cars, the operation of level 3–4 cars will involve exercising a dangerous activity under Colombian law, as they use "things or energies that may cause harm to third parties." As previously explained, an activity can be dangerous when it multiplies energy, when it contains a great possibility of causing harm, given the instruments used, and when its effects are uncertain (CSJ 1976). In determining the "dangerousness" of level 3–4 cars, it is helpful the case law distinction between the risk derived either from the structure or from the operation of the thing. In particular, there is danger in the structure when the thing has dynamism and retains the possibility of damage, given its location, design, construction, or materials used. On the contrary, the danger in the operation of the thing arises from the use given to it, regardless of whether it has dynamism[5] (Velásquez Posada 2015).

[5] The distinction of the two classes of dangerousness arose as a result of certain cases in which the courts found that there were people responsible for the potential danger of the structure of certain

In the case of level 3–4 cars, a dangerous multiplication of energy can be observed in the way they are designed and built (structure) and the risk they can create when driven (operation). The danger in the *structure* refers to, among other reasons, the car's design that can lead to damage, the possibility of being hacked, and the AI's central control of the vehicle, which can lead to unforeseen effects. On the other hand, the *operation* of level 3–4 cars implies an alteration of forces and the corresponding risks that people must ordinarily endure in the ordinary course of life. When a level 3–4 car is operating, people are exposed to mechanical or functional errors, which increases the probability of injury to their integrity their assets, even when they act with complete diligence.[6]

3.2 Who Is Responsible for Level 3–4 Car Damages

Even if a self-driving car has a broad range of autonomy, the obligation to compensate damages derived from a dangerous activity does not relapse on the vehicle—of course, it is not bound—; on the contrary, it resides in the person who exercises its custody.

At the heart of the discussion on autonomous vehicles lies whether this artificially intelligent technology can or should be considered for the law as an entity capable of assuming the obligations of civil liability. In other words, whether the law can or should consider a self-driving car as civilly liable. The answer is negative. According to the courts' legal tradition and the Civil Code, only the human person and the legal entities have legal capacity. Under no circumstances can level 3–4 cars be considered entities capable of civil liability in Colombia.

Under Colombian law, the possibility of granting legal capacity to self-driving cars becomes unnecessary and inconvenient. It is unnecessary because current Colombian law allows for adequate protection of victims of self-driving cars accidents, while there is no discouragement of AI technology. It is inconvenient because granting legal capacity to self-driving cars would release manufacturers, owners, or holders from their responsibility, causing an imbalance to the detriment of potential victims.

---

things, and others who were responsible for the handling or use of them. Therefore, in order to find the true culprit, it was necessary to determine which danger was the cause of the damage (Velásquez Posada 2015).

[6]See case law guidelines in: CSJ. exp. 5012 of 1999.

3.3 *Liability for Defective Product (Manufacturer, Programmer)*

When attributing responsibility for damages caused by a self-driving car, an alternative is to determine that the car was a defective product "due to an error in design, manufacture, construction, packaging or information" (Consumer Statute 2011). It is also a kind of strict liability under which the affected party must only prove the defect of the self-driving car and the causal link of this defect with the damage (CSJ 2009; Navarro-Michel 2020)

As mentioned, the defect can arise in the different stages of production and distribution of the product. In particular, in the case of self-driving cars, a design error may occur in the vehicle as such or in some of its components, or in the manufacture following that design and, even, in the information that must be provided to the users of the vehicle.

If it generates damages to third parties, the general rule is that the third party may only request the repair of the damages from the keeper of the car and not from the manufacturer of the car. For a product liability action to proceed, there must be a consumer relationship between the affected party and the manufacturer or supplier. Therefore, the only one authorized to file a suit against the manufacturer will be the user/driver/owner of the car, either because he or she is the keeper of it or because he or she hired the services of a self-driving car, such as, for example, a self-driving taxi.

The manufacturer or supplier can claim any of the grounds for exoneration provided by law. Concerning self-driving cars, the "development risk defense" draws attention. According to this ground for exoneration, the manufacturer or supplier is not obliged to respond "when at the time the product was put into circulation, the state of scientific and technical knowledge did not allow the existence of the defect to be discovered." (Consumer Statute 2011). In other words, if state of the art made it possible to foresee the defect, the manufacturer would be liable; but if that state of the art did not allow foreseeing it, the manufacturer would be exonerated of its responsibility (Velandia 2013). For some, this ground may benefit self-driving cars manufacturers because, due to the new nature of the car and the scientific techniques at the time of its manufacture, the possible harmful consequences of the product were unknown to them (Coral Díaz et al. 2019).

Finally, it is worth a special mention of the self-driving car's software, an essential operation element. The one who has programmed the vehicle system, its manufacturer, may also be the car's keeper. A self-driving car is an information system within the Electronic Commerce Act (Law 527 of 1999) framework because it is "a system for generating, sending, receiving, storing or otherwise processing data messages." Any instructions or the car's operating movements can be considered a data message. This message will be attributed to the software developer (manufacturer) since it was sent by "an information system programmed by, or on behalf of, the originator to operate automatically." Self-driving car systems and, in general, artificial intelligence are programmed to operate automatically, so the operations carried out by these systems can be attributed to whoever programmed them.

For this reason, a liability regime for dangerous activities, in the case of self-driving cars, should be complemented with the defective product liability regime regulated in the Colombian Consumer Statute and the Electronic Commerce Act data message attribution guidelines.

4 Liability for Level 5 Cars

There is no specific legal regulation for level 5 cars in Colombia, and no bill has been submitted to amend the current law. However, in light of the Colombian legal framework, the liability for damages caused by this kind of vehicle is similar to that explained for level 3–4 cars. In particular, a fully automated or level 5 car will be a dangerous activity because its actions become uncontrollable or unpredictable due to energy and movement multiplication. The activity is dangerous due to its structure because the car's design does not allow the occupant to take control of it and because artificial intelligence can generate unexpected effects. On the other hand, the activity is dangerous in its operation due to the multiplication of energies and forces that can generate damage in a road environment such as the Colombian one. The person responsible for the damages caused by this dangerous activity will be the car's keeper (owner, transport company).

If a manufactory defect causes damages, the car's owner or the one who benefits from the services of the car under a consumer agreement (i.e., self-driving taxi) may request compensation for damages from the manufacturer or supplier of the vehicle (Consumer Statute 2011; Constitutional Court 2000). The manufacturer or supplier may be exempted from liability if he proves the "development risk defense" previously explained (Mantilla and Ternera 2014). In any case, that manufacturer or supplier of the fully automated car will have an obligation to monitor if there are defects in the cars sold.

Finally, the actions of the car's automated system—using data messages—are attributed to the manufacturer or programmer of the software according to the provisions of the electronic commerce law, as explained in the previous section.

5 Conclusions

Although there has not been a specific ruling on autonomous vehicles in Colombia, their activity will be considered dangerous since it is "carried out through the use of things or energies capable of causing damage to third parties." The car's keeper's fault will be presumed, who can only disprove this presumption by proving a fortuitous event, a third party's fact, or the victim's responsibility.

According to the standard described above, the car's keeper qualification will depend on the car's automation level. Thus, in the first two levels of automation (1, 2), the owner/driver will continue to be the vehicle's primary keeper since he or she is the one in control and principal supervisor of driving, even when he has the help of the automated functions. The same keeper qualification can be attributed to a Level 3 vehicle driver; even though he is not driving, the system may ask him to take over driving at certain times. There will be no driver in the last two levels (4 and 5); there may be no wheel or pedals to control. Therefore, occupants, often simple users of third-party services, do not have a keepers' role. In the words of the Supreme Court of Justice, the vehicle's keeper is the one who has the power of direction and control over it. A self-driving car occupant will only be considered a keeper and, therefore, responsible if he has had any involvement in its power, such as the owner or software programmer. Precisely, who can also be attributed responsibility for damage caused by a vehicle with high levels of autonomy is its manufacturer, who has been in charge of programming the corresponding system and attributed the effects of the data messages for the vehicle to operate automatically, under the terms of the Electronic Commerce Act.

In sum, the long-established responsibility regime for dangerous activities should be complemented in self-driving cars, with liability for faulty products and data message attribution rules.

Law

Concept of the Superintendency of Industry and Commerce. Settled No. 16-424467 of January 1, 2017

Concept of the Superintendency of Industry and Commerce. Settled No. 16-160560 of July 27, 2016

Corte Suprema de Justicia [2016] (Sala de Casación Civil)

Decree 679/2016

Corte Constitucional, T-609 [2014]

Consumer Statute, Law 1480/2011

Corte Suprema de Justicia [2009] (Sala de Casación Civil)

Corte Constitucional C-1141 [2000]

Electronic Commerce Act, Law 527 [1999]

Corte Suprema de Justicia [1995] (Sala de Casación Civil)

Corte Suprema de Justicia [1992] (Sala de Casación Civil)

Corte Suprema de Justicia [1976] (Sala de Casación Civil)

Corte Suprema de Justicia [1972] (Sala de Casación Civil)

Corte Suprema de Justicia [1962] (Sala de Negocios Generales)

Corte Suprema de Justicia [1938a] (Sala de Casación Civil)

Corte Suprema de Justicia [1938b] (Sala de Casación Civil)

Corte Suprema de Justicia [1938c] (Sala de Casación Civil)

Colombian Commerce Code
Colombian Civil Code

References

Anderson J, Kalra N, Stanley K, Sorensen P, Samaras C, Oluwatola O (2016) Autonomous vehicle technology: a guide for policymakers. RAND Corporation, Santa Monica

Beale A (2018) Who's Coffers Spill when autonomous cars kill? A new tort theory for the computer code road. Widener Commonwealth Law Review

Coral Díaz D, Díaz Trujillo M, Macías Rodríguez Á (2019) Robótica y responsabilidad civil: reflexiones en torno al fundamento del deber de reparar, 1st edn. Universidad Externado de Colombia, Bogotá

Giuffrida I, Lederer F, Vermerys N (2018) A legal perspective on the trials and tribulations of AI: how artificial intelligence, the internet of things, smart contracts, and other technologies will affect the law. Case West Reserve Law Rev 3:761

Google Data Studio (2021) INFORMES INTERACTIVOS ANDEMOS. [online] Available at: <https://datastudio.google.com/reporting/ceb8deeb-3b00-4e08-8536-5a0f2ebb5cf2/page/yakzB>. Accessed 9 Nov 2021

Instituto Nacional de Medicina Legal y Ciencias Forenses (2021) Boletín estadísitica mensual. Centro de Referencia Nacional sobre Violencia-CRNV. [online] Available at: <https://www.medicinalegal.gov.co/documents/20143/628335/Boletin+octubre+2021.pdf>. Accessed 9 Nov 2021

Mantilla F, Ternera F (2014) El riesgo de desarrollo en la responsabilidad por productos del derecho colombiano. Ius et Praxis

Navarro-Michel M (2020) Application of road traffic regulations to accidents caused by automated and autonomous vehicles. Universitat de Barcelona

Rheinstein M (1952) Common law and civil law: an elementary comparison. Revista Jurídica Universidad de Puerto Rico

Runt (2021) Parque automotor registrado en RUNT. [online] Available at: <https://www.runt.com.co/runt-en-cifras/parque-automotor>. Accessed 9 Nov 2021

Santos Ballesteros J (1996) Instituciones de responsabilidad civil, vol I. Pontificia Universidad Javeriana, Bogotá, pp 165, 166

Steven E (1996) Emanuel law outlines: torts, 5th edn. Wolters Kluwer, New York

Tamayo Jaramillo J (2007) Tratado de la responsabilidad civil, vol I. Legis, Bogotá, pp 881–929, 935, 897, 921, 1039, 136

Velandia M (2013) Acciones derivadas de las fallas en los productos. In: Valderrama CL (dir.) Perspectivas del derecho del consumo. Universidad Externado de Colombia, Bogotá

Velásquez Posada O (2015) Responsabilidad civil extracontractual, 2nd edn. Temis, Bogotá

World Health Organization. Global status report on road safety. Geneva: 2018

World Health Organization (2021) Road traffic injuries. [online] Available at: <https://www.who.int/news-room/fact-sheets/detail/road-traffic-injuries>. Accessed 9 Nov 2021

Zuleta Angel E (1974) Estudios jurídicos. Ed. Temis, Bogotá, p 236

Part III
Asia

Autonomous Driving in China

Mingyan Nie and Yuan Shen

Abstract This chapter focuses on autonomous driving in China. Firstly, the recent development of China's autonomous vehicles (AVs) industry and policies released by the related governmental administrations to promote its growth are outlined. Then, the taxonomy of driving automation for vehicles of China (TDAV) is introduced as the basis for further discussion. The central and local government rules about AVs road testing and demonstration application are analyzed thereafter. The commercialization of AVs is still in its infancy in China, and a comprehensive legal framework has not yet been formulated, so these rules are the most relevant ones to AVs at the present stage. Lastly, the attribution of liability caused by AVs with different levels is discussed in accordance with the existing laws and regulations in China.

1 The Development of the AVs Industry in China and Relevant Policies

Since 2009 China has been the largest automobile market in the world both in terms of demand and supply.[1] The production of automobiles in China accounted for almost one- third of the global production in 2021, exceeding that of the United States, Japan and Germany combined.[2] Despite the impact of the coronavirus

[1]Ministry of Public Security (2021) China's car ownership surpassed the US in 2020 for the first time. https://www.sohu.com/a/523492883_191175. Accessed 8 Jan 2023.

[2]OICA (2021) 2021 production statistics. https://www.oica.net/category/production-statistics/2021-statistics/. Accessed 8 Jan 2023.

M. Nie
Law Faculty, Nanjing University of Aeronautics and Astronautics (P.R.C), Nanjing, China

Y. Shen (✉)
Luther Rechtsanwaltsgesellschaft mbH (Germany), Cologne, Germany
e-mail: yuan.shen@luther-lawfirm.com

H. Steege et al. (eds.), *Autonomous Vehicles and Civil Liability in a Global Perspective*, Data Science, Machine Intelligence, and Law 3,
https://doi.org/10.1007/978-3-031-41992-8_6

pandemic the sales in 2022 are expected to reach 26 million units, an increase of 1.7% compared to the sales in 2021.[3] The Chinese government has regarded new energy cars as essential for China's transformation from an automobile giant to an automobile power.[4] In the "New Energy Vehicle Industry Development Plan (2021-2035)" (hereinafter referred to as the "Development Plan") published by the State Council on October 20, 2020, in terms of self-driving cars the goal is to use highly self-driving cars (comparable to TDAV Level 4) in the commercial application of limited areas and specific scenarios by 2025 and to achieve its scale application by 2035. It is believed that achieving this goal should contribute to reducing emissions and improving social operation efficiency.[5] Furthermore, according to the "Development Plan", China is also going to improve the legal framework for road traffic, accident liability, and data utilization to meet the requirements of the development of intelligent connected vehicles.

To further promote the growth of the autonomous vehicles (AVs) industry in China, on February 24, 2020, the National Development and Reform Commission, as well as other ten Chinese national administrations announced the "Innovation and Development Strategy of AVs[6]" (hereinafter referred to as the "2020 Strategy") together. Compared to the "Development Plan", the "2020 Strategy" is concerned with a comprehensive growth of the AVs industry. It proposes to improve the system concerning the AVs industry which includes such aspects as innovative technology, industrial ecology, infrastructure, rules and standards, product supervision, and cyber safety. As a purpose of the initial stage, the "2020 Strategy" outlines a vision planning of scale production of the AVs, as well as marketization of the highly automated driving intelligent connected cars by 2025. Moreover, the "2020 Strategy" put forward that the comprehensive system of China's AVs could be finally established by 2050 to meet the requirements of China to get the status of an AVs power.[7]

With the goal of creating an AVs power, the "2020 Strategy" takes a number of measures. More specifically, it requires the relevant central and local administrations to (1) coordinate effectively to ensure the AVs industry growth; (2) establish supporting policies; (3) focus on the mechanism of talents introduction and

[3]China Association of Automobile Manufacturers (2023) Forecast of vehicle sales in 2022. https://www.workercn.cn/c/2023-01-06/7693533.shtml. Accessed 8 Jan 2023.

[4]Xi Jinping (2014) The development of new energy vehicles is the only way to become an automobile power. http://www.xinhuanet.com/politics/2014-05/24/c_1110843312.htm. Accessed 28 Feb 2021.

[5]New Energy Vehicle Industry Development Plan (2021–2035) [2020] State Council of the People's Republic of China.

[6]Innovation and development strategy of AVs in China [2020] National Development and Reform Commission, Cyberspace Administration of China, Ministry of Science and Technology of the People's Republic of China and Ministry of Industry and Information Technology of the People's Republic of China.

[7]Part 2(3), Innovation and Development Strategy of AVs in China.

exchanges; (4) deepen international cooperation in this field; and (5) optimize the environment for the growth of the AVs industry.[8]

With respect to the legal aspects of AVs, the "2020 Strategy" envisages addressing issues concerning the status of legal subjects of the AVs, the attribution of liabilities, cyber safety, and data management by formulating laws dealing with AVs testing, authorizing, utilizing and supervision, and improving the existing laws such as the "Law on Road Traffic Safety". In addition, the "2020 Strategy" calls for an improvement in technical standards for AVs.

The drawing of the legal framework for AVs in China is driven by technology development. Chinese internet giants such as Baidu and Tencent, etc. started their careers in AVs R&D in 2015 and 2016.[9] Until 2022, about 600 corporations in China have announced their plans in developing AVs or AVs-related technologies.[10] Stimulated by the rapid growth in the industrial community, the Chinese government released several policies/measures to administrate the forthcoming issues.

These policies/measures, briefly distinguished into categories of central and local ones, together with the existing laws and regulations, such as the "Law on Road Traffic Safety",[11] and the "Product Quality Law",[12] etc., constitute the fundamental legal framework for addressing the liability allocation pertain to AVs in China.

2 Taxonomy of Driving Automation for Vehicles (TDAV) of China

The Ministry of Industry and Information Technology (MIIT) of China proposed in March 2020 a draft of the recommended national standards "Taxonomy of Driving Automation for Vehicles (TDAV)[13]". On 1 March 2022, the TDAV became effective.[14] It is in principle consistent with SAE J3016[15] in terms of the overall grading

[8] Part 4(1)-(4), Innovation and Development Strategy of AVs in China.

[9] Tencent Research Institute (2017) The development status and trend of the autonomous driving industry, https://www.tisi.org/4896, the original language is in Chinese. Accessed 28 Feb 2021

[10] Netease (2022) Big data analysis of Chinese autonomous driving enterprises in 2021, https://www.163.com/dy/article/H70IMN86051481OF.html, the original language is in Chinese. Accessed 11 Jan 2023.

[11] Law of the People's Republic of China on Road Traffic Safety [2003] Standing Committee of the Tenth National People's Congress Order of the President of the People's Republic of China, No. 8.

[12] Product Quality Law of the People's Republic of China (amendment) [2000] Standing Committee of the National People's Congress Order of the President of the People's Republic of China No. 33.

[13] Taxonomy of Driving Automation for Vehicles [2020] Ministry of Industry and Information Technology of the People's Republic of China.

[14] Taxonomy of Driving Automation for Vehicles [2021] State Administration for Market Regulation, Standardization administration of the People's Republic of China.

[15] SAE International (2021) SAE J3016™ levels of driving automation, https://www.sae.org/binaries/content/assets/cm/content/blog/sae-j3016-visual-chart_5.3.21.pdf. Accessed 11 Jan 2023.

ideas and classification standards and classifies the degree of driving automation into six different levels. For the specific definition of each level of autonomous vehicles, the two standards are also roughly the same with only some differences in certain aspects.

The TDAV intends to define the different roles of drivers and the automatic systems in different AVs levels, which is of benefit to determine the liabilities allocation in accidents in the context of applying China's existing laws and regulations. In addition, the classification of AVs levels under the TDAV could also influence the required amendments of Chinese laws and regulations associated with AVs.

The TDAV is designed to apply to the cars belonging to M and N classes.[16] It illustrates the fundamental principles for classifying the AVs levels and is based on the consideration of the degree to which the driving automation system can perform dynamic driving tasks, the roles to which dynamic driving tasks are assigned, and whether the operational design conditions are limited. The driving automation is divided into 0~5 levels. Level 0 is defined as "emergency assistance", this level is not contained in the scope of automatic driving. The automation system in Level 0 can percept the driving environment and assist the driver by providing early warning service. Level 1 is "partial driver assistance", it refers to the condition that the driver and the automation system perform the dynamic driving task together, and the driver supervises the activities of the driving automation system and performs appropriate response and operation. The formulation of Level 2 is totally the same as Level 1, but entitled "combined driver assistance". Level 3 is conditionally automated driving, where the user performs dynamic driving task (DDT) to a reasonable extent. Level 4, highly automated driving, means that at this level if the passenger does not respond to the system's request to intervene, the system is able to reach a state of minimal risk. Level 5 means fully automated driving, i.e. the passenger does not have to respond to the request to intervene, the system is able to independently reach a state of minimal risk, and there are no operational design condition restrictions (except for constraints due to commercial and legal factors).

The relationships between the classification factors and the six levels are summarized in Table 1.

---

[16] According to the Chinese national standard "Classification of Power-Driven Vehicles and Tailers" (GB/T 15089-2001), M-Class refers to power-driven vehicles having at least four-wheels and used for the carriage of passengers; N-Class refers to power-driven vehicles with at least four wheels and used to carry cargo.

Table 1 AVs levels and classification factors under TDAV[a]

Level	Name	Lateral/ longitudinal vehicle motion control	Object and event detection and response	Dynamic driving task fallback	Operational design domain
0	Emergency assistance	Driver	Driver and system	Driver	Restrictions
1	Partial driver assistance	Driver and system	Driver and system	Driver	Restrictions
2	Combined driver assistance	System	Driver and system	Driver	Restrictions
3	Conditionally automated driving	System	System	Dynamic driving task fallback-ready user	Restrictions
4	Highly automated driving	System	System	System	Restrictions
5	Fully automated driving	System	System	System	No Restrictions (with exceptions)

[a]Annex A of the Taxonomy of Driving Automation for Vehicles

3 Central and Local Government Rules About the AVs Road Testing

Rules directly regulating AVs of Chinese central and local governments are mainly applied to road testing or demonstration application. In April 2018, MIIT together with the Ministry of Public Security and the Ministry of Transport issued the "Management Rules for Road Testing of Intelligent Connected Vehicles (Trial)[17]" (hereinafter referred to as the "Management Rules 2018"), which is a typical central governmental rule of AVs testing. On July 27, 2021, a new version of "Management Rules", i.e., the "Management Rules for Road Testing and Demonstration Application of Intelligent Connected Vehicles (Trial)[18]" (hereinafter referred to as the "Management Rules 2021") was issued which became valid on October 1, 2021, replacing the "Management Rules 2018". The "Management Rules 2021" is intended to allow road testing in a wider range of test scenarios, formulate rules for demonstration applications, and achieve mutual recognition of test results between different regions. Although these rules are not formulated for the allocation

[17]Management rules for road testing of intelligent connected vehicles (trial) [2018] Ministry of Industry and Information Technology, the Ministry of Public Security and the Ministry of Transport of the People's Republic of China.

[18]Management Rules for Road Testing and Demonstration Application of Intelligent Connected Vehicles (Trial) [2021] Ministry of Industry and Information Technology, the Ministry of Public Security and the Ministry of Transport of the People's Republic of China.

of liability in the event of an accident involving AVs, they nevertheless contain several relevant clauses that deserve to be analyzed.

3.1 *Management Rules for Road Testing and Demonstration Application of Intelligent Connected Vehicles (Trial)*

The "Management Rules 2021" belongs to the third hierarchical level of the Chinese legal system.[19] It consists of 38 provisions and applies to the road testing and demonstration application of AVs belonging to Level 3 to Level 5. It stipulates that the administrations of local governments may establish detailed rules for implementation based on the fundamental norms created by it.

Except for the preface and supplementary articles, the "Management Rules 2021" can be divided into the following parts:

3.1.1 The Party, Driver, and Vehicle for Road Testing or Demonstration Application

The party conducting the road testing is defined as an entity that conducts and organizes the testing, and is also responsible for damages caused by the testing. Specifically, an entity shall meet the following requirements to obtain the testing qualification:

(1) it is an independent legal entity registered in China;
(2) it is qualified to manufacture automobiles and parts, conduct research and development or test AVs, etc.;
(3) it is capable to compensate the damage caused by the testing;
(4) it has an assessment procedure for the road testing of AVs;
(5) it is capable to remotely monitor the test vehicles;
(6) it is capable to record, analyze and reproduce events on the test vehicles; and
(7) it is capable to guarantee the network security of the road testing vehicles and remote monitoring platforms.[20]

---

[19] According to the Chinese Constitution and the "Legislation Law", the national law are divided into three hierarchical levels below the Constitution: (1) laws enacted by the National People's Congress and its Standing Committee; (2) administrative regulations formulated by the State Council, the national administrative organ; (3) departmental regulations formulated by the Ministries and Commissions of the State Council, the People's Bank of China, the State Audit Administration as well as other organs endowed with administrative functions directly under the State Council. See: Legislation law of the People's Republic of China (amendment) [2015] National People's Congress of the People's Republic of China.

[20] Art. 5, Management Rules for Road Testing and Demonstration Application of Intelligent Connected Vehicles (Trial) [2021].

The demonstration application party means an entity that applies for the demonstration application of AVs, organizes demonstration application and assumes corresponding responsibilities. It shall meet the following requirements:

(1) it is an independent corporate entity or a consortium consisting of multiple independent corporate entities registered in China;
(2) it has the business capabilities related to AVs such as manufacturing of automobiles and their parts, technological research and development, experimentation and testing, or operation of demonstration application;
(3) it is a demonstration application party consisting of multiple independent corporate entities, at least one of which has the capacity to operate demonstration applications, and all legal entities sign relevant agreements on operation services and the sharing of relevant tort liabilities;
(4) it has sufficient capacity for civil compensation for personal injury and property damage that may be caused by the demonstration application of AVs;
(5) it has the demonstration application scheme for AVs;
(6) it has the capabilities to conduct real-time remote monitoring of demonstration application vehicles;
(7) it has the capabilities to record, analyze and reproduce events related to demonstration application vehicles; and
(8) it has the capabilities to guarantee network security of demonstration application vehicles and remote monitoring platforms.[21]

The driver for road testing or demonstration application of AVs is authorized by the qualified testing or demonstration party. She/he is required to

(1) conclude a valid labor or service contract with the testing or demonstration entity;
(2) obtain a driving license for the relevant vehicle types and have driving experiences of more than three years;
(3) have no full recorded marks within the last three consecutive scoring periods[22];
(4) have no records of serious traffic violations such as exceeding the speed limit by 50% or more, or violating the traffic signals in the recent one year;
(5) have no record of driving after drinking or drunk driving, or records for taking psychotropic drugs or narcotic drugs under state control;

---

[21] Art. 6, Management Rules for Road Testing and Demonstration Application of Intelligent Connected Vehicles (Trial) [2021].

[22] Art. 24 of "Law of the People's Republic of China on road traffic safety" indicates: "The traffic administrative department of the public security organ shall apply a system of keeping accumulative scores to the motor vehicle drivers who violate the road traffic safety laws and regulations in addition to imposing administrative penalties upon them in accordance with the law. The traffic administrative department of the public security organ shall detain the motor vehicle driving licenses of those whose accumulative scores have reached the prescribed limit, arrange for the education of road traffic safety laws and regulations and the corresponding re-examinations for them; the motor vehicle driving licenses shall be returned to those who are examined to be qualified."

(6) have no records of driving liability accidents for causing people serious injury or death; and
(7) have been trained by the testing or demonstration entity regarding operating AVs, been familiar with the testing procedure or demonstration application scheme, gained the skill of operating AVs testing or demonstration application, and had the capacity to deal with an emergency.[23]

With respect to the vehicles for testing or demonstration application, the "Management Rules 2021" excludes low-speed motors and motorbikes. Additionally, the testing or demonstration application AVs must meet the following requirements:

(1) it is not a registered motor vehicle;
(2) it meets mandatory inspection requirements for ensuring safety. If certain parts of the testing vehicle fail to meet these requirements for autonomous driving, the testing entity must prove the safety performance;
(3) vehicles must have both a manual and an automatic driving modes, and these two modes must be capable of being switched safely, quickly, and easily. The vehicle can be immediately converted to a manual operation mode under all circumstance; and
(4) vehicles must have the functions of recording, storage and online monitoring.

Crucial information such as vehicle control model, vehicle location, vehicle speed, etc. must be transmitted back in real time, and other relevant information must be automatically recorded and stored for at least one year.[24]

3.1.2 Application and Assessing of the Road Testing

According to the "Management Rules 2021", before conducting a test, the testing parties shall ensure that road testing vehicles have been fully tested in real-vehicle testing in specific areas such as test areas (fields), comply with relevant national and industry standards and specifications, the testing requirements issued by relevant competent departments of provincial and municipal governments, and the testing and evaluation procedures of road testing parties, and meet the conditions for conducting road testing.[25]

Moreover, the testing parties are required to provide an self-statement on the safety of road testing of intelligent connected vehicles, which shall be confirmed by the relevant competent departments of provincial and municipal governments. A testing period shall not exceed 18 months in principle, and the testing shall be

[23] Art. 7, Management Rules for Road Testing and Demonstration Application of Intelligent Connected Vehicles (Trial) [2021].

[24] Art. 8, Management Rules for Road Testing and Demonstration Application of Intelligent Connected Vehicles (Trial) [2021].

[25] Art. 9, Management Rules for Road Testing and Demonstration Application of Intelligent Connected Vehicles (Trial) [2021].

conducted during the validity of the certificate of safety and technical inspection, and the validity of the insurance certificate. The self-statement on the safety of road testing shall be submitted to the relevant competent department of a provincial or municipal government along with certain supporting materials.[26]

3.1.3 Application for Demonstration Application

Demonstration application, which is defined as "the operation of carrying passengers or goods by AVs within a pilot program or trial implementation, which is conducted on the designated sections of various roads for the passage of social motor vehicles within the scope of highways (including expressways), urban roads, or areas, among others[27]", is a new content in the "Management Rules 2021". For an initial application, the vehicles must have been subject to a total of not less than 240 h or 1000 km of road testing under the self-driving mode on the road sections or areas for demonstration application, and during the testing there is no traffic violation or traffic accident for which the test vehicles are liable.[28]

3.1.4 Management of the Road Testing and Demonstration Application

For the road testing or demonstration application, the relevant competent departments of provincial and municipal governments are required to select several typical road sections and areas that support the realization of self-driving and network connection functions.[29] Furthermore, they shall by multiple means, release the periods, items, and safety precautions, among others, of the road testing and demonstration application of intelligent connected vehicles to the public and especially the surrounding areas of the road sections and areas for road testing and demonstration application.[30]

---

[26] The materials include, for example, basic information on the road testing party, the testing driver and the road testing vehicle; descriptions of self-driving functions and the certificate proving that the safety performance of the vehicle is not degraded; the compulsory traffic accident liability insurance certificate and the certificate of the traffic accident liability insurance of not less than RMB 5 million per vehicle or the letter of guarantee of compensation for self-driving road testing accidents of not less than RMB 5 million, etc. See Art. 10, Management Rules for Road Testing and Demonstration Application of Intelligent Connected Vehicles (Trial) [2021].

[27] Art. 2, Para. 2, Management Rules for Road Testing and Demonstration Application of Intelligent Connected Vehicles (Trial) [2021].

[28] Art. 16, Management Rules for Road Testing and Demonstration Application of Intelligent Connected Vehicles (Trial) [2021].

[29] Art. 22, Management Rules for Road Testing and Demonstration Application of Intelligent Connected Vehicles (Trial) [2021].

[30] Art. 23, Management Rules for Road Testing and Demonstration Application of Intelligent Connected Vehicles (Trial) [2021].

For managing and supervising the test, all testing or demonstration application vehicles must obtain temporary licenses.[31] The testing entities and the drivers shall observe the existing rules and regulations concerning traffic safety, and perform the testing in accordance with the testing time, section, and project.

During the testing or demonstration application, drivers shall monitor the operating condition of the of vehicles and the surrounding environment, and take appropriate action in a timely manner if they find that the vehicles are in a condition that is not suitable for self-driving or if the system indicates that manual operation is necessary to avoid accidents.[32]

Further, other than professionally trained testers and counterweights used to simulate cargo, the vehicle shall not carry other personnel or goods not associated with the testing. During the demonstration application, the vehicles may carry personnel and goods needed for exploring business models, and the personnel and owners of the transported goods shall be informed in advance of the associated risks and take necessary safety measures. The vehicles shall not be operated for transport business activities or carry dangerous goods during the testing and demonstration application.[33]

3.1.5 Traffic Violations and Accident Handling

Traffic violations during the testing or demonstration application must be addressed by the relevant authorities in accordance with the existing traffic rules and regulations. In the event of a traffic accident, the liability of the parties concerned shall be determined in accordance with the "Road Traffic Safety Law"; the liability for damages is to be determined under the applicable civil law including judicial interpretations; in cases that constitute a criminal offense, criminal liability shall be assigned.

As a procedural requirement, the parties conducting road testing or demonstration application must report the information on traffic accidents that occur during the road testing or demonstration application to the relevant competent authorities each month. In the case of serious injury, death, or vehicle destruction, the parties must report it to the departments of provincial or municipal governments within 24 h, and the departments report it to MIIT, MPS, and MOT in three working days.[34] After the liability of an accident is determined, the parties must submit documentation in writing to the departments of provincial or municipal governments within five

[31] Art. 24, Management Rules for Road Testing and Demonstration Application of Intelligent Connected Vehicles (Trial) [2021].

[32] Art. 26, Management Rules for Road Testing and Demonstration Application of Intelligent Connected Vehicles (Trial) [2021].

[33] Art. 27, Management Rules for Road Testing and Demonstration Application of Intelligent Connected Vehicles (Trial) [2021].

[34] Art. 35, Management Rules for Road Testing and Demonstration Application of Intelligent Connected Vehicles (Trial) [2021].

working days with the reasons for the accident, the result of the liability determination, and the complete accident analysis report, and the departments shall forward this documentation to MIIT, MPS, and MOT in five workdays.[35]

The "Management Rules 2021" outlines the basic legal framework for the testing and demonstration application of AVs in China, the rights and obligations of the participants in the testing or demonstration application and address the liability for accidents and damages from several crucial perspectives. Although these rules are not directly related to the allocation of liability for the automated system, they are nonetheless important for the formulation of specific rules in the future.

As a basic rule in the field of testing and demonstration application of AVs issued by the relevant ministers of the Chinese central government, the "Management Rules 2021" authorizes local governments to release specific measures for implementation, which should be highlighted.

3.2 Local Governmental Rules of AVs Road Testing

In the context of the "Management Rules", more than ten Chinese cities have released specific implementing measures for the road testing of AVs.[36] The measures announced by Beijing are particularly weighted among them.

Beijing is the first Chinese city that released the implementing measures. As early as 2017, the "Beijing Rules for the Implementation of Road Test Management of AVs" (hereinafter referred to as "Beijing Rules") was published. At the same time, 200 roads with a total length of 700 km were opened for AVs testing in Beijing.[37] The "Beijing Rules" is a typical local regulation for the road testing of Avs based on the "Management Rules". The implementation rules of the other cities are similar to those of Beijing.

After the announcement in 2017, the "Beijing Rules" has been amended three times, and the latest version was released on November 12, 2020.[38] Compared with the previous versions, the following differences should be highlighted to understand the recent developments of China's local rules concerning AVs road testing: firstly,

[35] Art. 36, Management Rules for Road Testing and Demonstration Application of Intelligent Connected Vehicles (Trial) [2021].

[36] For more details of the AVs road testing implementing measures of the main cities in China, see: China Electric Vehicle Association (2018) Analysis and suggestions of autonomous vehicle testing standards in major cities of China, https://auto.gasgoo.com/a/70070281.html, the original language is in Chinese. Accessed 01 Mar 2021.

[37] For more details, please see: Sun Hongyang, Cao Zheng (2020) Beijing has 200 roads for autonomous driving tests, with 45 vehicles approved for manned tests, https://news.bjd.com.cn/2020/12/16/36701t100.html, the original language is in Chinese. Accessed 01 Mar 2021.

[38] Beijing rules for the implementation of road test management of AVs [2020] Beijing Municipal Commission of Transport, Beijing Traffic Management Bureau and Beijing Municipal Bureau of Economy and Information Technology.

the new version extends the validity period of the testing license from one year to two years; secondly, the new version coordinates the types of technologies to be tested by supporting the testing of general technology, enrichening the content of the specific technical test and advancing the operational test; thirdly, for the specific technical testing, the new version continues to support the formation driving test and adds new testing items such as special weather environment, high-speed environment, unmanned technology testing scenarios; and forth, the "Beijing Rules" continues to sustain the operation service model, the testing of manned and cargo driving technologies.[39]

To sum up, at the present stage, these rules for AVs road testing are more significant to AVs as an industry, as they enable and encourage the development of AVs technologies. Through road testing, more experiences concerning the operations of AVs could be obtained. In this regard, local rules may also serve as a testing bed for the development of future laws and regulations at the national level. In June 2020, DiDi, China's leading mobile transportation platform company, began using AVs (Level 4) to provide its traditional ride services to the public in assigned test zones in Shanghai.[40] This is a positive signal to the AVs applications; accordingly, more specific rules at the central and local levels are expected.

4 Existing Laws and Regulations and the Attribution of Liability Caused by AVs

The emerging national policies, central/local measures and rules aiming at regulating potential issues arising from the development of AVs will obviously impact the application of existing laws and regulations in transportation areas in China, particularly traditional civil or tort law governing the liability with regard to AVs.

4.1 Liability for Cars of Level 0-2

According to the classification of AVs levels under the TDAV, the current laws and regulations in China apply to the attribution of liability and disputes caused by vehicles up to Level 2. Within these levels (Level 0 to Level 2), the liability is allocated between the parties involved in the accident.

[39]Beijing Municipal Commission of Transport (2020) Interpretations of "Beijing rules for the implementation of road test management of AVs", http://jtw.beijing.gov.cn/xxgk/zcjd/202011/t20201119_2140422.html, the original language is in Chinese. Accessed 01 Mar 2021.

[40]More specifics of the AVs application in driving services, please see: http://www.shanghai.gov.cn/nw17239/20200920/0001-17239_1456791.html, the original language is in Chinese. Accessed 01 Mar 2021.

According to Art. 76 of the “Law on Road Traffic Safety” of China, where a motor vehicle is involved in a traffic accident and causes personal injury or property damage, the insurance company shall pay compensation within the limit of the compulsory third-party liability insurance for the motor vehicle. This compulsory insurance for vehicle accident liability is the first compulsory insurance regulated by law in China,[41] the premium is prescribed uniformly. On September 19, 2020, the total liability limit was increased from 122,000 to 200,000 CNY.[42]

Aside from insurance coverage, the liability must be allocated by considering the following different situations:

(1) in a traffic accident between motor vehicles, the party at fault is liable; if both parties are at fault, each shall bear its proportionate share of the liability;
(2) in a traffic accident between a motor vehicle and a non-motorist or a pedestrian, the party of the motor vehicle (also referred to as the motor vehicle party) shall compensate the damage if the non-motorist or the pedestrian was not at fault; if there is evidence that the non-motorist or the pedestrian was at fault, the compensation of the party of the motor vehicle may be reduced in accordance with the degree of fault. If the party of the motor vehicle was not at fault, the share of its compensation liability shall not exceed 10%.

If the damage is caused due to an intentional act of the non-motorist or the pedestrian, the party of the motor vehicle shall not be liable for the damage.

The situation may be more complicated in certain cases where the driver is liable or partly liable. Art. 76 of the “Law on Road Traffic Safety” generally provides that the motor vehicle party at fault is liable for damages in a traffic accident. However, the motor vehicle party is generally considered to be composed of the vehicle holder (including the owner, the lessee and the borrower, etc.) and the driver.

Chapter five of China’s Civil Code (“Civil Code”) regulates liability in motor vehicle accidents in more detail and, in particular, differentiates between the liability of the owner, the caretaker, and the user of a motor vehicle. According to Art. 1209 “Civil Code”, where the owner or caretaker of a motor vehicle is not the same person as the user in the case of leasing or borrowing, the user of the motor vehicle shall be liable for compensation; if the owner or caretaker of the motor vehicle is at fault for the damage, he or she shall bear the corresponding liability.[43]

---

[41] China Banking and Insurance News (2014) The first compulsory insurance in China- compulsory insurance for vehicle accident liability was launched in 2006, http://www.sinoins.com/zt/2014-01/27/content_87809.htm, the original language is in Chinese. Accessed 01 Mar 2021.

[42] See: Xue Yingjun (2020) Guidance on comprehensive reform of auto insurance was formally implemented on September 19, https://www.thepaper.cn/newsDetail_forward_9041995, the original language is in Chinese. Accessed 01 Mar 2021.

[43] Civil Code of the People’s Republic of China [2020] National People’s Congress of the People’s Republic of China.

4.2 Liability for Cars of Level 3 and Level 4

In Level 3 and Level 4, both the driver and the automation system can dominate the vehicle. From this point of view, the status of the driver and the status of the system in the driving process are equivalent. If an accident occurs, both the driver (including the relevant persons as mentioned above) and the automation system are potential liability holders.

Under current Chinese law, the automation system cannot independently assume liability. When an accident occurs due to a system-related problem, it is deemed as a product quality defect, so that the manufacturer or seller shall bear the related liability. If a traffic accident can be attributed to a problem with the automation system of the vehicle, the driver or the other persons involved, after paying the compensation to the injured party, can claim compensation from the manufacturer or seller of the vehicle.

The "Product Quality Law" stipulates that the manufacturer is liable for personal injury or property damage caused by a product defect.[44] If the product defect was caused by the seller, or the seller cannot clearly identify the manufacturer or supplier of the defective product, the seller is liable for compensation.[45] For the victim of the product, the seller and the manufacturer are jointly liable.[46]

To sum up, at the present stage, the liability of Level 3 and Level 4 AVs is still judged under the traditional laws and regulations in China. The driver and the other persons involved are still the main parties when it comes to compensating the damage caused by a traffic accident, in particular when the mandatory insurance amount is not enough to cover the damage. If the defect of the automatic system can be determined as the main cause of the traffic accident, the manufacturer or the seller of the vehicle is, in the context of recourse claims, liable for the compensation in accordance with the "Product Quality Law".

4.3 Liability for Cars of Level 5

Level 5 is fully automated driving. The system fully dominates the driving process and there are no restrictions due to operational design conditions. At this level, liability for accidents is always (mainly) attributed to the system. Therefore, the system itself or the manufacturer of the system (or the vehicle that installed the system) should be liable for damages. At the present stage, the system itself is not qualified as a liability subject under Chinese law. From the perspective of AV testing regulations, the driver should be allowed to take over the driving of the vehicle at any time, regardless of the extent to which the automation system can make decisions

[44] Art. 41, Product Quality Law of the People's Republic of China.

[45] Art. 42, Product Quality Law of the People's Republic of China.

[46] Art. 43, Product Quality Law of the People's Republic of China.

while driving. From this point of view, it can be seen that if Level 5 AVs are marketed in China in the future, the liability among the "subjects" of the driver (including other relevant persons), the non-motorist or pedestrian, the manufacturer (or seller), or even the relevant administrations for approving the marketing of the (defective) vehicle system should be examined and allocated. The automation system, even if it partially or fully controls driving at Level 3 to Level 5, can be only treated as an "object" under the current framework of Chinese law. On 23 June 2022, the "Regulation on the Administration of Intelligent Connected Vehicles in Shenzhen Special Economic Zone" ("Shenzhen ICVs Regulation") was adopted.[47] It is the first comprehensive (local) regulation on the administration of automated driving in China. The "Shenzhen ICVs Regulation" does not include details concerning the allocation of liability by AVs from Level 0 to Level 5, but confirms that drivers, vehicle owners, or caretakers are qualified responsible parties in the event of accidents. This is a local try for legislation on AVs.

5 Conclusion

This chapter first summarizes the rapid development of the AVs industry in China, then outlines the TDAV, and describes the classification of AVs levels in China. Generally, China's AVs levels are established in accordance with the internationally recognized standards. Next, the central and local policies and legal measures for AVs testing are summarized. These legal measures for testing indicate the specific rules that China might adopt for regulating AVs in the future. In general, the dominance of the driver is must always be supported, regardless of the level. It can be seen that, in order to ensure road safety, the automation system will, in terms of liability, be regarded as an "object", not a "subject". Damages directly caused by system defects should be attributed to the vehicle manufacturer or seller. However, a dilemma arises when it is difficult to judge the system as defective when an accident is caused; under the current legal framework, manufactures or sellers remain liable. Thus, rules on the burden of proof and its consequences that adequately take into account the interests of the manufacturer or seller should be introduced.

In a word, for addressing the issues of liability allocation that might be raised by the application of AVs in China, the current general rules such as the provisions in the "Civil Code", the "Law on Road Traffic Safety" and the "Product Quality Law" are applicable. If the existing laws are applied, the liability for damages caused by vehicles of Level 0 to Level 2 could be determined appropriately. Although there is still ambiguity about the relevant provisions, most cases can be clarified in judicial practice. Level 5 AVs are still in the minds of developers, which might not challenge the current legal framework in the near future. Challenges come with Level 3 and

[47] Regulation on the Administration of Intelligent Connected Vehicles in Shenzhen Special Economic Zone [2022] Standing Committee of Shenzhen Municipal People's Congress.

Level 4 AVs and require lawmakers to come up with an effective legal regulation soon.

Laws, Regulations and Other Rules

1. Rules for the Implementation of Road Test Management of AVs of Beijing [2020] Beijing Municipal Commission of Transport, Beijing Traffic Management Bureau and Beijing Municipal Bureau of Economy and Information Technology.
2. Civil Code of the People's Republic of China [2020] National People's Congress of the People's Republic of China.
3. Innovation and Development Strategy of AVs in China [2020] National Development and Reform Commission, Cyberspace Administration of China, Ministry of Science and Technology of the People's Republic of China and Ministry of Industry and Information Technology of the People's Republic of China.
4. Law on Road Traffic safety of the People's Republic of China [2003] the Standing Committee of the Tenth National People's Congress, Order of the President of the People's Republic of China, No. 8.
5. Legislation Law of the People's Republic of China (amendment) [2015] National People's Congress of the People's Republic of China.
6. Management Rules for Road Testing of Intelligent Connected Vehicles (trial) [2018] Ministry of Industry and Information Technology, the Ministry of Public Security and the Ministry of Transport of the People's Republic of China.
7. Management Rules for Road Testing and Demonstration Application of Intelligent Connected Vehicles (Trial) [2021] Ministry of Industry and Information Technology, the Ministry of Public Security and the Ministry of Transport of the People's Republic of China.
8. Management Specification for Intelligent Networked Vehicle Road Testing and Demonstration applications (for trial implementation) (draft for comments) [2021] the Ministry of Industry and Information Technology.
9. New Energy Vehicle Industry Development Plan (2021–2035) [2020] State Council of the People's Republic of China.
10. Product Quality Law of the People's Republic of China (amendment) [2018] Standing Committee of the National People's Congress, Order of the President of the People's Republic of China No. 33.
11. Regulation on the Administration of Intelligent Connected Vehicles in Shenzhen Special Economic Zone [2022] Standing Committee of Shenzhen Municipal People's Congress.
12. Taxonomy of Driving Automation for Vehicles [2020] Ministry of Industry and Information Technology of the People's Republic of China.

References

Literatures and Online Resources

Beijing Municipal Commission of Transport (2020) Interpretations of "Beijing rules for the implementation of road test management of AVs". http://jtw.beijing.gov.cn/xxgk/zcjd/202011/t20201119_2140422.html, the original language is in Chinese. Accessed 01 Mar 2021

China Association of Automobile Manufacturers (2023) Forecast of vehicle sales in 2022. https://www.workercn.cn/c/2023-01-06/7693533.shtml. Accessed 8 Jan 2023

China Automotive Technology and Research Center Co., Ltd., China Intelligent Transportation Systems Association (ed) (2020) Annual report on the development of autonomous driving industry in China (2020). Social Sciences Academic Press, Beijing

China Banking and Insurance News (2014) The first compulsory insurance in China- compulsory insurance for vehicle accident liability was launched in 2006. http://www.sinoins.com/zt/2014-01/27/content_87809.htm, the original language is in Chinese. Accessed 01 Mar 2021

China Electric Vehicle Association (2018) Analysis and suggestions of autonomous vehicle testing standards in major cities of China, https://auto.gasgoo.com/a/70070281.html, the original language is in Chinese. Accessed 1 Mar 2021

Hongyang S, Zheng C (2020) Beijing has 200 roads for autonomous driving tests, with 45 vehicles approved for manned tests. https://news.bjd.com.cn/2020/12/16/36701t100.html, the original language is in Chinese. Accessed 01 Mar 2021

Jinping X (2014) The development of new energy vehicles is the only way to become an automobile power. http://www.xinhuanet.com/politics/2014-05/24/c_1110843312.htm. Accessed 28 Feb 2021

Ministry of Public Security (2021) China's car ownership surpassed the US in 2020 for the first time. https://www.sohu.com/a/523492883_191175. Accessed 8 Jan 2023

Netease (2022) Big data analysis of Chinese autonomous driving enterprises in 2021, https://www.163.com/dy/article/H70IMN86051481OF.html, the original language is in Chinese. Accessed 11 Jan 2023

OICA (2021) 2021 production statistics. https://www.oica.net/category/production-statistics/2021-statistics/. Accessed 8 Jan 2023

Tencent Research Institute (2017) The development status and trend of the autonomous driving industry, https://www.tisi.org/4896, the original language is in Chinese. Accessed 28 Feb 2021

Yingjun X (2020) Guidance on comprehensive reform of auto insurance was formally implemented on September 19, https://www.thepaper.cn/newsDetail_forward_9041995, the original language is in Chinese. Accessed 01 Mar 2021

Autonomous Driving and Civil Liability in Japan

Antonios Karaiskos

Abstract In Japan, no legislation specifically regulating the issue of civil liability for autonomous driving exists. Therefore, this issue is dealt within the framework of the currently existing legislation applying to conventional driving. The basic pieces of legislation applying are the Civil Code (mainly its provisions on torts, and secondarily its provisions on contractual liability), the Act on Securing Compensation for Automobile Accidents and the Product Liability Act. In most cases, damage will be covered within the framework of the Act on Securing Compensation for Automobile Accidents by an insurance. Therefore, the Product Liability Act will be of limited use, and tort liability (as well as contractual liability) will be of secondary importance in practice. Since the currently existing legal framework is not structured to specifically deal with issues resulting from autonomous driving, various problems and discussions arise, and future legislative developments are anticipated.

1 General Framework

1.1 Legislative Framework

In Japan, no legislative rules exist for the time being that especially address the issue of autonomous driving and civil liability. As will be mentioned later in this paper, recent legislative developments related to autonomous driving were the amendments of the Road Traffic Act and the Road Transport Vehicle Act. The purpose of these amendments was to develop a legal framework which allows for the realization of level 3 and level 4 autonomous driving in Japan, an aim set by the Japanese

This work was supported by JSPS KAKENHI Grant Number 21K01269.

A. Karaiskos (✉)
Faculty of Law, Ryukoku University, Kyoto, Japankaraiskos@ymail.ne.jp

H. Steege et al. (eds.), *Autonomous Vehicles and Civil Liability in a Global Perspective*, Data Science, Machine Intelligence, and Law 3,
https://doi.org/10.1007/978-3-031-41992-8_7

Government already back in 2018.[1] However, the provisions of these two laws are not related to the issue of civil liability.

Therefore, the general legislative framework related to conventional driving and civil liability in Japan, applies also to autonomous driving. This framework comprises of the Civil Code as a general law (especially its provisions about torts), and the Act on Securing Compensation for Automobile Accidents as well as the Product Liability Act as major special laws. This chapter will present an outline of these laws and analyze the liability issues for each level of automatization under them. Further, an overview of the recent amendments of the Road Traffic Act will also be provided.[2]

1.2 Levels of Automation

In the past, the Japanese government was using an original categorization of automation, comprising of four levels. However, to avoid confusion occurring from the differences between this original categorization when compared to the international trends, the six-levels categorization by the Society of Automotive Engineer (SAE) in the US[3] was adopted, translated into Japanese in the form of JASO TP18004. [4] This chapter will also rely on the same categorization.

2 Tort Liability

2.1 General

The Japanese Civil Code (JCC)[5] is the general law for civil transactions. With regard to civil liability occurring from driving, important are the provisions about torts, contained in Part III (Claims), Chapter V.[6] This Chapter has 17 provisions, starting

[1] Kanmin ITS-Koso Rodomappu 2018 [Public-Private ITS Concept/Roadmap 2018] drafted by the IT Comprehensive Strategy Headquarters in June 2018.

[2] The Road Transport Vehicle Act (Act No. 185 of 1951) dealing with registration of vehicles and securities standards for them has also been amended to cope with level 3 autonomous driving. Details of this amendment will not be treated in this chapter.

[3] SAE International J3016 (September 2016).

[4] Kanmin ITS-Koso Rodomappu 2018 [Public-Private ITS Concept/Roadmap 2018].

[5] Act No. 89 of April 27, 1896. The provisions related to the law of obligations were largely amended in 2017, and the new content came into effect as of April 1, 2020. Regarding this recent amendment, see among others Wrbka (2014); Karaiskos (2018); Kihara (2012), pp. 72–84.

[6] In case the owner of the automobile incurs damage due to the autonomous driving system, the owner would also be able to pursue, as a purchaser, contractual liability of the seller for

from art. 709 and finishing with art. 724-2.[7] A basic feature of Japanese tort law system which can be seen in many legal systems worldwide is that it adopts the fault liability principle.

2.2 *Regulation of Torts*

2.2.1 Requirements for Liability

Article 709 JCC

According to the basic provision of art. 709 JCC, a person that has intentionally or negligently infringed the rights or legally protected interests of another person is liable to compensate for damage resulting in consequence.[8] This means that, in order for a person who has inflicted a tort to bear liability for this, at least the existence of negligence on its behalf will need to be proven. The requirements for liability set by this provision are as follows: (i) intention or negligence,[9] (ii) infringement of rights or legally protected interests of another person, (iii) occurrence of damage,[10] and (iv) existence of causal relation. The burden of proof lies on the victim of the tortious act.

According to the commonly accepted theory, the negligence mentioned above, as a requirement for tort, is understood as being a breach of an objective duty to avoid risk. Based on this understanding, which conceives negligence in an objective manner, negligence consists of a breach of a duty, imposed on the person who acts, to avert an objective result which could be foreseen by that person (objective fault).[11] Thus, the existence of negligence in concrete cases is judged based on whether, (i) it was possible to foresee the risk (*foreseeability*), and (ii) there was a duty to avoid the occurrence of the result (*duty to avoid risk*). Regarding (ii), if the person has taken proper measures to avoid the occurrence of the result, or it is

non-performance based on non-conformity or breach of the duty to explain. These possibilities will not be treated in this paper.

[7]For an overall analysis and presentation of the Japanese tort law system, see Yamamoto (2019), Oda (2021), pp. 191–211; Kihara (2015).

[8]The English translations of Japanese law provisions in this paper are based on the unofficial English translations in the website "Japanese Law Translation" (http://www.japaneselawtranslation.go.jp/?re=01) by the Japanese Ministry of Justice, with minor adjustments when considered necessary.

[9]From a tort law point of view, there is no meaning in drawing a line between intention and negligence as a requirement for torts, since the legal effects do not differ depending on whether there is intention or negligence on behalf of the person who committed the tort.

[10]There are no provisions setting any kind of limits to the liability (extent of damages) of a person who has committed a tort.

[11]For the content of objective fault, as well as the concept of subjective fault which suggests a different conception of fault, see Yamamoto (2019), p. 104 ff.

impossible to avoid the occurrence of the result regardless of the measures taken, no breach of such duty will exist.

Article 715 JCC

Art. 715 JCC provides for the liability of a person who employs another person for a business undertaking. According to the same provision, such a person is liable to compensate for damage inflicted on a third party by that person's employees with respect to the execution of that business. However, this does not apply if the employer exercised reasonable care in appointing the employee or in supervising the business, or if the damage could not have been avoided even if the employer had exercised reasonable care. In the past, this provision was thought to be providing for a personal liability of the employer based on its failure to exercise reasonable care in appointing the employee or supervising the business. On the contrary, today, it is understood that this provision stipulates for a vicarious liability of the employer, which is based on the existence of tort liability on behalf of the employee. Such vicarious liability is understood to be justifiable either by the fact that the employer makes profit and expands the range of its activities by employing another person,[12] or by the increase of danger through the usage of such a person.[13] In the analysis which follows below regarding liability under the tort regime, whenever the driver bears liability, it will also be possible to establish liability of the driver's employer, based on art. 715 JCC.

2.2.2 Scope of Damage

The scope of damage compensated in cases of torts is broad. In lack of an express provision, it is commonly accepted that the provision of art. 416 (scope of compensation for loss or damage) regarding contracts applies to torts by analogy.[14] According to the same provision, the purpose of the claim for compensation for the loss or damage for failure to perform an obligation is to have the obligor to pay the compensation for loss or damage which would ordinarily arise from the failure (para. 1). Further, the obligee can also claim the compensation for damage which has arisen from any special circumstances if the party did foresee or should have foreseen the circumstances (para. 2). As already mentioned, the same will apply to the scope of damage in cases of torts, by analogy.

[12] Supreme Court decision of July 1, 1988, Mishu, Vol. 42, No. 6, p. 451 ff.

[13] Supreme Court decision of April 30, 1958, Minshu Vo. 11, No. 4, p. 646 ff. For details about these perceptions of this liability, see Oda (2021), p. 193 ff.; Nomi and Kato (2019b), p. 407 ff.

[14] Regarding court practice and theoretical debate related to this issue, see Oda (2021), p. 199 ff.

2.2.3 Liability in Cases of Autonomous Driving

Levels 1 and 2 SAE

Since in levels 1 and 2, the driver is expected to pay attention to the surroundings and react properly, the driver's liability will be judged based on the same framework as in cases of conventional driving, and this framework will not be affected by malfunctions of the autonomous driving system. Such malfunctions will be taken into consideration when evaluating the existence or not of negligence on behalf of the driver, depending on whether the requirements of foreseeability and duty to avoid risk are fulfilled.[15]

Levels 3 and 4 SAE

In cases of levels 3 and 4, inside the operational design domain (ODD),[16] driving tasks are performed by the autonomous driving system. In level 3, if the continuance of the autonomous operation is difficult, a takeover request (TOR)[17] will be made by the system, driving operations will be passed to the driver, and the driver will be thereafter responsible for safe driving. Therefore, it would seem natural to consider that no negligence on behalf of the driver can be perceived inside the ODD (levels 3 and 4), or before a TOR is made (level 3). However, in the light of the above-mentioned content of negligence as conceived by the common understanding in Japan, the possibility of negligence of the driver being admitted also inside the ODD before a TOR is made (level 3) cannot be excluded.[18]

Accordingly, there is an opinion which asserts that the driver is responsible for incidents during autonomous driving also when no TOR has been made, as long as the requirements of foreseeability and duty to avoid risk are fulfilled.[19] According to a second opinion, the only responsibility borne by the driver before the TOR is to properly react once it is made.[20] Finally, according to a third opinion, no negligence of the driver can in principle be admitted for incidents before the TOR.[21]

Outside the ODD (levels 3 and 4), as well as after a TOR has been made (level 3), responsibility belongs to the driver.

Level 5 SAE

Since driving operations in level 5 are fully performed by the autonomous driving system, no negligence of the driver can be conceived.

---

[15] Kurita (2018), p. 28.

[16] For the discussions related to ODD, see Lee et al. (2020), pp. 1317–1322.

[17] As a recent analysis of TOR, see DeGuzman et al. (2021).

[18] Kurita (2018), p. 28.

[19] *Ibid*. Kurita suggests that the existence of foreseeability should be admitted, only in cases where the relevant incident could be foreseen by the driver, had the driver paid the level of attention required so as to notice a TOR.

[20] Fujita (2018a), p. 117.

[21] Konnai (2016), p. 1610. Kurita (2018), p. 28 notes that if negligence of the driver is to be admitted even exceptionally, the criteria for admitting so need to be clarified.

Other Issues Related to Liability Under the Tort Regime

An important issue related to tort liability, is that of inadequate maintenance of the car, including maintenance of the autonomous driving system. This is of especial relevance in cases of inadequate updates of the autonomous driving system. In such cases of inadequate maintenance (including inadequate updates), the driver or any other person in charge of maintenance may bear tort liability for the accident, if all requirements for tort liability are fulfilled.[22]

In cases of autonomous driving, depending on the level of automation, apart from the driver, tort liability of the manufacturer and the seller[23] of the car can also be established, as long as the requirements are fulfilled. However, proving negligence and causal relation with regard to the manufacturer or the seller will be difficult in general.[24]

Although tort liability is theoretically important, and it can be used as basis for compensation claims in automobile accidents, its function in practice is limited. Due to the wide scope of accidents covered by the Act on Securing Compensation for Automobile Accidents and the Product Liability Act, as will be analyzed below, it becomes practically important mostly in cases of death or injury to persons when the requirements set in the Act on Securing Compensation for Automobile Accidents are not fulfilled,[25] and of damage to property which does not fall under the scope of the Product Liability Act.[26]

3 Act on Securing Compensation for Automobile Accidents

3.1 General

The purpose of the Act on Securing Compensation for Automobile Accidents (ASCAA)[27] is to protect injured parties and contribute to the sound progress of automobile transportation by establishing a system to guarantee compensation for damage in the event of a person's death or bodily injury due to the operation of an automobile (art. 1 ASCAA).

[22] *Idem*: p. 29.

[23] The seller might also bear contractual liability against the purchaser.

[24] Kurita (2018), p. 29.

[25] Itagaki (2018), p. 5.

[26] Kubota (2020), p. 21.

[27] Act No. 97 of 1955.

3.2 *Liability Regime*

3.2.1 Requirements for Liability

According to art. 3 of the ASCAA, a person that puts an automobile into operational use for that person's own benefit is liable to compensate for damage arising from the operation of the automobile if this results in the death or bodily injury of another person. However, this does not apply if the person and the driver prove that they have exercised due care in connection with the operation of the automobile, that the injured party or a third party other than the driver has acted intentionally or negligently, and that there was no defect in automotive structure or function.

The requirements for liability based on art. 3 of the ASCAA are as follows: (i) falling under the notion of a "person that puts an automobile into operational use for that person's own benefit", (ii) existence of "operation" of the automobile, (iii) death or injury resulting "from the operation", (iv) death or injury resulting to "another person".[28]

It is evident that liability in the ASCAA is restricted to death or bodily injury of another person. Further, contrary to the JCC, which is based on the fault liability principle, the ASCAA provides for a strict liability regime. Since the person that puts an automobile into operational use is exempted from its liability if it proves (shift of burden of proof) the above-mentioned three exemption requirements in art. 3, liability under this provision is a *relative strict liability*.

3.2.2 The Notion of the "Person That Puts an Automobile Into Operational Use"

General

According to the provision above, the person who bears liability for an automobile accident causing death or bodily injury of another person, is the "person that puts an automobile into operational use for that person's own benefit".[29] The Supreme Court decision of September 24, 1968,[30] judged that liability of this person is justified by the fact that it creates the relevant risk and that it enjoys the benefits from using the automobile, and that the scope of this notion is therefore determined based on who the person is, which (i) has control over the use and (ii) enjoys the benefits from the use.[31]

As long as the "person in possession" becomes liable to compensate for damage under art. 3, the victim is entitled to compensation against the insurer or the mutual insurance cooperative (art. 16 para. 1 ASCAA). The term "person in possession"

---

[28] For details, see Nomi and Kato (2019a), p. 144 ff. [by Oshima Sh].

[29] For details of this notion and the liability borne by it, see Fujita (2018a).

[30] Hanreijiho, No. 539, p. 40 ff.

[31] Urakawa (2018), p. 132.

means not only the owner of an automobile, but also any other person with the right to use an automobile, that puts an automobile into operational use for that person's own benefit (art. 2 para. 3 ASCAA). Unless the driver is also the owner of the car, evaluation of whether the driver falls under the notion of the person that puts an automobile into operational use for that person's own benefit needs to be made. If the damage has occurred to the person who puts an automobile into operational use itself, that person cannot claim for compensation under the ASCAA, and will therefore have to use other legal ground for compensation claims, such as torts or the PLA.[32]

Control Over and Profit from Use

According to the Supreme Court decisions of September 24, 1968[33] and July 16, 1970,[34] as well as other court decisions where similar judgments have been made, the notion of "control over the use" of the automobile refers to the "right to control over the use of the automobile" or the "position which allows to instruct or manage the operation of the automobile". Thus, if the owner or any other person having the right to use the automobile has allowed another person to put the automobile in operational use, the owner or such other person will bear liability. Further, according to Supreme Court decision of September 12, 2008,[35] such "approval" to another person for putting the automobile in operational use exists also if it can be reasonably evaluated that the case in question falls within the limits of approval, judged objectively based on the appearance. Therefore, the owner or other person having the right to use the automobile can bear liability for an accident even if these persons were not in the automobile and were not even aware of the existence of the driver.

The notion of "benefit from use" is judged objectively, based on for whose benefit the automobile was set into operation, when evaluated overall.[36] Therefore, in the abstract analysis of the state of liability depending on the level of automation which follows in this chapter, no specific reference will be made to this criterion.

If the driver was driving the automobile for the benefit of another person, neither control over nor profit from use belong to it, and it bears no liability under the ASCAA. Further, art. 3 of the ASCAA does not cover material damage. In both these cases, tort liability or liability under the PLA might be established.

Exemption from Liability

According to the above-mentioned proviso of art. 3 of the ASCAA regarding exemption from liability, the person who should bear liability based on the same

[32] Regarding damage incurred by the driver in cases of levels 4 and 5, Yamaguchi (2019), p. 57 asserts that apart from tort or PLA claims, claims against the owner etc. as the person who puts the automobile into operational use for that person's own benefit might also be possible, since the driver could be characterized more a passenger rather than an actual driver of the automobile.

[33] Hanreijiho no. 539, p. 40 ff.

[34] Hanreijiho no. 600, p. 89 ff.

[35] Hanreijiho no. 2021, p. 38 ff.

[36] Kurita (2018), p. 31.

provision is exempted from it, if it proves that (i) that person and the driver have exercised due care in connection with the operation of the automobile, (ii) that the injured party or a third party other than the driver has acted intentionally or negligently, and (iii) that there was no defect in automotive structure or function.[37] These three requirements need to be proven cumulatively. This means that the person who put the automobile into operational use bears liability also for defects in the structure or function of the automobile.[38] Although this structure creates issues regarding the relation between the ASCAA with the PLA, it is in general understood that this wide liability established by the ASCAA is justifiable, in light of the above-mentioned purpose of the ASCAA to protect injured parties.[39]

In general, these grounds for exemption apply to autonomous driving in the same manner as for conventional driving, with minor adjustments. For example, due care ((i)) will also include the duty to maintain the functions of the autonomous driving system,[40] such as applying all updates available. Further, in cases a third party intruding into the autonomous system without authorization (such as hacking), usually intent on behalf of that third party will be admitted. Moreover, the notion of a "defect in automotive structure or function", also includes defects in the autonomous system.

3.2.3 Insurance Coverage

A characteristic of the liability under the ASCAA is that, there exists a legal system providing support for the performance of such liability, by means of automobile liability insurance. The existence of such coverage balances the strict liability provided by the ASCAA. It needs however to be emphasized that this insurance system for liability under the ASCAA does not cover all liability arising from the ASCAA.[41] The part of the compensation due which exceeds the maximums stipulated by the law, will need to be covered by means of optional insurance. Although the purchase rate for optional insurance is quite high in Japan, there are significant differences in the rate depending on the area.[42]

---

[37] According to Shiomi (2017), p. 239, defects as a reason for exemption, do not include "warning/instruction defects", which are included in the notion of "defects" under the PLA, as is mentioned below.

[38] Urakawa (2018), p. 136.

[39] Yamaguchi (2019), p. 54. Otherwise, the victims would bear the burden to select the most proper person against whom and legal ground based on which they would claim for compensation.

[40] Itagaki (2018), p. 11.

[41] According to the Order for Enforcement of the Act on Securing Compensation for Automobile Accidents (Cabinet Order No. 286 of October 18, 1955), the maximum compensation covered by compulsory insurance is 30 million yen in case of death, and 40 million yen for grave severe residual disabilities.

[42] Kubota (2020), p. 20.

3.3 *Liability in Cases of Autonomous Driving*

3.3.1 Levels 1 and 2 SAE

In levels 1 and 2, there is always a human driver in the vehicle. Therefore, the owner of the car or any other persons eligible for liability under the ASCAA, will bear liability, even if they are not the driver of the car, as long as they have "approved" the operation of the car by the driver.

3.3.2 Levels 3 and 4 SAE

What was mentioned above for levels 1 and 2, also applies to levels 3 and 4. An issue arises in the case of level 4, because in this level, when the autonomous system operates, it fully takes over the driving of the car. There is debate regarding whether it is possible to apply the legal framework described above for conventional driving, also in cases of autonomous driving.

According to one opinion, the person who has activated the autonomous system and set the destination should be considered to be in a "position which allows to instruct or manage the operation of the automobile". The reason for this is that such a person would also bear liability in case another person is driving, and there is no necessity to differentiate when driving is performed not by another person but by the autonomous system.[43]

According to a second opinion, these two situations are not comparable, and it might be non-justifiable to admit that the person who has activated the autonomous system and set the destination is in a "position which allows to instruct or manage the operation of the automobile". The main arguments drawn by this opinion are as follows. Firstly, the owner of the car or any other persons eligible for liability under the ASCAA can select and supervise the driver but not the autonomous system (and even if they could select the system, the nature of such choice would be different from the choice of the driver). Secondly, the benefits of a car can be enjoyed even without selection of another person as a driver, whereas the benefits of autonomous driving cannot be enjoyed unless operation of the autonomous system is admitted. Thirdly, Supreme Court decision of December 20, 1973[44] and other similar decisions have judged that the owner of a car or any other persons eligible for liability under the ASCAA do not bear liability in cases of accidents caused after the car has been stolen. However, if one admits that one bears liability simply by admitting the operation of the car by the autonomous system, this would also apply to cases of stolen cars, contrary to the case-law mentioned above.[45]

[43] Fujita (2018a), p. 134.

[44] Minshu, vol. 27, no. 11, p. 1611.

[45] Kurita (2018), p. 30.

If one adopts the second opinion, the outcome is that under the currently existing system, the victim will need to pursue liability under torts or the PLA.

3.3.3 Level 5 SAE

What was mentioned above for level 4 when the autonomous system operates, also applies to level 5.

4 Product Liability Act

4.1 Requirements for Liability

The Product Liability Act (PLA),[46] has as purpose to protect victims by establishing liability against manufacturers, regarding compensation for loss or damage if a defect in a product causes damage to life, body, or property of another person, and thereby to contribute to the stabilization and improvement of the lives of the citizenry and to the sound development of the Japanese economy (art. 1 PLA).[47]

Same as the ASCAA, the PLA also provides for a strict liability regime regarding defective products (however, not a *relative* strict liability as under the ASCAA). According to art. 3 of the PLA, the "manufacturers, etc."[48] are liable to compensate for loss or damage, if death or bodily injury to others or infringement of property of others are caused by a defect in the delivered product, which was manufactured, processed, imported, or on which indications including a name referred to in items (ii) or (iii) of art. 2 para. 3 of the PLA are used. However, this does not apply if the damage occurs to the product alone, namely there is no consequential damage. This means that contrary to the ASCAA, the PLA also covers cases where damage was incurred by the driver of the automobile itself.

According to the provision of art. 3 of the PLA, the requirements for liability based on the same provision are as follows: (1) the product concerned was manufactured, processed, imported, or indications including a name referred to in items (ii) or (iii) of article 2 para. 3 are used , (2) the manufacturer etc. "delivered" the product, (3) there was a "defect" in the product, (4) death or bodily injury to others or infringement of property of others was caused, and (5) there is causal relationship between the defect and the damage.[49] The fulfillment of these requirements needs to be proven by the victim.

---

[46] Act No. 85 of 1994.

[47] For details of the Japanese product liability law system, see Nottage (2012), as well as Nagano and Karaiskos (2018), pp. 59–74, where recent case law relevant to the PLA is also presented.

[48] The word "etc." is mentioned in the relevant provisions of the PLA, since the notion "manufacturer, etc." does not include only manufacturers, but other persons such as importers too, as will be explained later in this chapter.

[49] Nomi and Kato (2019b), p. 179 [by Asami Y].

4.2 *Definitions of Basic Notions*

4.2.1 Manufacturer Etc

The above-mentioned provision of art. 2 para. 2 of the PLA provides a definition of the term “manufacturer, etc.”. According to this provision, a manufacturer, etc. is a person who (i) manufactured, processed, or imported the product in the course of business (hereinafter simply referred to as “manufacturer”), (ii) uses a name, trade name, trademark or other kinds of indications on the product as the manufacturer of the product, or any person that uses indications including a name on the product which may mislead others into believing that the person is the manufacturer thereof, or (iii) uses indications including a name on the product which can be recognized by others as that of the real manufacturer, taking the manner of the manufacturing, processing, importation or sale of the product, and other circumstances into consideration.

4.2.2 Product

The term “product” means movables which are manufactured or processed (art. 2 para. 1 PLA). Movables are only tangible objects (art. 86 para. 2, art. 85 JCC). The autonomous driving system in itself is not tangible, but since the automobile to which the autonomous driving system has been installed is tangible, it is possible to evaluate the defects of the autonomous driving system as being defects of the automobile.[50] If the manufacturer of the system and the manufacturer of the car are different persons, the latter will bear liability against the victim, and can claim reimbursement or pursue non-performance liability against the former. In practice, reimbursement claims by the manufacturer of the car against the manufacturer of components of the car are quite rare even in cases of non-autonomous driving automobile accidents.[51] Therefore, it is questionable whether this situation will change under the autonomous driving regime.

[50] Urakawa (2017), p. 33. Kurita (2018), p. 32 questions the conformity of this practice, where accidents caused by defects of an intangible autonomous driving system are treated within the framework of the PLA, the structure of which requires a defect in a movable.

[51] Kurita (2018), p. 32. Fujita (2018a), p. 205. According to Fujita, the main reasons for this are that in most cases the reason for the accident lies with the driver, that it is difficult to prove existence of defect and causal relationship with the accident, that the person who puts the automobile into operation for that person’s own benefit does not incur economic loss thanks to insurance coverage, and that the insurance companies do not exercise their rights to reimbursement taking into consideration the load and cost for this.

4.2.3 Defect

The term "defect" means a lack of safety which a product should normally have, taking into account the characteristics of the product, the normally foreseeable usage manner, the time at which the manufacturers, etc. delivered the product, and other circumstances of the product (art. 2 para. 2 PLA). In general defects under the PLA are divided into three categories: (i) manufacturing defects, (ii) design defects, and (iii) warning/instruction defects.[52]

Whether a defect of the autonomous driving system also constitutes a defect of the automobile can be questionable if weight is given to the nature of the autonomous driving system as assistance of driving. From this perspective, it might become possible to assert that an autonomous system defect does not constitute a defect of the automobile, as long as the other ordinary functions of the automobile are free of defects.[53]

In general, the evaluation of defects under the PLA will function similarly for conventional and autonomous driving. Among the above-mentioned categories of defect, the one which might need special treatment for autonomous driving, depending on the case, is that of design defects. In Japan, there is debate about what the criterion is for defining whether a design defect exists. The question here is whether the existence of a defect should be based on a statistical average comparison, or on specific aspects of the driving function.[54]

4.3 Liability in Cases of Autonomous Driving

Due to the above-mentioned content of the requirements for liability under the PLA, the structure or issues of such liability does not in general differ depending on the level of automation. Issues with the liability under the PLA can occur because the burden of proof about the existence of a defect falls on the victim. Especially in levels 0 to 4, the driver's participation to driving actions might complicate the proof of causal relationship between defect and damage.[55]

A specific issue arising under levels 3 and 4 is that, there might be cases where the person that puts the automobile into operational use will need to compensate under

[52] Shohishacho Shohishaanzenka [Consumer Affairs Agency, Consumer Safety Division] (2018). This categorization is influenced by the relevant one in the USA.

[53] Yamaguchi (2019), p. 54.

[54] According to Urakawa (2017), p. 34, the criterion for this should be an "average human driver". Kurita (2018), p. 32, criticizes this, since there might be cases where although autonomous driving is statistically far safer than driving of an average human driver, comparisons regarding specific aspects might lead to the affirmation of existence of a defect in the system. Kurita further questions the properness of using a PL system designed for tangibles, to intangible autonomous driving systems.

[55] Yamaguchi (2019), p. 54.

the ASCAA also for damage resulting from defects of the autonomous driving system. Normally, such damage should be covered by the manufacturer of the autonomous driving system, based on the PLA.[56] Hence, the existence of insurance systems related to the ASCAA shifts part of the liability which should ordinarily be borne by the manufacturer to the person that puts the automobile into operational use.[57]

5 Road Traffic Act

The Road Traffic Act[58] has as purpose to prevent road hazards and otherwise ensure the safety and fluidity of traffic, as well as to contribute to preventing blockages arising from road traffic (art. 1 Road Traffic Act). According to art. 70 of the Road Traffic Act (safe driving obligation), the driver of a vehicle must work the vehicle's steering wheel, brakes, and other equipment in a consistent manner, and must drive at a speed and in a manner that poses no hazard to others in consideration of road conditions, traffic conditions, and the condition of the vehicle. In the past, the wording of this provision was creating concerns related to autonomous driving, since it was evident that the Road Traffic Act was based on the assumption of a driver steering the automobile, and bearing the obligation to drive safely.[59] Eventually, the Act was amended in 2019, to ensure a secure autonomous level 3 driving,[60] and the new provisions came into force in April 2020.[61]

The main content of this amendment was as follows. Firstly, provisions that are related to definitions of autonomous operation equipment were inserted. Secondly, provisions related to the obligations of drivers using autonomous operation equipment were established. Thirdly, provisions related to records by operation status recorders were added. Further, the Road Traffic Act was amended in 2022, to establish a permission system for autonomous level 4 driving, and the new provisions came into force in April 2023.[62]

---

[56] Kubota (2020), p. 21; Nomura (1969) p. 131.

[57] For details of the automobile insurance structure in Japan and the influences brought to it by autonomous driving, see Ikeda (2018).

[58] Act No. 105 of 1960.

[59] For details of such concerns, see Urakawa (2018), p. 139.

[60] As already mentioned, the Road Transport Vehicle Act has also been amended to cope with level 3 autonomous driving.

[61] For details of the amended Road Traffic Act and Road Transport Vehicle Act, see Sato (2019), pp. 4–11; Imai (2019).

[62] Details of the respective provisions are not treated in this paper, since, this Act, same as the Road Vehicles Act, is not directly related to civil liability, but rather aims at preventing automobile accidents *ex ante*.

6 Future Perspectives

This chapter has demonstrated the outline and basic issues related to autonomous driving and civil liability in Japan. As it has been analyzed in this chapter, in Japan, there is no specific legislation regulating civil liability for autonomous driving. Therefore, the relevant issues are dealt within the framework of the currently existing legislation applying to conventional driving. As explained in this chapter, the ASCAA and the PLA will be used as legal grounds for relevant claims in most cases, and damage will be usually covered by the insurance. Tort liability (as well as contractual liability) will be in practice of secondary importance. Since the currently existing legal framework about civil liability is not specifically structured so as to deal with issues resulting from autonomous driving, various issues and discussions arise regarding its application to autonomous driving.

To close this chapter, the author would like to present some basic features of the future perspectives. A major issue is, whether the current structure where civil liability is primarily borne by the person that puts an automobile into operational use for that person's own benefit (ASCAA) should be maintained. Options other than changing the current structure would be maintaining it and (i) having insurance fees being shared by the person that puts an automobile into operational use for that person's own benefit and the car manufacturer, (ii) requesting the car manufacturer to provide information relevant to the liability of the person that puts an automobile into operational use for that person's own benefit, or (iii) enhancing liability borne by the car manufacturer.[63]

There is a suggestion that an alternative could be a system where the provider of the autonomous system bears such liability primarily.[64] Here, a further option would be enlightening the burden of proof borne by the victim, either by shifting the burden of proof on the manufacturer, or by obliging the manufacturer to equip autonomous driving cars with drive recorders.[65]

In any case, taking into consideration the theoretical and practical issues arising from the application of the currently existing legal framework for civil liability on autonomous driving, and the perspective of high level autonomous driving been used in practice in Japan, legislative developments are to be expected in the near future.

[63] Fujita (2018b), p. 284 ff. Regarding case (ii), Fujita suggests that if the manufacturer refuses to provide such information and civil liability of the person that puts an automobile into operational use for that person's own benefit is admitted, the manufacturer should not be able to use the existence of a defect as a defense against reimbursement claims by the person that puts an automobile into operational use.

[64] *Idem*, p. 281 ff., p. 287 ff. See also Kubota (2018), p. 188 ff. who refers to future perspectives related to reimbursement claims.

[65] Daira (2019), p. 80 ff. also emphasizes the necessity to ensure accurate information about the reasons causing an accident by means of drive recorders or event data recorders (EDR).

Laws

- Product Liability Act (Act No. 85 of July 1, 1994)
- Road Traffic Act (Act No. 105 of June 25, 1960)
- Act on Securing Compensation for Automobile Accidents (Act No. 97 of July 29, 1955)
- Road Transport Vehicle Act (Act No. 185 of 1951)
- Civil Code (Act No. 89 of April 27, 1896)

References

Literature

Daira Y (2019) Jidounten ni muketa Songaihokengyokai no Taio – Jibaisekihoken/Jidoshahoken ni kansuru Taio [Response of the non-life insurance industry for autonomous driving: the response regarding automobile liability insurance and automobile insurance]. In: Nakayama K and others (eds) Jidounten to Shakaihenkaku [Autunomous driving and social changes]. Shojihomu, Tokyo

DeGuzman CA, Kanaan D, Hopkins SA, Donmez B (2021) Takeover request (TOR) effects during different automated vehicle failures. J Intell Transp Syst. https://doi.org/10.1080/15472450.2021.1891536

Fujita T (2018a) Jidounten to Unkokyoyosha no Sekinin [Autonomous driving and liability of the person that puts an automobile into operational use]. In: Fujita T (ed) Jidounten to Ho [Autonomous driving and the law]. Yuhikaku, Tokyo

Fujita T (2018b) Jidounten o meguru Minjisekininhosei no Shoraizo [Future images of civil liability legislation over autonomous driving]. In: Fujita T (ed) Jidounten to Ho [Autonomous driving and the law]. Yuhikaku, Tokyo

Ikeda Y (2018) Jidounten to Hoken [Autonomous driving and insurance]. In: Fujita T (ed) Jidounten to Ho [Autonomous driving and the law]. Yuhikaku, Tokyo

Itagaki T (2018) Jidounten ni okeru Minjijo no Sekinin ni kansuru Kosatsu [A study on civil liability in autonomous driving]. Shin PL Kenkyu [J New Prod Liabil] 3:3–13

Karaiskos A (2018) Civil code reform in Japan: is the new regulation of standard contract terms a desirable one? In: Heidemann M, Lee J (eds) The future of the commercial contract in scholarship and law reform. Springer, Berlin

Kihara H (2012) Japan's civil code reform and consumer protection. Asia Univ Law Rev 47(1): 72–84

Kihara H (2015) Tort law in Japan. Kluwer Law International, Alphen aan den Rijn

Konnai K (2016) Jidountenjidosha ni yoru Kotsujiko no Hotekisekinin [Legal liability for traffic accidents caused by autonomous driving cars]. Kokusaishojihomu [J Jap Inst Int Bus Law] 44(11):1609–1619

Kubota A (2018) Jidounten to Hanbaiten/Meka no Sekinin [Autonomous driving and liability of dealers/manufacturers]. In: Fujita T (ed) Jidounten to Ho [Autonomous driving and the law]. Yuhikaku, Tokyo

Kubota A (2020) Jidounten ni kansuru Genjo to Kadai (1) – Minjisekinin no Kanten kara [Current state and issues related to autonomous driving: from the viewpoint of civil liability]. Horitsu no Hiroba 73(2):18–27

Kurita M (2018) Jidountensha no Jiko to Minjisekinin [Autonomous driving car accidents and civil liability]. Horitsujiho 91(4):27–33

Lee ChW, Nayeer N, Garcia DE, Agrawal A, Liu B (2020) Identifying the operational design domain for an automated driving system through assessed risk. In: 2020 IEEE Intelligent Vehicles Symposium (IV):1317–1322

Nagano F, Karaiskos A (2018) Pharmaceuticals products liability in Japan. Pharm Policy Law 20: 59–74

Nomi Y, Kato S (eds) (2019a) Rontentaikei Hanreiminpo 8 Fuhokoi I [Points at issue: civil law cases 8 – Torts I], 3rd edn. Daiichihoki, Tokyo

Nomi Y, Kato S (eds) (2019b) Rontentaikei Hanreiminpo 9 Fuhokoi II [Points at issue: civil law cases 9 – Torts II], 3rd edn. Daiichihoki, Tokyo

Nomura Y (1969) Unkokyoyoshasekinin ni tsuite no Ichikosatsu [A consideration on the liability of the person that puts an automobile into operational use]. Jurist 431:131–141

Nottage L (2012) Product safety and liability law in Japan. Routledge, London

Oda H (2021) Japanese law, 4th edn. Oxford University Press, London

Sato N (2019) Jidounten no Jitsugen ni muketa Dorounsosharyoho oyobi Dorokotsuho no Kaisei no Gaiyo [Outline of the reform of the road transport vehicle act and of the road traffic act towards a realization of autonomous driving]. NBL 1149:4–11

Shiomi Y (2017) Kihonkogi Saikenkakuron II Fuhokoiho [Basic lectures of special part of law of obligations II: torts], 3rd edn. Shinseisha, Tokyo

Shohishacho Shohishaanzenka [Consumer Affairs Agency, Consumer Safety Division] (2018) Seizobutsusekininho (PL-ho) no Chikujokaisetsu [Article-byArticle Commentary of the Product Liability Act], 2nd edn, available at the website of the Japanese Consumer Affairs Agency (https://www.caa.go.jp/policies/policy/consumer_safety/other/product_liability_act_annotations/)

Urakawa M (2017) Jidounten to Minjisekinin [autonomous driving and civil liability]. NBL 1099: 30–36

Urakawa M (2018) Jidounten ni yoru Jidoshajiko to Minjisekinin [Autonomous driving accidents and civil liability]. In: Nakajima S and others (eds.), Jidounten no Genjyo to Kadai [Current state and issues of autonomous driving], Shakaihyoronsha, Tokyo

Wrbka S (2014) Japan's Civil Code Reform Plan – Seen from a Western Perspective. Kyoto University Legal Research Bulletin. http://www.law.kyushu-u.ac.jp/programsinenglish/kulrb/stefan2.pdf. Accessed 8 Jan 2022

Yamaguchi N (2019) Jidounten to Hotekisekinin [Autonomous driving and legal responsibility]. Gendaishohishaho [Curr Consum Law] 42:52–59

Yamamoto K (2019) Basic features of Japanese tort law. Jan Sramek Verlag, Vienna

En-Route to a Driverless City-State: Examining the Current Legal Landscape for Autonomous Vehicles in Singapore

Thomas Hufnagel and Jaryl Lim Zhi Wei

Abstract As part of Singapore's Smart Nation movement, research and development trials for autonomous vehicles are already well underway in Singapore. However, Singapore's laws relating to autonomous vehicles remain in their nascent stages as the Singapore government is presently still studying the legal consequences of autonomous vehicles and their related technologies. In 2017, the government had set up a 'regulatory sandbox' to regulate the trial and use of autonomous vehicles for a period of 5 years, and, in 2019, an industrial standard known as Technical Reference 68 was introduced by the Singapore Standards Council to set out best practices for the deployment of autonomous vehicles on Singapore roads.

This chapter will examine the various liability addressees involved in Level 0 to 5 autonomous vehicles (as defined under SAE J3016) based on existing legal principles in Singapore. It will also discuss recent legal trends—including the aforementioned 'regulatory sandbox' and Technical Reference 68—and their impact on the future regulation of autonomous vehicles in Singapore.

1 Introduction: Turning Vision Into Reality

Since 2014, the Singapore government has been paving the way for the development of a driverless vehicle mobility system as part of its 'Smart Nation' movement. The 'vision', as it were at the time, was to design a "*totally new mobility concept for a green-field, future town*" where there are green and open spaces for residents and pedestrians to walk and cycle as a mode of intra-town travel, supplemented by environmentally friendly, autonomous vehicle-enabled point-to-point type of public transport that is free of the smoke, noise, congestion and safety concerns posed by

T. Hufnagel (✉)
Partner, Luther LLP (Singapore), Singapore, Singapore
e-mail: Thomas.hufnagel@luther-lawfirm.com

J. L. Z. Wei
Associate, I.N.C. Law LLC, Singapore, Singapore

H. Steege et al. (eds.), *Autonomous Vehicles and Civil Liability in a Global Perspective*, Data Science, Machine Intelligence, and Law 3,
https://doi.org/10.1007/978-3-031-41992-8_8

vehicles today.[1] A Committee on Autonomous Road Transport for Singapore ("**CARTS**") was thus set up in 2014 to chart the strategic direction for autonomous vehicles ("**AV**") enabled land mobility concepts in Singapore, and a Memorandum of Understanding was signed with the Agency for Science, Technology and Research, a research & development statutory board, to set up the Singapore Autonomous Vehicle Initiative for the purposes of exploring the technological possibilities AVs can create.

Since that time, several AV research and development trials have begun in Singapore and are well underway. These include, for example, driverless taxis developed by US-headquartered company nuTonomy, on-demand driverless shuttle buses deployed in Sentosa Island, driverless shuttle buses connecting student halls in the Nanyang Technological University ("**NTU**") campus, and even self-driving golf-buggies developed jointly by the National University of Singapore ("**NUS**") and the Singapore-MIT Alliance for Research and Technology.

In November 2017, Singapore's Transport Minister announced that self-driving buses will be deployed in 3 towns during off-peak periods starting from 2022—Punggol, Tengah and Jurong Innovation Districts—and autonomous shuttles will also provide first-last mile connection for residents and workers in these districts.[2] In October 2019, the Land Transport Authority further announced that it would be expanding its test-bed from the said 3 towns to now include all of western Singapore, spanning more than 1000 km of public roads and including neighbourhoods such as Bukit Timah, Clementi and Jurong.[3]

In the midst of these developments however, the legal implications of AV was first thrown into the public spotlight in October 2016 when a self-driving car by nuTonomy was reported to have collided into a lorry during a test-drive.[4] nuTonomy later announced publicly that the accident was caused by an "*extremely rare combination of software anomalies*" that nuTonomy claimed to have since

[1]Ministry of Transport (27 August 2014) Speech by Mrs Josephine Teo at Autonomous Transport: Paving the Road for Future Mobility Event on 27 August 2014. Available via Government of Singapore. https://www.mot.gov.sg/news-centre/news/Detail/Speech-by-Mrs-Josephine-Teo-at-Autonomous-Transport-Paving-the-Road-for-Future-Mobility-Event-on-27-August-2014/. Accessed 2 May 2020.

[2]Kevin Kwang (22 November 2017) Self-driving buses, shuttles to be tested in 3 towns from 2022. Available via Channel News Asia, Mediacorp Pte Ltd. https://www.channelnewsasia.com/news/technology/self-driving-buses-shuttles-test-punggol-jurong-tengah-9427850. Accessed 2 May 2020.

[3]Toh Ting Wei (25 October 2019) Western Singapore set to become test bed for self-driving vehicles. Available via The Straits Times, Singapore Press Holdings Ltd. https://www.straitstimes.com/singapore/transport/western-spore-set-to-become-test-bed-for-self-driving-vehicles. Accessed 2 May 2020.

[4]Melissa Lin (19 October 2016) Driverless car his lorry during test drive. Available via The Straits Times, Singapore Press Holdings Ltd. https://www.straitstimes.com/singapore/driverless-car-hits-lorry-during-test-drive. Accessed 2 May 2020.

eliminated following extensive testing both in simulation and on private roads.[5] Although no one was injured during the collision, the incident highlighted the need to consider the apportionment of liabilities for accidents involving AVs. This topic has been the subject of detailed studies ever since CARTS was first set up back in 2014, but the collision drew greater public attention and focus on how future accidents involving AVs should be dealt with, especially given the increase in the number and scope of AV trials in Singapore as well as the projected deployment of AVs in public towns in 2022.

These rapid developments in the AV field also prompted discussions on the difficulties and deficiencies posed by existing laws to address AV-related liability issues. Such difficulties were duly acknowledged in the Singapore Parliament in 2017 by the then Second Minister for Transport Mr Ng Chee Meng, who noted:[6]

> The current Act (i.e. Singapore's Road Traffic Act) assumes a human driver to be in control of the vehicle, and thereby responsible for monitoring the surroundings and making driving decisions. Where automated vehicle technology is installed in a motor vehicle, computer systems control the vehicle, replacing the driver's role. Where automated vehicle technology is installed in a motor vehicle, computer systems control the vehicle, replacing the drier's role.
>
> AVs thus challenge the very notion of human responsibility which lies at the core of our current road and criminal laws.

However, the Minister noted that the Singapore courts would adopt an incremental approach to deal with such situations by drawing analogies to other similar technologies. The Minister also hinted at the possibility of licensing AVs as a separate class of vehicles, but noted that these are presently the subject of detailed study:[7]

> On liability, the traditional basis of claims for negligence may not work so well – as Mr Dennis Tan has said – where there is no driver in control of a vehicle. When presented with novel technologies, courts often try to draw analogies to legal constructs in other existing technologies.
>
> In the case of AVs, the courts have autopilot systems for airplanes and autopilot navigational systems for maritime vessels, and product liability law to draw references from. As with accidents involving human-driven vehicles, it is likely that issues of liability for AVs will be resolved through proof of fault, and existing common law. We will require all test AVs to log travel data to facilitate accident investigations and liability claims.
>
> But having said this, these are early days yet. A lot of detailed studies will have to be done, including the point raised by Mr Dennis Tan on the future framework, whether this class of vehicles requires licensing. I think this will be an area to be studied.

[5] Zhaki Abdullah (25 November 2016) 'Software anomalies' to blame for driverless car accident. Available via The Straits Times, Singapore Press Holdings Ltd. https://www.straitstimes.com/singapore/software-anomalies-to-blame-for-driverless-car-accident. Accessed 2 May 2020.

[6] Parliamentary Debates Singapore: Official Report, vol 94 at pages 64 to 65 (7 February 2017) (Second Minister for Transport Mr Ng Chee Meng).

[7] Parliamentary Debates Singapore: Official Report, vol 94 at page 87 (7 February 2017) (Second Minister for Transport Mr Ng Chee Meng).

2 Recent Legal Developments in Singapore

2.1 Amendments to Singapore's Road Traffic Act

A few months after the nuTonomy incident, the Singapore Parliament passed a slew of amendments to Singapore's *Road Traffic Act*[8] in February 2017 to, among other things, establish a regulatory framework for the undertaking of trials and use of AVs on Singapore roads.

To be clear, the amendments were not intended to provide a comprehensive prescription of liability for AV-related incidents. Rather, the amendments were intended to give the Land Transport Authority (which is the local regulatory authority for land transport matters) the flexibility to impose or amend rules and requirements on AV trials and use as and when required given the myriad of possibilities and unforeseen circumstances that may occur, as the nuTonomy incident had aptly demonstrated. The Minister who moved the amendments likened them to a "*regulatory sandbox*" where the Land Transport Authority could study the legal consequences of AV trials for a period of 5 years to assess the appropriate regulatory response to AVs.[9]

And, for the first time ever in Singapore parliamentary proceedings, the Minister also made reference to the Society of Automotive Engineers ("SAE") International J3016 and said that the legislative amendments would apply to Levels 3, 4 and 5 of J3016, because "*[a]t levels 3, 4 and 5, it is the autonomous vehicle which monitors the driving environment, and the human driver is no longer required to actively monitor the vehicle. The amendments are not meant and not intended to capture vehicles outfitted with just driver assistance systems like cruise control and assisted parking functions. These systems are regulated through other means.*"[10]

New definitions[11] were introduced, primarily in order to complement the enactment of new rules for the trial and use of AVs in Singapore:

> "automated vehicle technology" means any particular technology that —
>
> (a) relates to the design, construction or use of autonomous motor vehicles; or
> (b) otherwise relates to advances in the design or construction of autonomous motor vehicles;
>
> "autonomous motor vehicle" means a motor vehicle equipped wholly or substantially with an autonomous system (also commonly known as a driverless vehicle), and includes a trailer drawn by such a motor vehicle;

---

[8] *Road Traffic Act* (Cap 276, 2004 Rev Ed Sing).

[9] Parliamentary Debates Singapore: Official Report, vol 94 at page 65 (7 February 2017) (Second Minister for Transport Mr Ng Chee Meng).

[10] Parliamentary Debates Singapore: Official Report, vol 94 at page 65 (7 February 2017) (Second Minister for Transport Mr Ng Chee Meng).

[11] *Road Traffic Act* (Cap 276, 2004 Rev Ed Sing), section 2(1).

Table 1 Comparison of the definitions of "*motor vehicles*" before and after the 2017 legislative amendments to the *Road Traffic Act*

Old definition of "*motor vehicles*"[a]	New definition of "*motor vehicles*"[b]
"*motor vehicle*" *means a mechanically propelled vehicle intended or adapted for use on roads;*	"*motor vehicle*" *means a vehicle that —* (a) *is propelled wholly or partly by a motor or by any means other than human or animal power; and* (b) *is used or intended to be used on any road;*

[a]Before 14 August 2017
[b]With effect from 14 August 2017

> "autonomous system", for a motor vehicle, means a system that enables the operation of the motor vehicle without the active physical control of, or monitoring by, a human operator;

In addition, the definition of motor vehicle was amended to take into account the definition of "*autonomous motor vehicles*". The definitions of "*motor vehicles*" before and after the 2017 amendments are provided in Table 1 by way of comparison.

The 2017 legislative amendments do not prescribe any rules regarding attribution of liability for AV-related accidents. Instead, the amendments implement a regulatory framework in order to minimise incidents during AV trials, and also impose certain duties on AV developers/operators, including the procurement of liability insurance.

Under the new *Road Traffic Act* rules, persons who wish to undertake the trial of AV Levels 3 to 5 must first obtain an approval from the Land Transport Authority.[12] In the course of obtaining such approval, the Land Transport Authority has put in place a "*robust regulatory framework*" that AV trials must meet in order to minimise the possibility of accidents, which include the following:[13]

> First, AVs must demonstrate basic roadworthiness and capabilities by passing a safety assessment, before they can even be trialled on the roads.
>
> Second, AVs developers must have robust accident mitigation plans for the trials, including having a safety driver who is trained to swiftly take control of the autonomous vehicle, whenever necessary. In answer to Mr Dennis Tan, these drivers have a Class 3 licence and they have no demerit points. This need for a safety driver can be waived after the AV developers have proven the competency of their AVs to LTA's full satisfaction of navigating our infrastructure and roads.
>
> Third, trials are allowed to start only on lightly used roads like the one we have at one-north. When the AVs are able to demonstrate higher competencies, they will be allowed to trial in more complex environments, including some of our public roads. The proposed RTA provisions will give LTA the flexibility to impose additional requirements or amend requirements, as the need arises, because in many of these trials, there may be unforeseen circumstances. Therefore, giving LTA the flexibility is the sensible thing to do.

[12]*Road Traffic (Autonomous Motor Vehicles) Rules 2017* (S 464/2017), rules 4, 5 & 7.

[13]Parliamentary Debates Singapore: Official Report, vol 94 at page 86 (7 February 2017) (Second Minister for Transport Mr Ng Chee Meng).

To summarize, the above legislative amendments were intended to set up a regulatory framework for AV trials and use for a period of 5 years in order to analyse the legal and practical issues faced with AV deployment. The 'regulatory sandbox' is expected to end by the year 2022,[14] at which time the Singapore government will then re-assess as to whether a further 'regulatory sandbox' period is required, or to move on to the next phase of AV deployment in Singapore.

2.2 Technical Reference 68 ("TR 68")

Subsequently in January 2019, the Singapore Standards Council released a 4-part industry standard known as Technical Reference 68 ("TR 68") to guide the local AV industry in the development and deployment of fully autonomous vehicles.[15] TR 68 is an industry-led effort comprising representatives from the AV industry, research institutions, institutes of higher learning and governing agencies aimed to promote the safe deployment of fully driverless vehicles in Singapore. In a joint press release by Enterprise Singapore, Land Transport Authority, Standards Development Organisation and Singapore Standards Council, TR 68 was touted as the "*first such national standard in world*" based on industry feedback.[16]

TR 68 covers 4 key areas of AV deployment: vehicle behaviour, vehicle functional safety, cybersecurity and data formats, and it is applicable to "*a vehicle designed to be operated exclusively by a level 4 or 5 automated driving system for all trips within its given operational design domain ("ODD") where the vehicle is designed to perform the sustained dynamic driving task ("DDT") and its fallback without the presence of a human operator on board*".[17]

It must be noted that TR 68 is only a provisional standard and therefore does not presently have any force of law or binding effect in Singapore. However, it is anticipated that as AV technology matures and more feedback is gathered from the AV industry, TR 68 will continue to be refined and form the backbone for further legislative development in the field of AV.

---

[14]Parliamentary Debates Singapore: Official Report, vol 94 at page 65 (7 February 2017) (Second Minister for Transport Mr Ng Chee Meng).

[15]Land Transport Authority (31 January 2019) Joint Media Release by the Land Transport Authority (LTA), Enterprise Singapore, Standards Development Organisation & Singapore Standards Council – Singapore Develops Provisional National Standards to Guide Development of Fully Autonomous Vehicles. Available via Government of Singapore. https://www.lta.gov.sg/content/ltagov/en/newsroom/2019/1/2/joint-media-release-by-the-land-transport-authority-lta-enterprise-singapore-standards-development-organisation-singapo.html. Accessed 2 May 2020.

[16]Gwyneth Teo (31 January 2019) Singapore's driverless vehicle ambitions reach next milestone with new national standards. Available via Channel News Asia, Mediacorp Pte Ltd. https://www.channelnewsasia.com/news/technology/singapore-driverless-vehicle-autonomous-national-standards-11190498. Accessed 2 May 2020.

[17]Technical Reference 68 (2019), Part 1, page 8, section 3.2.

Table 2 Summary of levels of driving automation in SAE J3016[a]

Level	Name	Narrative definition	DDT: *Sustained lateral and longitudinal vehicle motion control*	DDT: *OEDR*	*DDT fallback*	*ODD*
Driver performs part or all of the *DDT*						
0	No Driving Automation	The performance by the *driver* of the entire *DDT*, even when enhanced by *active safety systems*.	*Driver*	*Driver*	*Driver*	n/a
1	Driver Assistance	The *sustained* and *ODD*-specific execution by a *driving automation system* of either the *lateral* or the *longitudinal vehicle motion control* subtask of the DDT (but not both simultaneously) with the expectation that the *driver* performs the remainder of the *DDT*.	*Driver* and *System*	*Driver*	*Driver*	Limited
2	Partial Driving Automation	The *sustained* and *ODD*-specific execution by a *driving automation system* of both the *lateral* and *longitudinal vehicle motion control* subtasks of the *DDT* with the expectation that the *driver* completes the *OEDR* subtask and *supervises* the *driving automation system*.	*System*	*Driver*	*Driver*	Limited
ADS ("*System*") performs the entire *DDT* (while engaged)						
3	Conditional Driving Automation	The *sustained* and *ODD*-specific performance by an *ADS* of the entire DDT with the expectation that the *DDT fallback-ready user* is receptive to *ADS*-issued *requests to intervene*, as well as to *DDT performance-relevant system failures* in other *vehicle* systems, and will respond appropriately.	*System*	*System*	*Fallback-ready user (becomes the driver during fallback)*	Limited
4	High Driving Automation	The *sustained* and *ODD*-specific performance by an *ADS* of the entire *DDT* and *DDT fallback* without any expectation that a *user* will respond to a *request to intervene*.	*System*	*System*	*System*	Limited
5	Full Driving Automation	The *sustained* and unconditional (i.e., not *ODD*-specific) performance by an *ADS* of the entire *DDT* and *DDT fallback* without any expectation that a *user* will respond to a *request to intervene*.	*System*	*System*	*System*	Unlimited

[a]SAE International (June 2018) Surface Vehicle Recommended Practice J3016, page 19

3 Autonomous Driving in Singapore

Our article will explain the existing legal position in Singapore with regard to the different levels of driving automation as described in SAE J3016, ranging from Level 0 to Level 5.

A tabular description of these levels adopted from SAE J3016 is provided in Table 2 by way of summary.

The explanation of liability in respect of AVs from a Singapore legal perspective will be covered in 4 sections.

The first section will analyse the distinction between Level 0 to 2 cars on the one hand, and Levels 3 to 5 cars on the other hand, and the impact of this distinction on legal liabilities under Singapore law.

The second section will cover apportionment of liabilities for Level 0 to 2 cars. These are situations where the driver performs part or all of the dynamic driving task, or "**DDT**", the meaning of which will be explained in the next section.

The third and fourth sections will cover the apportionment of liabilities for Levels 3 to 5 AVs, which involve an interplay between the human driver and the automated driving system. As already mentioned above, the 2017 amendments to the Road Traffic Act apply to Levels 3 to 5 AVs, while the 2019-released TR 68 apply exclusively to Levels 4 to 5 AVs, so a closer examination will be conducted on these two recent developments.

3.1 Distinction Between Levels 0 to 2 Cars, and Levels 3 to 5 Cars

In a Level 0 car (No Driving Automation), the driver fully performs the DDT (see Table 1 above). Dynamic driving task, or "**DDT**", refers to:[18]

> All of the real-time operational and tactical functions required to operate a vehicle in on-road traffic, excluding the strategic functions such as trip scheduling and selection of destinations and waypoints, and including without limitation:
>
> Lateral vehicle motion control via steering (operational);
>
> Longitudinal vehicle motion control via acceleration and deceleration (operational);
>
> Monitoring the driving environment via object and event detection, recognition, classification, and response preparation (operational and tactical);
>
> Object and event response execution (operational and tactical);
>
> Maneuver planning (tactical); and
>
> Enhancing conspicuity via lighting, signaling and gesturing, etc. (tactical).

In the case of Level 1 cars (Driver Assistance) and Level 2 cars (Partial Driving Automation), the automated driving system may perform a part or all of the "*sustained longitudinal and/or lateral vehicle motion control subtasks of the DDT*" (*i.e.* control of the vehicle's motion). However, it is ultimately the driver who completes the rest of the DDT. The rest of the DDT would include (a) the monitoring of the driving environment and the execution of an appropriate response to objects and events by the driver, also known as "Object and Event Detection and Response" or "**OEDR**" for short, as well as (b) the response by the driver to achieve

[18] SAE International (June 2018) Surface Vehicle Recommended Practice J3016, section 3.13, pages 6 to 7.

a minimal risk condition (such as to prevent car crashes), also known as "Dynamic Driving Task Fallback" or "**DDT Fallback**" for short.

Underpinning these definitions for Levels 0 to 2 cars is the common element of human intervention during DDT. Put simply, the distinction between Levels 0 to 2 cars on the one hand, and Levels 3 to 5 cars on the other, is that the human driver performs part or all of the DDT while driving a Level 0, 1 or 2 car, whereas in a Level 3, 4 or 5 car it is the automated driving system which performs the entire DDT while engaged (see pages 21 to 22 of SAE J3016).

This distinction is critical in considering the apportionment of liability between man and machine. In the case of Levels 0 to 2 cars, the driver remains the responsible actor for driving the vehicle, and therefore any criminal or civil liability arising from the task of driving would fall on the driver, and not the driving system. Singapore's existing road traffic or penal laws currently do not explicitly recognise the responsibility of the autonomous driving system or the people behind that system, so, depending on the liability involved, the usual responsible actors in respect of Level 0 to 2 cars would be its driver, manufacturer or owner, as the case may be.

For Levels 3 to 5 cars, 3 other actors come into the frame:

(i) the programmer, who programs the automated driving system of an AV;
(ii) the developer, who creates or assembles an AV; and
(iii) the operator, who manages and operates one or several AVs.

The liabilities and responsibilities for these 3 additional actors will be covered in greater detail in the sections on liability for Levels 3 to 5 cars.

3.2 *Liability for Levels 0 to 2 Cars*

As mentioned, in the case of Level 0 to 2 cars, the automated driving system does not perform the entire DDT. Even in a Level 2 car, the automated driving system is only limited to lateral and longitude vehicle motion control which can be immediately disengaged upon the driver's request. A strong argument can thus be made that liability for Level 0 to 2 cars closely mirrors the current principles governing motor vehicles in Singapore.

3 liability addresses will therefore be considered below: (i) the driver, (ii) the manufacturer, and (iii) the owner.

3.2.1 Driver

3.2.1.1 Civil Liability

Where a road incident occurs involving a car without an automated driving system performing the DDT, the driver will normally be held responsible for any civil liabilities incurred in the course of driving, such as negligent driving resulting in personal injury. Negligence is a tortious claim in Singapore common law that requires the claimant to prove that (i) a duty of care existed on the part of the driver, (ii) that duty of care was breached, and (iii) the breach caused damage to the claimant and the damage was not too remote.

The Singapore courts have previously held that road users have a duty to each other to take care that they do not injure each other.[19] Thus, a breach of such a duty to take due care for the safety of others resulting in personal injury is claimable in the Singapore courts by way of damages. The extent of such damages, however, may be reduced if there was contributory negligence involved, *i.e.* the victim himself had contributed to its injury by its own actions or omissions (such as by failing to keep a proper lookout).

3.2.1.2 Criminal Liability

It is a criminal offence not to comply with the provisions of Singapore's *Road Traffic Act* or its accompanying rules thereunder.[20] This includes not complying with the provisions of the Highway Code,[21] which is a code of conduct for road users and the responsibilities of road users towards each other.

Singapore's *Penal Code*[22] also contains several criminal offences relating to road use, such as rash driving,[23] or, in cases where death occurs, causing death by rash or negligent act (*e.g.* road rage resulting in the victim's death).[24]

Since the driver remains responsible for the DDT in a Level 0 to 2 car, the driver is normally charged with the relevant criminal offence if an injury or death results from the use of a Level 0 to 2 car.

3.2.2 Manufacturer

3.2.2.1 Civil Liability

For Level 0 to 2 cars the key liability to be considered is the product liability for accidents that were caused by a malfunction of a component of the motor vehicle.

There is no single comprehensive legislation in Singapore that governs the manufacturing, import or supply of defective products. Product liability claims in Singapore are governed instead by common law.

---

[19] *Ng Swee Eng (administrator of the estate of Tan Chee Wee, deceased) v Ang Oh Chuan* [2002] 2 SLR(R) 321.

[20] *Road Traffic Act* (Cap 276, 2004 Rev Ed Sing), section 131.

[21] *Highway Code* (Cap 276, R 11, 1990 Rev Ed Sing).

[22] *Penal Code* (Cap 224, 2008 Rev Ed Sing).

[23] *Penal Code* (Cap 224, 2008 Rev Ed Sing), section 279.

[24] *Penal Code* (Cap 224, 2008 Rev Ed Sing), section 304A.

The common law rules on tortious liability for drivers are equally applicable to manufacturers. In order to establish a tort claim in negligence against a manufacturer of a vehicle, the claimant would similarly have to prove vis-à-vis the manufacturer that (i) a duty of care existed on the part of the manufacturer, (ii) that duty of care was breached, and (iii) the breach caused damage to the claimant and the damage was not too remote.

3.2.2.2 Criminal Liability

In terms of criminal liability, certain provisions in Singapore's criminal statutes are drafted widely such that legal entities (in addition to individuals) could also be found liable for contravening road traffic laws. For example, section 131 of the *Road Traffic Act* provides that any "*person*" who does not comply with the provisions of the *Road Traffic Act* or its rules without lawful excuse is guilty of a criminal offence:

> ***Offences and penalties***
> ***131.***—*(1) A person who without lawful excuse —*
> (a) *refuses or neglects to do anything he is by this Act or the rules required to do;*
> (b) *fails to comply with the requirements of any notice served on him under this Act or the rules; or*
> (c) *acts in contravention of any provision of this Act or the rules, shall be guilty of an offence.*

Under the *Interpretation Act*,[25] the word "*person*" is interpreted to not only mean an individual, but also includes "*any company or association or body of persons, corporate or unincorporate*" unless otherwise expressly provided. This statutory definition of "*person*" applies to every written law in Singapore, including the *Road Traffic Act*.

Thus, in a criminal provision such as section 131 of the *Road Traffic Act* where the liability addressee is described in the section as a "*person*", such a criminal provision may be extended to cover corporate or legal entities such as the manufacturer (unless the context or a purposive interpretation of the provision would otherwise suggest that it should only be restricted to individuals).

3.2.3 Owner

The owner of a motor vehicle is normally not liable for any accidents caused by the actions of the driver of a motor vehicle.

However, if the driver was driving a company vehicle in the course of his employment, then it is possible, in addition to suing the employee-driver, to also sue the employer (as owner of the company vehicle) for vicarious liability.

[25] *Interpretation Act* (Cap 1, 2002 Rev Ed Sing).

The key legal test for proving vicarious liability is whether there is a 'close connection' between the employee's tortious act and his employment so as to make it fair and just to hold his employer vicariously liable for the accident. This is known as the 'close connection' test. In applying the 'close connection' test, some of the factors that a Singapore court would consider in the context of road accidents is whether or not the employee-driver was driving the company vehicle within the course of his employment, and whether or not his conduct in question resulting in the accident was within the express and implied authority granted by the employer-company.

In the Singapore High Court case of *Chong Kwai Choy v Lee Fung Onn James alias James Lee Fung Oon and Another (Marliese Kathy French, Third Party)* [1998] SGHC 63, an employee-driver and his employer had been sued as co-defendants for a road accident that caused severe injuries to the plaintiff-victim, but the employee-driver was not served with the writ because he could not be found so the employer was effectively the only defendant from which the plaintiff-victim could obtain compensation if his case is successfully proven. The Singapore High Court held that the employee-driver was solely liable for the accident, but also found that the employer was vicariously liable for the injuries suffered by the plaintiff because the employee-driver was acting within the course of his employment when the accident occurred. The Singapore High Court came to this conclusion after considering the following factors: (a) the employee-driver was transporting his cousin at the material time, which the employer's principal officers knew about, (b) it was an employment perk for the employee-driver to provide transportation to his cousin using the company's vehicle, and (c) the accident happened during the normal working hours of the employee-driver. Thus, the Singapore High Court found that it was fair and just for the employer to be held fully liable for the damages claimed by the plaintiff due to his injuries. Vicarious liability is thus a form of strict liability, as an employer may be found liable even if the employer is or is not at fault.

Aside from vicarious liability arising from an employer-employee relationship as described above, any liability incurred by the driver would remain his own; an owner would generally not be liable for the accident in addition to the driver.

3.3 Liability for Levels 3 and 4 Cars

As alluded to earlier, the liabilities involved in the case of Levels 3 and 4 AVs are more complex because an automated driving system in such vehicles takes over a certain amount of control, including some of the DDT, from the driver.

There are presently no Singapore cases or legislation that deal with the liability of Level 3 or 4 cars. It is therefore not yet determined whether or not a programmer, developer or operator may be held responsible for any civil, criminal or product liability claims arising out of incidents involving the use of AVs. This is hardly surprising given that AV regulations are still in their infancy while the Singapore government continues to assess the role that AVs will play in Singapore.[26] Legal and insurance experts alike also observe that liability issues involving AV technology are presently unclear, with one observing that the introduction of driverless vehicles on Singapore roads would likely create "*one of the greatest Legal Conundrums in modern history when it comes to the Blame Game*".[27]

In the meantime however, the Land Transport Authority has implemented a regulatory framework for AV trials, which also deals with the liability for accidents. This includes the mandatory purchase of liability insurance before the trial and/or use of AVs can be carried out. The interim regulatory framework will remain in place until the completion of the 'regulatory sandbox' and clearer rules are developed for the attribution of liability in AV-related incidents.

The ensuing section will discuss the possible ways in which liability in Level 3 and 4 AV-related accidents may be attributed, taking into account the existing legal framework in Singapore, the legislative amendments made in 2017 as well as the recently released TR 68 in 2019.

3.3.1 Driver

In Level 3 to 4 cars, the DDT is completely performed by the automated driving system and not the driver. The key difference between a Level 3 and a Level 4 car lies in the DDT fallback: in a Level 3 car, the driver is expected to respond to an automation-issued request to intervene and take over the car in a timely manner in order to minimise the risk of an accident, whereas in a Level 4 car the DDT Fallback is fully performed by the automated driving system without any expectation of the human driver to respond to a request to intervene.

Thus, given the possibility of human intervention in a Level 3 car, it is still possible for the human driver to be liable for accidents if the driver does not respond to the system's request to intervene in a timely manner, or indeed if he fails to respond to the request at all. Alternatively, if the driver's handling of the DDT Fallback was rash, reckless or otherwise negligent and this resulted in the death of the victim, the driver could also be charged with a criminal offence, such as an offence under section 304A of the *Penal Code* for causing death by a rash or negligent act.

That is however not to say that a driver could escape liability altogether if the DDT Fallback is handled by the automated driving system, as is the case of a Level 4 car. Although there is no expectation for a human driver to respond to a request to

[26] Christabel Seah (5 February 2018) Autonomous vehicles: are we there yet? Available via Ince. https://www.incegd.com/en/knowledge-bank/autonomous-vehicles-are-we-there-yet. Accessed 29 May 2020.

[27] Zhaki Abdullah (13 December 2016) Driverless vehicles could change laws, insurance policies. Legal, insurance experts say liability issues involving such technology are still unclear. Available via The Straits Times, Singapore Press Holdings Ltd. https://www.straitstimes.com/singapore/transport/driverless-vehicles-could-change-laws-insurance-policies. Accessed 29 May 2020.

intervene in a Level 4 car, an *omission* by the driver to intervene (despite him knowing or having ought to have known that he should have done so) could still amount to negligence as this would be a breach of the legal duty imposed on road users to take care not to injure each other.[28]

In fact, the Singapore courts have even gone so far as to say that an *omission* could even constitute *rashness* if the driver, by his omission, took a conscious and deliberate risk without due regard to the consequences. In the Singapore Court of Appeal case of *Jali bin Mohd Yunos v Public Prosecutor* [2014] 4 SLR 1059, the Court of Appeal explained this legal concept (at paragraph 21) in the following manner:

> With respect, there is no reason in principle why an omission to do something (here, to check the traffic lights at a signalised traffic junction) could not constitute rashness. The concept of rashness connotes a heedlessness or indifference towards risk. In the present case (and, indeed, in all such similar cases), the risk is clear: If a driver drives into the junction when the traffic lights are not in his favour, there would be a clear possibility of an accident as well as other undesirable (even horrendous) consequences ensuing (including damage to property and/or personal injury or even (as was the case here) death). If a driver drives in such a manner, how can it be said that he had not taken a conscious and deliberate risk that the consequences just mentioned might in fact (and, in this case, did) result?

Thus, in a situation where the automated driving system of a Level 4 car malfunctions and drives against a traffic light, and the human driver in the AV could have intervened but chose not to (whether because he expected the automated driving system to perform the DDT Fallback or otherwise), he may not be able to escape liability simply by relying on the system. The Singapore courts will look at all the facts and circumstances surrounding the driver's omission to determine whether or not his omission could amount to negligence, or whether it could even be elevated to rashness if his omission was coupled with a heedlessness or indifference towards risk. If either situation is made out, the driver could be sued by the victim for damages and, additionally, he could also be prosecuted for a criminal offence if his negligence or recklessness endangered human life, or resulted in actual injuries or death.

3.3.2 Manufacturer

Where an AV-related accident is caused by a malfunction of or defect in an AV component or autonomous equipment, the Singapore courts will adopt the same common law elements on negligence to determine whether or not an AV manufacturer ought to be liable for the accident (*i.e.* whether or not a duty of care ought to be established on the part of an AV manufacturer, finding that that duty of care was breached, and finding that the breach caused damage which was not too remote). This is because the common law framework on negligence fashioned by the

[28] *Ng Swee Eng (administrator of the estate of Tan Chee Wee, deceased) v Ang Oh Chuan* [2002] 2 SLR(R) 321.

Singapore courts in the last decade or so is intended to be one of 'universal applicability'[29] to determine whether or not a defendant ought to be liable for negligence, having regard to the particular facts of each case.

However, an issue remains as to whether or not the Singapore courts would establish a duty of care on the part of an AV manufacturer, and what the standard of care ought to be if such a duty of care is established. One commentator has noted that there is currently a lack of global unanimity as to the standard of care that AV manufacturers ought to be subjected to, and no global industry standard for the manufacturing of AVs.[30] Thus, the specific duty and standard of care to be imposed on AV manufacturers within the common law framework for negligence, if any, still remains an open question to be determined by the Singapore courts.

3.3.3 Owner

In the case of AV owners, our views in the previous section continue to apply here. An owner would not be liable for a motor vehicle-related incident if no fault can be attributed to him. However, if an employer-employee relationship arises between owner and driver, and the vehicle in question it is a company-owned vehicle, then it is still possible for the Singapore courts to impose vicarious liability on the employer if it determines that it would be fair and just to do so.

3.3.4 Software Programmer

3.3.4.1 Civil Liability

There are presently no statutory rules or Singapore case law that explicitly considers whether or not AV software programmers may be subject to civil liability for AV-related accidents. The 2017 legislative amendments to the *Road Traffic Act*, the newly enacted the *Road Traffic (Autonomous Motor Vehicles) Rules 2017* (S 464/2017) and the 2019-released TR 68 do not specify any duties, obligations, responsibilities or liabilities on the part of an AV programmer. It is not certain whether or not the omitted reference to an AV programmer in these rules or industry standards is deliberate, nor has there been any active discussion by the Singapore government or industry players elsewhere as to whether or not software programmers ought to be liable for AV-related incidents.

Thus, if there are any AV-related incidents that may be attributed to negligence in programming, then the only recourse at present would be to the common law framework on negligence. Under the framework, the Singapore courts would have to determine whether or not an AV programmer ought to owe a duty of care to the

[29] Tan and Goh (2013), p. 510.

[30] Wong (2017/18).

victim and therefore ought to be liable to the victim for damages in a civil claim. The Singapore common law however evolves on an incremental basis, so until a suitable case arises before the courts for determination, these remain open questions that presently do not have a definitive answer.

Besides common law, there is also a possibility that the Singapore Parliament may introduce new legislation to regulate the responsibilities and liabilities of the various actors who are involved in the deployment of AVs. This may include the enactment of specific statutory duties, obligations and/or responsibilities for AV manufacturers, programmers, developers, operators and drivers. Given the close attention that the Singapore government has already given to the AV field (*e.g.* by way of the 'regulatory sandbox' and the setting up of CARTS), it is likely that the government will proactively regulate the attribution of liability in the AV context via the enactment of new statutory legislation, but such legislative action may take some time since the AV 'regulatory sandbox' is still well underway and is subject to review only in the year 2022. Currently, no plans have been announced by the Singapore government to release new legislation for AV-related liability issues as at the date of this article.

3.3.4.2 Criminal Liability

Similarly, there are presently no criminal cases dealing with the use of AVs and its attendant liabilities in Singapore as criminal laws relating to AVs are still relatively in its infant stages.

However, the broad statutory definitions of who may be held liable for road-related criminal offences under the *Road Traffic Act* and the *Penal Code* suggests that the Singapore courts do have leeway to consider whether a person other than the driver of an AV may be subject to prosecution as well.

For example, section 304A of the *Penal Code* states that "*Whoever causes the death of any person by doing any rash or negligent act not amounting to culpable homicide, shall be punished*...". On a broad interpretation, the word "*Whoever*" may suggest that the programmer of an AV fleet could also be held liable if he had been the person responsible for programming an AV in a way that resulted in a passer-by's death.

Similarly, under section 131 of the *Road Traffic Act*, "[a] *person who without lawful excuse*... *acts in contravention of any provision of* [the *Road Traffic Act* or its rules]" would be guilty of a criminal offence. Although the *Road Traffic Act* has a specific definition for "*driver*",[31] the offence under section 131 uses the word "*person*" rather than the "*driver*". Using the rules of statutory interpretation, one could therefore argue that, in specifically omitting of use of the word "*driver*" and

[31] A "*driver*" is defined in the *Road Traffic Act* (Cap 276, 2004 Rev Ed Sing) to mean:

(a) in relation to a trailer, includes a driver of a vehicle by which the trailer is drawn, and "drive" shall be construed accordingly;

using the broader term "*person*" instead, the Singapore Parliament intended for section 131 to also cover other persons than the driver.

Given these observations, an argument could be made that these road-related offences are potentially broad enough to even cover programmers behind the automated driving system or an operator company managing and operating an AV fleet.

Whether or not the Singapore courts would agree with such a broad interpretation of these existing criminal law provisions remains to be seen. One commentator has observed that it would "*stretch the realm of reality*" to apply existing *Penal Code* provisions to a scenario where an AV runs over a pedestrian.[32] There are also cogent arguments to be made that criminal laws should be interpreted very strictly because they invariably curtail fundamental liberties, statutory rights and common law freedoms, and therefore current criminal legislation should not be expanded so liberally to include AV programmers unless and until Parliament decides to enact specific legislation to deal with them.

Additionally, other legal issues must be considered before liability may be attributed to persons other than the AV driver. These include establishing the accused's intention and state of mind (for example, whether or not an AV programmer intended or would have intended to cause the accident via his software/ programme) and establishing a chain of causation between the AV programmer and the passer-by's death.

These competing legal issues therefore prompt for further development of existing criminal laws to cover the field of AV. The Singapore government may similarly take a proactive approach to clarify the scope of existing AV-related criminal laws or to introduce new criminal offences dealing with the manufacture, supply, deployment and/or operation of AVs. Such legislative plans, however, have yet to be announced as at the date of this article.

3.3.5 Developer and Operator

3.3.5.1 Definitions of Developer and Operator

The TR 68 defines "AV developer" and "AV operator" as follows:

- An AV developer is "*[a]n individual or legal entity responsible for creating an AV or its components*". Additionally, it goes on to say, "*In the case of developing*

(b) where a separate person acts as a steersman of a motor vehicle, includes that person as well as any other person engaged in the driving of the vehicle, and "drive" shall be construed accordingly; and

(c) includes the rider of a bicycle and any person propelling a tricycle or trishaw or pushing or pulling a cart, and "drive" shall be construed accordingly.

[32] Wong (2017/18).

a whole AV this may be done either by assembling the whole vehicle or adapting a vehicle platform manufactured by others."[33]
- An AV operator is "*[a]n individual or legal entity (business, government agency or other organisation) that manages and operates one or several regulated autonomous vehicles.*"[34]

We will first consider the possible civil and criminal liabilities of AV developers and operators, followed by the obligations of AV developers and operators relating to the trial and use of AVs under the 2017 amendments to the *Road Traffic Act*, and then finally we will cover the duties and obligations that TR 68 seeks to prescribe to AV developers and operators.

3.3.5.2 Civil Liability

Though the role of an AV developer appears to be wider in scope than that of an AV manufacturer, both AV developers and AV manufacturers perform a similar function in that they are responsible for "*creating*" an AV or its components. Thus, for the AV developer the same legal considerations apply as for the AV manufacturer. If an AV-related accident is caused by a malfunction of or defect in an AV component or autonomous equipment, any legal recourse against an AV developer would similarly be evaluated in accordance with the common law negligence framework to determine, whether or not the AV developer is liable for negligence.

As for an AV operator, we have mentioned above that there is presently no Singapore statutory legislation that deals with the civil liability of Level 3 or 4 cars (whether it is for software programmers, AV developers or in this case AV operators), nor have the Singapore courts discussed the extent of liability for AV operators under Singapore's common law negligence framework to date. The key legal question that remains to be answered, thus, is whether or not AV operators, like drivers, also have a legal duty to other road users to take care not to injure each other since they are also operating a motor vehicle on the road (in some cases, alongside a human driver). If such a duty is imposed on AV operators, then a further legal issue would arise in the area of contributory negligence: for instance, if a Level 3 or 4 AV is operated remotely by an AV operator with a human operator onboard, to what extent would the AV operator and the human operator be liable respectively for negligence? This would call for further consideration into the extent of the involvement of the AV operator (such as whether or not the AV operator can intervene in the DDT Fallback in a Level 4 AV), the possibility and the extent of human intervention in the AV, whether there was an omission on the part of the AV operator or the human driver, *etc*. The imposition of a duty of care unto AV operators, as well as the rules regarding contributory liability on the part of AV operators, remain open legal

[33] Technical Reference 68 (2019), Part 1, page 8, section 3.3.

[34] Technical Reference 68 (2019), Part 1, page 8, section 3.4.

questions to be determined either by the Singapore courts or by way of statutory legislation.

3.3.5.3 Criminal Liability

As with manufacturers, AV developers could also be found liable for contravention of criminal provisions under the *Road Traffic Act* or its accompanying rules insofar as the role of AV developers are akin to that of manufacturers. For example, car manufacturers (or car dealers) have a legal duty to notify the Land Transport Authority and motor vehicle owners if they become aware of any safety-related defects in any vehicle manufactured or sold by them, and a failure to comply with this duty of notification is a criminal offence.[35] If AV developers perform the functions of a manufacturer, then such a provision could be extended to cover AV developers as well.

As for AV operators, the question that arises is whether or not existing criminal provisions relating to road use, which are normally used for human drivers (such as rash driving under section 279 of the *Penal Code*), could be purposively interpreted to extend to AV operators. In analysing this legal issue, the Singapore courts will generally look at the legislative history of the provision in question, the parliamentary intention at the time when the provision was enacted, whether the Singapore Parliament has subsequently reformulated or re-stated the parliamentary intention behind the provision, and whether or not there are any other extrinsic historical documents or evidence that could shed light on whether or not the provision could be readily extended and applied to AV operators. AV technology, however, is a relatively new form of technology that rapidly gained prominence over the last couple of decades or so, so it is unlikely that the legislative history of Singapore's existing criminal provisions relating to road use would have contemplated AVs in mind during their enactment—in fact, the 2017 legislative amendments to the *Road Traffic Act* and the parliamentary debates surrounding those amendments were a first attempt by Parliament to establish a regulatory framework for AVs and AV technology, so it is unlikely that there is any historical parliamentary material prior to 2017 suggesting that existing criminal provisions could be readily extended to AV operators.

Thus, moving forward, Singapore legislators and courts would have to consider whether or not the criminal liability of AV operators could be shoehorned into existing criminal provisions (which could result in an artificial interpretation of these provisions), or whether new criminal provisions ought to be enacted to deal specifically with the liability of AV operators. The 2017 legislative amendments, for instance, introduce new criminal offences relating specifically to the trial and use of AVs on Singapore roads (which will be discussed in the next section), so a similar legislative effort could be undertaken after the Singapore government completes its

[35] *Road Traffic Act* (Cap 276, 2004 Rev Ed Sing), section 23A.

5-year review in 2022 to enact further criminal provisions relating to road use that are applicable specifically to AV operators.

3.3.5.4 Duties and Obligations Under the 2017 Legislative Amendments to the *Road Traffic Act*

The *Road Traffic (Autonomous Motor Vehicles) Rules 2017* would be applicable to AV developers and AV operators if they wish to conduct the trial or use of an AV in Singapore.

First, if an AV developer or AV operator wishes to (i) undertake any trial of automated vehicle technology or an autonomous motor vehicle on any road, or (ii) use an autonomous motor vehicle on a road, then the AV developer or AV operator (as the case may be) must make an application to the Land Transport Authority first for authorisation.[36]

The Land Transport Authority may also impose conditions in the course of granting an authorisation for the trial or use of an AV.[37] These may include conditions such as the geographical area to undertake the approved trial or use of the AV, requiring a qualified safety driver or qualified safety operator to monitor the operation of the AV, prohibiting the AV from carrying passengers, prohibiting the AV from being used for hire or reward, and/or stating the persons who may participate in the approved trial or use of the AV.

Once authorisation is given for the trial or use of an AV, the specified person must also comply with the following obligations and duties:

- Liability insurance must be obtained before an approved trial or approved special use (as the case may be) can commence, and such liability insurance must be in force throughout the duration of the approved trial/special use.[38] If liability insurance cannot be obtained, a security deposit of no less than S$1.5 million may be provided to the Land Transport Authority in lieu of liability insurance.[39]
- The specified person also has a duty to:
  - ensure maintenance of the AV;[40]
  - ensure every AV used in the approved trial/special use is installed with an operational data recorder;[41]
  - keep records of the approved trial/special use;[42]

[36] *Road Traffic (Autonomous Motor Vehicles) Rules 2017* (S 464/2017), rule 4.
[37] *Road Traffic (Autonomous Motor Vehicles) Rules 2017* (S 464/2017), rule 9.
[38] *Road Traffic (Autonomous Motor Vehicles) Rules 2017* (S 464/2017), rule 14.
[39] *Road Traffic (Autonomous Motor Vehicles) Rules 2017* (S 464/2017), rule 15.
[40] *Road Traffic (Autonomous Motor Vehicles) Rules 2017* (S 464/2017), rule 16.
[41] *Road Traffic (Autonomous Motor Vehicles) Rules 2017* (S 464/2017), rule 17.
[42] *Road Traffic (Autonomous Motor Vehicles) Rules 2017* (S 464/2017), rule 18.

- notify incidents and accidents[43] involving:

  the malfunction of the autonomous system of an AV during the approved trial/special use; and/or

  any accident involving the death, bodily injury or damage to property, of any person caused by, or arising out of, the use of an AV in the approved trial/special use; and

- comply with the Land Transport Authority's notice to produce any AV for testing.[44]

3.3.5.5 Duties and Obligations Under TR 68

Since no human operator will be onboard the AV, TR 68 makes the assumption that "*the AV developer/operator of the autonomous vehicle will maintain the responsibility for providing a safe vehicle deployment, including appropriate validation and verification of the AV and associated systems*" [emphasis in underline].[45] The set of guidelines laid out in TR 68 is therefore specifically applicable only to AV developers and AV operators, and not to any other liability addressees.

To be clear, TR 68 does not prescribe specific rules or guidance on the attribution of liability for AV-related incidents. What it does is that it lists a series of guiding principles relating to industry-accepted standards, responsibilities and obligations that AV developers and/or operators ought to hold themselves to in the context of vehicle functional safety and cybersecurity. If TR 68 is adopted as a starting point for the enactment of future legislation involving AV liability issues, then it is possible to prescribe that a breach of the duties, responsibilities and/or obligations listed in TR 68 could give rise to civil liability, criminal liability or regulatory penalties in the future.

The specific responsibilities and obligations of AV developers and/or operators set out in TR 68 are summarised below.

Vehicle functional safety (Part 2 of TR 68)

- Before deployment of the AV, the AV developer is responsible for providing sufficient proof that the AV is safe and fit for its intended purpose.[46]
- At the deployment level, the AV developer and AV operator apportion their own respective responsibilities when deploying the AV. The AV developer/operator is expected to provide a clear allocation of competency, roles, responsibilities and accountabilities of the AV developer/operator, the manufacturers and the operator

[43] *Road Traffic (Autonomous Motor Vehicles) Rules 2017* (S 464/2017), rule 19.

[44] *Road Traffic (Autonomous Motor Vehicles) Rules 2017* (S 464/2017), rule 20.

[45] Technical Reference 68 (2019), Part 1, page 11, section 4(e).

[46] Technical Reference 68 (2019), Part 2, page 12, section 5.2.1.

as well as the suppliers. This is to ensure clear traceability in the legal structure of the AV developer/operator.[47]

- The AV developer/operator is required to record and make available relevant documentation for driving test results and reviews and provide them upon request.[48]
- The AV developer/operator is expected to demonstrate that a change management plan is in place and is followed.[49]
- The AV developer/operator is required to describe aspects related to the functionality of the AV and demonstrate a thorough understanding of the ODD of the Singapore operations for the AV.[50]
- The AV developer/operator is required to have a hazard analysis and risk assessment method in place.[51]
- The AV developer/operator is required to demonstrate how hazardous events are mitigated in the design or in the ODD of the AV and develop a risk mitigation strategy.[52]
- The AV developer/operator is required to demonstrate that the validation practices are both appropriate and complete for the intended AV functionality, risk mitigation strategy and operational scope in Singapore.[53] Validation means the examination and tests used by the AV developer/operator to provide assurance that the goals are sufficient and have been achieved.
- The AV developer is responsible for managing system software updates in a transparent and verifiable manner.[54]
- The supplier of AV equipment shall define the rules, conditions and constraints relevant to safety in the application of the AV system or its equipment.[55]
- Where there are exceptional conditions, it would be necessary for a remote takeover to be possible without the involvement of a human on board the vehicle.[56]
- If artificial intelligence ("AI") is involved in an autonomous driving system feature, the AV developer/operator is expected to provide a functional architectural description of such AI, the requirements on the needed performance to be achieved by such a feature, and a safety argument for acceptable risk of safety-related AI based functionality.[57]

[47] Technical Reference 68 (2019), Part 2, page 13, section 5.2.5.

[48] Technical Reference 68 (2019), Part 2, page 13, section 5.2.6.

[49] Technical Reference 68 (2019), Part 2, pages 14 to 15, section 5.4.

[50] Technical Reference 68 (2019), Part 2, pages 17 to 18, section 6.2.

[51] Technical Reference 68 (2019), Part 2, pages 18 to 21, sections 6.3 and 6.4.

[52] Technical Reference 68 (2019), Part 2, pages 21 to 22, section 6.5.

[53] Technical Reference 68 (2019), Part 2, pages 22 to 23, section 6.6.

[54] Technical Reference 68 (2019), Part 2, pages 23 to 24, section 6.7.

[55] Technical Reference 68 (2019), Part 2, page 24, section 6.8.

[56] Technical Reference 68 (2019), Part 2, page 25, section 7.7.

[57] Technical Reference 68 (2019), Part 2, pages 25 to 26, section 8.

Cybersecurity (Part 3 of TR 68)

- Cybersecurity systems are a key part of AV deployment, and so the AV developer/operator is expected to identify and apply the standards and codes which provide the cybersecurity process best suited for their products and organisation.[58]
- Generally speaking, the key cybersecurity principles referenced and in-line with industry best practices are as follows: (1) Security-by-design; (2) Defence-in-depth; (3) Continuous operational management and oversight; and (4) Resiliency.[59]
- In addition, the AV developer/operator is responsible for providing a cybersecurity safeguarding framework, and the framework is to be provided to an independent third-party for a cybersecurity assessment of the AV.[60] The independent third-party assessment is intended to discover and feedback on cyber threats that may have been overlooked by the AV developer/operator, as well as to validate the preparedness of the AV against cyber threats. However, it is to be noted that any safety-related responsibility stipulated in TR 68 will always remain with the AV developer/operator and will not be transferred to the independent assessor.[61]

With the introduction of these responsibilities and obligations, TR 68 is intended to provide greater guidance to AV developers and operators alike so as to support the development of AV technology and deployment. It bears reiterating nevertheless that TR 68 is only a provisional standard that does not have any legal or binding effect, nor does it currently prescribe any civil liabilities, criminal punishments or regulatory penalties for anyone who fails to adhere to TR 68.

The working committee has also specified that TR 68 is presently only made available for application over a period of 3 years. At the end of the three years, the working committee will review TR 68, taking into account any feedback or other considerations, to further its development into a Singapore Standard if found suitable.[62]

3.4 *Liability for Level 5 Cars*

3.4.1 Driver

A Level 5 vehicle is defined as the "*sustained and unconditional (i.e. not ODD-specific) performance by an* (automated driving system) *of the entire DDT*

[58] Technical Reference 68 (2019), Part 3, pages 14 to 15, section 5.2.

[59] Technical Reference 68 (2019), Part 3, pages 13 to 14, section 5.1.

[60] Technical Reference 68 (2019), Part 3, pages 16 to 23, section 6.

[61] Technical Reference 68 (2019), Part 3, page 13, section 4(e).

[62] Technical Reference 68 (2019), Part 1, page 6.

and DDT fallback without any expectation that a user will respond to a request to intervene".[63] In other words, the automated driving system performs the DDT and DDT Fallback fully without intervention of a human driver. The operational design domain, or "**ODD**" for short, is also unlimited, meaning that the automated driving system is designed to function without limitations as to environmental, geographical and time-of-day restrictions, and/or the requisite presence or absence of traffic or roadway characteristics.[64]

Thus, the scope of liability that may be imposed on a human driver in a Level 5 car situation becomes even more limited, depending on the extent of human intervention inside the car (if human intervention is even possible in the first place). Although a Level 5 car is not commercially available at this point in time, it is conceivable that a fully-deployed Level 5 car in the future may not even require a human onboard at all (such as the transportation of goods using an AV that is fully controlled by an AV operator off-site).

If a human driver is onboard a Level 5 car, that does not mean that the driver may be completely excluded from liability. According to SAE J3016, a driver is still required to (i) verify the operational readiness of an AV equipped with an automated driving system, and (ii) determine whether to engage the automated driving system.[65] The driver only becomes a passenger while the automated driving system is engaged. Thus, if the human (as driver) is negligent in his duties to verify the operational readiness of an AV, for instance, then the possibility of him being sued or being prosecuted in the event of an accident occurring as a result of his negligence cannot be ruled out. The extent of his liability would vary, depending on his culpability and the apportionment of liability between the driver and the automated driving system.

3.4.2 Manufacturer

The same common law test for negligence in Singapore applies to the manufacturers of Level 5 cars. The laws of negligence in Singapore presently do not distinguish or apply different tests among the different levels of AVs.

3.4.3 Owner

Similar as in the case of Level 0 to 4 cars, an owner would generally not be liable for motor vehicle-related incidents (including AV-related incidents) unless a special

[63] SAE International (June 2018) Surface Vehicle Recommended Practice J3016, page 25, section 5.6.

[64] The definition of ODD may be found at SAE International (June 2018) Surface Vehicle Recommended Practice J3016, section 3.22, page 14.

[65] SAE International (June 2018) Surface Vehicle Recommended Practice J3016, page 23.

relationship arises between owner and driver so that vicarious liability may be imposed on an owner, such as in an employer-employee relationship.

3.4.4 Software Programmer, Developer and Operator

As a general point, our observations regarding the liabilities of AV software programmers, developers and operators mentioned in Levels 3 and 4 cars above apply equally in the context of Level 5 cars. Specifically, these would include the following.

- Our observations regarding the potential liabilities of AV software programmers, developers and operators for Levels 3 and 4 cars equally apply to Level 5 cars. (Please refer to the above sub-sections titled "*Software programmer*" and "*Developer and operator*" for Levels 3 and 4 cars.)
- The duties and liabilities imposed on AV developers and operators via the 2017 legislative amendments are also applicable to Level 5 cars. The Minister who moved the legislative amendments expressly noted that those laws cover Levels 3, 4 and 5 cars under the SAE J3016:

  > Clause 6 read with the amendments in clause 3(a), applies to motor vehicles with conditioner (sic) high and full automation, as defined in levels 3, 4 and 5 of the Society of Automotive Engineers (SAE) International J3016. The SAE levels of automation are commonly used in the motor industry to define and distinguish the various levels of automation. At levels 3, 4 and 5, it is the autonomous vehicle which monitors the driving environment, and the human driver is no longer required to actively monitor the vehicle.[66]

  (Please therefore refer the above sub-section titled "*Liabilities for trial and use of AVs in Singapore by developers and operators under the 2017 legislative amendments to the Road Traffic Act*" for Levels 3 and 4 cars.)
- Additionally, the responsibilities and liabilities for AV developers and operators mentioned in TR 68 that we have set out above would also apply to Level 5 cars. This is because the scope of TR 68 specifically covers Levels 4 and 5 cars:

  > For the purposes of this TR, an AV is defined as a vehicle designed to be operated exclusively by a level 4 or 5 automated driving system...[67]

  (Please therefore refer to the above sub-section titled "*Further guidance under TR 68 on the responsibilities and obligations of developers and operators*" for Levels 3 and 4 cars.)

There are no other specific statutory provisions or common law rules that are only applicable to Level 5 cars to the exclusion of all the other levels of AVs.

---

[66] Parliamentary Debates Singapore: Official Report, vol 94 at page 87 (7 February 2017) (Second Minister for Transport Mr Ng Chee Meng).

[67] Technical Reference 68 (2019), Part 1, page 8, section 3.2.

4 Conclusion

At the turn of this decade, the world races towards the widespread implementation of driverless vehicles, including Singapore. A key issue will be whether Singapore's legal regulations will be able to keep pace with the rapid technological developments in the field of AV. In the authors' view, regulation of AVs in Singapore is still in its infant stages. There is presently no omnibus legislation in Singapore dealing with AV-related liability issues, nor have the Singapore courts to date encountered specific legal issues relating to civil, criminal or product liability in the context of AV or considered the liability of additional actors that are introduced therefrom (specifically, AV programmers, developers and operators). The government and industry players are also presently still assessing the role that AVs will realistically play in Singapore, and this is currently being done via the 'regulatory sandbox' which is expected to last at least until 2022.

Despite its present stage of development, the authors nevertheless note that Singapore appears to be well-poised to spearhead rapid progress in the AV sector in time to come, having been ranked first in Asia and second in the world by KPMG in its readiness to adopt AV due to the strength of Singapore's government, infrastructure and consumer acceptance of AVs.[68] Singapore also led the global AV arena by formulating the first-of-its-kind industrial standard for AVs in 2019 via TR 68.

The regulatory road towards full AV deployment no doubt would involve many thorny legal issues. However, with its strong industrial expertise, regulatory infrastructure and government support, Singapore seems to be amply prepared to confront the legal challenges that lie ahead in its quest towards its 'Smart Nation' vision of a driverless vehicle mobility system.

Cases

Road Traffic (Autonomous Motor Vehicles) Rules 2017 (S 464/2017)
Jali bin Mohd Yunos v Public Prosecutor [2014] 4 SLR 1059
Penal Code (Cap 224, 2008 Rev Ed Sing)
Road Traffic Act (Cap 276, 2004 Rev Ed Sing)
Ng Swee Eng (administrator of the estate of Tan Chee Wee, deceased) v Ang Oh Chuan [2002] 2 SLR(R) 321
Interpretation Act (Cap 1, 2002 Rev Ed Sing)
Highway Code (Cap 276, R 11, 1990 Rev Ed Sing)

[68] KPMG International Cooperative (March 2019) 2019 Autonomous Vehicles Readiness Index. Available at KPMG Services Pte Ltd. https://home.kpmg/sg/en/home/insights/2019/02/2019-autonomous-vehicles-readiness-index.html. Accessed 2 May 2020.

Chong Kwai Choy v Lee Fung Onn James alias James Lee Fung Oon and Another (Marliese Kathy French, Third Party) [1998] SGHC 63

References

Christabel Seah (5 February 2018) Autonomous vehicles: are we there yet? Available via Ince. https://www.incegd.com/en/knowledge-bank/autonomous-vehicles-are-we-there-yet. Accessed 29 May 2020

Gwyneth Teo (31 January 2019) Singapore's driverless vehicle ambitions reach next milestone with new national standards. Available via Channel News Asia, Mediacorp Pte Ltd. https://www.channelnewsasia.com/news/technology/singapore-driverless-vehicle-autonomous-national-standards-11190498. Accessed 2 May 2020

Kevin Kwang (22 November 2017) Self-driving buses, shuttles to be tested in 3 towns from 2022. Available via Channel News Asia, Mediacorp Pte Ltd. https://www.channelnewsasia.com/news/technology/self-driving-buses-shuttles-test-punggol-jurong-tengah-9427850. Accessed 2 May 2020

KPMG International Cooperative (March 2019) 2019 Autonomous Vehicles Readiness Index. Available at KPMG Services Pte Ltd. https://home.kpmg/sg/en/home/insights/2019/02/2019-autonomous-vehicles-readiness-index.html. Accessed 2 May 2020

Land Transport Authority (31 January 2019) Joint Media Release by the Land Transport Authority (LTA), Enterprise Singapore, Standards Development Organisation & Singapore Standards Council – Singapore Develops Provisional National Standards to Guide Development of Fully Autonomous Vehicles. Available via Government of Singapore. https://www.lta.gov.sg/content/ltagov/en/newsroom/2019/1/2/joint-media-release-by-the-land-transport-authority-lta-enterprise-singapore-standards-development-organisation-singapo.html. Accessed 2 May 2020

Melissa Lin (19 October 2016) Driverless car his lorry during test drive. Available via The Straits Times, Singapore Press Holdings Ltd. https://www.straitstimes.com/singapore/driverless-car-hits-lorry-during-test-drive. Accessed 2 May 2020

Ministry of Transport (27 August 2014) Speech by Mrs Josephine Teo at Autonomous Transport: Paving the Road for Future Mobility Event on 27 August 2014. Available via Government of Singapore. https://www.mot.gov.sg/news-centre/news/Detail/Speech-by-Mrs-Josephine-Teo-at-Autonomous-Transport-Paving-the-Road-for-Future-Mobility-Event-on-27-August-2014/. Accessed 2 May 2020

Parliamentary Debates Singapore: Official Report, vol 94 (7 February 2017) (Second Minister for Transport Mr Ng Chee Meng)

SAE International (June 2018) Surface Vehicle Recommended Practice J3016.

Tan D, Goh YH (2013) The Promise of Universality: The Spandeck Formulation Half a Decade On. SAcLJ 510

Technical Reference 68 (2019)

Toh Ting Wei (25 October 2019) Western Singapore set to become test bed for self-driving vehicles. Available via The Straits Times, Singapore Press Holdings Ltd. https://www.straitstimes.com/singapore/transport/western-spore-set-to-become-test-bed-for-self-driving-vehicles. Accessed 2 May 2020

Wong K (2017/18) Considerations in an Autonomous Era. Juris Illuminae Vol 9

Zhaki Abdullah (13 December 2016a) Driverless vehicles could change laws, insurance policies. Legal, insurance experts say liability issues involving such technology are still unclear. Available via The Straits Times, Singapore Press Holdings Ltd. https://www.straitstimes.com/singapore/transport/driverless-vehicles-could-change-laws-insurance-policies. Accessed 29 May 2020

Zhaki Abdullah (25 November 2016b) 'Software anomalies' to blame for driverless car accident. Available via The Straits Times, Singapore Press Holdings Ltd. https://www.straitstimes.com/singapore/software-anomalies-to-blame-for-driverless-car-accident. Accessed 2 May 2020

Civil Liability Regime on Autonomous-Driving Vehicle Accident in Korea

Jeung-Jun Park and Jun-Kyu Ahn

Abstract With the recent legal discussions on autonomous-driving vehicles that are becoming more active, the amendment to the Motor Vehicle Accident Compensation Guarantee Act Compensation Act was promulgated on April 7, 2020 and took effect on October 8, 2020, governing the legal liability arising from traffic accident involving the SAE Level 3 AVs.

Based thereon, we will discuss on which concerned party is liable in case of a relevant vehicle accident, the underlying laws and regulations, relevant requirements, and the difference between the liabilities at each stage according to the SAE Level 0-5, mainly focusing on the differences from conventional legal liabilities in the case of SAE Level 3-5 AVs.

Thereby, we will briefly explain the concept of AVs under Korean law and the scenarios that assume legal liability, dividing them into (1) SAE Level 0-2 to deal with casualty/property damages by offenders and victims, (2) SAE Level 3 to understand where the liability for casualty damages of victims lies under the amended Compensation Act, and (3) SAE Level 4-5 to examine whether the liability under the amended Compensation Act will be applied as it is by referring to the current academic discussions.

1 Introduction: Concept of Autonomous Driving Under Korean Law

In August 2015, the Motor Vehicle Management Act (the "**MVM Act**") was amended to include autonomous-driving vehicle(s) (the "**AV(s)**") under the definition of motor vehicle. AVs are classified as a type of motor vehicle and are distinguished from regular motor vehicles (Article 2 No. 1-3 of the MVM Act). Furthermore, the Act on Promotion and Support of Commercialization of

J.-J. Park (✉) · J.-K. Ahn
Attorney-at-Law, Kim & Chang, Seoul, Korea
e-mail: jeungjun.park@KimChang.com

H. Steege et al. (eds.), *Autonomous Vehicles and Civil Liability in a Global Perspective*, Data Science, Machine Intelligence, and Law 3,
https://doi.org/10.1007/978-3-031-41992-8_9

Autonomous Vehicles (the "**AV Act**") was enacted on April 30, 2019, coming into effect on May 1, 2020. Both said laws define AV as "*a motor vehicle which can self-operate without any operation by its driver or passengers*".

According to Article 2(2) of the AV Act, AVs are divided into "*partially-autonomous vehicles*" and "*fully-autonomous vehicles*". A partially-autonomous vehicle refers to an "*autonomous vehicle that requires intervention by a driver or a passenger as the vehicle cannot be operated with an autonomous driving system alone or requires continuous attention from its driver,*" which is deemed to be a SAE Level 3 AV. On the other hand, a fully-autonomous vehicle refers to an "*autonomous vehicle that is operated driverless or does not require intervention by a driver or a passenger as the vehicle can be operated on an autonomous driving system alone,*" which is deemed to be a SAE Level 4 or 5 AV.

Under Korean law, SAE Level 0-2 AVs may be interpreted as regular motor vehicles. Pursuant to Article 48(1) of the Road Traffic Act (the "**RT Act**"), a motor vehicle driver is obligated to properly operate the steering system, brakes and other related devices while driving. Consequently, as for legal liability in traffic accidents, discussions based on laws and court precedent on regular vehicles will apply to SAE Level 0-2 vehicles, while new discussions are still emerging regarding SAE Level 3 or higher AVs.

2 Proposed Scenario

For the sake of our discussion, liability for damages caused by a traffic accident will be explained by proposing the following scenario:

> A vehicle owner driving his/her car causes a traffic accident due to negligent driving. His/her car crashes into another car and results in casualty/property damage to the other driver.
>
> The fault between the vehicle owner and the other driver is 100:0.
>
> This traffic accident results in casualty/property damages to the vehicle owner.
>
> This traffic accident is attributable, to a certain degree, to an already existing defect in the car.

3 Liability in Traffic Accidents Involving SAE Level 0-2 AVs

Depending on the type of damage to the victim and the offender, the following liability issues may exist.

3.1 *Compensation for Casualty Damage to Victim*

3.1.1 Claim for Damages Against the Vehicle Owner/Insurance Company Based on an Operator's Liability Under the Motor Vehicle Accident Compensation Guarantee Act

A traffic accident is a matter of an unlawful act caused mainly by negligence. In Korea, regulations on civil liability for damages attributable to an unlawful act may be applicable. However, civil legal provisions on tort are not sufficient to regulate liability for damages in traffic accidents due to difficulty in proving negligence, etc. For that reason, the Motor Vehicle Accident Compensation Guarantee Act (the "**Compensation Act**") was enacted, prescribing an operator's liability without fault. In other words, under the Compensation Act, a vehicle owner bears primary liability for damages (Article 3) and must purchase a liability insurance policy that covers casualty damage to a third party (Article 5(1)), which allows the victim to seek the prompt remedies.

A vehicle owner refers to a person who owns a motor vehicle or a person entitled to use a motor vehicle, who operates the motor vehicle for personal interest (Article 2 No. 3). As for recognizing such an operator, precedent court decisions rendered rulings based on the concepts of "*interests in operation*" and "*control in operation*." Interests in operation indicate that interests generated from operating a motor vehicle belong to an operator. The interests include economic interests gained directly and indirectly from said operation of motor vehicle as well as psychological benefits, such as a sense of satisfaction (*see* the Supreme Court Decision No. 86*Daka*1807 rendered on January 20, 1987; Supreme Court Decision No. 86*Daka*556 rendered on December 23, 1986; Supreme Court Decision No. 2002*Da*47181 rendered on November 26, 2002, etc.).In the case of SAE Level 0-2 AV, vehicle owners are recognized as operators. As interests in operation and control in operation are not recognized for motor vehicle manufacturers, they do not bear any liability as an operator.

For a vehicle owner to be exempt from liability, it needs to be proven that (i) he/she was not negligent in operating the vehicle; and that (ii) a victim or a third party, other than the person operating the motor vehicle and the driver thereof, committed an intentional or negligent act and that there were no structural defects or functional disorders in the motor vehicle (Article 3[1]). Therefore, even if there was a

[1] **Article 3 (Liability for Damage Caused by Motor Vehicles)**

Where any person who operates a motor vehicle for personal use has killed or injured another person by such operation, he/she shall be liable to compensate the damages therefrom: Provided, that the same shall not apply to any of the following cases.

3.1 In cases of the death or injury of any person other than a passenger, when it has been proven that the person who has operated the motor vehicle and the driver thereof were not negligent in its operation, that a victim or a third party, other than the person who has operated the motor vehicle and the driver thereof, commits an intentional or negligent act, and that there were no structural defects or functional disorders in the motor vehicle;

technical defect in a motor vehicle, its owner cannot be exempted from liability. This is a type of "*liability without fault.*" The law is prescribed for a vehicle owner to be liable in principle, while the vehicle owner can be exempted from liability only in exceptional cases.[2] Hence, this differs significantly from tort liability, which is established only where intent or negligence is proven.[3]

Due to the ease in making a claim for damages, for most traffic accidents in Korea, the insurance companies that sell liability insurance policies to vehicle owners primarily bear no-fault liability, and in cases where vehicle owners do not want to use insurance policies to cover damages, the vehicle owners compensate damages out of their own pocket.

3.1.2 Claim for Damages Against the Driver Based on Unlawful Acts Under Civil Law

In cases where the vehicle owner and the driver do not match (e.g., in the case where a vehicle was temporarily borrowed and driven by a friend), the actual driver may not be covered by liability insurance. In the event that such a driver committed an intentional or negligent act in a traffic accident, the victim can prove the driver's intention or negligence and claim for damages caused by an unlawful act against the victim pursuant to Article 750 of the Civil Code. However, even in this case, the vehicle owner still bears no-fault liability as an operator. Therefore, in practice, the victim is compensated only for damages attributable to the operator's liability and may find it difficult to claim compensation from the driver for damages caused by an unlawful act. A vehicle owner or an insurance company that compensated the victim for such damages may subsequently prove the driver's negligence and exercise the right to indemnity.

3.1.3 Claim for Damages Against the Vehicle Manufacturer Based on Product Liability

In cases where a traffic accident is attributable not only to a driver's negligence but also a vehicle defect, a vehicle manufacturer may also bear liability. Product liability, which is a special tort liability under the Product Liability Act (the "**PLA**"), will pose an issue (Article 3(1)[4]). In theory, a vehicle manufacturer is jointly liable for damages with the driver and/or vehicle owner. However, even if a consumer's

---

3.2 . In cases of the death or injury of a passenger, when the passenger has died or been injured due to his/her intentional act or suicidal act.

[2]Lee and Hwang (2018), p. 11.

[3]Operator's liability under the Compensation Act is recognized for an "*accident involving casualty damages*," civil liability for damages attributable to tort is applied to an "*accident involving property damages*."

[4]**Article 3 (Product Liability)**

burden of proof is lessened for product liability compared to regular tort liability, the litigant still has the burden of proof. In practice, under the Compensation Act, a victim's claims for damages are basically only against a vehicle owner or the insurance company. As for the compensating vehicle owner or insurance company, in case they believe that the traffic accident is attributable to a defect in the vehicle, the vehicle owner or insurance company makes claims for damages against vehicle manufacturers under the civil principle of the subrogation based on the PLA (Article 682(1) of the Commercial Code) or seek to subrogation right for the exempted portion if they are joint tort-feasors.

Requirements for liability for damages under the PLA are further explained in Section C. 1) below.

3.2 *Compensation for Property Damage to Victim*

3.2.1 Claim for Damages Against Vehicle Owner

Liability for damages under the Compensation Act does not apply to damage to property. Therefore, as for property damage, a victim needs to claim compensation from an actual driver for damage caused by unlawful acts.

A person that caused damage to other person's property by a traffic accident is subject to criminal sanction (Article 151 of the RT Act). However, if said person has an insurance policy covering property damages, he/she is exempted from criminal sanction (Article 4(1) of the Special Act on the Settlement of Traffic Accidents). Given the foregoing, in practice, nearly all vehicle owners have comprehensive insurance policies that cover both casualty and property damages to other persons. Accordingly, in practice, insurance companies primarily compensate the victim for property damage.

3.2.2 Claim for Damages Against Driver Based on Unlawful Acts

In cases where a full coverage insurance policy a vehicle owner purchased does not cover the actual driver, a victim needs to claim compensation from the actual driver for damages caused by his/her unlawful act.

6.1 A manufacturer shall compensate for damages to the life, body or property of a person caused by a defect of a product (excluding damages inflicted only to the relevant product).

(Omitted)

3.2.3 Claim for Damages Against Vehicle Manufacturer Based on Product Liability

As stated above, in case an accident is attributable to not only a driver's negligence in operation but also a vehicle defect, a vehicle manufacturer may also bear product liability under the PLA. However, in practice, since vehicle owners usually have full coverage insurance policies, victims rarely claim for damages against vehicle manufacturers. In fact, it is the vehicle owner or insurance company that exercises the right to indemnity against vehicle manufacturers on the grounds of a vehicle defect.

3.3 Compensation for Casualty Damages Inflicted on an Offender Himself

3.3.1 Claim for Damages Against Vehicle Manufacturer Based on Product Liability

In cases where an accident is attributable to not only a driver's negligence in operation but also a vehicle defect, a driver or vehicle owner may claim damages against a vehicle manufacturer for product liability under the PLA. However, the PLA is established only in case a vehicle defect resulted in "*extended damage*" (e.g., in case a vehicle defect caused a fire and other property was burned or where a defect caused an accident and resulted in bodily injury). Therefore, a vehicle owner can claim only for casualty damages to themselves and casualty/property damages to other persons. Property damage to the owner's vehicle need to be claimed under other grounds.

Under the PLA, "*defects*" are divided into defects in (i) manufacturing, (ii) design, and (iii) indication. "*Defect in manufacturing*" means the lack of safety caused by manufacturing or processing of any product not in conformity with the originally intended design, regardless of whether the manufacturer manufactured or processed the product with the requisite duty of care and diligence. "*Defect in design*" means the lack of safety caused by failure of a manufacturer to adopt a reasonable alternative design in a situation where any damage or risk caused by the product would otherwise have been reduced or prevented if an alternative design had been adopted. "*Defect in indication*" refers to cases where damage or risks caused by a product could have been reduced or avoided if a manufacturer had given reasonable explanations, instructions, warnings or other indications on the product (Article 2 of the PLA).

Meanwhile, it may be difficult for a customer to ascertain the existence of a defect. The PLA, as amended on April 18, 2017, coming into effect on April 19, 2018, introduced a legal provision that if the following conditions were proven, it is presumed *per se* that a vehicle had a defect and that such defect caused damage: (i) damage was caused while the vehicle was being used normally; (ii) such damage is attributable to a cause practically controllable by the manufacturer; and (iii) the damage would not ordinarily be caused if it were not for the relevant defect of the vehicle (Article 3-2 of the PLA).

Since the PLA does not mandate manufacturers to be insured, quick relief may be difficult.[5]

3.4 Compensation for Property Damages Inflicted on an Offender Himself

3.4.1 Claim for Damages Against Vehicle Seller Based on Warranty Liability Against Defect/Liability for Non-Performance of Obligation

For the distribution of imported vehicles in Korea, importers and sellers are usually separated as independent entities. A person that purchased a vehicle can claim for damages against the motor vehicle seller based on contractual liability.

Liability for warranty against defect is based on Article 580 of the Civil Code. It refers to a seller's liability against a buyer in case a defect is found, and thus a buyer can claim compensation from a seller for damages caused by a defect. Regarding the concept of such defects, court precedents ruled that a defect is found when the subject matter of a sale lacks properties or performance commonly expected in transactions or lacks properties expected from or guaranteed by the parties.[6] As for said right, it must be exercised within six (6) months from the time when the buyer was first aware of the defect (Article 582 of the Civil Code).

Next, liability for non-performance of obligation may be claimed for failing to supply a vehicle without a defect (Article 390 of the Civil Code). If a vehicle purchase agreement entered into by and between a vehicle seller[7] and a buyer has a separate clause on claim for damages, the relevant provision will prevail. In the case of such contractual liability, the liability remains ten (10) years from the date such a claim could have first been exercised. (Article 162(1) of the Civil Code).

In theory, claim for damages attributable to an unlawful act may be indeed be possible against a seller. However, lower court precedents have denied the establishment of tort liability, except for warranty liability against defect, in case a product with a defect has been supplied, on grounds that "*incomplete performance*" cannot be deemed an intentional or negligent unlawful act. Therefore, in practice, it would not be easy to hold a seller liable for tort.[8]

---

[5] Lee and Hwang (2018), p. 13.

[6] Supreme Court No. 98*Da*18506 rendered on January 18, 2000.

[7] However, if the vehicle is an imported vehicle, since there is no direct contractual relationship between the manufacturer of the vehicle and the consumer, it will be difficult to hold the manufacturer liable. In practice, a vehicle dealer bears liability based on a warranty.

[8] *Seoul* Central District Court No. 2008*Gahap*57857 rendered on May 8, 2009; *Gwangju* District Court No. 2010*Gadan*1048 rendered on September 7, 2010.

In such a case, whether an offender can claim for damages to his own vehicle resulting from a traffic accident should be decided by considering the causal relationship between a defect, a traffic accident and the degree of a driver's negligence in operation.

3.4.2 Claim for Damages Against Vehicle Manufacturer Based on Unlawful Acts

As explained above, the PLA is applied to "*extended damage*" only. In order to claim for property damages inflicted on an offender's own vehicle against a vehicle manufacturer, the claim should be based on Article 750 of the Civil Code. In order to claim for damages attributable to an unlawful act, requirements such as an "*unlawful act,*" "*intention or negligence,*" "*damages,*" and "*causation*" need to be proven, which would be challenging.

3.4.3 Vehicle Manufacturer's Liability Based on Warranty

Pursuant to Article 32-2(1) of the MVM Act and Article 49-3(1) of the Ordinance of the Ministry of Land, Infrastructure, and Transport (the "**MOLIT**"), motor vehicle manufacturers are obligated to provide free repair services on vehicles they have sold for a period of three (3) years from the date of sale (and mileage of up to 60,000 km) for motors and power transfer devices, and for a period of two (2) years from the date of sale (and mileage of up to 40,000 km) for other devices. In practice, motor vehicle manufacturers offer free repair warranties on vehicles they have sold for the aforementioned period or a longer period. Accordingly, motor vehicle manufacturers indeed have liability on vehicle defects based on their warranties, but it is uncertain whether such warranty liability also includes claim for damages resulting from an accident.

4 Liability in Traffic Accidents Involving SAE Level 3 AVs

4.1 Differences from Regular Motor Vehicles

SAE Level 3 AVs feature autonomous driving enabled by an autonomous driving system in a certain section without any operation by a driver. As such, in case an accident occurs during autonomous driving enabled by an autonomous driving system, since a driver was not involved in operating the vehicle, it is more likely that damages are attributable to a vehicle defect. In this context, there has been much discussion on who should be liable.

Meanwhile, academics expected that since the existence of a manufacturer's defect and its severity may vary depending on whether an accident occurred during a full autonomous driving mode or in a situation where control had to be taken over by a driver, a manufacturer's product liability attributable to a motor vehicle defect would be more critical in determining liability, unlike the earlier liability structure that was based on the premise of "*control by a human driver*."[9]

However, from a perspective of protecting victims, there has been discussion on whether motor vehicle manufacturers should be deemed operators within the meaning of the Compensation Act and whether they should bear no-faulty liability accordingly.

When test driving AVs, the manufacturers are the vehicle owners. Therefore, vehicle manufacturers are required to purchase a liability insurance policy and bear a driver's liability under the Compensation Act.[10] The AV Act requires any person conducting research and pilot operation in a pilot operation district to purchase an insurance policy to cover damages in case an accident involving AVs occurs (Article 19). Article 4 of the Regulation on Safe Operation Requirements and Test Operation of AVs also sets forth that any person seeking to obtain a temporary operation permit on an AV for test and research purposes must be covered by an insurance policy prescribed in Articles 5(1) and 5(2) of the Compensation Act.[11] Therefore, according to these legal provisions, in case a traffic accident occurs, its operator primarily bears liability set forth in Article 3 of the Compensation Act.

As the commercialization of AVs became more feasible, amendments to the Compensation Act were proposed a few times at the National Assembly. Finally, the amended Compensation Act was promulgated on April 7, 2020, and took effect on October 8, 2020.

4.2 Liability for Casualty Damages to Victim Under the Amended Compensation Act

Under the amended Compensation Act,[12] just like in the past, the driver's liability of a vehicle owner is maintained the same. Therefore, a vehicle owner or an insurance

---

[9]Lee and Hwang (2016), p. 93.

[10]Park (2019), pp. 179–207: 190–191.

[11]**Article 4 (Liability for Damages and Purchase of Insurance)**

13.1 A vehicle owner or a person entitled to use a vehicle and seeking to obtain a temporary operation permit on an AV for a test and research purpose (an "applicant for a temporary AV operation permit") shall be liable for damages prescribed in Article 3 of the Compensation Act for any traffic accidents damages resulting from operating the relevant vehicle.

13.2 An applicant for a temporary AV operation permit must purchase an insurance policy prescribed in Articles 5(1) and 5(2) of the Compensation Act in order to appropriately cover damages resulting from a traffic accident.

[12]However, there are comments that it is not clear which level of AVs is referred by Article 2-1(3) of the MVM Act that defines AVs and whether SAE Level 3 AVs are included [Lee (2019), pp. 1–26: 15].

company would primarily bear no-fault liability for damage resulting from an accident occurred while an AV is in operation. In order for them to be exempted from the liability, they must fulfill every criteria for exemption set forth in Article 3 of the Compensation Act. A vehicle defect alone would not be enough for exemption.

Instead, the Compensation Act states that in case "*a person has died or been injured*" or "*property has been destroyed or damaged*" due to an accident caused by a defect in an AV and an insurance company, etc. has already paid out insurance money, the insurance company, etc. may exercise the right to indemnity against a person liable for the damages (Article 29-2 of the amended Compensation Act). Under the previous Compensation Act, if an insurance company, etc. paid insurance money, the insurance company was allowed to exercise the right to indemnity against a person liable for damages only in the following cases: a driver did not have a driver's license authorized under the RT Act (Article 29(1) No. 1); a driver operated a vehicle under the influence (Article 29(1) No. 2); or a driver failed to take appropriate measure when an accident occurred (Article 29(1) No. 3). However, the amended Compensation Act sets forth explicit grounds for an insurance company to exercise the right to indemnity against vehicle manufacturers, etc.

As explained above, in the past, a vehicle owner or an insurance company claimed for product liability against a manufacturer on behalf of the insured through a subrogation clause because there were no explicit grounds to claim for the right to indemnity. However, the amended Compensation Act (i) introduced legal grounds to facilitate the direct exercise of the right to indemnity, (ii) required vehicle manufacturers to install devices recording autonomous driving information on AVs, and (iii) mandated the creation of the Autonomous Vehicle Accident Investigation Committee (the "**Investigation Committee**") in order to identify the cause of an accident by collecting and analyzing recording devices (Articles 39-14 through 39-17 of the Compensation Act). Going forward, in case of an AV accident, investigations by the Investigation Committee may become more important than the decisions of the courts.

4.3 Other Legal Relationship

Except for the introduction of a legal provision in the Compensation Act for a vehicle owner and an insurance company to exercise the right to indemnity against a vehicle manufacturer, the discussions for SAE Level 0-2 AVs provided above may be applied the same for any other legal relationship.

5 Liability in Traffic Accidents Involving SAE Level 4-5 AVs

5.1 *Whether Driver's Liability Under the Amended Compensation Act is the Same*

The amended Compensation Act is to introduce driver's liability for accidents that occur while AVs are in operation and to allow the exercise of the right to indemnity after the fact. However, the amended Compensation Act regulates only "*autonomous vehicles*." Based on the literal interpretation of the legal provisions of the amended Act, the same principle may apply to SAE Level 4-5 AVs as well.

However, the MOLIT announced that the Compensation Act is applicable to SAE Level 3 AVs and that there would be a separate legislation for SAE Level 4-5 AVs. In light of the foregoing, the Compensation Act is likely to be amended to account for SAE Level 4-5 AVs going forward.

It may not be reasonable to maintain driver liability under the Compensation Act for SAE Level 4-5 AVs as well. It is difficult to justify driver liability because in the case of SAE Level 4-5 AVs, a vehicle owner, which is the main agent of operation, could also be in a position of a "*passenger*" or the "*other person*," which is intended to be the object of protection.[13] In particular, SAE Level 5 AVs are "*fully autonomous*" and do not require any intervention by a driver. Therefore, in case an accident occurs, it is indeed highly likely for the accident to be attributable to a "*vehicle defect*." There is an issue as to whether it is reasonable to regulate such circumstances based on a driver's liability. In particular, this may be unconstitutional as it is in violation of the "*principle of liability with fault*" under the Korean Constitution.

In an accident involving an SAE Level 4-5 AV, a victim may file a claim for damages against the owner, driver and/or manufacturer of the vehicle based on the existing legal principles of the Civil Code or the PLA. However, in order to address the issue fundamentally, the Compensation Act needs to be amended to reflect the aforementioned discussions.

5.2 *Current Discussions on Liability in Accidents Involving SAE Level 4-5 AVs*

5.2.1 SAE Level 4 AV

SAE Level 4 AVs requires significantly less intervention by a driver compared to SAE Level 3 AV. However, even if the vehicles are operated in an autonomous

[13] Kwon and Lee (2016), pp. 482–483.

driving mode, the fact that there is an operator is still recognized[14] and the duty of care still exists. Therefore, they are still distinguishable from fully autonomous SAE Level 5 AVs.

Due to such a difference, some view that the driver's duty of care must be clearly defined for the commercialization of SAE Level 4 AVs. Under that perspective, since a driver may be required to intervene if an abnormal situation is perceived due to a system defect in SAE Level 4 AVs, the driver's duty to perceive, determine and intervene in such a situation should be clearly set forth in the law.[15]

Opinions are divided, with those acknowledging a driver's liability as just the same as it is under the current regulatory regime and those denying such in accidents involving SAE Level 4 AVs. The former is based on the fact that driver's intervention still exists (since a driver has control over the AV in an emergency situation, the vehicle owner can conceptually regulate and control operation by having control in operation), while the latter is mainly based on logic that an autonomous driving system has full control in operation in an autonomous driving mode and thus cannot be regulated with a system of driver liability.[16] Furthermore, a legislative bill proposed in February 2018 for the amendment of the Compensation Act was based on the premise that primary liability for damage in AV accidents was to be jointly borne by vehicle owners and manufacturers and also mandated manufacturers be covered by liability insurance. Therefore, the possibility of the Compensation Act being amended again to account for SAE Level 4-5 AVs cannot be fully ruled out. However, there has been discussion that liability will vary depending on driving mode and that speedy relief measures to victims will be difficult because lawsuits may be prolonged due to issues such as it being relatively difficult for victims to prove a vehicle defect.[17]

For now, the amended Compensation Act that took effect has adopted the same driver liability for AVs as is prescribed in the past legal provisions. This is reasonable in the sense that a consumer's burden of proof is lessened while quick relief for victims is possible. If vehicle manufacturers are the party primarily burdened with no-fault liability, it may be too harsh and adversely affect their business operations. As a result, it is not recommended that vehicle owners bear primary liability as an owner in terms of fair distribution of damages, and unless the Compensation Act is newly amended in the future, SAE Level 4 AVs are also likely to be regulated based on a driver's liability under the Compensation Act, similarly to SAE Level 3 AVs.

[14] Seo and Choi (2017), p. 365.

[15] Lee (2017), pp. 108–110.

[16] Park (2019), pp. 179–207: 193–194.

[17] Ibid., 187.

5.2.2 SAE Level 5 AV

SAE Level 5 VAs are fully autonomous vehicles and, in principle, cannot be interrupted by a driver's intervention. In particular, in the case of SAE Level 5 AV, operation of driverless AVs is expected to become common and thus operational control will eventually completely disappear. Therefore, it would not be reasonable to regulate SAE Level 5 AVs based on a driver's liability as prescribed in the Compensation Act.

Many believe that when an accident occurs while a SAE Level 5 AV is in operation, the duty of care to manage and maintain vehicles such as updating systems and checking vehicle functions would be considered far more important than a driver's duty of care.[18] The government has announced that it would legislate a law introducing mandatory system maintenance by a vehicle owner. On November 8, 2018, the government announced a roadmap to pre-emptively abolish restrictions in the field of AVs by setting up short-term goals to introduce mandatory vehicle system maintenance, streamline civil and criminal liability and insurance regulations, improve regulations on collecting and utilizing video information in the field of infrastructure and, in the mid-to-long term, outlining issues in the areas of a principal agent driving, vehicles/devices, and operation regarding SAE Level 4 and 5 AVs. [19]

In the end, discussion should focus on a manufacturer's legal liability under the PLA rather than the liability under the Compensation Act. Therefore, in addition to the Compensation Act, legislation is necessary. Such legislation should include legal provisions that make product liability insurance mandatory for AV accidents since accidents involving the operation of AVs are likely to be caused by a vehicle defect and are therefore likely to be attributable to a manufacturer's liability.[20]

Under the PLA, in principle, a victim has the burden of proof on the causal relationship between the existence of a "*defect*" and damages, and thus there are limits in claiming for product liability in the case of an accident. Accordingly, the possibility of establishing product liability is examined, including who is liable in the case of a software defect and what the scope of a defect is. There are also discussions on interpretations to apply legal principles through a perspective of protecting consumers.[21] In other words, with respect to accidents involving AV computer systems, legal liability may vary depending on whether the computer system controls individual vehicles through a centralized computer operated and managed by the vehicle manufacturer or operates individual vehicles independently through a computer system installed in each vehicle. In the former case, the vehicle manufacturer will be legally liable for accidents, while in the latter case, the program operator

[18] Seo and Choi (2017), p. 366.

[19] http://www.opm.go.kr/opm/news/press-release.do?mode=view&articleNo=118993.

[20] Ibid., 368.

[21] Kwon and Lee (2016), pp. 469–477; Ryu (2016), pp. 62–63.

is likely to be liable.[22] Going forward, not only the liability of vehicle manufacturers but also sharing of the liability with program developers should be discussed.

In this regard, the key task will be amending the relevant legal provisions to shift the focus on liability from the operator to the motor vehicle manufacturer. As an example, in the case of a recent legislative bill[23] to partially amend the Compensation Act, which was proposed in February 2018, it required AV manufacturers to have mandatory insurance policies and partly share liability via the insurance policies even from the stage of compensating victims for damages. In the case of full AVs, there is a chance that a legislative bill may be proposed against such a change. We will need to wait and see how legal provisions will be amended regarding SAE Level 4-5 in the future.

6 Conclusion

As for the liability in traffic accidents involving SAE Level 0-2 AVs, based on an operator's liability and vehicle owner's mandatory liability to subscribe insurance policy under the Compensation Act, a driver (vehicle owner) and insurance company bear the primary liability for damages without fail in traffic accident caused by the motor vehicles. In case where a driver is not an operator nor the insured, the driver may be held liable to compensate damages attributable to an unlawful act. In cases where a traffic accident is attributable to a vehicle defect, a vehicle manufacturer may also bear liability for damages to victims based on unlawful acts under the Civil Code, or damages based on product liability under the PLA. In such a case, it is not easy for the victim to be compensated from the vehicle manufacturer as the victim should be held liable for burden of proof to a certain extent. In fact, it is not frequent that a victim claims compensation from the vehicle manufacturer for damages based such grounds of purport.

As for the liability in traffic accidents involving SAE Level 3 AVs, there has been theoretical discussion on who will be considered as an operator under the Compensation Act among a vehicle manufacturer, vehicle owner and driver. However, according to the recently amended Compensation Act governing the liability for traffic accidents involving SAE Level 3 AVs, an operator's liability and vehicle owner's mandatory liability to subscribe insurance policy under the Compensation Act shall equally apply to the SAE Level 3 AVs, just like SAE Level 0-2 AVs. Therefore, it is now available for a victim to be compensated for damages from the vehicle owner who is a driver. Provided, in case of an autonomous driving of SAE Level 3 AVs, causes attributable to vehicle manufacturers are highly likely to be existed. Due to such reason, the amended Compensation Act newly introduced any

[22] Kim (2015), p. 251.

[23] Bill No. 2011832 was proposed by the Assemblyman Hee Hwang and others, but has been recently repealed.

Table 1 Overview of major legal liabilities in case of accidents involving the operation of AVs

SAE Level	Vehicle Owner/ Insurance Company	Driver	Manufacturer
0-2	Compensation Act (no-fault liability)	Liability for damages attributable to an unlawful act	Victim: Product liability Vehicle Owner/Insurance Company: seek a subrogation claim and exercise the right to indemnity against manufacturer
3	Compensation Act (no-fault liability)	Liability for damages attributable to an unlawful act	Victim: Product liability Vehicle Owner/Insurance Company: exercise the right to indemnity against manufacturer (Compensation Act) – setting up the Investigation Committee and installing a device to record accidents are mandatory
4-5	Compensation Act (no-fault liability)	Liability for damages attributable to an unlawful act	Victim: Product liability Vehicle Owner/Insurance Company: exercise the right to indemnity against manufacturer (Compensation Act) – setting up the Investigation Committee and installing a device to record accidents are mandatory
	Regarding the question of whether to recognize a driver's liability as it is will have to be determined by the legislature.		

vehicle manufacturers' liability for accidents that occur while AVs are in operation to allow the exercise of the right to indemnity after the fact. Under the existing legal principle, an insurance company is allowed to claim for product liability against a manufacturer on behalf of the insured through a subrogation clause or claim for the right to indemnity. However, the amended Compensation Act introduced legal grounds to facilitate the direct exercise of the right to indemnity, required vehicle manufacturers to install devices recording autonomous driving information on AVs, and mandated the establishment of the Investigation Committee. In such case, if a driver is not an operator nor the insured and is negligent to a certain degree, the driver should be held liable to compensate damages attributable to an unlawful act, separately with an operator, insurance company and manufacturers.

As for the liability in traffic accidents involving SAE Level 4-5 AVs, whether or not the current rules of the amended Compensation Act can be applied to SAE Level 4-5 AVs as well are still under consideration. In particular, in case of SAE Level 5 AVs, it is full autonomous driving, which do not require any invention by a driver. Therefore, it can be seen as contrary to the principle of self-responsibility to hold the driver liable. Discussion on this part has developed further, and it remains as an issue that must be resolved through legislation.

The legal liability discussed above with regard to accidents involving the operation of AVs is summarized in Table 1.

References

Kim YK (2015) Legal issues concerning accidents during operation of autonomous vehicles and insurance coverage (자율주행 자동차의 운행 중 사고와 보험적용의 법적 쟁점). Legal Theory Pract. Korea Soc Legal Theory Pract 3(2):251

Kwon YJ, Lee SE (2016) Autonomous vehicle accidents and civil liability (자율주행자동차 사고와 민사책임). Civil Code Rev 75:469–477, 482–483

Lee BS (2019) Several legislative challenges surrounding concept of autonomous vehicle under the Motor Vehicle Management Act – in light of discussion on civil liability (자동차관리법상 이른바 '자율주행자동차'라는 것의 개념을 둘러싼 몇몇 입법과제 –민사책임 논의에 비추어-). Korean Law Legis Rev 56(1–26):15

Lee JK, Hwang CK (2016) Necessity to overhaul liability legislation and liability insurance system to prepare for operation of autonomous vehicles (자율주행자동차 운행에 대비한 책임법제와 책임보험업제의 정비 ?요성). Financ Law Rev 13(1):93

Lee JK, Hwang CK (2018), Structure of Level 3 autonomous vehicle accidents and liability: focusing on uber autonomous vehicle accidents (3단계 자율주행차 사고와 책임의 구조: 우버 자율주행자동차 사고를 중심으로), Chung-Ang Law Rev 20(3):11: 13

Lee JS (2017) Improvement of the road traffic act for commercialization of level 3 autonomous vehicles (level 3의 자율주행자동차 상용화를 위한 도로교통법의 개선방안). Ajou Law Rev 11(1):108–110. Ajou University Law Research Institute

Park SM (2019) Study on interpretation of legal principles on civil liability for accidents in level 3 autonomous vehicles under autonomous driving mode (레벨3 자율주행자동차의 자율주행 모드시 사고에 따른 민사상 책임법리의 해석에 대한 연구), Corp Law Rev 33(1):179–207, 190–191, 193–194

Ryu CH (2016), Study on application of product liability to autonomous vehicles (자율주행자동차에 대한 제조물책임의 적용에 관한 연구). Ajou Law Rev 10(1):62-63. Ajou University Law Research Institute

Seo KS, Choi KJ (2017) Civil law review on liability for damages for accidents involving autonomous vehicles (자율주행자동차 사고 시 손해배상책임에 관한 민사법적 검토). Gachon Law Rev 10(4):365–366

Court Decisions

Supreme Court No. 98Da18506 rendered on January 18, 2000

Seoul Central District Court No. 2008*Gahap*57857 rendered on May 8, 2009

Gwangju District Court No. 2010Gadan1048 rendered on September 7, 2010

Web Site

http://www.opm.go.kr/opm/news/press-release.do?mode=view&articleNo=118993

Part IV
Australia

Steering the Transition to Automated Vehicles in Australia: Current Status and Regulatory Changes Ahead

Angelika Yates and Samuel Siskovic

Abstract Automated vehicles (**AV**) (The term "automated vehicle", abbreviated to "AV" is used throughout this chapter to refer to vehicles that are equipped with an automated driving system, abbreviated to "ADS". Australian government agencies and regulators use the term "automated vehicle" rather than "autonomous vehicle".) will drastically alter the experiences in transport and road safety in Australia. Not only will driverless technology remove human errors such as driver fatigue, speeding and driver distraction on Australian roads, it can also improve efficient road use and minimise congestion. Looking towards the existing success of driverless technology in the Australian mining industry (Darren Gray, Sydney Morning Herald (27 April 2019), No one behind the wheel: The new workforce driving Australia's mines.) and the introduction of driverless trains as part of the Sydney Metro project in 2019 (NSW Government (19 February 2019), What does embracing technology mean for transport? https://www.sydneymetro.info/article/what-does-embracing-technology-mean-transport. Accessed 13 January 2023.), it is clear that the adoption of AV will bring significant advantages. However, Australian regulatory frameworks are not yet equipped to address all levels of AV. Existing laws regarding road safety are based on the concept of a human driver being in control of the vehicle. This fails to recognise the reality that in vehicles operating at high levels of automation a human driver will no longer control the vehicle.

Currently only Level 0 to 2 AV (as defined under SAE J3016) (SAE International (April 2021), Surface Vehicle Recommended Practice J3016. An overview of the SAE Levels and their features is set out in the Table 1.1) are allowed to be driven on public roads in Australia. Even though Level 3 to 5 AV (as defined under SAE J3016) (SAE International (April 2021), Surface Vehicle Recommended Practice

The authors wish to thank William Tong, Law Student at University of Sydney and Alina Kuehn, Legal Trainee, for their valuable support with the preparation of this chapter during their internship with Addisons. This article is current as at 31 January 2023.

A. Yates (✉) · S. Siskovic
Addisons, Sydney, NSW, Australia
e-mail: angelika.yates@addisons.com

H. Steege et al. (eds.), *Autonomous Vehicles and Civil Liability in a Global Perspective*, Data Science, Machine Intelligence, and Law 3,
https://doi.org/10.1007/978-3-031-41992-8_10

J3016.) are increasingly trialled in Australia, a comprehensive legislative reform is required before these AV can be commercially deployed at scale.

This chapter gives an overview of the current regulatory framework and proposed legislative reforms in Australia with respect to conduct of trials, safety at first supply and in-service safety of Levels 0 to 5 AV.

1 Introduction: Are We There Yet?

In KPMG's *2020 Autonomous Vehicle Readiness Index* (**AVRI**), Australia is one of only four countries that received the highest AVRI score for its AV regulations.[1] However, Australian laws are not quite there yet when it comes to regulating AV at higher automation levels.

The legal reforms are coordinated by the *Transport and Infrastructure Council of Australian Governments* (**TIC**) which consists, amongst others, of the Federal Minister of Transport and the Ministers for Transport of the various Australian States and Territories.[2] TIC sets priorities and makes decisions on the legislative reforms for AV and is supported by the *National Transport Commission* (**NTC**).[3]

The NTC has largely adopted the levels of automation set out in the *Society of Automotive Engineers* (**SAE**) *International Standard J3016* (**Levels**).[4] AV at Levels 0 to 2 can be encompassed within existing legislations as the human driver remains the dominant, controlling party during vehicle operation. However, Australian laws currently struggle to accommodate AV at Levels 3 to 5 as these vehicles can operate without a human driver.

Table 1 gives an overview of the Levels and their specific features as they are used in this chapter.

NTC released its first discussion paper on AV in May 2016.[5] During this initial review, over 700 barriers in Australia's current Federal and State legislation to the

[1] KPMG International Cooperative, 2020 Autonomous Vehicles Readiness Index, page 26. https://home.kpmg/xx/en/home/insights/2020/06/autonomous-vehicles-readiness-index.html. Accessed 13 January 2023. However, Australia's overall AVRI score ranks it at number 15 of 30 countries participating in the AVRI, see KPMG International Cooperative, 2020 Autonomous Vehicles Readiness Index, page 3.

[2] The *Transport and Infrastructure Council* under the direction of the *Council of Australian Governments*, brings together Commonwealth, State, Territory and New Zealand Ministers with portfolio responsibilities for transport, infrastructure, roads, and planning, as well as the Australian Local Government Association.

[3] The *National Transport Commission* is a corporate body established as a national transport reform agency by the *National Transport Commission Act* 2003 (Cth).

[4] SAE International (April 2021), Surface Vehicle Recommended Practice J3016.

[5] National Transport Commission (May 2016), Regulatory Options for Automated Vehicles Discussion Paper. https://www.ntc.gov.au/sites/default/files/assets/files/Regulatory-options-for-automated-vehicles-May-2016.pdf. Accessed 13 January 2023.

Table 1 Overview of Levels of automation as referred to in this chapter

Level 0	Level 1	Level 2	Level 3	Level 4	Level 5
No Automation	Driver Assistance	Partial Automation	Conditional Automation	High Automation	Full Automation
Human driver performs all aspects of dynamic driving task (**DDT**) with very limited or no driver support features (**DSF**).	Human driver performs DDT and DSF provide steering **or** brake / acceleration support.	Human driver performs DDT and DSF provide steering **and** brake / acceleration support.	Automated driving system (**ADS**) undertakes entire DDT under limited conditions specified in the operational design domain (**ODD**).	ADS undertakes entire DDT under conditions specified in the ODD. No human driver is required.	ADS undertakes entire DDT under all conditions at all times. No human driver is required.
Human driver must monitor driving environment.			ADS monitors driving environment.		
Human driver has control of vehicle at all times.			Human driver must take control upon request by ADS or when ADS fails (**Fallback-ready user**).	Human driver not required / able to take control. ADS continues to control vehicle following ADS failure or ODD exit.	

deployment of AV at higher levels of automation were identified.[6] In an AV, control of the vehicle will gradually transfer from a human driver to an automated driving system (**ADS**), depending on the level of automation. Accordingly, a comprehensive overhaul of prevailing legislative frameworks is required to accommodate the use of AV at Levels 3 to 5.

Since 2016, NTC has received and analysed submissions from multiple stakeholders, issued various policy and discussion papers, including a comprehensive '*In-service safety for automated vehicles, decision regulation impact statement*'[7] and its '*Automated Vehicle Program Approach*'[8] setting out a clear path for legislative reform. In 2020 NTC issued a discussion paper with respect to '*A national in-service safety law for automated vehicles*' which asked for submissions by relevant

[6]National Transport Commission (July 2019), Automated Vehicle Program, page 4. https://www.ntc.gov.au/sites/default/files/assets/files/Automated%20Vehicle%20Program%20Approach_0.pdf. Accessed 13 January 2023.

[7]National Transport Commission (June 2020), In-service safety for automated vehicles, decision regulation impact statement. https://www.ntc.gov.au/sites/default/files/assets/files/NTC-Decision-RIS-In-service-safety-for-AVs.pdf. Accessed 13 January 2023.

[8]National Transport Commission (September 2020), Automated Vehicle Program Approach. https://www.ntc.gov.au/sites/default/files/assets/files/Automated%20vehicle%20approach.pdf. Accessed 13 January 2023.

stakeholders by 11 December 2020[9] and updated the '*Guidelines for Trials of Automated Vehicles in Australia*' (**December 2020 Trial Guidelines**[10]). Most recently, NTC issued another discussion paper with respect to '*On-road enforcement for automated vehicles*' to examine whether state and territory law enforcement officers have sufficient powers to interact with and respond to the road safety risks of AV.[11]

In Australia, one of the challenges regarding legal reforms for AV is the division of power between Federal and State Governments. The lawmaking power of the Federal Government is confined to matters expressly specified in the Australian constitution.[12] The Federal Government has the legislative power to set requirements for the introduction of new vehicles, including AV, into Australia. However, the specified matters allocated to the Federal Government do not include road transport, use of motor vehicles or road safety.[13] Consequently, each Australian State and Territory currently has its own laws concerning the use of motor vehicles on public roads.

In November 2018, the Australian transport ministers endorsed a nationwide end-to-end regulatory system to support the safe, commercial deployment of AV at all levels of automation.[14] Since then, a clear path has been developed by the NTC to a new national *Automated Vehicle Safety Law* (**AVSL**). In 2020, the infrastructure and transport ministers agreed to key elements of the in-service framework for AV based on the NTC's recommendations. In February 2022, it was decided to implement the AVSL through Commonwealth law by 2026.[15]

---

[9] National Transport Commission (October 2020), A national in-service safety law for automated vehicles. https://www.ntc.gov.au/sites/default/files/assets/files/NTC-Discussion-Paper-national-in-service-safety-law-for-AVs.pdf. Accessed 13 January 2023.

[10] National Transport Commission (December 2020), Guidelines for trials of automated vehicles in Australia. https://www.ntc.gov.au/sites/default/files/assets/files/AV-trial-guidelines-2020.pdf. Accessed 13 January 2023.

[11] National Transport Commission (July 2022), On-road enforcement for automated vehicles. NTC Discussion Paper - On-road enforcement for automated vehicles_0.pdf. Accessed 17 January 2023.

[12] *Commonwealth of Australia Constitution Act 1900* (Cth), section 51 of the Constitution sets out the legislative powers of the Federal parliament which comprises 39 specific matters.

[13] *Commonwealth of Australia Constitution Act 1900* (Cth), section 51 (xxxii) only refers to "*the control of railways with respect to transport for the navel and military purposes of the Commonwealth*".

[14] National Transport Commission (September 2020), Automated Vehicle Program Approach, page 4.

[15] National Transport Commission homepage, The regulatory framework for automated vehicles in Australia (February 2022). Accessed 17 January 2023.

2 Key Stages of AV Deployment

In examining Australia's existing laws and ongoing legislative reforms, we have focused on the following key stages of the introduction of AV:

- on-road trials for AV;
- market entry, referred to as 'safety at first supply'; and
- on road use, referred to as 'in-service safety'.

Due to the unprecedented nature of AV at higher levels of automation, legal reform is required at each of these key stages. The current regulatory regimes regarding 'safety at first supply' and conduct of trials are also suitable for Levels 3 to 5 AV with only minor modifications, whereas major reforms are required to cover 'in-service safety' of AV at Levels 3 to 5.

3 Regulations for On-Road Trials of AV

In Australia, AV technology trials are much smaller in scale when compared to countries with large automotive and IT industries, such as European countries and the US.[16] As at December 2020, there have been 32 automated vehicle trials across Australia,[17] with the majority of these trials involving low-speed automated shuttle buses operating on set routes.

As mentioned above, AV of higher automation levels cannot legally operate on Australian roads. Therefore, organisations seeking to run AV trials must first obtain a permit or exemption from the road transport agency of the relevant State or Territory in which the trial is proposed to be conducted.

The legislative power to regulate on-road trails vests with the Australian States and Territories.[18] South Australia was the first Australian State to regulate trials of AV.[19]

To create regulatory certainty, clarify industry expectations and create minimum safety standards for AV trials throughout Australian States and Territories, NTC and Austroads issued their first '*Guidelines for trials of Automated Vehicles in Australia*'

[16]National Transport Commission (December 2020), Lessons learned from automated vehicle trials in Australia, page 31. https://www.ntc.gov.au/sites/default/files/assets/files/AV-lessons-learned-2020_0.pdf. Accessed 13 January 2023.

[17]National Transport Commission (December 2020), Lessons learned from automated vehicle trials in Australia, page 10. https://www.ntc.gov.au/sites/default/files/assets/files/AV-lessons-learned-2020_0.pdf. Accessed 13 January 2023.

[18]Under the Australian constitution, the Federal Government only has legislative powers with respect to specified matters (see section 51 of the *Commonwealth of Australia Constitution Act 1900* (Cth)). The use of public roads is not one of them.

[19]*Motor Vehicles (Trials of Automotive Technologies) Amendment Act* 2016 (SA).

Table 2 National Trial Guidelines—Summary of Conditions

Permit/Exemption	A permit must be obtained from the road transport agency in the relevant State or Territory exempting the trialling organisation from compliance with road rules.[a]
Trial Management	Certain key management criteria, including the purpose of the trial, trial location, description of technology being trialled and a traffic management plan must be prepared and submitted with the road transport agency in the relevant State or Territory.[b]
Safety Management Plan	A safety management plan must be developed by the trialling organisation which outlines all relevant key safety risks for the trial and how these risks will be mitigated or eliminated.[c]
Insurance	Appropriate insurance coverage must be taken out by the trialling organisation to ensure any road users injured by an AV during the trial are no worse off than if they were injured by a human-operated vehicle.[d]
Trial data	Details of data that will be recorded by the ADS and stored by the trialling organisation and how such data will be supplied to relevant parties must be provided.[e]

[a]National Transport Commission (December 2020), Guidelines for trials of automated vehicles in Australia, pages 5 to 6
[b]National Transport Commission (December 2020), Guidelines for trials of automated vehicles in Australia, pages 7 to 8
[c]National Transport Commission (December 2020), Guidelines for trials of automated vehicles in Australia, pages 11 to 12
[d]National Transport Commission (December 2020), Guidelines for trials of automated vehicles in Australia, page 9
[e]National Transport Commission (December 2020), Guidelines for trials of automated vehicles in Australia, pages 13 to 14

(“**National Trial Guidelines**”) in May 2017[20] which were replaced by the updated December 2020 Trial Guidelines.[21] After the National Trial Guidelines became first available in May 2017, New South Wales and Victoria chose to enact their own legislation with respect to AV trials.[22]

Notwithstanding their differences, the overall approach to AV trials is similar under the National Trial Guidelines and each relevant State legislation. For example, the National Trial Guidelines recommend that all trials must satisfy the conditions given in Table 2.

[20]National Transport Commission (May 2017), Guidelines for Trials of Automated Vehicles in Australia. https://www.ntc.gov.au/sites/default/files/assets/files/AV_trial_guidelines.pdf. Accessed 13 January 2023.

[21]National Transport Commission (December 2020), Guidelines for trials of automated vehicles in Australia. https://www.ntc.gov.au/sites/default/files/assets/files/AV-trial-guidelines-2020.pdf. Accessed 13 January 2023.

[22]*Transport Legislation Amendment (Automated Vehicle Trials and Innovation) Act* 2017 (NSW); *Road Safety (Automated Vehicles) Regulations* 2018 (VIC).

4 Regulation of 'Safety at First Supply'

4.1 Self-Certification and Type Approval

The lawmaking power with respect to 'safety at first supply' vests with the Federal Government of Australia. The current system for introducing new vehicles to the Australian market is based on self-certification and type approval, wherein a prototype vehicle of a particular design undergoes tests to demonstrate compliance with the applicable *Australian Design Rules* (**ADR**) which set out safety standards for road vehicles.[23] If the test vehicle complies with the ADR then the manufacturer will be granted type approval and may supply vehicles of the approved type in Australia in unlimited numbers.

Australian transport ministers agreed in November 2018 to incorporate a self-certification system for automated driving systems (**ADS**) as a component and AV containing an ADS into the federal *Road Vehicles Standards Act* 2018 (Cth).[24] Under the proposed regime, an automated driving systems entity (**ADSE**), i.e. the organisation that produces or is seeking to import ADS into Australia, will need to self-certify against a set of certain criteria to demonstrate that the ADS complies with applicable safety standards set out in the ADR and that the ADSE will adequately manage safety risks.

The proposed eleven self-certification criteria for ADS include:[25]

- Safe system design and validation process
- Operational design domain
- Human-machine interface
- Compliance with relevant road traffic laws
- Interaction with enforcement and other emergency services
- Minimal risk condition
- On-road behavioural competency
- Installation of system upgrades
- Verifying for the Australian road environment
- Cybersecurity
- Education and training.

Once testing is successfully completed and all further corporate obligations[26] are met, type approval for the ADS as a component, or the AV containing the ADS, will then be published in an online database, the Register of Approved Vehicles (**RAV**),

[23] *Road Vehicles Standards Act* 2018 (Cth), section 22; *Vehicle Standard (Australian Design Rule – Definitions and Vehicle Categories)* 2005.

[24] National Transport Commission (September 2020), Automated Vehicle Program Approach, page 12.

[25] National Transport Commission (June 2020), In-service safety for automated vehicles, decision regulation impact statement, page 27.

[26] See Section 4.2 Further Corporate Obligations for ADSE.

and will identify the ADSE for the approved ADS or AV. This is legally significant because the ADSE named in the RAV will be responsible for the continuous safety of the ADS. The type approval will also state the ADS' operational design domain (**ODD**), which outlines the limits of the ADS' on-road capabilities, and it will verify that an ADS can comply with the relevant road rules.

4.2 Further Corporate Obligations for ADSE

In order to ensure that the ADSE will be able to manage its liability for events such as road traffic law breaches and crashes caused by its ADS, the ADSE must:

- enable data recording, sharing and storing in Australia;
- have a corporate presence in Australia; and
- satisfy minimum financial and asset requirements.[27]

4.3 Personal Liability of Executive Officers of the ADSE

Executive officers of the ADSE will be subject to the executive officer liability regime under the *Road Vehicles Standards Act 2018* (Cth) (**RVSA**) under which they will be personally liable if the ADSE commits an offence against criminal provisions or contravenes a civil penalty provision of the RVSA and all of the following conditions are satisfied:[28]

- the officer knew that the offence would be committed or the contravention would occur;
- the officer was in a position to influence the conduct of the ADSE in relation to the commission of the offence or the contravention; and
- the officer failed to take reasonable steps to prevent the commission of the offence or the contravention.

This liability of the executive officer of the ADSE will be civil or criminal, depending on whether the ADSE committed a criminal offence or a contravention of a civil penalty provision.

[27] National Transport Commission (June 2020), In-service safety for automated vehicles, decision regulation impact statement, page 27.

[28] *Road Vehicles Standards Act* 2018 (Cth), section 33.

5 Regulation of 'In-Service Safety'

With regards to the in-service safety regulation of AV, existing regulations are sufficient to cover AV at Levels 0 to 2, whereas AV at Levels 3 to 5 are currently not permitted on Australian roads as they are inconsistent with existing requirements that a licensed human driver must be in control of the vehicle.

The 'in-service safety' of vehicles is the responsibility of the States and Territories and, originally, each State and Territory enacted its own road safety laws. In an attempt to unify road safety laws across Australia, the NTC developed the *Australian Road Rules* (**ARR**) as a model law.[29] The ARR do not have legislative force on their own. However, the ARR form the basis of road rules that have been enacted by the Australian States and Territories.[30]

The ARR (as adopted by the States and Territories) are not suitable for AV at Levels 3 to 5 as they are not considering that driving could be performed without a human driver. For example, section 297(1) of the **ARR** requires that the driver must have 'proper control' of the vehicle. Furthermore, existing laws are unable to address the roles and potential liabilities of new parties involved in the operation of AV. For example, these new parties include the ADSE and remote drivers. Comprehensive legal reforms are necessary to properly address the in-service safety of Levels 3 to 5 AV. The approach to these reforms is well progressed. The NTC has presented its final recommendations on in-service safety to TIC in May 2021 which then were agreed to in February 2022.

The approach taken by the NTC with respect to 'in-service safety' regulation of AV at Levels 3 to 5 is based on the concept of influence over the ADS which will replace the current concept of driver control. Effectively, this means that multiple parties may be liable for accidents caused by AV, depending on their level of influence over the operation of the ADS at the relevant time. In particular, the ADSE self-certifying that its ADS can safely perform the driving task in place of a human driver may be liable for accidents caused by a defect or failure of the ADS. Other persons with 'major influence' over the safe operation of the ADS may also be liable, including the human driver, manufacturer, importer, repairer and road manager.[31]

The following sections will examine in more detail the current liability regime for AV at Levels 0 to 2 and then analyse the proposed new laws regarding 'in-service safety' for AV at Levels 3 to 5.

---

[29] *National Transport Commission (Road Transport Legislation – Australian Road Rules) Regulations* 2006 (Cth).

[30] See for example, *Road Rules 2014* (NSW) *Road Safety Rules 2017* (Victoria) and *Road Transport (Road Rules) Regulation 2017* (ACT).

[31] National Transport Commission (June 2020), In-service safety for automated vehicles, decision regulation impact statement, page 45 to 46.

6 In-Service Safety: Liability for AV at Levels 0 to 2 Under Existing Laws

AV at Levels 0 to 2 are currently permitted on Australian roads. In Australia, each State and Territory has its own statutory regime which applies in respect of the operation of motor vehicles on the road. [32] A substantial portion of these statutes are based on the ARR,[33] which are a national model law and have been adopted (sometimes in modified form), by the States and Territories to varying degrees. Where the term "ARR" is used in this chapter, it refers to the ARR as they have been adopted in the relevant State or Territory where the breach occurs.

Relevantly, it remains a requirement in each State and Territory that, with the exception of approved AV trials, a human driver must remain in control of a vehicle, and, consequently, is potentially liable in respect of a breach of relevant laws or an act of negligence.[34]

6.1 Human Driver

In AV at Levels 0 to 2, the human driver retains predominant control in performing the vehicle's dynamic driving task (**DDT**). Under the ARR and its various State and Territory implementations, the human driver must have 'proper control' of the vehicle and is liable for his/her actions whilst in control of the vehicle.[35]

Where the human driver causes an accident, the injured person may make a civil liability claim under common law negligence. To establish that the human driver is civilly liable in respect of a motor vehicle accident, the plaintiff is required to establish that:

- the human driver owed the plaintiff a duty of care;
- the human driver breached their duty of care;
- the plaintiff suffered an injury or other damage as a result of the human driver breaching their duty of care owed to the plaintiff.

[32] Australian Capital Territory: *Road Transport (Safety and Traffic Management) Regulation 2000* (ACT); Northern Territory: *Traffic Regulations 1999* (NT); New South Wales: *Road Rules 2014* (NSW); Queensland: *Transport Operations (Road Use Management-Road Rules) Regulation 2009* (QLD); South Australia: *Road Traffic Act* 1961 (SA); Tasmania, *Road Rules 2009* (TAS); Victoria: *Road Safety Road Rules 2009* (VIC); Western Australia: *Road Traffic Code 2000* (WA).

[33] *Australian Road Rules* approved by the Transport and Infrastructure Council as at 22 November 2019, https://www.pcc.gov.au/uniform/2019/Australian-Road-Rules-22November2019.pdf. Accessed 13 January 2023.

[34] For example, see the *Australian Road Rules* at section 297(1), which is mirrored in the *Road Rules 2014* (NSW) at section 291(1).

[35] *National Transport Commission (Road Transport Legislation – Australian Road Rules) Regulations* 2006 (Cth), section 297(1).

Most States and Territories have a Civil Liability Act or other statutory instruments in which the relevant principles that apply to the breach of a duty of care are set out.[36]

Personal injury claims are subject to a statutory regime in the relevant State or Territory in which the accident occurred. Notwithstanding that a statutory regime will apply, a personal injury claim is a tortious claim based on the principles of common law negligence.

More severe breaches of the ARR may also result in the driver incurring criminal liability. The words '*Offence Provision*' referred to in a rule of the ARR indicate that a contravention of that rule is a criminal offence. For example, the following ARR are offence provisions:

- rule 20 ARR (*Obeying Speed Limits*);
- rule 37 ARR (*U-turns*); and
- rule 48 ARR (*Change of Direction Signal*).

Furthermore, a human driver commits a statutory criminal offence under the applicable Crimes Act or Criminal Code Act as in force in the relevant State or Territory if he/she causes the death of, or bodily harm to, another person where the driver is driving either under the influence of alcohol or another drug, or if the vehicle is driven at a speed or in a manner dangerous to others.[37]

6.2 Registered Owner

Before a vehicle can legally operate on the road, the owner must present the vehicle for registration and certification of its roadworthiness.[38] However, in-service safety obligations in current legislations are predominantly addressed to the human driver, with limited statutory liabilities and obligations for the registered owner. Having said that, under common law principles, it is possible for the owner to be liable for actions of the human driver based on 'vicarious liability' if some form of agency can be established.[39] For example, if the driver can be construed as acting on the authority of the owner in an employment context or otherwise, the owner, as the employer or the principal, may be held liable for the actions of the driver in his /her capacity as employee or agent.[40]

[36] For example, see *Civil Liability Act 2002* (NSW) at section 5B.

[37] *Crimes Act 1900* (ACT), section 29; *Criminal Code Act 1983* (NT), section 174F; *Crimes Act 1900* (NSW), section 52A; *Criminal Code Act 1899* (QLD) section 328A; *Criminal Law Consolidation Act* 1935 (SA), section 19A; *Criminal Code Act 1924* (TAS), section 167A; *Crimes Act 1958* (VIC), sections 318 and 319; *Road Traffic Act 1974* (WA) section 59.

[38] Service NSW, Vehicle Registration. https://www.service.nsw.gov.au/transaction/renew-vehicle-registration. Accessed 13 January 2023.

[39] *Prince Alfred College Incorporated v ADC* [2016] HCA 37.

[40] *Sobulsky v Ega* [1960], 103 CLR 215.

6.3 *Manufacturers, Importers and Suppliers*

Both, the manufacturer (which includes the importer) and the supplier, may be liable for injury or damages caused by defects in AV at Levels 0 to 2. In Australia, the manufacturer's, and/or supplier's liability for defective products is derived from a combination of common law principles and legislation.[41]

6.3.1 Product Liability

Under common law, it is well-established that manufacturers and suppliers owe a duty of care in negligence to protect purchasers and users of the product from foreseeable risks of injury when using the product as intended.[42] Manufacturers and suppliers of AV will consequently have a duty of care to passengers and road users to prevent the AV manufactured or supplied by them from causing harm.

In addition to the common law tort of negligence (which is fault based), the *Australian Consumer Law* (**ACL**)[43] imposes a strict liability regime for manufacturers of goods that have a safety defect.[44] The term "*manufacturer*" is defined broadly under the ACL and includes, amongst others, a person who imports goods into Australia, where the manufacturer of the goods does not have a place of business in Australia.[45] Accordingly, for AV that are not manufactured in Australia, the importer is also liable for safety defects in the goods they import into Australia.

Under the relevant provisions of the ACL, an individual who suffers injury, loss or damage because of a safety defect of a good can recover such loss or damages from the manufacturer without the need to prove fault on the part of the manufacturer.[46] Goods have a safety defect if their safety is not of a standard that persons are generally entitled to expect.[47] In determining whether goods are defective, regard must be given to all relevant circumstances, such as the purpose for which they are marketed and instructions or warnings provided with the goods.[48]

Under the statutory product liability regime, the manufacturer (which includes the importer) of AV at Levels 0 to 2 is liable to any individual (including passengers and

[41] Global Legal Group, "*The International Comparative Legal Guide to: Product Liability 2015*", 13th Edition, page 51.

[42] *Donoghue v Stevenson* [1932] UKHL 100; Global Legal Group, "*The International Comparative Legal Guide to: Product Liability 2015*", 13th Edition, page 51.

[43] The Australian Consumer Law is set out in Schedule 2 to the *Competition and Consumer Act* 2010 (Cth).

[44] The product liability provisions are set out in Part 3-5 (*Liability of manufacturers for goods with safety defects*) of Schedule 2 to the *Competition and Consumer Act* 2010 (Cth).

[45] Section 7(1)(e) of Schedule 2 to the *Competition and Consumer Act* 2010 (Cth).

[46] Section 138 of Schedule 2 to the *Competition and Consumer Act* 2010 (Cth).

[47] Section 9(1) of Schedule 2 to the *Competition and Consumer Act* 2010 (Cth).

[48] Section 9(2) of Schedule 2 to the *Competition and Consumer Act* 2010 (Cth).

Table 3 Overview Consumer Guarantees

Consumer guarantee	Liability
AV is of acceptable quality[a]	Manufacturer and supplier
AV corresponds with description[b]	Manufacturer and supplier (with respect to any description of the AV provided by them)
AV will satisfy any express promises made about them (express warranties)[c]	Manufacturer and supplier (with respect to any express warranty for the AV given by them)
AV is fit for the purposes disclosed by the customer to the supplier before sale[d]	Supplier
AV matches any demonstration model shown to the purchaser[e]	Supplier
Reasonable supply of spare parts or repair facilities for the AV will be available after purchase[f]	Manufacturer

[a]Section 54 of Schedule 2 to the *Competition and Consumer Act* 2010 (Cth)
[b]Section 56 of Schedule 2 to the *Competition and Consumer Act* 2010 (Cth)
[c]Section 59 of Schedule 2 to the *Competition and Consumer Act* 2010 (Cth)
[d]Section 55 of Schedule 2 to the *Competition and Consumer Act* 2010 (Cth)
[e]Section 57 of Schedule 2 to the *Competition and Consumer Act* 2010 (Cth)
[f]Section 58 of Schedule 2 to the *Competition and Consumer Act* 2010 (Cth)

other traffic participants) who suffers a loss or damage because the AV has a safety defect.

6.3.2 Liability for Breach of Consumer Guarantees

The ACL also sets out certain statutory guarantees with respect to consumer goods which cannot be excluded.[49] Consumer goods include motor vehicles, regardless of price and whether the vehicle is acquired for personal or business use.[50] Under the ACL, manufacturers (which includes importers) and suppliers of AV are liable to the purchaser for a breach of the consumer guarantees given in Table 3.

In the case of AV manufactured overseas, importers will be liable to Australian consumers in respect of a breach of consumer guarantees relating to products for which the manufacturer would be responsible if the manufacturer were in Australia.[51]

Where one or more of these consumer guarantees are breached with respect to the AV, the purchaser is entitled to certain statutory remedies, including compensation

[49]Part 3-2 (*Consumer Transactions*), Division 1 (*Consumer Guarantees*), Subdivision A (*Guarantees relating to the supply of goods*) of Schedule 2 to the *Competition and Consumer Act* 2010 (Cth).

[50]Section 2 and section 3(1) of Schedule 2 to the *Competition and Consumer Act* 2010 (Cth).

[51]Section 7(1)(e) of Schedule 2 to the *Competition and Consumer Act* 2010 (Cth).

for damages, costs, refunds and replacement.[52] The type of remedy will depend on the seriousness of the breach of the consumer guarantee and whether the claim is made against the manufacturer or the supplier. For example, the purchaser of an AV will only be entitled to a full refund of the purchase price, or a replacement AV, if the defect was a 'major failure'[53] and if the claim is made against the supplier.[54] In addition to the remedies available against the supplier, the consumer is also entitled to make a direct claim against the manufacturers (which includes the importer) for breach of a consumer guarantee, to recover damages for any reduction in the value of the goods or any loss or damages suffered because of the breach of the guarantee.[55] If the defect is a manufacturing defect, for which the consumer has made a claim against the supplier, the supplier may seek recourse against the manufacturer or importer under a statutory indemnity.[56]

7 Liability for AV at Levels 3 to 5

Currently, AV at Levels 3 to 5 are not permitted on Australian roads as the existing in-service safety laws are insufficient to adequately address the risks and implications of these AV. However, legal reforms in Australia with respect to AV at Levels 3 to 5 are well advanced and there is a clear roadmap to the development of a nationally consistent regulatory framework to support the safe operation of AV at higher automation levels in Australia.[57]

In October 2020, the NTC released a comprehensive discussion paper setting out a proposal for a national in-service safety law for AV, the AVSL, which will establish a general safety duty on ADSE and place due diligence obligations on their executive officers.[58] Given that the constitutional legislative power with respect to the use of motor vehicles vests with the State and Territories, it was proposed that the AVSL will either be implemented through the use of complementary Federal and State and Territory laws which would ensure greater national consistency[59] or State

[52] Part 5-4 (*Remedies relating to guarantees*) of Schedule 2 to the *Competition and Consumer Act* 2010 (Cth).

[53] See section 260 of Schedule 2 to the *Competition and Consumer Act* 2010 (Cth) as to when a failure to comply with a guarantee is a major failure.

[54] Section 259(3) of Schedule 2 to the *Competition and Consumer Act* 2010 (Cth).

[55] Sections 271 and 272 of Schedule 2 to the *Competition and Consumer Act* 2010 (Cth).

[56] Section 274 of Schedule 2 to the *Competition and Consumer Act* 2010 (Cth).

[57] National Transport Commission (October 2020), A national in-service safety law for automated vehicles, page 8.

[58] National Transport Commission (October 2020), A national in-service safety law for automated vehicles, page 8.

[59] National Transport Commission (June 2020), In-service safety for automated vehicles, decision regulation impact statement, page 115; National Transport Commission (October 2020), A national in-service safety law for automated vehicles, page 9. Under this model, the Federal Government

and Territory applied law.[60] However, based on the NTC's final recommendations on the AVSL, the infrastructure and transport ministers agreed to implement the AVSL through Commonwealth law.[61] In terms of enforcement, it is proposed to establish an 'in-service safety' national regulator with statutory powers to monitor, investigate and issue penalties for contraventions of the AVSL.[62] It is expected that States and Territories would remain responsible for managing matters such as road access, roadside enforcement and human users of AV.

Under the proposed AVSL, the extent of liability which may be imposed on relevant parties will depend on their level of 'influence' on the in-service operation of the ADS.[63]

Figure 1 below illustrates the shift in the level of influence on the in-service safety of AV from the human driver to the ADSE with increased automation levels.

7.1 ADSE

7.1.1 Definition and Characteristics of ADSE

The ADSE is the self-nominated party that will certify that the ADS can safely perform the driving task in place of a human driver.[64] The ADSE will self-nominate at first supply when applying for type approval of the ADS.[65]

---

would pass a national law that regulates ADSE, its executive officers and remote drivers in reliance on its corporations and communications heads of power under section 51(v) and (xx) of the *Commonwealth of Australia Constitution Act 1900* (Cth). Whereas the State and Territories would regulate other participants, such as fallback ready users and AV occupants under 'complementary' legislation.

[60] National Transport Commission (June 2020), In-service safety for automated vehicles, decision regulation impact statement, page 131. Under this model, one State or Territory (so called 'host' jurisdiction) would enact a law regulating in-service safety of AV that is then applied by other participating jurisdictions. An example of such an 'applied law' model is the Heavy Vehicle National Law (HVNL). However, the HVNL has not achieved nationally consistent law as Western Australia and the Northern Territory have chosen not to apply it.

[61] National Transport Commission (February 2022), The regulatory framework for automated vehicles in Australia. https://www.ntc.gov.au/sites/default/files/assets/files/NTC%20Policy%20Paper%20-%20regulatory%20framework%20for%20automated%20vehicles%20in%20Australia.pdf. Accessed 24 January 2023.

[62] National Transport Commission (October 2020), A national in-service safety law for automated vehicles, Chapter 6 (*Functions of the in-service safety regulator*), page 58 et seq.

[63] National Transport Commission (June 2020), In-service safety for automated vehicles, decision regulation impact statement, Chapter 3 (*Parties with an influence on the in-service safety of automated vehicles*), page 43 et seq.

[64] National Transport Commission (October 2020), A national in-service safety law for automated vehicles, page 20, Table 1 (*Key Concepts*).

[65] See Section 4.1 Self-Certification and Type Approval.

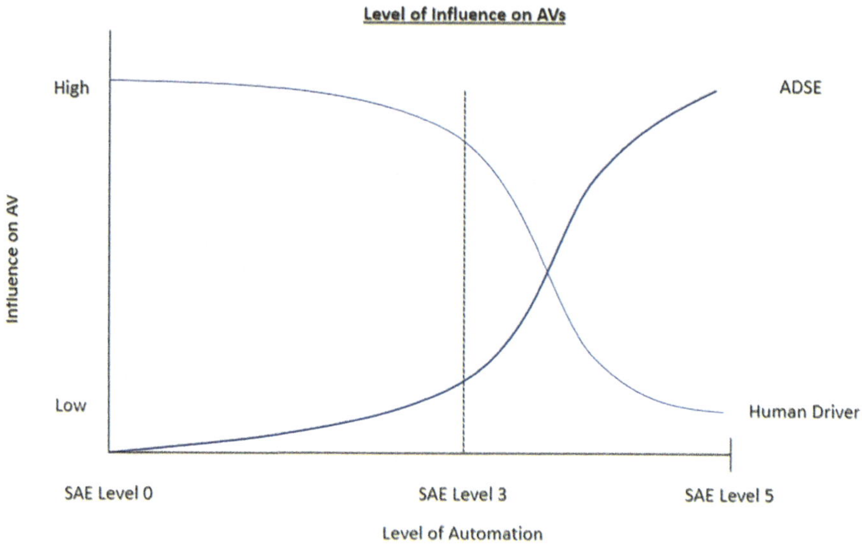

Fig. 1 Level of Influence of ADSE and Human Driver on AVs

For example, the ADSE could be the manufacturer of the ADS, the manufacturer of the vehicle or any other organisation importing the ADS into the Australian market. As the ADSE is required to self-certify at first supply, give an undertaking to manage safety risks when the ADS is engaged, and is responsible for maintaining and updating the ADS throughout its product life cycle, the ADSE is deemed to have major influence over the in-service operation of the AV.

Whilst the human driver has predominant control in more traditional vehicles (Levels 0 to 2 AV), this control shifts to the ADSE when the AV is operating at higher levels of automation and the ADS is assuming the dynamic driving task. This shift is depicted in the image above and, at higher levels of automation, the ADSE will have significant influence on the safe performance of the AV and the ADS.[66]

Being a new category of stakeholders within the Australian motor industry, current laws are ill-equipped to address the unique role of the ADSE.[67] For example, even though the ADSE may have duties under the common law tort of negligence to take reasonable care to prevent foreseeable harm that may be caused by the ADS, the qualitative standard of 'reasonable care' has yet to be determined in relation to AV

[66] National Transport Commission (June 2020), In-service safety for automated vehicles, decision regulation impact statement, page 46.

[67] National Transport Commission (June 2020), In-service safety for automated vehicles, decision regulation impact statement, page 66.

and their ADS.[68] In addition, the tort of negligence is a reactive, private right which requires the plaintiff to suffer damages before it may be enforced in court.[69]

7.1.2 General Safety Duty

Under the proposed AVSL, the ADSE will have a general duty of care to ensure the safe operation of the ADS ("**general safety duty**").[70] The ADSE must identify and address reasonably foreseeable safety risks and develop procedures to mitigate these risks.[71] To limit the broad concept of a general safety duty, it is proposed that in order to discharge that duty, the ADSE must ensure safety only to the extent it is 'reasonably practicable'.[72] The 'reasonably practicable' test is designed to give flexibility to accommodate the ADSE's unique business models and different contexts and will develop over time as technology and practices evolve.[73] The following was concluded with respect to the 'reasonably practicable' qualification in a review of a comparable general safety duty under the Australian work, health and safety laws (**WHS Laws**):

> The standard of 'reasonably practicable' is a high one, requiring the duty holder to consider all of the circumstances and take measures that are commensurate to the likelihood and seriousness of the harm which may result from the relevant activities, and relieved only by consideration of what is not possible or what is clearly unreasonable in the circumstances. A duty holder must clearly understand that this standard must be met.[74]

To discharge its general safety duty, the ADSE must have systems in place to enable it to identify and respond to reasonably foreseeable safety risks that may be caused by their ADS. For example, this may include:[75]

- having policies in place that ensure systemic software errors are quickly identified and addressed;

[68] National Transport Commission (June 2020), In-service safety for automated vehicles, decision regulation impact statement, page 66.

[69] *Civil Liability Act* 2002 (NSW), section 5A.

[70] National Transport Commission (October 2020), A national in-service safety law for automated vehicles, Chapter 3 (*ADSE Duties and enforcement framework*), page 26 et seq.

[71] National Transport Commission (October 2020), A national in-service safety law for automated vehicles, page 28.

[72] National Transport Commission (October 2020), A national in-service safety law for automated vehicles, page 28.

[73] National Transport Commission (June 2020), In-service safety for automated vehicles, decision regulation impact statement, page 78.

[74] National Review into Model Occupational Health and Safety Laws, First Report, October 2008, page 40.

[75] National Transport Commission (October 2020), A national in-service safety law for automated vehicles, page 28.

- designing an ADS so that the ADSE can detect if it is unable to function safely or whether it has been interfered with (such as through hacking or modification) and then prevent it from engaging;
- having effective communications strategies to alert consumers to safety issues (for example, through a smartphone app, email or advertising campaign);
- encouraging employees of the ADSE to escalate safety concerns, and generally promoting a safety culture (including disciplining employees who fail in their duty to prioritise safety); and
- making prompt disclosures to the regulator and other relevant authorities when the safety of their ADS is no longer assured.

The specific actions that will need to be taken by the ADSE to discharge its general safety duty will become clearer over time and will be further defined by case law, once it becomes available. In the NTC's view, the ADSE itself is best place to identify the steps necessary to ensure the ADS's safety.[76]

7.1.3 Prescriptive Duties

To support the general safety duty (which is limited by the 'reasonably practicable' test), it is proposed that the AVSL will include certain prescriptive duties. Given the novelty of AV at higher automation levels, it is impossible to predict all areas where prescriptive duties may be beneficial.[77] Therefore, in the NTC's opinion, it is important that the AVSL provides sufficient authority for additional prescriptive duties to be introduced in a timely way.[78] It is further proposed that the prescriptive rules will be agreed in conjunction with State and Territory governments.[79]

The NTC has suggested including the following prescriptive rules in the AVSL under which the ADSE must ensure that:[80]

- internal systems are developed, used and maintained to carry out its general safety duty;
- system upgrades to the ADS are installed safely and do not result in the operation of an unsafe ADS;
- the ADS software is without risks to the health and safety of users;

[76] National Transport Commission (October 2020), A national in-service safety law for automated vehicles, page 28.

[77] National Transport Commission (October 2020), A national in-service safety law for automated vehicles, page 30.

[78] National Transport Commission (October 2020), A national in-service safety law for automated vehicles, page 18.

[79] National Transport Commission (October 2020), A national in-service safety law for automated vehicles, page 18.

[80] National Transport Commission (October 2020), A national in-service safety law for automated vehicles, Table 2 (*Potential prescriptive duties under the general safety duty in the AVSL*), page 29 et seq.

- education and training is provided to relevant parties such as users of its ADS;
- the operation of an ADS is prevented when the ADSE is aware the ADS is unsafe;
- the ADS can comply with relevant road traffic laws that may change from time to time; and
- the ADS cannot operate outside its approved ODD.

In addition, under the proposed prescribed duties, the ADSE must:[81]

- notify the in-service regulator and users of any systemic safety issues affecting the ADS; and
- record and store data relevant to compliance with the general safety duty.

Furthermore, a rule-making power for prescriptive duties will be included in the AVSL.[82]

7.1.4 Consequences of Breach

7.1.4.1 Criminal and Civil Penalty Regime

It is proposed that a criminal and civil penalty regime will be introduced for breaches of obligations under the AVSL that follows the approach in the *Heavy Vehicles National Law* 2013 (**HVNL**).[83] This means that penalty provisions will be included within the AVSL and will be categorised to reflect the degree of culpability and the severity of potential damages caused.

The penalty provisions will be enforced by the in-service regulator.

Table 4 gives an overview of the proposed penalty regime.[84]

Amounts for the monetary penalties have not yet been proposed. However, looking at the comparable penalty regime under the HVNL legislation, it is to be expected that the monetary penalties will be quite severe. For example, a Category 1 offence under the HVNL legislation attracts a maximum monetary penalty of AUD 300,000 for an individual or AUD 3,000,000 for a corporation.[85]

---

[81] National Transport Commission (October 2020), A national in-service safety law for automated vehicles, Table 2 (*Potential prescriptive duties under the general safety duty in the AVSL*) page 29 et seq.

[82] National Transport Commission (February 2022), The regulatory framework for automated vehicles in Australia, page 32. https://www.ntc.gov.au/sites/default/files/assets/files/NTC%20Policy%20Paper%20-%20regulatory%20framework%20for%20automated%20vehicles%20in%20Australia.pdf. Accessed 18 January 2023.

[83] *Heavy Vehicles National Law* 2013 (NSW).

[84] National Transport Commission (October 2020), A national in-service safety law for automated vehicles, Appendix B (*Illustrative obligations and offence provisions*), page 134 et seq.

[85] Section 26F of the Heavy Vehicle National Law.

Table 4 Proposed penalty regime

Breach of proposed obligation	Proposed penalties
General safety duty: The ADSE must ensure, so far as is reasonably practicable, that its ADS is safe when used for a purpose for which it was designed, manufactured, supplied or installed.	A breach of the general safety duty is a criminal offence and will be divided into 3 categories: **Category 1:** ADSE acts reckless **and** exposes an individual to whom a duty of care is owed to the risk of death, serious illness or injury: **Maximum penalty**: • Individuals: prescribed monetary penalty or up to 5 years imprisonment or both • Corporations: prescribed monetary penalty × 5 **Category 2:** ADSE does not act reckless but exposes an individual to whom a duty of care is owed to the risk of death, serious illness or injury: **Maximum penalty:** • Individuals: prescribed monetary penalty • Corporations: prescribed monetary penalty × 5 **Category 3:** ADSE does not act reckless **and** does not expose any individual to whom a duty of care is owed to the risk of death, serious illness or injury: **Maximum penalty**: • Individuals: prescribed monetary penalty • Corporations: prescribed monetary penalty × 5
Prescriptive Duties	Civil penalty offence

7.1.4.2 Statutory Cause of Action for Injured Persons

At this stage, the NTC has decided against recommending a statutory cause of action which would give injured persons a direct claim against the ADSE for damages caused by a breach of the general safety duty. If the AVSL would include such a cause of action, those injured by an ADSE's breach of the general safety duty would be able to bring a direct claim against the ADSE for damages they have suffered as a consequence of that breach. Without such a statutory right of action, the injured person must rely on the tort of negligence and other statutory actions, for example under the product liability laws which are set out in the Australian Consumer Law.

Even though the NTC acknowledges that allowing injured persons to litigate safety duty breaches under the AVSL could increase the likelihood of compliance by the ADSE,[86] it has considered it premature to recommend a statutory cause of action until it transpires that the tort of negligence, motor accident injury insurance schemes

[86] National Transport Commission (October 2020), A national in-service safety law for automated vehicles, page 34.

and the Australian Consumer Law are insufficient for injured individuals to recover damages suffered from defective ADS.[87]

7.2 ADSE Executive Officers

7.2.1 Executive Officers

Under the proposed AVSL, executive officers of the ADSE may also be responsible for the in-service safety of AV at Levels 3 to 5.[88]

The term 'executive officer' refers to an individual who has decision-making authority within the ADSE.[89] Typically, these individuals are senior officers of the ADSE who are in a position to have major influences over the safe operation of the ADS as they have the power to make decisions which affect the design or operation of the ADS.

7.2.2 Due Diligence Obligation

It is proposed that the AVSL will impose a due diligence obligation on executive officers of the ADSE to ensure that the ADSE as an organisation complies with its safety obligations.[90] While it may seem harsh to impute personal liability on executive officers, pecuniary penalties imposed on the ADSE would penalise the ADSE's shareholders rather than the management team who is responsible for the breach.[91] The concept of personal liability of officeholders is well-established in Australian law where a corporation's breach puts individuals at risk of injury or death.[92]

The proposed due diligence obligation will be linked to the ADSE's general safety duty. Only those executive directors who are in a position to influence the ADSE's compliance with its general safety duty will be subject to a due diligence

[87] National Transport Commission (October 2020), A national in-service safety law for automated vehicles, page 35.

[88] National Transport Commission (October 2020), A national in-service safety law for automated vehicles, page 26.

[89] National Transport Commission (October 2020), A national in-service safety law for automated vehicles, page 35.

[90] National Transport Commission (October 2020), A national in-service safety law for automated vehicles, page 36.

[91] National Transport Commission (June 2020), In-service safety for automated vehicles, decision regulation impact statement, page 47.

[92] National Transport Commission (June 2020), In-service safety for automated vehicles, decision regulation impact statement, page 47. Other examples are the Heavy Vehicles National Law and the Work Health and Safety Laws.

obligation and only to the extent of their own personal influence.[93] In order to discharge the due diligence duty, the relevant executive officer must take reasonable steps to:[94]

- acquire knowledge and keep up to date about ADS safety matters;
- ensure the ADSE has the right resources and processes in place, and uses those resources and processes to eliminate or minimise ADS safety risks;
- ensure the ADSE has the right processes to receive and respond to reports of safety related incidents, hazards or issues, and processes to comply with the general safety duty; and
- verify that the processes and resources set out above are being used.

Most recently, the NTC has suggested to include the right of the executive directors to refer to a defence of 'reasonable reliance' in case of prosecution.[95]

7.2.3 Consequences of Breach

A breach of the due diligence obligation by the executive officer is a criminal offence punishable with a monetary penalty or imprisonment or both. The executive officer may be convicted of an offence even if the ADSE has not been proceeded against for, or convicted of, a breach of its general safety duty.[96]

7.3 Human Driver/Fallback Ready User

As shown in the image above, the human driver still retains influence over the operation of AV at Level 3 as the human driver is expected to resume control of the AV when demanded by the ADS. Here, the human driver is referred to as the 'fallback ready user'. Because the fallback ready user must respond to the ADS' requests and be capable of driving, Australian transport ministers have agreed that the fallback ready user in an AV at Level 3 should remain subjected to prescriptive

[93] National Transport Commission (October 2020), A national in-service safety law for automated vehicles, page 36; COAG Principles on Directors' Liability Provisions, adopted December 2009.

[94] National Transport Commission (October 2020), A national in-service safety law for automated vehicles, page 36.

[95] National Transport Commission (February 2022), The regulatory framework for automated vehicles in Australia, page 35. https://www.ntc.gov.au/sites/default/files/assets/files/NTC%20Discussion%20Paper%20-%20On-road%20enforcement%20for%20automated%20vehicles_0.pdf. Accessed 18 January 2023.

[96] National Transport Commission (October 2020), A national in-service safety law for automated vehicles, page 137.

rules similar to those imposed on human drivers of AV at Levels 0 to 2 under the ARR and their implementation laws at State level.[97]

As the ADS may require the fallback ready user to resume control of the AV at any time during operation, the fallback ready user must be licenced, be of minimum age and must be physically able to drive the AV. This means the fallback ready user must at all times comply with intoxication and fatigue requirements.[98] Although these matters are already addressed by existing legislation with respect to human drivers, current laws were not consciously drafted in consideration of fallback ready users. Therefore, it is anticipated that the existing road and traffic laws at State level may be modified to include specific duties of a fallback ready user, such as a duty to remain sufficiently vigilant to respond to ADS requests or mechanical failure and regain control of the vehicle without undue delay when required.[99] To ensure a uniform approach, it is expected that the NTC will develop model provisions for inclusion in the ARR that set out specific duties for fallback ready users and the State and Territories will then implement these provisions into their road traffic laws.

In AV at Levels 4 to 5, the control of the AV lies predominantly with the ADS and the human driver becomes a mere occupant of the vehicle. Hence, the human driver will have only minor or no influence and significantly lower responsibility for ensuring the safe operation of the AV.[100]

7.4 Remote Driver

The remote driver is a human who takes control of the AV upon the request of the ADS but unlike the fallback ready user, the remote driver is typically not an occupant within the vehicle. As mentioned above, current laws assume that the person in control of the vehicle is physically present within the vehicle. In relation to AV at Levels 4 and 5, this assumption is undoubtedly outdated as the remote driver may not even be in Australia. Although the NTC indicated that various current criminal negligence offences[101] may apply to remote drivers, the national inconsistency and ill-suited nature of these provisions to remote drivers mean that these laws will unlikely be effective in practice.

---

[97] National Transport Commission (June 2020), In-service safety for automated vehicles, decision regulation impact statement, page 49.

[98] National Transport Commission (June 2020), In-service safety for automated vehicles, decision regulation impact statement, page 49.

[99] National Transport Commission (June 2020), In-service safety for automated vehicles, decision regulation impact statement, page 49.

[100] National Transport Commission (June 2020), In-service safety for automated vehicles, decision regulation impact statement, page 46.

[101] For example, *Criminal Code Act* 1899 (QLD), section 289; *Crimes Act 1958* (VIC), sections 318 and 319.

Currently, there is no intention to include specific regulations for remote driver licensing in the AVSL. However, the ministers have agreed to include a head of power to make prescriptive rules to cover remote driving.[102]

Unless and until, remote drivers are regulated under the AVSL, their liability will be confined to the common law tort of negligence. Where the remote driver is not based in Australia, jurisdictional issues arise, and it may be difficult for persons that were injured in Australia due to a fault of the remote driver to enforce their rights against the remote driver in a foreign jurisdiction.

7.5 *Registered Owner*

As mentioned above, vehicles must be registered before they can be legally driven on public roads.[103] After registration, the registered owner, whether it may be an individual or corporation, will be legally responsible for the registered vehicle.

Under the NTC analysis, the registered owner has major influence over the safe operation of the ADS because the registered owner is imputed the duty to conduct regular servicing and maintenance of the vehicle, including the ADS.[104] The registered owner must also oversee ADS software upgrades, and this is a critical duty as outdated or malfunctioning software may cause otherwise avoidable collisions and incidents.

Under the current proposal for Levels 3 to 5 AV, the registered owner will be regulated under existing State and Territory legislation which may need to be modified to address the unique responsibilities of a registered owner of an AV that no longer requires a human driver. In particular, it is anticipated that infringement notices that are currently issued to the human driver may in the future be issued to the registered owner.

7.6 *Manufacturer, Importer and Supplier*

The manufacturer or importer will be the ADSE and as such will be subject to the general safety duty and supportive prescriptive duties under the AVSL.[105]

---

[102] National Transport Commission (February 2022), The regulatory framework for automated vehicles in Australia, page 42. https://www.ntc.gov.au/sites/default/files/assets/files/NTC%20Policy%20Paper%20-%20regulatory%20framework%20for%20automated%20vehicles%20in%20Australia.pdf. Accessed 18 January 2023.

[103] Service NSW, Vehicle Registration. https://www.service.nsw.gov.au/transaction/renew-vehicle-registration. Accessed 13 January 2023.

[104] National Transport Commission (June 2020), In-service safety for automated vehicles, decision regulation impact statement, page 52.

[105] See Section 7.1ADSE.

In addition to being regulated under the proposed AVSL, the manufacturer and importer of an AV, including its ADS, will be liable under the existing Australian product liability regime. Accordingly, if damages or injuries are suffered because of a safety defect in an ADS, the injured person may have causes of action under the common law tort of negligence (which is fault based) or the product liability provisions of the ACL (which impose a strict liability).[106]

Similarly, the manufacturer, importer and supplier of AV are liable to Australian consumers if a statutory consumer guarantee is breached in respect of the AV, including the ADS.[107]

7.7 Repairers and Modifiers

NTC considers repairers and modifiers as persons with major influence over the safe operation of AV as they have direct access and control over AV components when repairing, replacing, or modifying hardware and software.[108] Parties such as vehicle manufacturers, dealers, and independent mechanics could fall under this category.

However, despite their major influence on the in-service safety of AV, the NTC does not propose to regulate repairers or modifiers under the AVSL. Rather, the substantive regulation of repairers and modifiers will remain the responsibility of State and Territory governments. However, most States and Territories do not have a certification process for modifications, and existing regulations address the relationship between repairers and their customers rather than public safety.[109]

In addition, repairers and modifiers will continue to be liable to their customers under the ACL for breach of consumer guarantees as they apply to the provision of services.[110]

7.8 Road Managers

Road managers are responsible for ensuring that roads are in proper condition and to take reasonable care in avoiding foreseeable risk of harm to road users when

[106] See section 6.3.1 6.3(a) Product Liability.

[107] See section 6.3.2 6.3(b) Liability for breach of consumer guarantees.

[108] National Transport Commission (June 2020), In-service safety for automated vehicles, decision regulation impact statement, pages 50 and 51.

[109] *Motor Dealers and Repairers Act* 2013 (NSW), section 3.

[110] Part 3-2 (*Consumer transactions*), Division 1 (*Consumer guarantees*), Subdivision B (*Guarantees relating to the supply of services*) of Schedule 2 to the *Competition and Consumer Act* 2010 (Cth).

exercising their powers.[111] However, as technology advances and connected AV, i.e. AV which are able to digitally communicate with road infrastructures and other vehicles (**CAV**), are introduced, the liability of road managers will increase. Indeed, road managers must indirectly contribute to the on-road safety of AV by ensuring the proper digital and physical operation of road infrastructure. Examples of the road manager's duty include ensuring that road surface markings are visibly painted on roads, electronic devices can promptly communicate changes in traffic lights or speed limits to CAV, and the ADS or ADSE is swiftly notified of any road works or road closures.[112] These duties apply to both public road managers, such as State, Territory, and local governments, and private road managers, i.e. companies responsible for maintaining particular roads such as tollways.

7.9 Vehicle Inspectors

Vehicle inspectors are regulatory or licenced entities who certify a vehicle's roadworthiness before the vehicle is legally permitted to be driven on roads. The NTC considers that vehicle inspectors only possess minor influence on the in-service safety of AV because their role is to merely test for faults and certify repairs rather than actually conducting repairs.[113]

However, this conclusion may be challenged, as vehicle inspectors are the final checkpoint before an AV is released onto public roads. Although States and Territories differ in their determination of when vehicles must be inspected, all jurisdictions largely concur in that the vehicle must be presented for inspection at sale of the vehicle and annual registration.[114] Hence, as the vehicle inspector will likely have access to the vehicle at key moments of the vehicle's operational lifespan, this demonstrates the significant role and greater liability vehicle inspectors should have with respect to the safe in-service operation of AV.

[111] National Transport Commission (June 2020), In-service safety for automated vehicles, decision regulation impact statement, page 53; *Road Management Act* 2004 (VIC), section 40; *Brodie v Singleton Shire Council* (2001) 206 CLR 512.

[112] *Roads Act* 1993 (NSW), section 38B.

[113] National Transport Commission (June 2020), In-service safety for automated vehicles, decision regulation impact statement, page 54.

[114] Transport for NSW, Safety checks and vehicle inspections. https://www.rms.nsw.gov.au/roads/vehicle-registration/renewing-vehicle/safety-checks-and-vehicle-inspections/index.html. Accessed 13 January 2023.

7.10 Telecommunication Service Providers

Telecommunication service providers are responsible for facilitating the communication between the ADS, ADSE, and road infrastructures for CAV. For ADS that rely upon constant connectivity with the ADSE or road infrastructures for operation, steady and reliable telecommunication connections are crucial for safe operation of the AV. Despite this, the NTC expects the ADSE, at the design stage of the ADS, to anticipate the risk of communication network failure such that if required, the ADS must be capable of making decisions without information from communication networks.[115] This shifts the majority of potential liability to the ADSE, which the NTC argues as reasonable since failures such as outages to telecommunication infrastructures are foreseeable.[116] Furthermore, telecommunication service providers have not made any guarantees that their services are fit for ADS use, and the ADSE ultimately undertook at first supply to ensure the safe operation of its ADS. This may appear as overly onerous for the ADSE, and it is contentious that, at least within Australian metropolitan regions, it may be reasonable to expect steady network connections from telecommunication service providers.

Currently, telecommunication service providers are overseen by the *Australian Communications and Media Authority* (**ACMA**) and regulated by legislations such as the *Telecommunications Act* 1997 (Cth). As neither the regulatory body nor the legislation were created with governing the service of telecommunication providers to keep ADS in mind, it is possible that there will also be legislative reform on this front. As the legislative powers with respect to telecommunication vests with the Federal Government, it would be either possible to regulate the duties and responsibilities of the telecommunication service providers in the AVSL or to adapt the Commonwealth telecommunication laws.[117]

7.11 Other Road Users

Other road users, such as human drivers, passengers, motorcyclists, cyclists, and pedestrians, may also be liable for the in-service safety of AV as road users owe a general duty of care to not cause harm to other road users. This duty is also owed to road users who disobey road rules by acting carelessly or even being intoxicated.[118]

---

[115] National Transport Commission (June 2020), In-service safety for automated vehicles, decision regulation impact statement, page 56.

[116] National Transport Commission (June 2020), In-service safety for automated vehicles, decision regulation impact statement, page 56.

[117] The last option was proposed by the NTC, cf. National Transport Commission (February 2022), The regulatory framework for automated vehicles in Australia, page 40.

[118] *March v E & MH Stramare Pty Ltd* [1991], 171 CLR 506, 527 (McHugh J).

In addition, certain road users may also be accountable to rules specific to that category of road users. For example, pedestrians are prohibited from causing traffic hazards by moving into the path of a driver.[119] However, the ARR were not developed in consideration of AV and may be incomprehensive in addressing new situations or undesirable behaviours in relation to AV. States and Territories will need to consider if additional regulation is required.[120]

8 Conclusion

The reforms to shape the Australian regulatory landscape for AV are well advanced. Since 2016, the NTC has worked closely with government and industry to develop a national framework and published its final recommendations in 2021. Based on the NTC's recommendations, in February 2022, the infrastructure and transport ministers agreed on a clear path for a regulatory framework for AV in Australia. The drafting of the AVSL is about to begin and it is expected that it will be passed by the Federal Parliament by 2026. The AVSL will define duties and liabilities of the ADSE, its executive officers and remote drivers whereas the fallback ready users and other persons with influence on in-service safety of the AV will remain subject to existing State and Territory legislation with necessary modifications.

With respect to 'safety at first supply' it is proposed that the existing self-certification framework contained in the *Road Vehicle Standards Act* will be modified to extend to ADS of Levels 3 to 5 AV. The ADSE will be required to self-certify and seek type approval of their AV before they may be driven on Australian roads.

Considering the complex technology of AV at higher automation levels and the interaction between the ADS with other vehicle components and telecommunication systems, it is to be expected that a malfunction leading to an accident will rarely be due to a single cause but rather a combination of multiple failures. Therefore, in practice, the difficulty will be to allocate the liability between the various persons with influence on the in-service safety of the AV. Given the division of regulations of the various persons with influence on the in-service safety of the AV between Federal (ADSE, its executive directors and remote driver) and State laws (fallback ready user, repairers and modifiers) the challenge will be to identify the responsible person(s) and to apportion the liability amongst them.

---

[119] *National Transport Commission (Road Transport Legislation – Australian Road Rules) Regulations* 2006 (Cth), section 236.

[120] National Transport Commission (February 2022), The regulatory framework for automated vehicles in Australia, page 45.

Legislation

Road Vehicles Standards Act 2018 (Cth)
Road Safety (Automated Vehicles) Regulations 2018 (VIC)
Transport Legislation Amendment (Automated Vehicle Trials and Innovation) Act 2017 (NSW)
Road Safety Rules 2017 (VIC)
Road Transport (Road Rules) Regulation 2017 (ACT)
Motor Vehicles (Trials of Automotive Technologies) Amendment Act 2016 (SA)
Road Rules 2014 (NSW)
Motor Dealers and Repairers Act 2013 (NSW)
Heavy Vehicles National Law 2013 (NSW)
Competition and Consumer Act 2010 (Cth)
Transport Operations (Road Use Management-Road Rules) Regulation 2009 (QLD)
Road Rules 2009 (TAS)
Road Safety Road Rules 2009 (VIC)
National Transport Commission (Road Transport Legislation – Australian Road Rules) Regulations 2006 (Cth)
Vehicle Standard (Australian Design Rule – Definitions and Vehicle Categories) 2005 (Cth)
Road Management Act 2004 (VIC)
National Transport Commission Act 2003 (Cth)
Civil Liability Act 2002 (NSW)
Road Traffic Code 2000 (WA)
Road Transport (Safety and Traffic Management) Regulation 2000 (ACT)
Traffic Regulations 1999 (NT)
Roads Act 1993 (NSW)
Criminal Code Act 1983 (NT)
Road Traffic Act 1974 (WA)
Road Traffic Act 1961 (SA)
Crimes Act 1958 (VIC)
Criminal Law Consolidation Act 1935 (SA)
Criminal Code Act 1924 (TAS)
Commonwealth of Australia Constitution Act 1900 (Cth)
Crimes Act 1900 (ACT)
Crimes Act 1900 (NSW)
Criminal Code Act 1899 (QLD)

Case Law

Prince Alfred College Incorporated v ADC [2016] HCA 37
Brodie v Singleton Shire Council [2001], 206 CLR 512

March v E & MH Stramare Pty Ltd [1991], 171 CLR 506
Sobulsky v Ega [1960], 103 CLR 215
Donoghue v Stevenson [1932] UKHL 100

References

Darren Gray, Sydney Morning Herald (27 April 2019) No one behind the wheel: The new workforce driving Australia's mines

Global Legal Group, The International Comparative Legal Guide to: Product Liability 2015, 13th Edition

KPMG International Cooperative (July 2020) Autonomous Vehicles Readiness Index. https://home.kpmg/xx/en/home/insights/2020/06/autonomous-vehicles-readiness-index.html. Accessed 24 January 2023

National Transport Commission (May 2016) Regulatory Options for Automated Vehicles Discussion Paper. https://www.ntc.gov.au/sites/default/files/assets/files/Regulatory-options-for-automated-vehicles-May-2016.pdf. Accessed 24 January 2023

National Transport Commission (May 2017) Guidelines for Trials of Automated Vehicles in Australia. https://www.ntc.gov.au/sites/default/files/assets/files/AV_trial_guidelines.pdf. Accessed 24 January 2023

National Transport Commission (July 2019) Automated Vehicle Program. https://www.ntc.gov.au/sites/default/files/assets/files/Automated%20Vehicle%20Program%20Approach_0.pdf. Accessed 24 January 2023

National Transport Commission (June 2020a) In-service safety for automated vehicles, decision regulation impact statement. https://www.ntc.gov.au/sites/default/files/assets/files/NTC-Decision-RIS-In-service-safety-for-AVs.pdf. Accessed 24 January 2023

National Transport Commission (September 2020b) Automated Vehicle Program Approach. https://www.ntc.gov.au/sites/default/files/assets/files/Automated%20vehicle%20approach.pdf. Accessed 24 January 2023

National Transport Commission. (October 2020c) A national in-service safety law for automated vehicles. https://www.ntc.gov.au/sites/default/files/assets/files/NTC-Discussion-Paper-national-in-service-safety-law-for-AVs.pdf. Accessed 24 January 2023

National Transport Commission (December 2020d) Guidelines for trials of automated vehicles in Australia. https://www.ntc.gov.au/sites/default/files/assets/files/AV-trial-guidelines-2020.pdf. Accessed 24 January 2023

National Transport Commission (December 2020e) Lessons learned from automated vehicle trials in Australia. https://www.ntc.gov.au/sites/default/files/assets/files/AV-lessons-learned-2020_0.pdf. Accessed 24 January 2023

National Transport Commission (February 2022a) The regulatory framework for automated vehicles in Australia. https://www.ntc.gov.au/sites/default/files/assets/files/NTC%20Policy%20Paper%20-%20regulatory%20framework%20for%20automated%20vehicles%20in%20Australia.pdf. Accessed 24 January 2023

National Transport Commission (July 2022b) On-road enforcement for automated vehicles. https://www.ntc.gov.au/sites/default/files/assets/files/NTC%20Discussion%20Paper%20-%20On-road%20enforcement%20for%20automated%20vehicles_0.pdf. Accessed 24 January 2023

National Transport Commission Submissions. https://www.ntc.gov.au/submission/540. Accessed 24 January 2023

National Transport Commission homepage, In-service safety for automated vehicles, timeline. https://www.ntc.gov.au/transport-reform/ntc-projects/in-service-safety-AVs. Accessed 24 January 2023

NSW Government. (19 February 2020) What does embracing technology mean for transport? https://www.sydneymetro.info/article/what-does-embracing-technology-mean-transport. Accessed 24 January 2023

SAE International (April 2021) Surface Vehicle Recommended Practice J3016

Service NSW., Vehicle Registration. https://www.service.nsw.gov.au/transaction/renew-vehicle-registration. Accessed 24 January 2023

Transport for NSW, Safety checks and vehicle inspections. https://www.rms.nsw.gov.au/roads/vehicle-registration/renewing-vehicle/safety-checks-and-vehicle-inspections/index.html. Accessed 24 January 2023

Part V
Europe

Autonomous Vehicles and Civil Liability in Belgium

Orian Dheu, Jan De Bruyne, Peggy Valcke, and Ilse Samoy

Abstract In the absence of an *ad hoc* liability regime for autonomous vehicles, victims will have to rely on existing Belgian legal frameworks. Whereas certain road victims, the so-called 'vulnerable users', will presumably continue to benefit from facilitated compensation stemming from the no-fault regime of Article 29*bis*, other road victims, including drivers, will have to resort to general fault liability, liability for things or product liability as a legal basis for bringing a claim. The following analysis identifies potential legal hurdles that may arise when applying these regimes and which may vary according to the vehicle's automation level. Furthermore, the analysis provides an overview into the potential for autonomous vehicles to 'disrupt' existing liability paradigms, notably through the foreseen digitalisation and 'servitisation' of mobility.

The following chapter does not touch upon developments at the European level such as the Commission's recent proposals on an AI Liability Directive or a revised Product Liability Directive as they are not yet binding. See for a first analysis of these proposals: Dheu O, De Bruyne J, Ducuing C (2022) The European Commission's Approach To Extra-Contractual Liability and AI – A First Analysis and Evaluation of the Two Proposals. CiTiP Working Paper. Available on SSRN: https://papers.ssrn.com/sol3/papers.cfm?abstract_id=4239792.

Part of this research has received funding from the European Union's Horizon 2020 Research and Innovation - Marie Skłodowska-Curie actions program under the Safer Autonomous Systems (SAS) project, grant agreement n° 812.788: https://etn-sas.eu/. This publication reflects only the author's view, exempting the European Union from any liability.

O. Dheu · J. De Bruyne (✉)
KU Leuven Centre for IT & IP Law - imec, Leuven, Belgium
e-mail: jan.debruyne@kuleuven.be

P. Valcke
KU Leuven Centre for IT & IP Law, Leuven, Belgium

I. Samoy
KU Leuven Institute for the Law of Obligations, UHasselt, Leuven, Belgium

H. Steege et al. (eds.), *Autonomous Vehicles and Civil Liability in a Global Perspective*, Data Science, Machine Intelligence, and Law 3,
https://doi.org/10.1007/978-3-031-41992-8_11

1 Introduction

The world is preparing for autonomous vehicles, and so is Belgium.[1] In 2018, it introduced rules to allow the testing of such cars.[2] While autonomous vehicles are expected to increase the general levels of safety, ruling out all accidents is not a realistic scenario. In fact, incidents and accidents involving automated vehicles have already occurred. An example is the 2016 fatal Tesla car crash in the United States (US). The question of the allocation of liability is undoubtedly raised and in the absence of *ad hoc* frameworks, the adequacy of current civil liability mechanisms will be tested.

This chapter discusses the liability and compensation rules under Belgian civil law that are relevant in the event of an accident involving an autonomous vehicle. Various parties can be targeted under current Belgian civil law rules: the 'driver' or user of the vehicle, the owner/guardian of the vehicle, its manufacturer, the software developer, the service providers or infrastructure managers, etc. This chapter will map and explain the existing liability and compensation mechanisms applicable to more traditional vehicles (Sect. 2). We will subsequently examine how these liability and compensation regimes could apply to vehicles of SAE[3] levels 3 and 4 (Sect. 3). Finally, we will analyse how such regimes may apply to SAE level 5 vehicles that are fully autonomous (Sect. 4). The most important findings are briefly summarised in a conclusion (Sect. 5). The methodological approach used in our analysis is both descriptive and evaluative. The assessment is prospective as virtually no case law exists on the subject matter in Belgium. The analysis is therefore based on existing scholarship as well as relevant case law under the respective civil liability and compensation regimes.

[1] See for an extensive study on the societal impact of autonomous vehicles: De Bruyne (2021), 408 p.

[2] Royal decree (arrêté) of 18th March 2018 related to tests with automated vehicles. This statute inserts a new article, art. 59/1 in the royal decree (arrêté) of December the 1st 1975. This text establishes the possibility for the public authorities (Minister in charge of road traffic) to grant derogations to the Road traffic code in order for tests of autonomous vehicles to be carried out including in real time traffic. See: Cassart (2018), p. 136 (n°12–13). A code of practice for testing with autonomous vehicles on public roads was previously adopted in 2016 (however, it is non-binding guidelines).

[3] The Society of Automotive Engineers (SAE) designed a 6 level scale ranging from no automation (level 0) to full automation (level 5).

2 Liability for Level 0–2 Vehicles: Mapping the Existing Liability and Compensation Mechanisms for Accidents Involving 'Traditional' Vehicles

Level 0 vehicles have no driving automation and have already been on the roads for more than a century. In such cars, the user retains full control and is in charge of the vehicle's motion. Level 1 vehicles entail 'driver assistance' and constitute the lowest level of automation with features such as *lane keeping assistance* and *adaptive cruise control* assisting the user.[4] Level 2 vehicles with 'partial automation' correspond to cars that embed advanced driving technologies such as *park assist* and collision avoidance systems.[5] However, the human driver retains control and monitoring duties. This means that the existing liability mechanisms will apply. While certain categories of victims (so-called 'vulnerable' road users) may benefit from a protective regime (Sect. 2.1), fault-based liability based on the Articles 1382–1383 of the Old Belgian Civil Code (OBCC) remains the cornerstone of damage claims in traffic-related matters (Sect. 2.2). However, other legal bases for liability can also be relied upon by the victim of a car accident, namely liability of the custodian for defective things (Sect. 2.3) and product liability (Sect. 2.4).

*2.1 Article 29*bis *and Automatic Compensation of 'Vulnerable' Road Victims*

In Belgium, there is no specific liability regime applicable to road traffic accidents.[6] General fault-based liability pursuant to Article 1382 OBCC usually applies. The effectivity of the liability regime is strengthened by a system of mandatory third-party motor vehicle insurance.[7] This regime requires the vehicle owner to contract an insurance policy that covers liability incurred by the vehicle.[8] The law provides that such an insurance policy must cover the liability of the owner, of any keeper of the vehicle and the driver who injured a party.[9] However, proving liability through fault may sometimes be challenging. In that respect, a specific act establishes a compensation scheme for certain categories of road victims in which they are not required to prove fault from the driver. The victim has a direct claim for compensation against

[4] Malengreau (2019a), p. 15.578 (n°6).

[5] *Ibid.*

[6] Kruithof (2018), p. 99 (n°163).

[7] *Ibid.* p. 99 (n°164).

[8] Article 2 §1 paragraph 1 and 2 of the Act of 21 November 1989 on Mandatory Liability Insurance for Auto Motor Vehicles (la Loi du 21 novembre 1988 relative à l'assurance obligatoire de la responsabilité en matière de véhicules automoteurs).

[9] Article 3 §1 paragraph 1 of the Act of 21 November 1989 on Mandatory Liability Insurance for Auto Motor Vehicles.

the liability insurer of any motor vehicle that is involved in the accident.[10] This compensation scheme can be found in Article 29*bis* of the Act of 21 November 1989 on Mandatory Liability Insurance for vehicles.[11]

2.1.1 Conditions of Application

In order to apply this regime, the plaintiff only needs to prove that he/she was the victim of road accident (1) where a motor vehicle (2) was involved (3). First, such a traffic accident (*accident de la circulation*) must have happened on public roads, terrains open to the public or terrains not open to the public but open to certain categories of persons authorised to ride through them.[12] The notion of 'accident' is not defined by the law but one can refer to its common understanding, which is a sudden and unforeseen event that causes a damage.[13] The law also refrains from defining the notion of 'traffic' (*circulation*). The notion of 'traffic accident' has, however, been extensively interpreted by judges.[14] Second, the accident needs to involve an 'motor vehicle' (*véhicule automoteur*), which is defined as a vehicle intended to circulate on the ground and that can be actioned by a mechanical force without being linked to a railway.[15] Third, the notion of the vehicle's involvement in the accident is a flexible yet complex condition that is broadly interpreted.[16] It is usually defined as the participation of the vehicle in the accident.[17] The Belgian *Court of Cassation* states that a vehicle is involved when any link between the vehicle and the accident can be demonstrated.[18] Such a link can be demonstrated if

---

[10] Kruithof (2018), p. 101 (n°167).

[11] Article 29bis of the Act of the 21 November 1989 on Mandatory Liability Insurance for Auto Motor Vehicles (as amended). For more information on Article 29*bis*, see: Schoubroeck and Meurs (2013), 102 p; Malengreau (2019b), p. 15582 (n°57).

[12] Article 2 §1 of the Act of 21 November 1989 on Mandatory Liability Insurance for Auto Motor Vehicles.

[13] De Coninck and Dubuisson (2015), p. 28.

[14] See: Court of Cassation, 6 January 2012, as reported in (2012) RGAR (7), p. 14892 (R.G. n° C.10.0343.F); Court of First Instance, Brussels 70e ch., 26 January 2006, as reported in (2008) RGAR (3), p. 14373; Court of First Instance, Brussels 87 e ch., 5 December 2016, as reported in (2017) RGAR (6), pp. 15399–15405; Also see: Bocken et al. (2014), p. 251 with references.

[15] Article 1 paragraph 1 of the Act of 21 November 1989 on Mandatory Liability Insurance for Auto Motor Vehicles ("véhicules destinés à circuler sur le sol et qui peuvent être actionnés par une force mécanique sans être liés à une voie ferrée").

[16] See for instance: Police Court Charleroi, 4e ch., 2 November 2006, as reported in (2007) JLMB (10), p. 401.

[17] See for instance: Court of First Instance, Brussels, 27 January 2006, as reported in (2007) RGAR (5), p. 14262; Fagnart (1994), p. 12388 (n°51–54).

[18] Court of Cassation, 15 November 2012, as reported in (2013) RGAR (4), p. 14972; Court of Cassation, 30 September 2010, as retrieved from Juportal.be (R.G. n° C.09.0545.F); This requirement of involvement, however, cannot be confused with the required 'causal link', which is one of

there was a contact between the vehicle and the victim.[19] However, the proof of a causal link between the vehicle and the accident is not strictly required for the vehicle to be involved.[20]

2.1.2 Protected Persons and Covered Damage

Article 29*bis* applies to so-called 'vulnerable road users' which include, among others, cyclists, pedestrians and passengers. It, therefore, excludes drivers that incurred damages from the benefit of this protective regime.[21] The evidentiary burden of proving the quality of 'vulnerable user' rests on the victim.[22] Moreover, Article 29*bis* applies to damages resulting from bodily harm or death and includes medical costs but also loss of income.[23] However, Article 29*bis* excludes damages to goods or property from its scope with the exception of damages to clothes that the victim was wearing and functional prosthetic devices (*i.e.* devices used by the victim to compensate physical deficiencies).[24] Article 29*bis* provides that the rules on civil liability remain applicable for everything that is not expressively governed by this article.[25] This means that damages to victims not covered by this compensation regime as well as damages to property must be dealt through the channel of general civil law provisions, namely fault-based liability (and accessorily liability for things

---

the requirements to impose liability pursuant to Article 1382 of the OBCC. See also *infra* for more information.

[19] A contact between the vehicle and the victim is not always strictly required. See case law mentioned by De Rode (2023), pp. 27–28 (n°32–33), footnote 120 & 124–125; on the influence of a vehicle in the accident's occurrence see: Police Court, Brussels, 15 April 2002, as reported in Dr. circ., 2002, p. 358; On the involvement of a bus whose presence obscured the view of a pedestrian and a vehicle, which prevented them from correctly evaluating the situation, see: Court of First Instance, Civ., Brussels, 27 January 2006, as reported in RGAR, 2007, p. 14262.

[20] Court of Cassation, 28 April 2011, as reported in (2011) RGAR (6), p. 14765: the link between the vehicle and the accident exists as long as there was a contact between the vehicle and the victim; Court of Cassation, 3e ch. N, 9 January 2006, as reported in (2006) Pasicrisie Belge (Pas), I, p. 87 (RG n° C.04.0519.N): a motor vehicle is involved in a traffic accident when there is any link between the vehicle and the accident. The existence of a causal link between the vehicle and the accident is not required for the vehicle to be involved.

[21] However, there is one exception to this exclusion: the driver ('*ayant droit*') beneficiary of a victim who was not a driver and if such victim did not intentionally cause the damage. See: Article 29bis §2 of the Act of the 21 November 1989 on Mandatory Liability Insurance for Auto Motor Vehicles.

[22] Court of Cassation, 1ère ch. F, 26 October 2007, Les Assurances Fédérales c/ R.F, as retrieved from Juportal.be (RG n° C.06.0341.F).

[23] Article 29bis §1 of the Act of the 21 November 1989 on Mandatory Liability Insurance for Auto Motor Vehicles.

[24] *Ibid.*

[25] Article 29bis §5 of the Act of the 21 November 1989 on Mandatory Liability Insurance for Auto Motor Vehicles.

or product liability). Finally, the liability insurer can be subrogated in the victim's rights of recourse against liable third parties.[26]

2.2 *The Driver's Fault-Based Liability as a Cornerstone for Compensation*

In Belgium, extra-contractual fault liability lies in the Articles 1382 and 1383 of the OBCC.[27] Under these articles, the victim needs to prove the wrongful act, the damage and the causal link between both.[28] However, these two articles refrain from defining the requirement of fault, which has been extensively developed by Belgian case law. A wrongful act consists of two elements, namely an objective element and a subjective element.[29] Whereas the objective element refers to the wrongful act itself, the subjective element relates to a person's 'culpability' and the imputability of the wrongful act to that person.[30] The wrongful act is assessed against two alternative criteria,[31] the violation of a specific legal rule of conduct on the one hand (Sect. 2.2.1) and negligence or violation of a general duty of care on the other hand (Sect. 2.2.2).[32]

2.2.1 Violation of Specific Legal Rule of Conduct

The first criterion used to determine personal fault rests in the violation of a statutory provision or legal rule of conduct[33] that imposes a determined behaviour or

[26] Article 29bis §4 of the Act of the 21 November 1989 on Mandatory Liability Insurance for Auto Motor Vehicles.

[27] See in general: Bocken et al. (2014), 270 p.; Stijns and Samoy (2020), 192 p.

[28] In Belgium, the theory of equivalence of conditions usually prevails with regards to causality (Vansweevelt and Weyts 2009, pp. 763–873, with further references; Fagnart 2006, p. 14.080 (n°3); Van Quickenborne 2007, pp. 31–46 with further references). See, however: Kruithof (2015), pp. 139–208 (concluding that another rule can better describe Belgian positive law on causation in civil liability. Under this rule a fact is a cause of a loss if (i) the loss would in the given circumstances not have occurred as it specifically occurred without the fact and (ii) the fact has increased the specific risk of which the specific loss occurrence was a realisation). In some cases, the theory of adequate causality is relied upon as well (Glansdorff 2013, pp. 119–120 (n°22–23)).

[29] Stijns and Samoy (2020), p. 52 (n°8).

[30] *Ibid.;* See also: Kruithof (2018), p. 45 (n°60).

[31] Glansdorff (2013), p. 104 (n°3).

[32] See on the violation of a legal provision: Court of Cassation, 16 May 2011, as retrieved from Juportal.be (R.G. n° C.10.0664.N). See on the violation of the general duty of care: Court of Cassation, 4 November 2010, as retrieved from Juportal.be (R.G.n° C.09.0214.F).

[33] Court of Cassation, 10 April 1970 as reported in (1970) Pasicrisie Belge (Pas), I, p. 682, as mentioned by Van Ommeslaghe (2010), p. 1189 (n°830); Dubuisson et al. (2009), p. 30 (n°13); De Bruyne and Tanghe (2017), p. 337.

abstention.[34] According to the Belgian *Court of Cassation*, the violation of a statutory or regulatory legal rule of conduct committed without a ground of justification is *per se* wrongful.[35] Negligence is thus not required. [36] In this regard, the Belgian Highway Code[37] is of particular importance as it imposes several obligations to the driver of the vehicle and other traffic participants. The violation of these obligations is rather easy to determine for vehicles up to level 2. For instance, a driver may have crossed a red light (Article 61.1.1°) or a solid white line (Article 72.2) or not have yielded priority (Article 12). Such violations, if having resulted in damage, can lead to the civil liability of the driver.

2.2.2 Violation of the General Duty of Care

The objective element of fault can also consist of the violation of the general duty of care, which is formulated as the reasonable amount of care that a normal, foreseeing and prudent person (*bonus pater/mater familias*) would have applied when placed in the same concrete circumstances.[38] As noted by Kruithof, a "behaviour is negligent if it falls short of the standard level of care".[39] Moreover, the violation of the general duty of care also requires that it was reasonably foreseeable for the defendant that his/her conduct could result in some kind of damage.[40]

[34] Van Ommeslaghe (2010), p. 1189 (n°830).

[35] Court of Cassation, 9 February 2017, as retrieved from Juportal.be (R.G. n° C.13.0143.F), as reported in (2017) Pasicrisie Belge (Pas), p. 337; Court of Cassation, 16 May 2011, as retrieved from Juportal.be (R.G. n° C.10.0664.N); See case law as reported by Kruithof (2018), p. 50 (n°68).

[36] See case law reported by De Bruyne and Tanghe (2017), p. 337, footnote n°56: Court of Cassation, 10 April 2014 (C.11.0796.N), as reported in (2014) Arr Cass, p. 962, (2014) Pasicrisie Belge (Pas), p. 25; Court of Cassation, 16 May 2011, as retrieved from Juportal.be (R.G. n° C.10.0664.N), and reported in (2011) Pasicrisie Belge (Pas), p. 1339; Also see: Vandenberghe et al. (2000), p. 1564; Vansweevelt and Weyts (2009), p. 138.

[37] Arrêté royal du 1er décembre 1975 portant règlement général sur la police de la circulation routière [et de l'usage de la voie publique].

[38] Court of Cassation., 9 December 2011, as reported in (2017) RGDC (10), p. 545; Bocken et al. (2014), p. 89 and further; Vansweevelt and Weyts (2009), p. 125 and further; Stijns and Samoy (2020), pp. 54–55 (n°11–12).

[39] Kruithof (2018), p. 46 (n°62); On the requirement of a general duty of care which is assessed against the behavior of a reasonable person placed in the same circumstances, see: Court of Cassation, 25 November 2002, as reported in (2002) Pasicrisie Belge (Pas), p. 2230; Dubuisson (2009), p. 23 (n°3); Van Ommeslaghe (2010), p. 1190 (n°830).

[40] See: Court of Cassation, 5 May 1971, as reported in (1971) Pasicrisie Belge (Pas), p. 802 and note, as mentionned by Van Ommeslaghe (2010), p. 1190 (n°830); Court of Cassation, 12 November 1951, as reported in (1951) Pasicrisie Belge (Pas), I, p. 128, as mentionned by Stijns and Samoy (2020), p. 55 (n°12), footnote n°57; Vansweevelt and Weyts (2009), pp. 134–136; Bocken et al. (2014), p. 91.

2.3 *Liability of the Custodian as an Alternative Ground for Compensation*

The victim of level 0 to 2 vehicles can alternatively hold the custodian of the vehicle liable. Indeed, Belgium has a general liability regime for defective things[41] pursuant to Article 1384, first paragraph, OBCC.[42] This provision entails that the person who has the custody of a defective thing is liable for the damage caused by it, irrespective of whether that person committed any fault.[43] The two essential conditions for applying this regime are a defective thing causing damage (Sect. 2.3.1) and its custodian (Sect. 2.3.2).[44]

2.3.1 Defective Thing

Article 1384, first paragraph, OBCC, applies to 'things' (*choses*). Even though part of the doctrine nuances this stance,[45] most authors commonly agree that this article only applies to material things, therefore excluding intangible objects from its scope of application.[46] In order to apply the liability regime of Article 1384, first paragraph, OBCC, the claimant has to prove that the object that caused the damage was defective (*vice de la chose*). In the absence of any textual definition, the notion of "defect" and its characterisation have been subject to an abundant case law. The defect of the thing has been defined by the *Court of Cassation* as an abnormal characteristic or state of the object that makes it capable of causing harm in certain circumstances.[47] Judges seem to mainly refer to the conformity to the model as well

---

[41] Court of Cassation, 26 May 1904, as reported in (1904) Pasicrisie Belge (Pas), I, p. 246, as mentioned by Van Zuylen (2015a), pp. 7–8, (n°1) footnote n°5.

[42] See for more information: Fagnart (2017), 196 p; Cornelis and Bützler (1982), 728 p; Stijns and Samoy (2020), p. 101 and following (n°69 and following).

[43] See also: Van Zuylen (2015a), p. 9 (n°3); Kruithof (2018), pp. 91–92 (n°149) with further references.

[44] For recent case law on these conditions, see: Court of Cassation, 12 November 2015, as retrieved from Juportal.be (R.G. n°C.14.0468.N, C.14.0469.N.).

[45] Fagnart (2017), pp. 19–20 (n°34); Ninane and Van Zuylen (2015), pp. 229–230 (n°12).

[46] Van Ommeslaghe (2010), p. 1347 (n°941). See case law mentioned by author: Court of Cassation, 21 April 1972, as reported in (1972) Pasicrisie Belge (Pas), I, p. 773 and note; Bocken et al. (2014), p. 184; Vansweevelt and Weyts (2009), p. 456; Vandenberghe and Van Quickenborne (2011), p. 387.

[47] See for example: Court of Cassation, 8 March 2018 as retrieved from Juportal.be (R.G. n° C.17.0248.N); Court of Cassation, 7 October 2016, as retrieved from Juportal.be (R.G. n° C.15.0314.N-C.16.0060.N); Court of Cassation, 4 January 2016, as retrieved from Juportal.be (R.G. n° C.15.0191.F); Court of cassation, 13 March 2015 as retrieved from Juportal.be (R.G. n° C.14.0284.N).

as that of legitimate expectations as criteria to assess the abnormality.[48] The first criterion requires a comparison between the thing that caused the damage and its usual model.[49] Following the second criterion, the thing is deemed defective if it does not answer the safety requirements that a prudent and diligent man could legitimately expect.[50] The defect does not have to be intrinsic to the thing in the sense that it is not caused by an external factor, nor needs the structure of the object to be affected.[51] A judge can infer the defect of the thing from the thing's behaviour if (s)he excludes any other cause other than the thing's defect (cf. negative proof).[52] Finally, the cause of the defect is not relevant for applying this regime.[53]

2.3.2 Custody

The victim also needs to establish that the thing was under the custody of a person. Such custody or guard has not been defined in the law. For this reason, one has to refer to case law that has described custodianship as the person who uses the thing on his or her own behalf, who enjoys it or who keeps it with a power of supervision, lead and control.[54] The first three elements (using, enjoying or keeping the thing) are alternative, not limitative and not sufficient to establish custody.[55] However, the custodian must be able to cumulatively exercise a power of supervision, lead and control over the thing,[56] which denote a power of commandment (*pouvoir de*

[48] See for instance: Court of Cassation, 4 January 2016 as retrieved from Juportal.be (R.G. n° C.15.0191.F); Court of Cassation, 11 March 2010, as retrieved from Juportal.be (R.G. n° C.09.0186.N); Ninane and Van Zuylen (2015), pp. 251–254 (n°32–35).

[49] On recent case law, see: *op. cit.*, Court of cassation, 4 January 2016, as retrieved from Juportal.be (R.G. n° C.15.0191.F); Court of cassation, 11 March 2010 as retrieved from Juportal.be (R.G. n° C.09.0186.N), as reported in (2010) Pasicrisie Belge (Pas), p. 777; Kruithof (2018), p. 91 (n°147); Stijns and Samoy (2020), pp. 108–109 (n°74).

[50] See: Court of Cassation, 25 April 2005, as reported in (2005) Pasicrisie Belge (Pas), p. 924; Court of appeals (Civ.) Brussels (9e ch.), 2 November 2010, as reported in (2012) RGAR (2), p. 14831 and note; Court of appeals Liège (20e ch.), 20 May 2010, as reported in (2011) RGAR (1), p. 14704 ("Pour apprécier si le vice de la chose qui est invoqué constitue une caractéristique anormale, il convient de se référer à la destination normale du bien et de fixer, sur cette base, le degré de sécurité qu'il doit normalement presenter") (as mentioned by, *op. cit.*, Van Zuylen 2015a, pp. 40–41 (n°33)).

[51] Kruithof (2018), p. 92 (n°147) with further references.

[52] Court of Cassation, 11 September 2008, as retrieved from Juportal.be (R.G. n° C.07.0200.F); Court of Cassation, 1 December 2006, as reported in (2006) Pasicrisie Belge (Pas), pp. 2535–2539; Court of Cassation, 24 February 2006, as retrieved from Juportal.be (R.G. n° C050112F).

[53] Kruithof (2018), p. 92 (n°149); Lambert (2016), p. 9 (n°20).

[54] See for instance: Court of Cassation, 22 February 2018, as retrieved from Juportal.be (R.G. n° C.17.0313.N); Court of Cassation, 13 September 2012, as retrieved from Juportal.be (R.G. n° C.10.0226.F); Court of Cassation, 22 January 2009, as retrieved on Juportal.be (R.G. n° C.06.0418. F).

[55] Dubuisson (2006), p. 18; Lambert (2016), pp. 5–6 (n°8–9).

[56] Ninane and Van Zuylen (2015), p. 234 (n°15).

commandement) over the thing.[57] The custodian needs to have 'intellectual' power (*pouvoir intellectuel*) over the thing. This refers to the possibility for the custodian to exercise his or her power of direction over the thing,[58] which does not necessarily need to be exercised.[59] As Kruithof notes, a custodian is the person who "had the thing under his detention or used it, and who had factual control over the object".[60] With vehicles levels 0 to 2, the custodian would usually be the owner of the vehicle. However, the user, in case (s)he is different from the owner, could also be held liable on the basis of Article 1384, first paragraph, OBCC, as custody can be transferred.[61]

2.4 Product Liability as a Residual Ground for Compensation

Product liability constitutes another ground for claiming compensation for damages related to a defect in the vehicle.[62] This regime stems from the Product Liability Directive[63] and is transposed in the Belgian Product Liability Act.[64] Under this Act, the producer is liable for defective products[65] that cause damages to persons and moral loss and, under certain conditions, damages to property.[66] It establishes a (semi) strict no fault liability mechanism where the plaintiff needs to prove that a product (Sect. 2.4.1) was defective (Sect. 2.4.2) in order to obtain compensation from the producer (Sect. 2.4.3).

---

[57] Lambert (2006), pp. 5–6 (n°9); Dubuisson (2006), p. 18 (n°6).

[58] Ronneau (2019), p. 97.

[59] Van Zuylen (2015b), pp. 20–21 (n°15); Dubuisson (2006), p. 19 (n°7).

[60] Kruithof (2018), p. 92 (n°148). See mentioned case law at footnote n°474.

[61] The owner is not necessarily the custodian. Ownership over the thing is not a sufficient condition. See: Court of cassation, 22 March 2004, as reported in (2004) Pasicrisie Belge (Pas), p. 487; Stijns and Samoy (2020), p. 102 (n°70); On the transfer of custody, see: Ninane and Van Zuylen (2015), p. 243 (n°23).

[62] Delforge (2015), p. 295 and further.

[63] Council Directive 85/374/EEC of 25 July 1985 on the approximation of the laws, regulations and administrative provisions of the Member States concerning liability for defective products [1985] OJ L210/29.

[64] Act of 25 February 1991 Relating to Liability for Defective Products ("Loi relative à la responsabilité du fait des produits defectueux"), hereafter referred to as the 'Product Liability Act'.

[65] Article 1 Product Liability Act.

[66] Article 11 §1 and §2 Product Liability Act; Kruithof (2018), p. 96 (n°157) ("Damage to goods is only covered for goods other than the defective product itself and if the damaged goods are of a type ordinarily intended for private use or consumption").

2.4.1 Product

A product is defined as any tangible movable good, even incorporated in another movable or immovable good.[67] Therefore, the liability regime covers material objects[68] and either applies to the vehicle as a whole or to one of its material components. For instance, if the vehicle suffers a mechanical failure, linked to one of its components, the victim may target the producer. However, the plaintiff will need to prove the product's defectiveness.[69]

2.4.2 Defect

Pursuant to Article 5 of the Product Liability Act, a product is defective when it does not offer the safety that one can legitimately expect (from it) taking into account all circumstances.[70] As Montero notes, the criteria for determining the defectiveness resides in the potentially damageable character of the product to physical integrity of individuals or for their private consumer goods.[71] If proving the defect may not be as straightforward, the demonstration of causation may be even more daunting as illustrated in certain fields of litigation such as medical products.

2.4.3 Liable Parties

According to the Product Liability Act,[72] the producer can either be the manufacturer of the finished product, the manufacturer of a component of the finished product or of a raw material or any other person that presents itself as the manufacturer or producer by adding his name, trademark, or any other distinctive sign.[73] The manufacturer of the vehicle as well as the component manufacturers may, therefore, be held liable for defects in the vehicle. The Act also provides that when several persons are jointly liable in accordance with this law, they are held jointly and

[67] Article 2 Product Liability Act.

[68] The Product Liability Act refers in Article 2 to products as tangible items ("biens meubles corporels"), whereas the Product Liability Directive is not as straightforward. It only mentions "movables" without strictly referring to their tangibility.

[69] Article 1 Product Liability Act.

[70] Article 5 Product Liability Act.

[71] Montero (2006), p. 626 (n°6) ("Il ressort de cette définition que la notion de défaut est étroitement liée au concept de sécurité. Le critère réside dans le caractère potentiellement dommageable du produit pour l'intégrité physique des individus ou pour leurs biens de consommation privés").

[72] Article 3 Product Liability Act.

[73] In some cases, the product supplier can also be qualified as the producer: see Article 4 § 2 Product Liability Act; The importer can also be held liable under the conditions of article 4 § 1 of the Product Liability Act.

severally liable,[74] without prejudice to recourse actions. This means in practice that if the failure comes from a vehicle component, the victim may sue either the component manufacturer or most likely the end manufacturer. The Act also stipulates that when the damage is caused both by a defect in the product and a third party's act, the producer's liability to the victim cannot be limited or excluded.[75] However, the producer may have recourse action against this liable third party. Finally, the injured party's contributory negligence may limit or exclude the liability of the producer.[76]

3 Liability for Level 3 and Level 4 Vehicles: Unequal Applicability of Existing Liability and Compensation Mechanisms

SAE level 3 vehicles with 'conditional automation' embed technologies that allow them to be partially autonomous, meaning that the user can delegate most driving functions to the vehicle's systems. However, the user must be ready to regain control when the on-board system requires him/her to do so.[77] In SAE level 4 vehicles with 'high automation', this autonomy is further advanced and the user has very little, if any, monitoring or fall-back duties. The vehicle can move and proceed to its journey without any human intervention,[78] provided that it is within its operational domain.[79] If the existing no-fault compensation scheme may continue to apply to vulnerable road users (Sect. 3.1), implementing fault liability to users of autonomous vehicles will be made more difficult (Sect. 3.2). Alternatively, applying liability for things of the custodian (Sect. 3.3), the manufacturer's product liability (Sect. 3.4), as well as liability of service providers and public authorities (Sect. 3.5) may be possible, albeit challenging. Due to the restricted scope of the chapter, we will not develop on the problematic application of causation to AI enabled autonomous vehicles, which is a transversal source of concern when applying most liability regimes.[80]

---

[74] Article 9 Product Liability Act.

[75] Article 10 §2 paragraph 2 Product Liability Act.

[76] Article 10 §2 paragraph 1 Product Liability Act.

[77] See: Malengreau (2020), p. 54 (n°7).

[78] *Ibid.*

[79] Indeed, under certain circumstances (unfavorable weather conditions, complex environments, etc.), the vehicle may come to a halt or the user may have to resort to driving manually (if that is possible).

[80] On the issue of causation, see for instance: Martin-Casals (2019), pp. 201–228.

*3.1 Applicability of Article 29*bis *to Certain Victims*

3.1.1 Continued Compensation of Vulnerable Road Users

As mentioned above, several conditions must be met in order to apply the regime under Article 29*bis*. First, proving the traffic accident involving an autonomous vehicle should not be problematic. Second, this provision applies to motor vehicles, which are defined in very broad terms,[81] therefore allowing autonomous vehicles to fall within the scope of application of Article 29*bis*.[82] Thirdly, the notion of involvement (*implication*) allows for an extensive application to accidents in which autonomous vehicles would be involved. Indeed, this compensation scheme is disconnected from liability, precluding the victim from having to prove fault or negligence and, therefore, facilitating his or her claim for redress.[83] In short, Article 29*bis* will likely also apply to level 3–4 autonomous vehicles, provided that the conditions explained in Sect. 2.1.1 are met.

3.1.2 The Uncertain Characterisation of the User of the Vehicle

As seen in Sect. 2.1, this no-fault compensation regime only benefits 'vulnerable road users' and, therefore, excludes drivers which suffered damage from a road traffic accident. It is usually accepted that such a 'driver' is defined restrictively since his/her exclusion from the benefit of Article 29*bis* is a derogation to the general rule.[84] In accordance with the *Court of Cassation's* recent case law, the 'driver' is defined as the person who has the control (*maitrise*) of the vehicle when the accident occurs.[85] Malengreau and Sauvage, referring to case law, suggest that the sole 'control' criterion is currently employed for determining the existence of a driver. Tanghe and De Bruyne, basing themselves on case law from the Belgian *Court of Cassation* and the Benelux Court of Justice,[86] argue that both a factual criterion, that

[81] Article 1 of the Act of 21 November 1989 on Mandatory Liability Insurance for Auto Motor Vehicles.

[82] Malengreau (2019b), p. 15582 (n°59).

[83] Van Ommeslaghe (2010), p. 1470 (n°1035).

[84] Van Schoubroeck and Meurs (2013), pp. 20–23; De Rode (2023), p. 32; Bogaert (2004), pp. 54–55; Malengreau (2020), p. 118 (n°61); However, two authors disagree: Tanghe and De Bruyne (2018), p. 15 (n°22). According to them, the driver should not be restrictively defined. Such a restrictive definition does not give much clues as to how the driver may be characterized.

[85] Court of Cassation, 13 May 2016, as retrieved from Juportal.be (R.G. n° C.15.0415.F); Court of Cassation, 7 September 2015, as retrieved from Juportal.be (R.G. n° C.14.0209.F).

[86] See case law reported by Tanghe and De Bruyne (2018), p. 16 (n°23), footnote n°61: Court of Cassation, 13 April 2007, as reported in (2007) Arr.Cass. p. 776, (2007) Pasicrisie Belge (Pas), p. 681, (2007–2008) RW, p. 1078, note by Simoens, (2007) TBH, n°8, p. 797, note by Van Schoubroeck C; T.Pol. 2007, n°4, p. 203, note; Cass. 7 June 2012, T.Verz. 2012, n°4, 477, note by Van Den Hout E; Benelux Court of Justice, 8 December 1994, n° A 93/5, Benelux Jur. 1994, 109, consideration 17.

of actually and independently driving the vehicle, and a normative criterion, that of being responsible for driving, should allow to qualify the user as a driver.[87] In any case, such definitions require to examine the specific circumstances surrounding the accident since determining who had the control of the vehicle may not be straightforward. With level 4 vehicles, it is highly likely that the user will not be legally qualified as a driver as he or she relinquishes control to the on-board system.[88] With level 3 vehicles, things are more complex as the driving functions are shared between the user and the on-board computer. As such, the criterion of the user controlling the vehicle in an 'independent' manner may not be fulfilled. Tanghe and De Bruyne consider that such a criterion will not likely prevent the user from being characterized as a driver under the alleged normative criterion.[89] He or she has the possibility of regaining control of the vehicle.[90] However, Malengreau argues that when the vehicle is in automated mode, and abiding to the strict definition of control, the mere supervision of the vehicle will not be sufficient to characterize the user's effective control over such a vehicle.[91] In this situation, the victim should be characterized as a 'vulnerable user', therefore benefiting from the protective regime.[92] However, questions may arise in the 'grey zone' when the car's systems transfer the control to the user.

3.2 Challenged Applicability of Fault Liability to Users of Semi-Autonomous Vehicles

Victims not covered by Article 29*bis* will have to resort to liability grounds such as fault liability. It has already been mentioned that fault-based liability relies on two alternative criteria: the violation of a specific statutory rule or the general duty of care (negligence). The application of such criteria seems challenging in the context of SAE level 3–4 vehicles.[93] For instance, it needs to be determined whether there still is a 'driver' in an autonomous vehicle to whom the violation of traffic regulations initially committed by the vehicle can be attributed (Sect. 3.2.1). Moreover, proving that the user of the vehicle acted negligently will not be straightforward because the dynamicity and autonomy of software systems and the rigid interaction with users may decrease the reasonable foreseeability and avoidability of the damage (Sect. 3.2.2). Finally, exoneration causes may be relied upon by the user (Sect. 3.2.3).

---

[87] Tanghe and De Bruyne (2018), p. 16 (n°23–24); De Bruyne and Tanghe (2017), pp. 339–341.

[88] Malengreau (2019b), p. 15582 (n°63).

[89] Tanghe and De Bruyne (2018), p. 17 (n°25–26).

[90] *Ibid.* p. 17 (n°26).

[91] Malengreau (2019b), p. 15582 (n°63).

[92] *Ibid.*

[93] See also: Swinnen (2018), p. 1091.

3.2.1 The User's Violation of Specific Legal Rules of Conduct

The violation of a legal rule or statutory provision committed without any justification grounds, is wrongful by itself and the plaintiff is not required to prove negligence.[94] The Highway Code contains several statutory obligations the 'driver' has to comply with. The notion and characterisation of the user as a 'driver' is, therefore, essential with regards to the associated legal consequences, namely establishing a road traffic violation. The driver is defined in Article 2.13 of the Highway Code as any person that assures the direction of the vehicle.[95] De Bruyne and Tanghe deduct two alternative criteria from case law, namely a factual and a normative criterion.[96]

The *Court of Cassation* established that any person who puts the vehicle in motion can be defined as the driver.[97] The Court further held that a person drives a vehicle even if it was not his or her intention and/or he or she did not use the steering wheel.[98] This is the factual criterion according to De Bruyne & Tanghe.[99] Sauvage argues that this definition will be sufficient to characterise the user as a 'driver' whatever the degree of autonomy may be.[100] Regarding the normative criterion, the Court established that the driver is also the person responsible for the driving even though he or she did not perform any act that had an actual impact on the movement of the vehicle.[101] When applying these criteria to autonomous vehicles, De Bruyne and Tanghe note that the person "whose task is to supervise the operating systems"[102] is the person responsible for driving. This could therefore concern level 3 vehicles where the user has some degree of monitoring duties but

---

[94] Court de Cassation, 9 February 2017, as reported in (2017) Pasicrisie Belge (Pas), p. 337 (R.G. n° C.13.0143.F); Also see case law reported by: De Bruyne and Tanghe (2017), p. 337 (footnote n°56).

[95] Article 2.13 of the Belgian Highway Code ("Arrêté royal du 1er décembre 1975 portant règlement général sur la police de la circulation routière et de l'usage de la voie publique"): "Le terme "conducteur" désigne toute personne qui assure la direction d'un véhicule ou qui guide ou garde des animaux de trait, de charge, de monture ou des bestiaux".

[96] De Bruyne and Tanghe (2017), pp. 339–340.

[97] See: Court of Cassation, 30 October 1967, Pas. 1968, I, 303, as mentioned by De Bruyne and Tanghe (2017), p. 339, footnote n°62.

[98] See: Court of Cassation, 23 October 1973, as reported in (1974) Arr Cass p. 221, (1973) I Pasicrisie Belge (Pas.) p. 203 and (1974) RGAR n°9330 with notes by Fagnart J.-L. Malengreau explains that this decision was given on the basis of the previous version of the Highway Code (Article 2-6° of the 14 March 1968 Royal decree "portant règlement general sur la police de la circulation routière). However, he adds that the definition of the driver under this article was very similar to that of the current Article 2.13 of the Highway Code (Malengreau 2019a, p. 15578 (n°16), footnote n°88). See also concurring: Sauvage (2020), p. 276 (n°41), footnote 211.

[99] De Bruyne and Tanghe (2017), pp. 339–340.

[100] Sauvage (2020), p. 276 (n°41): "le niveau d'autonomie du véhicule ne devrait pas remettre en question la qualité de conducteur de l'utilisateur". However, he nuances this assertion when referring to vehicles without a steering wheel.

[101] Court of Cassation, 4 June 1973, as reported in (1973) Arr Cass, p. 961 and (1973) I Pasicrisie Belge (Pas), I, p. 918, as mentioned by De Bruyne and Tanghe (2017), pp. 339–340.

[102] De Bruyne and Tanghe (2017), p. 341.

less so for vehicles level 4 where the user is mainly passive. Malengreau argues that in level 3 and 4 vehicles, there still may be a driver on-board the car either because the user must be able to regain control of the vehicle (level 3) or because the user pushed the start button (level 4).[103] In the absence of the normative criterion, De Bruyne & Tanghe contend that the mere pushing of the start button—which corresponds to the factual criterion—would not necessarily amount to the user being qualified as a 'driver' for the entire journey.[104] However, Malengreau argues that the user who engages the automatic system will remain a 'driver' as long as there is a seat behind the steering wheel and control pedals,[105] therefore implying some possibility of regaining control. In the contrary situation, as noted by Sauvage, it would be difficult to qualify the user as a 'driver'.[106]

Notwithstanding the uncertain qualification of the user as a 'driver', questions as to the attribution to the user of the violation of a rule of conduct by the computer system remain unanswered.[107] With such self-driving technologies, the user of the vehicle may not have directly violated a rule of conduct if the vehicle was driving in automated mode. For instance, the system could have misread a traffic sign that caused the vehicle not to yield a priority at a certain intersection, eventually leading to a collision with other road users. However, a 1973 decision from the *Court of Cassation*[108] may shed light on the possible attribution of an infringement of the Highway Code to the user of the vehicle. In this decision, a driving instructor was held liable for the violation of a highway rule by his pupil since it was assumed he was responsible for the driving as he had the possibility to retain control and avert the violation of the traffic rule. Applied by analogy, several authors note this solution could mean that users of autonomous vehicles could be held liable for the vehicle's violation of road traffic rules,[109] at least in level 3 cars where the user retains some form of control over the vehicle. This solution may not straightforwardly apply to level 4 vehicles where the car drives entirely autonomously within its specified operational design domain. One may also wonder whether the user's inaction or failure to prevent the infringement may allow to attribute him/her the violation of the rule of conduct.[110] The answer to this question will depend on whether or not the user of the autonomous vehicle had the possibility of overriding the driving system.

According to De Bruyne and Tanghe, when the violation of a traffic rule cannot be directly and personally attributed to the driver, should there still be one, the user

[103] Malengreau (2019a), p. 15578 (n°17).

[104] De Bruyne and Tanghe (2017), p. 341.

[105] Malengreau (2019a), p. 15578 (n°17).

[106] Sauvage (2020), p. 276 (n°41).

[107] De Bruyne and Tanghe (2017), p. 341; See also: Swinnen (2018), pp. 1092–1093.

[108] *Op. cit.*, Court of Cassation, 4 June 1973, as reported in (1973) Arr Cass, p. 961 and in (1973) I Pasicrisie Belge (Pas), p. 918.

[109] De Bruyne and Tanghe (2017), p. 342; Malengreau (2019a), p. 15578 (n°18); Sauvage (2020), p. 276 (n°41).

[110] De Bruyne and Tanghe (2017), p. 341.

could be held liable for the violation of Article 8.3 of the Highway Code. This provision stipulates that the driver needs to have his/her vehicle well under control at all times, which is a general obligation of control.[111] This argument could work for level 3 vehicles in which the user is supposed to maintain some form of control through his/her capacity to regain control of the vehicle should he/she be required to do so (or when driving manually). Nonetheless, the control argument is less persuasive for level 4 vehicles driving in automated mode. Indeed, the user no longer has any control over the vehicle since he is not required to supervise the driving. One could question whether the user would still be at fault for violating the obligation of control of Article 8.3, even though the user had no factual control over the vehicle. It seems unlikely, and in any case, amendments to the Highway Code may be necessary in order to accommodate the arrival of highly autonomous vehicles (especially on the control requirement).[112]

3.2.2 The User's General 'Duty of Care'

The victim may want to invoke the user's fault through a negligent or imprudent behaviour under Article 1382 and 1383 OBCC. As mentioned before, such a behaviour is assessed against the standard of care, which is formulated as the reasonable amount of care that a normal, foreseeing and prudent person would have applied when placed in the same concrete situation.[113] Moreover, the *Court of Cassation* added that negligence supposes that damage must have been reasonably foreseeable by the liable party.[114] However, applying negligence based liability to the user of an autonomous vehicle may be challenging and will differ according to the level of autonomy and driving mode.[115]

More specifically, attributing a negligent behaviour will be complex in level 3 vehicles and a distinction must be made between driving modes. Indeed, if an accident occurs when the vehicle is in automated mode, the user will likely not be held liable for negligence since (s)he had no factual control over the vehicle.[116] Some argue, though, that the user could still be held liable for failing to overrule certain computer decisions (or on the contrary for having overruled such decisions), should that be possible.[117] If the accident occurred when the system passed over

[111] See also: De Bruyne and Tanghe (2017), pp. 342–343.

[112] Sauvage (2020), p. 276 (n°41–42); Malengreau (2019a), p. 15578 (n°20).

[113] See: *op. cit.*, Court of Cassation., 9 December 2011, as reported in (2017) RGDC (10), p. 545 (R.G. n° C.11.0015.F); Court of Cassation, 5 June 2003, as reported in (2003) Pasicrisie Belge (Pas), I, p. 1125 (R.G. n° C.01.0252.F).

[114] Court of Cassation, 5 May 1971, as reported in (1971) Pasicrisie Belge (Pas), p. 802 and note (case law mentioned by Van Ommeslaghe 2010, p. 1190 (n°830)); Jacquemin and Hubin (2017a), p. 118 (n°43).

[115] Sauvage (2020), p. 274 (n°39).

[116] Malengreau (2019a), p. 15578 (n°13).

[117] De Bruyne and Tanghe (2017), pp. 344–345; Sauvage (2020), p. 277 (n°44).

control to the user, the latter could be held personally liable under applicable tort law. The grey zone appears when the accident occurs during the transitional phase, when the on-board computer passes over the driving functions to the vehicle user.[118] First, it may not be clear who had control at the time of the accident.[119] Second, in such a situation, the user could be held liable for different actions or omissions such as failing to regain control of the vehicle or improperly regaining control of the vehicle.[120] In level 4 vehicles, the user can no longer be qualified as a driver and becomes a mere passenger who has almost no (or very little) control over the vehicle when the automated mode is on. This means that attributing a negligent behaviour to the user of the vehicle may be virtually impossible.[121]

However, Jacquemin and Hubin have asked whether the user could be held liable for not being vigilant enough to anticipate the actions of the on-board computer that could potentially cause damage.[122] The foreseeability criterion in negligence is rather problematic when applied to autonomous vehicles as the embedded technologies can be intrinsically unpredictable.[123] Indeed, this is the case for machine learning algorithms found in software that are employed for perception and decisional tasks. Moreover, autonomous vehicle's driving software will rely on continuous software updates, which may further blur the foreseeability and predictability of the damage.[124]

It should be noted that in situations where the user cannot be held liable for negligent driving, the victim could still hold the user liable for failing to properly maintain the vehicle or for failing to update a safety critical software and/or adding unauthorized third party software.[125] The user could also be held liable for using the vehicle unreasonably in such a way that the accident's occurrence was foreseeable.[126]

3.2.3 Force Majeure as an Exoneration Cause for the User?

In order to escape liability, users of autonomous vehicles could potentially invoke a ground of justification.[127] Force majeure, for instance, could be alleged in claims involving the violation of a road traffic rule. Force majeure is usually defined as any

[118] Malengreau (2019a), p. 15578 (n°13).

[119] However, reading and decrypting the on-board data event recorder may shed light to who was in control at the time of the accident, but may not always be sufficient.

[120] Malengreau (2020), pp. 60–63 (n°13).

[121] Sauvage (2020), p. 275 (n°39).

[122] Jacquemin and Hubin (2017a), p. 117 (n°43).

[123] De Bruyne and Tanghe (2017), pp. 345–346.

[124] De Bruyne and Tanghe (2017), p. 346.

[125] Sauvage (2020), p. 277 (n°44).

[126] Malengreau (2019a), p. 15578 (n°13); See also: Swinnen (2018), p. 1092.

[127] Sauvage (2020), p. 278 (n°46).

irresistible, non-foreseeable event that is beyond the control of the person that violates a rule of conduct.[128] As Malengreau notes, when negligence is invoked, the irresistibility that characterises force majeure questions whether the user could have (re)acted differently and whether he or she behaved as a normally prudent and diligent person.[129] This may be difficult to sort out in level 3 vehicles because of the challenges for the user of promptly and effectively regaining control of the vehicle in an emergency.[130] Moreover, in the case of the violation of a road traffic rule, the user could also be tempted to use force majeure as a ground of justification, defined as "an event not caused by the defendant that he could not reasonably avoid and that makes it impossible to comply with the legal rule".[131] As Tanghe and De Bruyne note, the non-foreseeability of the violation of a road traffic rule by the machine could be established as the system would have normally been programmed to abide to the Highway Code, meaning that any deviation could not have been expected by the user.[132] However, proving the irresistibility of the road traffic rule violation would imply that the user establishes that he could not have reasonably acted differently, which may be complex in practice to demonstrate.[133]

3.3 *Challenged Liability of the Custodian of the Autonomous Vehicle*

As a result of the identified difficultics in applying fault liability, victims may want to bring a claim on another legal basis such as the liability of the custodian of a defective thing under Article 1384, first paragraph, OBCC. This regime imposes a strict liability on the custodian of a defective thing. Applying this regime to autonomous vehicles may, however, prove difficult. The uncertain characterisation of

[128] On the general concept of justification cause see: Court of Cassation, 22 March 2002, as retrieved from Juportal.be (R.G. n° C.00.0401.F) ("une cause de justification ne peut résulter que d'un événement indépendant de la volonté humaine et que celle-ci n'a pu ni prévoir ni conjurer"); Court of Cassation, 8 January 1982, as retrieved from Juportal.be (R.G. n° 3210–3216); See: Van Ommeslaghe (2010), pp. 1381–1384 (n°966) and further; Delforge et al. (2019), p. 715 (n°262 and further); Van Zuylen (2013), p. 273 and further.

[129] Malengreau (2019a), p. 15578 (n°13). However, for a nuanced approach on force majeure and negligence, see: Kruithof (2018), p. 116 (n°188).

[130] Malengreau (2019a), p. 15578 (n°13 and 19).

[131] Kruithof (2017), p. 252.

[132] Tanghe and De Bruyne (2018), p. 40 (n°62); Malengreau (2019a), p. 15578 (n°19).

[133] Malengreau (2019a), p. 15578 (n°19). However, the author differentiates between fully autonomous vehicles, where the irresistibility issue is not relevant since the user has no control over the driving process, and partially autonomous vehicles where demonstrating such irresistibility will be more delicate.

software as a defective 'thing'/'good' (Sect. 3.3.1) coupled to the difficult identification of a custodian may be challenging (Sect. 3.3.2).[134]

3.3.1 The Delicate Characterisation of the Defective Thing

As mentioned in Sect. 2.3, in order to bring a claim on the basis of Article 1384, first paragraph, OBCC, the victim must prove that the thing was defective. In the case of an autonomous vehicle causing damage, the victim could either prove that the vehicle[135] or one of its components were defective (wheel, motor, sensor, software, etc.).[136] Difficulties could arise when the victim tries to prove that the damage originated from the defective software. Indeed, following a 1972 decision from the *Court of Cassation*,[137] a large part of doctrine usually considers that Article 1384, first paragraph, only applies to material things[138] (although nuances may exist).[139] While software are intangible items which would therefore *a priori* exclude them from the scope of article 1384, some authors such as De Bruyne and Tanghe consider them as objects that should fall within the scope of liability for things.[140] Malengreau and others argue that the victim could overcome this difficulty by proving the defectiveness of the physical support of the software (i.e. hardware) and not that of the software itself.[141] In any case, the victim would have to prove that the thing was defective, which has been defined by case law as an abnormal characteristic susceptible of causing damage.[142]

The abnormality of the characteristic is usually assessed against two alternative criteria as identified in Sect. 2.3.1. In the first criterion, the judge assesses the abnormality of the characteristic by comparing the thing to a standard model.[143]

[134] For a thorough analysis, see: Swinnen (2018), pp. 1078–1084.

[135] Vansweevelt and Weyts (2009), p. 454.

[136] Malengreau (2019b), p. 15582 (n°48). In the same direction, see: Sauvage (2020), p. 281 (n°53); Tanghe and De Bruyne (2018), p. 45 (n°71–72).

[137] Court of Cassation, 21 April 1972, as reported in (1972) Pasicrisie Belge (Pas), I, p. 773 and note, as mentioned by Van Ommeslaghe (2010), p. 1347 (n°941).

[138] See for instance: Van Ommeslaghe (2010), p. 1347 (n°941); Kruithof (2018), p. 91 (n°146).

[139] Ninane and Van Zuylen (2015), pp. 228–230 (n°10–12); Lambert (2016), pp. 4–5 (n°5).

[140] De Bruyne and Tanghe (2017), p. 349.

[141] Malengreau (2019b), p. 15582 (n°48); Van Zuylen (2015a), p. 14 (n°11); Jacquemin and Hubin (2017b), p. 124 (n°50).

[142] Definition given by Van Ryn Jean in 1946 (Van Ryn 1946, p. 166). See: Court of Cassation, 4 January 2016, as retrieved from Juportal.be (R.G. n°C.15.0191.F); Court of Cassation, 13 March 2015, as reported in (2015) Pasicrisie Belge (Pas), p. 721 (R.G.n° C.14.0284.N); Court of Cassation, 17th January 2014, as reported in (2014) Pasicrisie Belge (Pas), p. 143 (R.G.n° C.12.0510.F).

[143] Van Zuylen (2015a), p. 40 (n°32). See: Court of Cassation, 4 January 2016, as retrieved from Juportal.be (R.G. n° C.15.0191.F); Court of Cassation, 25 April 2005, as reported in (2005) Pasicrisie Belge p. 924 (R.G. n° C.03.0400.N).

Assuming that we consider software as an object falling under the scope of Article 1384, first paragraph, OBCC, determining the abnormal character of a software that embeds a machine learning algorithm may prove very challenging.[144] Indeed, it is not clear to what model one can refer to when the algorithm itself can self-learn and evolve over time. According to the second 'functional' criterion,[145] which builds on the destination and normal usage of the thing, the object is deemed defective if it no longer meets the safety requirements that a diligent and prudent man could legitimately expect.[146] In the case of autonomous vehicles, the demonstration of the thing failing to meet these safety expectations could be possible.[147] However, as noted by Jacquemin and Hubin, the self-learning capacity of machine learning algorithms could challenge the foreseeability of such safety expectations.[148] *De facto,* Belgian courts usually combine both approaches[149] and characterise the thing's abnormality through its comparison to things of the same kind and type in order to determine the thing's qualities that the victim was entitled to expect.[150]

Such abnormality traditionally relates to the intrinsic characteristics of an object affecting its physical structure.[151] This could be problematic when applied to autonomous vehicles as they are highly connected mobility artefacts, which cannot be apprehended as pure standalone objects but rather as 'systems' that communicate with the exterior. The damage could have therefore resulted from a defective communication of data or from an erroneous off-vehicle software.[152] In theory, this could result in making a claim based on liability for defective things difficult due to the exteriority of the data or software. However, as observed by some authors,[153] Belgian case law seems to be broadening the notion of defect by not necessarily requiring that the defect constitute an intrinsic characteristic of the object nor does it need to affect its structure.[154] This could allow the victim to bring a claim

---

[144] Malengreau (2019b), p. 15582 (n°54).

[145] As reported by Jacquemin and Hubin (2017b), p. 126 (n°51). This criterion has been advocated by: Van Zuylen (2015a), pp. 40–45 (n°32 36); Marchetti (2005), pp. 340–341 (n°7).

[146] Van Zuylen (2015a), pp. 40–41 (n°33).

[147] Malengreau (2019b), p. 15582 (n°55).

[148] Jacquemin and Hubin (2017b), p. 126 (n°51).

[149] Van Zuylen (2015a), pp. 42–43 (n°34).

[150] Court of Cassation, 4 January 2016, as retrieved from Juportal.be (R.G. n° C.15.0191.F); Court of Cassation, 11 March 2010, as retrieved from Juportal.be (R.G. n° C.09.0186.N); Van Zuylen (2015a), pp. 42–43 (n°34).

[151] See: Court of Cassation, 27 November 1969, as reported in (1970) Pasicrisie Belge (Pas), I, p. 277 as mentioned by: Van Ommeslaghe (2010), p. 1353 (n°944); De Bruyne and Tanghe (2017), p. 350.

[152] De Bruyne and Tanghe (2017), p. 350; Malengreau (2019b), p. 15582 (n°53).

[153] De Bruyne and Tanghe (2017), pp. 350–351; Kruithof (2017), p. 264; Jacquemin and Hubin (2017b), p. 125 (n°51).

[154] Court de Cassation, 13 Mars 2015, as retrieved from Juportal.be (R.G. n° C.14.0284.N), reported in (2015) Pasicrisie Belge (Pas), p. 721; Court of Cassation, 31 October 2013, as retrieved from Juportal.be (R.G. n° C.12.0628.N), reported in (2013) Pasicrisie Belge (Pas), p. 2115; Court of

against the custodian of the vehicle, even though the defect is not intrinsic to the vehicle itself.[155]

In any case, the victim could surpass some of the identified difficulties by proving the defectiveness of the vehicle as a whole and not that of a particular component. As Malengreau states, it would be sufficient to "identify a characteristic, without necessarily isolating it from the material [component] it affects" (*own translation*).[156] Case law seems to be leaning in this direction. For instance, one could refer to a Belgian Court of Appeal decision from Liège which ruled that, in the case of automatic doors closing on the victim, the defective thing is not the faulty sensor taken in isolation but the door taken as a whole.[157] Such a solution could therefore be applied to the autonomous vehicles, even if the failures originated from an erroneous software.

3.3.2 The Delicate Characterisation of the Custodian(s)

Under Article 1384, first paragraph, OBCC, the person liable for the defective thing is the custodian of the thing. As discussed in Sect. 2.3.2, the custodian is the defined as the person who uses the thing on his or her own behalf, who enjoys it (benefits from it) and who keeps a power of supervision, lead and control over the thing.[158] The determination of the custody is examined at the time of the damage's occurrence.[159] Though not unanimously shared, some authors such as Fagnart[160] question whether there is a presumption of custody that falls upon the owner. In any case, the owner usually gathers the attributes of a custodian, however, ownership rights are not sufficient to qualify the owner as the custodian. Applied to autonomous vehicles, the victim could continue to bring a claim against the vehicle owner as would have been the case with conventional vehicles.[161] However, the owner could prove that

cassation, 2 January 2009, (R.G. n° C.08.00774.N-C.08.0217.N), as reported in (2009) Pasicrisie Belge (Pas), p. 2. For instance, public road authorities have been held liable on the basis of article 1384, first paragraph, OBCC, for accidents which were caused by certain external elements that were present on the road (e.g. a rock or a concrete bloc). See: Marchetti (2005), pp. 338–339 (n°5); Kruithof (2018), p. 92 (n°147).

155 De Bruyne and Tanghe (2017), p. 351.

156 Malengreau (2019b), p. 15582 (n°48): "il suffit d'identifier une caractéristique, sans devoir nécessairement l'isoler du matériel qu'elle affecte".

157 Cour of Appeal Liège, 21 Novembre 2019, as reported by Fruy (2020), p. 1.

158 See, *e.g.*: Court of Cassation, 13 September 2012, as reported in (2012) Pasicrisie Belge (Pas), p. 1647 (R.G. n° C.10.0226.F); Court of Cassation, 28 May 2010, as reported in (2010) Pasicrisie Belge (Pas), p. 1649 (RG n° C.09.0233.F).

159 Jacquemin and Hubin (2017b), p. 128 (n°53).

160 Fagnart (2017), p. 61 (n°130).

161 Sauvage (2020), p. 280 (n°51).

the power of command,[162] which characterises the custodianship was transferred[163] to a third party enabling him to escape liability. [164] Indeed, Kruithof explains that such liability "attaches to the person who had the thing under his detention or used it, and who had factual control over the object".[165] The owner, even if qualified as a custodian, could have recourse action against the liable party for the defect.[166]

It has been questioned whether the vehicle's operating software could have a custodian distinct from that of the vehicle.[167] Sauvage argues affirmatively by asserting that the 'complex objects' premise, according to which a thing can be composed of numerous elements that form a whole, could allow to dissociate the vehicle software from the vehicle itself.[168] This would mean in practice that the custodian of the software could be the software designer/producer and not only its user.[169] Furthermore, as noted by Malengreau and others, most software programs are subject to an operating license given to the user. That license remains the property of the producer.[170] This could entail that the software producer keeps a certain control over the computer program (*maitrise*), meaning that he could be held liable on the basis of liability for defective things.[171] But this would only be possible if we consider software as a thing falling under the scope of Article 1384, first paragraph, OBCC, which is far from obvious as seen in Sect. 3.3.1. Further questions on how one would differentiate between the owner/user's custodianship of the vehicle from the software producer's custodianship remain. However, it may be possible that the vehicle have two custodians that would (jointly) share the guard of the vehicle and therefore be co- responsible. [172]

---

[162] Ninane and Van Zuylen (2015), p. 234 (n°15).

[163] Ninane and Van Zuylen (2015), p. 243 (n°23).

[164] Malengreau (2019b), p. 15582 (n°50).

[165] Kruithof (2018), p. 92 (n°148).

[166] See case law as reported by Kruithof (2018), p. 92 (n°149): Court of Cassation, 4 February 2011 as retrieved from Juportal.be (R.G. n° C.10.0236.N), (2011) Arr.Cass. p. 401, (2011) Pasicrisie Belge (Pas.), p. 429; Court of Cassation, 14 February 2013, as retrieved from Juportal.be (R.G. n° C.11.0793.F), as reported in (2013) Pasicrisie Belge (Pas), p. 429, (2013) Arr.Cass. p. 443.

[167] Malengreau (2019b), p. 15582 (n°51); Sauvage (2020), p. 280 (n°51); De Bruyne and Tanghe (2017), p. 354.

[168] Sauvage (2020), p. 280 (n°51); The guard of the thing is indivisible, however as noted by some, the thing itself can be divisible. See: Ninane and Van Zuylen (2015), pp. 240–243 (n°21–22).

[169] See: Malengreau (2019b), p. 15582 (n°50–51) who agrees that the software programmer could be held liable as the custodian of the software, provided that the software is considered a thing.

[170] Malengreau (2020), p. 109 (n°51); De Bruyne and Tanghe (2017), p. 354.

[171] Tanghe and De Bruyne (2018), pp. 52–53 (n°84).

[172] On the notion of joint custodianship, see: Court of Cassation, 25 March 1943, reported in (1948) RCJB, p. 157, as mentioned by Dubuisson (2006), p. 34 (n°20) footnote n°64; Lambert (2016), p. 6 (n°11); Sauvage (2020), p. 280 (n°51); Also see: Swinnen (2018), pp. 1083–1084. Case law has shown that 'cumulative' or joint custody of a thing can be possible (see: Fagnart 2017, p. 70 (n° 144)).

3.4 Challenged Applicability of Product Liability

Product liability could constitute another legal basis for the victim of an autonomous vehicle. However, applying this regime to autonomous vehicles may be complex.[173] If the damage originates from the car's software, questions in relation to the characterisation of software as a product falling under the scope of product liability can be raised (Sect. 3.4.1). Moreover, issues in relation to establishing the defectiveness of the autonomous vehicle may be brought forward as well (Sect. 3.4.2). Finally, we will see that the product liability regime provides defences that the manufacturer may successfully apply, eventually preventing the victim from claiming compensation on this legal ground (Sect. 3.4.3).

3.4.1 Uncertain Characterisation of Software as a 'Product'

The Product Liability Act applies to material items[174] and there is no doubt that the autonomous vehicle itself is a tangible item that qualifies as a product falling under this regime. However, questions arise when the damage originates from a software component. While the lack of materiality may exclude software from its scope of application, some scholars disagree or nuance this assertion.[175] In that respect, a differentiation could be made between embedded software and non-embedded software. On the one hand, non-embedded software is provided separately from the vehicle, which could make it difficult to label it as a product. The same would hold true for software updates provided over the air whose legal status remains unclear (although one could argue that the software is eventually stored on hardware). Embedded software, on the other hand, could be part of the vehicle considered as one standalone object and therefore a product taken in its entirety. Moreover, embedded software is fixed on a hardware.[176]

3.4.2 Uncertain Characterisation of the Defect

The problematic qualification of software is not the only challenge. In order to bring a claim against the manufacturer, the injured party needs to prove that the damage was caused by a defect in the product as well as the causal link between the damage and the defect, which may be burdensome to demonstrate.[177] A product is deemed

[173] See for an extensive analysis: Werbrouck (2018), pp. 529–603.

[174] Article 2 of the Product Liability Act.

[175] Coulon (2016), 6 p. (n°12); Enghelhard and De Bruin (2017), p. 56; Fairgrieve et al. (2016), pp. 46–47 (n°67–70); Fagnart and Boularbah (2000), pp. 132–134; De Bruyne and Tanghe (2017), p. 367; Sauvage (2020), pp. 265–266 (n°15).

[176] See: Werbrouck (2018), pp. 547–552.

[177] De Bruin (2016), p. 491.

defective if it fails to meet the legitimate safety expectations[178] taking into account all circumstances.[179] This definition entails a rather flexible criterion (also known as the 'consumer expectation test') for establishing the defect as it is not required for the victim to prove the exact nature of the defect.[180] However, the defect is assessed against the general public's expectations and not that of an individual.[181]

It remains uncertain how this broad criterion will be applied to defects in autonomous vehicles. De Bruyne and Tanghe have highlighted that this "unpredictability is strengthened by the fact that such vehicles are a new kind of product".[182] Because of the ground-breaking nature of these vehicles and the associated technologies (e.g. software, machine learning etc.) as well as the high level of (safety) benefits the public expects from it, it may not be easy to determine the legitimate level of safety that can be expected from such vehicles.[183]

The Product Liability Act provides several elements that can help determine these legitimate expectations such as the presentation of the product, the reasonably foreseen use of the product and the time at which the product was put into circulation.[184] The manufacturer will probably pay great attention to the presentation of the product. Indeed, if he commercialises a product as extremely safe, the legitimate safety expectations will be high. This in turn increases the manufacturer's liability exposure.[185] Conversely, if the manufacturer provides detailed information about the risks associated to the use of the vehicle, the public's legitimate safety expectations may be lower, therefore decreasing his liability exposure. However, this would limit the public's trust and acceptance of these vehicles and, therefore, prevent their expected social benefits from deploying. Several scholars add that implementing the Anglo-Saxon "risk utility test" in which the judge balances the overall benefits of the technologies with its risks could prove helpful in determining the legitimate safety expectations in order to characterise the defect.[186] The second element to determine the defect, namely the reasonably foreseen use of the product, raises some issues for the manufacturer as well. Indeed, as noted by De Bruyne and Tanghe, the manufacturer will have to design a vehicle "in such a way as to ensure that they are

[178] Article 5 of the Belgian Product Liability Act.

[179] Recent case law from the European Court of Justice has developed the notion of 'potential safety defect' which, if applied to mobility products, may impact the manufacturer (Emanuilov 2017, 8 p). See March the 5th 2015 ECJ judgment in joined Cases C-503/13 and C-504/13, the so-called *Boston Scientific* case.

[180] De Bruyne and Tanghe (2017), p. 359; Bocken et al. (2014), p. 197, footnote n°47.

[181] Van Ommeslaghe (2010), p. 1459 (n°1028); De Bruyne and Tanghe (2017), pp. 358–359.

[182] De Bruyne and Tanghe (2017), p. 359.

[183] *Ibid.*, pp. 359–361.

[184] Article 5 (a), (b), (c) of the Product Liability Act.

[185] De Bruyne and Tanghe (2017), p. 361.

[186] Malengreau (2020), pp. 86–87 (n°30); De Bruyne and Tanghe (2017), pp. 368–369.

adapted to (. . .) foreseeable misuse".[187] It is indeed highly possible that the vehicle will be used in way that the manufacturer(s) had not anticipated.

3.4.3 Potential Exoneration of the Producer Through Available Defences

Even if the defect is established, the Product Liability Act contains various defences which autonomous car producers may successfully invoke shielding them from liability. For instance, if the defect did not exist at the time the product was put into circulation or if it appeared after the product was put into circulation, the producer may not be held liable.[188] With artificial intelligence/machine learning driven systems, the damage could arise from an erroneous behaviour that the software self-learned. Indeed, machine learning techniques and algorithms can improve and evolve over time, therefore potentially leading to new defects. If such a situation could not have been (reasonably) foreseen at the time of designing the software application, the manufacturer may be immune from liability since the defectiveness appeared after its putting into circulation.[189] Notwithstanding their uncertain qualification as products, software updates also raise questions.[190] It is presumed that if these software updates provide substantial evolutions, they may be considered new products put on the market.[191] If the updates proved minor modifications, however, they may not be qualified as new products. As such, the defect would have to be assessed against its original putting in circulation.[192] Furthermore, the state-of-the-art defence[193] may also be relied upon by manufacturers to escape liability by proving that, given scientific knowledge at the time the product was put into circulation, it was not possible to detect the defect. The state of the art is assessed against the objective state of scientific and technical knowledge and thus not the specific individual knowledge of the producer.[194] Such a defence could potentially be successfully invoked by manufacturers should they demonstrate that the defect was not detectable when the product was put into circulation.[195]

[187] De Bruyne and Tanghe (2017), pp. 361–362.

[188] Article 8 (b) of the Product Liability Act.

[189] See: De Bruyne and Tanghe (2017), pp. 362–363; Sauvage (2020), p. 270 (n°28); Malengreau (2019b), p. 15582 (n°40–42); Jacquemin and Hubin (2017b), pp. 134–135 (n°59).

[190] De Bruyne and Tanghe (2017), p. 363; Sauvage (2020), p. 270 (n°28).

[191] Jacquemin and Hubin (2017b), p. 136 (n°59).

[192] Sauvage (2020), p. 270 (n°28).

[193] Article 8 (e) of the Belgian Product Liability Act.

[194] ECJ Judgement of 29 May 1997, *Commission of the European Communities v United Kingdom of Great Britain and Northern Ireland*. C-300/95, ECLI:EU:C:1997:255, paragraph 27; Court of Cassation, 6 April 2006 as reported in (2006) Pasicrisie Belge (Pas), p. 802 (R.G. n° C.05.0156.N).

[195] Jacquemin and Hubin (2017b), p. 137 (n°59); Malengreau (2019b), p. 15582 (n°43). See for an extensive analysis: Werbrouck (2018), pp. 580–591.

3.5 *Liability of Service Providers and Infrastructure Authorities*

Autonomous vehicles will increasingly rely on data, information and connectivity in order to function and will no longer resemble pure standalone objects. This new eco-system involves more actors that could be held liable for damages resulting from an accident. For instance, software, data and communication providers could be held liable for failure to provide quality data and communication (Sect. 3.5.1), though it is certain that contractual liability will play a predominant role in this regard. Depending on the operational setting, it is also possible that public authorities may be responsible for operating and maintaining smart road infrastructure. This could potentially result in their liability in case of failure (Sect. 3.5.2).

3.5.1 Liability of Software, Data and Communication Service Providers

Autonomous vehicles will heavily rely on software, data, network communications and other and services in order to safely operate. Such elements could have a causal role in the accident's occurrence but applying the corresponding liability regimes to software, data and communication providers may be difficult for several reasons.

First, while embedded software seems to be covered by the Product Liability Act, non-embedded software and specifically updates may not fall under its scope.[196] If the damage derives from a defective update (provided over the air), the victim may have to base his/her claim on another legal ground. Second, data providers (*e.g.*, mapping information which is essential for the vehicle to localise itself)[197] could see their liability invoked by the claimant (or by the vehicle liability insurer if recourse action is implemented), specifically for inaccurate data or if the data was breached by a cybersecurity attack. Third, communication providers may also be held liable for damages resulting from data loss and communication failures.

Contractual liability could play an important role, especially between the manufacturer and service providers or users (depending on the contractual setting). However, in the absence of contractual relations, extra-contractual liability between the victim and these providers may still be applied. In this case, the basis for bringing a claim may be Articles 1382 and 1383 OBCC. Yet, the victim may encounter several legal hurdles such as the proof of the different constitutive elements (*e.g.* wrongful act, causation, etc.). As such, a victim's claim may not be successful. Moreover, autonomous vehicles are extremely complex cyber physical artefacts and the damage may have had different causes resulting in a diffusion of responsibilities.

[196] See *supra*, Sect. 3.4.1.

[197] Sjafrie (2020), p. 58.

3.5.2 Liability of Public Authorities for Road Infrastructure

As explained by Ducuing, some technological settings, specifically Cooperative Intelligent Transport Systems (C-ITS)[198] are expected "to enable road vehicles to exchange messages in a peer-to-peer fashion, with one another and with other participants in their environment".[199] Depending on the technological setting that is eventually chosen, the vehicle could communicate with other vehicles (V2V[200]), other road users such as pedestrians (V2P[201]) and the road infrastructure itself (V2I[202]). In V2I scenarios, the autonomous vehicle communicates with smart road sensors and road traffic signs. Allegedly, such sensors and more generally road infrastructure could be managed by public authorities. Once level 3 and 4 vehicles are allowed to 'drive' on roads, liability challenges may therefore arise for the road authorities.[203]

In Belgium, such authorities vary according to the type of road (regional, provincial, communal).[204] The legal basis for claiming liability could either be fault-based liability under Articles 1382–1383 OBCC or the general liability for defective things of Article 1384, first paragraph, OBCC.[205] As infrastructure managers, public authorities are responsible for the safety of roads open to the public which means they have obligations to maintain and repair roads.[206] As such, they

[198] C-ITS, which stands for 'Cooperative Intelligent Transport systems', refers to systems where various technologies will allow "road users and traffic managers to share information and use it to coordinate their actions. This cooperative element [is] enabled by digital connectivity between vehicles and between vehicles and transport infrastructure" (European Commission website, "Cooperative, connected and automated mobility (CCAM)": https://ec.europa.eu/transport/themes/its/c-its_en). Such C-ITS solutions would enable the autonomous vehicles to communicate between each-other as well as with the smart road infrastructure.

[199] Ducuing (2021), p. 2.

[200] V2V: Vehicle to Vehicle.

[201] V2P: Vehicle to Pedestrians.

[202] V2I: Vehicle to Infrastructure.

[203] Enghelhard and De Bruin (2018), p. 96.

[204] Durviaux et al. (2010), pp. 189–190.

[205] Since the *Flandria* decision by the Court of Cassation, public entities are subject to the general rules of civil liability (Articles 1382 and following of the OBCC) (Court of Cassation, 5 November 1920, as reported in (1920) Pasicrisie Belge (Pas), I, p. 239, as mentioned by Durviaux 2010, pp. 225–226, footnote n°7).

[206] Road authorities must open roads that are sufficiently safe and must obviate any abnormal danger. See, *e.g.*: Court of cassation, 12 April 1984, as retrieved from Juportal.be (R.G. n° 7050): "Hormis le cas où une cause étrangère qui ne peut leur être imputée les empêche de remplir l'obligation de sécurité qui leur incombe, les pouvoirs publics doivent, par des mesures appropriées, obvier à tout danger anormal dans l'établissement des voies qu'ils ouvrent à la circulation publique". Article 135 § 2 of the New Communal Act (Nouvelle Loi Communale du 24 juin 1988) imposes a duty of safety (which amounts to a duty of care) for Communes which are responsible for the safety and maintenance of their roads. They must subsequently control such roads and ensure that they are not unsafe by taking adequate action: Durviaux et al. (2010), pp. 196–197/pp. 202–203; See also: Court of Cassation, 28 January 2005 (R.G. n° C.02.0272.N),

can be held liable for fault on the basis of Articles 1382 and 1383 OBCC. [207] For instance, a 2015 court decision held that communes must make sure that roads established in their territory do not present any abnormal dangers that could cause accidents.[208] This means that public authorities are required to manage roads that are sufficiently safe.[209] Public authorities may, therefore, possibly be held liable for failing to properly maintain the road sensors or smart traffic signals (once they are there), or for failing to repair them when they are not properly working, which would result in an abnormal danger for the road users. Public authorities managing road infrastructure are also responsible for the road in their quality of custodian. As such, they can be held strictly liable on the basis of Article 1384, first paragraph, OBCC, as the custodian of a defective thing.[210] Indeed, roads open to the public are understood as things falling under the scope of Article 1384, first paragraph, OBCC. The victim could prove that the road was intrinsically defective or could establish the abnormal behaviour of an element of the road caused the damage.[211] By extension, road sensors that would not be functioning properly eventually leading to an accident could possibly trigger the public authorities liability for things. However, the claimant may find it challenging to prove the defect and the causal link with the damage, precisely because autonomous vehicles operate within a complex socio-technological eco-system, which involves numerous cyber-physical components and

as reported in (2005) Pasicrisie Belge (Pas), p. 217: "En vertu de son obligation de ne construire des routes et de n'ouvrir celles-ci à la circulation routière que si elles sont suffisamment sûres, toute commune est tenue, sous réserve d'une cause étrangère qui ne lui est pas imputable et qui l'empêche de se conformer à cette obligation de sûreté, d'éviter par la prise de mesures adéquates tout danger anormal qui pourrait tromper la légitime attente des usagers, fût-il caché ou apparent".

[207] See for instance: Police Court of Brussels (9 e ch.), 12 September 2019, n° 15A388, as reported in (2020) VAVARC (2), pp. 26–31 (the road was known to be defective, and the region was held liable for fault for failing to take action); First Instance Court of Namur (division Dinant), 21 March 2017, n° 16/354/A, as reported in (2017) VAVARC (3), pp. 10 12; Durviaux et al. (2010), p. 200; Kruithof (2018), p. 102 (n°168): "fault liability can be used invoking a violation of the municipal government's statutory duty to maintain safe traffic on the public roads, streets and squares on its territory. The Court of Cassation has ruled that, while this statutory rule does not require the municipality to reach the result of safety at any cost, which would result in a quasi-strict liability of the municipality for traffic accidents caused by unsafe road conditions, but only imposes a duty of care on the government to take all reasonable measures to maintain safety, the government is presumed to be aware of all unsafe conditions on the roads on its territory".

[208] See: First Instance Court of Brussels, 13 January 2015, as retrieved from Juportal.be (R.G. n°12/4792/A); Court of Appeals of Liège, 17 September 2015, as reported in (2016) RGAR, p. 15268. Case law mentioned in: Delforge et al. (2019), p. 714.

[209] See for instance: Court of Cassation, 20 April 2001, as reported in (2001) Pasicrisie Belge (Pas), p. 655; Court of Appeals of Liège, 30 June 2010, as reported in (2011) RGAR (5), pp. 14743–14745; Durviaux et al. (2010), p. 200.

[210] See for instance: Court of Cassation, 23 May 2014, as reported in (2014) Pasiscrisie Belge (Pas), p. 1278 (R.G. n°C.12.0301.F); Durviaux et al. (2010), p. 197.

[211] For instance, road authorities have been held liable on the basis of Article 1384, first paragraph, OBCC for deteriorated road surfacing. See case law reported by: Durviaux et al. (2010), pp. 217–218.

various actors. Each of them could have had a causal role in the accident's occurrence.

4 Liability for Level 5 Vehicles: An Uncertain Future

The introduction of level 5 vehicles ('full automation') will go one step further, though their circulation on roads is not foreseen before some time (should that ever happen at all). In such cars, the user will no longer have any active role in the driving process. The embedded technologies will, therefore, shift the user from being a driver towards that of a mere passenger. This will impact the applicability of some liability regimes (Sect. 4.1). Moreover, new business and operational paradigms may challenge the current liability setting (Sect. 4.2).

4.1 Unequal Applicability of Existing Liability and Compensation Mechanisms

4.1.1 Applicability of Art. 29*bis* and the Insurance Impact of the Driver's Disappearance

Users of level 5 vehicles will become mere passengers and will, therefore, be covered by the protective provisions of Article 29*bis* as they are qualified as 'vulnerable users'. If this adds increased protection to all the users of autonomous vehicles, some have questioned whether this would challenge the current insurance setting and eco-system.[212] First, some scholars argue that it is important to keep Article 29*bis*' compensation scheme so that victims continue to have facilitated compensation.[213] With users sliding towards being mere passengers, some have questioned whether the driver's exclusion from this protective regime would still be relevant.[214] Secondly, with foreseen decreases in the number of car accidents (at least when there will be only fully autonomous vehicles on the roads), it is alleged that current vehicle centric liability insurance coverages may become less relevant. Insurance requirements may thus slide towards that of other parties such as the manufacturers. This points to fundamental questions that policymakers and legislators will have to tackle such as the relevance of changing the first line liability

[212] See: Malengreau (2019b), p. 15582 (n°65).

[213] Malengreau (2019b), p. 15582 (n°65). This author argues that the automatic compensation scheme for road users should continue to apply to autonomous vehicles; See also: Sauvage (2020), p. 264 (n°12).

[214] Malengreau (2019b), p. 15582 (n°64); Tanghe and De Bruyne (2018), p. 20 (n°31).

bearers. If such a change would be necessary, questions as to whom must be attributed primary liability and according to which criteria could be raised.

4.1.2 Inapplicability of the 'Driver's' Fault Liability and (Continued) Challenged Application of Liability for Things and Product Liability

As seen for level 4 vehicles, level 5 cars will be completely automated. The user will have virtually no control over the vehicle (except from inputting the destination in the computer system). Moreover, the user will not be required to monitor or supervise the driving in all circumstances. This obviously has an impact on the applicability of a fault-based liability regime.[215] First, as Sauvage notes, it will become nearly impossible to attribute the violation of a rule of conduct to the 'user' if the vehicle causes damage. [216] Secondly, the user cannot be held liable for negligently failing to control the vehicle or for monitoring the driving as all these functions fall upon the on-board computer system.[217] At the most, the user or owner may be held liable under negligence for having failed to update a safety critical software (of for having prevented it from occurring). Furthermore, the user could also be held liable for having installed an unauthorised third party software that caused damage.[218] Liability for things as applied to vehicles level 5 will face the same difficulties identified in Sect. 3.3. Moreover, product liability may also be confronted with the challenges that were mapped in Sect. 3.4.

4.2 Towards New Liability Paradigms?

The introduction of fully autonomous vehicles paves the road for new operational and liability paradigms. First, it appears important to nuance product liability as the main tool for dealing with damages related to autonomous vehicles (Sect. 4.2.1). Secondly, the digitalisation and interconnectedness of autonomous vehicles, which will question the current functional delineations of traditional actors (Sect. 4.2.2), and the technology induced servitisation of mobility (Sect. 4.2.3), may both impact liability.

---

[215] De Bruyne and Tanghe (2017), p. 336; Malengreau (2019a), p. 15578 (n°14); Sauvage (2020), p. 274 (n°39).

[216] Sauvage (2020), p. 274 (n°39).

[217] *Ibid.*

[218] Sauvage (2020), p. 277 (n°44).

4.2.1 Nuancing Product Liability as the Main Solution

On a general level, many authors[219] contend that product liability will be at the core of compensation claims. This is usually explained through the alleged liability shift from the human operator towards the manufacturer. However, different elements could help temper/nuance such an assertion.[220] First, *de lege lata*, implementing the national regime transposing the Product Liability Directive may prove challenging for reasons discussed in Sect. 3.4. Moreover, certain victims will prefer to rely on the no fault compensation regime of article 29*bis*. Secondly, and from a *de lege ferenda* perspective, questions as to the relevance of solely ascribing liability to the manufacturer can be raised. Indeed, other 'digital' actors create risks and could, therefore, also be held liable.[221] Autonomous vehicles increasingly rely on a socio-technological eco-system where a multiple and diverse set of cyber-physical components and services are needed for these vehicles to safely operate.[222] Moreover, new operational models may challenge even further the traditional liability paradigms. As such, product liability may not be seen as a "cure all tool"[223] for dealing with these novel mobility artefacts. Other remedies and legal solutions may have to be sought.

4.2.2 Digitalisation and the Blurred Functions of Traditional Actors

Several reports illustrate that actors' traditional functions may be extended/transformed by the introduction of these new technologies[224] which could therefore blur the ordinary functional delineations, thus potentially having an impact on liability attribution.[225] Autonomous vehicles are highly connected objects in which numerous data flows in and out of the vehicle. As the car deals with large amounts of data, the computational power and storage limitations of the vehicle may require that some of the data be processed and stored externally on back-end servers. There are competing models on who and how such servers would be managed.[226] Manufacturers could house and manage such servers. However, other stakeholders could also be involved. Putting the regulatory and liability focus on the manufacturer, or more

---

[219] Malengreau (2019b), p. 15582 (n°66); Kalra et al. (2009), p. 22; Lohmann (2017), p. 337 (n°2); Wagner (2018), p. 39; Smith (2017), p. 45.

[220] On this issue, see: Dheu et al. (2020), p. 13.

[221] Ducuing (2019), pp. 95–96 (n°8).

[222] *Ibid.*, p. 96 (n°9).

[223] Dheu et al. (2020), p. 14.

[224] See for instance: Expert Group on Liability, New Technologies formation (2019) Liability for artificial intelligence and other emerging digital technologies. Report for the European Commission: 70 p.

[225] Dheu et al. (2020), pp. 15–16.

[226] See, e.g.: McCarthy et al. (2017), 259 p.

specifically on the product, may be less relevant for these new vehicles.[227] As argued by some, the so-called "product liability paradigm",[228] according to which the manufacturer and the notion of product lie at the heart of regulatory intervention, may be challenged by the fact that these autonomous vehicles will be at the "crossroads of product and service".[229] This product versus service dichotomy "does not fit the 'new vehicle' reality where the manufacturing and operational phases are getting intertwined in time".[230] Moreover, autonomous vehicles can no longer be seen as standalone objects as one could classify them as 'systems' since they involve numerous cyber-physical components and actors for their functioning.[231]

The associated socio-technological interdependencies therefore challenge the characterisation of the vehicle as a pure product which may impact liability.[232] With such artefacts, it is not clear which regime would apply. Yet, delineating the boundaries between products and services is essential in order to determine the applicable liability regime, namely product liability or liability for fault. First, when implementing product liability, it appears unclear as to whether the claimant can successfully prove that the vehicle, taken as a product, was defective. Indeed, if the damage does not originate from the material inner workings of the vehicle, but rather from connectivity or communication issues, it may become difficult for the victim to rely on product liability. However, if actions are successfully brought against the manufacturer, this latter could possibly apply contractual liability provisions to allocate some of the costs to his subcontractors such as service providers. Secondly, when fault liability applies, identifying the liable party, proving its wrongful conduct and establishing causation may be extremely burdensome for the victim.

4.2.3 The Servitisation of Mobility and New Operational/Ownership Paradigms

The deployment of these complex mobility artefacts may also transform the existing automotive landscape and influence the way cars are operated and owned. Currently, most cars are used and owned at an individual level by private citizens. However, the

---

[227] Dheu et al. (2020), pp. 13–14.

[228] *Ibid.*, p. 14.

[229] *Ibid.*, p. 14.

[230] *Ibid.*, p. 14; Also see: Expert Group on Liability and New technologies (2019) Liability for artificial intelligence and other merging digital technologies. EC commissioned Report, pp. 39 and 44. The authors of the report seem to evoke the possibility of the manufacturer acting as an 'operator'. The 'frontend operator' is also distinguished from the 'back-end operator'. On the blurred product versus service dichotomy, also see: Evas Tatjana (2018) A common EU approach to liability rules and insurance for connected and autonomous vehicles. European Added Value Assessment accompanying the European Parliament's legislative own initiative report, p. 22.

[231] Herrmann et al. (2018), pp. 18–19.

[232] Ducuing (2019), pp. 95–96 (n°8).

autonomisation of vehicles opens new avenues for commercially operated cars and businesses as well as car sharing services. This servitisation process could impact current ownership paradigms, further fostering transport services involving the use of autonomous vehicles . For instance, autonomy enabled car sharing and fleet pooling would allow for the optimal use of vehicles, *inter alia* by new mobility service providers. Such providers could either own or lease vehicles to provide their services. There are currently many prospective business and operational models and it is difficult to foresee which ones will prevail.[233] Moreover, the development of Mobility as a Service (MaaS) will also play an important role within the context of enhanced mobility solutions.[234] In this complex and emerging eco-system, vehicle operations may move away from traditional ownership models. Automobile as a service may become the new norm.[235] New actors may come into the picture (think about fleet operators working for mobility service providers). All of this could impact the way risks associated with the operation of such vehicles are distributed and shared. This projected mobility eco-system, in which functions and responsibilities are less clearly delineated,[236] may have an impact on the implementation of liability mechanisms. When fully autonomous vehicles level 5 will be deployed (should they ever be deployed), the users could either own the vehicle individually, share it with other users through co-ownership, lease it from a company (the manufacturer), or even just use the car through a mobility service provider (taxi company).[237]

5 Concluding Remarks

Despite their (alleged) safety benefits, the deployment of autonomous vehicles will undoubtedly trigger accidents. In the absence of an *ad hoc* regime, the occurrence of such events will question the adequacy of existing liability and compensation mechanisms in dealing with the victim's damage(s). For vehicles up to level 2, traditional liability solutions will apply. However, when vehicles level 3 are introduced on public roads, the challenges will be more pressing. If it is assumed that some road victims, such as pedestrians, cyclists or passengers should continue to benefit from the no fault compensation regime of Article 29*bis*, other road victims, including drivers, will have to resort to general fault liability in order to obtain redress. The sharing of driving functions between the human user and the on-board computer will unmistakably trigger questions as to the fitness of applying fault liability solutions to partially autonomous vehicles. Moreover, attributing the

---

[233] Narayanana et al. (2020), pp. 255–293.

[234] Kostiainen and Tuominen (2019), pp. 239–254.

[235] At least in urban environments. This assertion may hold less true in the countryside.

[236] Kemp (2018), p. 35.

[237] On the notion of leasing vehicles: see (*op. cit.*) Sauvage (2020), pp. 268–269 (n°23).

violation of road traffic rules will become increasingly challenging. Alternatively, the victim could invoke liability for things and product liability as a basis for bringing a claim. However, applying these regimes may also prove difficult due to the dynamicity of technologies in and out of the vehicle. If with level 4 and 5 vehicles attributing personal liability to the user may become virtually impossible, some have alleged that the control shift from the human operator to the machine would translate into liability shifting away from the user towards the manufacturer. If this may partially hold true, it omits the fact the self-driving cars function and operate within a complex socio-technological eco-system which involve numerous actors (though, some of these actors may be contractually linked to one another). Moreover, the systemisation and digitalisation of road mobility may blur the traditional delineations, challenging the product versus service dichotomy. It remains to be seen how these technologies will deploy and what eco-system they will bring with them. It seems time for legislators and policymakers alike to step up their game in fostering an adequate legal framework which would balance conflicting interests. An *ad hoc* liability regime may be considered for autonomous vehicles, preferably at the European level.

References

Bocken H, Boone I, Kruithof M (2014) Inleiding tot het schadevergoedingsrecht: buitencontractueel aansprakelijkheidsrecht en andere schadevergoedingsstelsels. Die Keure, Bruges

Bogaert J (2004) Tien jaar praktijk artikel 29bis: de regeling ten voordele van zwakke weggebruikers. Kluwer, Mechelen, pp 54–55

Cassart A (2018) Législation. Bref point sur la situation Belge en matière de voiture autonome. Revue du droit des technologies de l'information 71:136

Cornelis L, Bützler R (1982) De buitencontractuele aansprakelijkheid voor schade veroorzaakt door zaken: rechtsvergelijkend onderzoek: België, Frankrijk, Nederland, Bondsrepubliek Duitsland en Engeland. Kluwer, Antwerp, p 728

Coulon C (2016) Du robot en droit de la responsabilité civile: à propos des dommages causés par les choses intelligentes. Responsabilité civile et assurances 4:6

De Bruin R (2016) Autonomous intelligent cars on the European intersection of liability and privacy: regulatory challenges and the road ahead. EJRR 3:491

De Bruyne J (2021) Autonome motorvoertuigen. Een multidisciplinair onderzoek naar de maatschappelijke impact. Vanden Broele, Bruges, p 408

De Bruyne J, Tanghe J (2017) Liability for damage caused by autonomous vehicles: a Belgian perspective. JETL 8(3)

De Coninck B, Dubuisson B (2015) L'indemnisation automatique des usagers faibles, victimes d'accidents de la circulation. Rapport belge. In: L'indemnisation des victimes d'accidents de la circulation en Europe, recueil des travaux du Groupe de Recherche Européen sur la Responsabilité civile et l'Assurance (GRERCA). Bruylant, Brussels, p 28

De Rode H (2023) L'indemnisation des victimes faibles d'accidents de circulation. Anthemis, Limal

Delforge C (2015) Le défaut de sécurité au sens de la loi du 25 février 1991 relative à la responsabilité du fait des produits défectueux. In: Delforge C, Van Zuylen J (coord) Les défauts de la chose, Responsabilités contractuelle et extracontractuelle. Anthemis, Limal, p 295

Delforge C, Delbrassinne C, Leleux A, Mortier S, Van Zuylen J, Vandenhouten L, Defosse M, Larielle S, Vandenberghe N (2019) Chronique de jurisprudence (2015 à 2016) - La responsabilité aquilienne (Articles 1382 et 1383 du Code civil). RCJB 4:715

Dheu O, Ducuing C, Valcke P (2020) The Emperor's new clothes: a roadmap for conceptualizing the 'new vehicle'. Revue Transidit 75

Dubuisson B (2006) La garde de la chose.... Pour des prunes? Note sous Cass., 20 mars 2003. RCJB 60(1):18

Dubuisson B (2009) Les faits générateurs de responsabilité. In: La responsabilité civile – chronique de jurisprudence 1996–2007. Larcier, Brussels, p 23

Dubuisson B, Callewaert V, Ganthem C, De Coninck B (2009) §1 Définitions et éléments de la faute civile. In: La responsabilité civile. Larcier, Brussels, p 30

Ducuing C (2019) Mobility, new technologies and innovation regulation. In: Rethinking IT and IP law – celebrating 30 years CiTiP. Intersentia, Cambridge

Ducuing C (2021) Understanding the rule of prevalence in the NIS Directive: C-ITS as a case study. Comput Law Secur Rev 60:105514

Durviaux A (2010) Chapitre 4 - La responsabilité civile des pouvoirs publics du fait de leurs agents et la responsabilité civile des agents des pouvoirs publics. In: Droit de la fonction publique. Larcier, Brussels, pp 225–226

Durviaux L, Kohl B, Fisse D (2010) Voirie: le point sur la responsabilité des pouvoirs publics. In: Lecocq P, Engels C (eds) Chronique de droit à l'Usage des Juges de Paix et de Police. Die Keure, Bruges

Emanuilov I (2017) Autonomous systems in aviation: between product liability and innovation. In: Proceedings of seventh SESAR innovation days, Belgrade, 28–30 November 2017, 8 p

Enghelhard EFD, De Bruin R (2017) EU Common Approach on the liability rules and insurance related to Connected and Autonomous Vehicules. Annexe I of the European Added Value Assessment Accompanying the European Parliament's legislative own-initiative report, p 56

Enghelhard EFD, De Bruin R (2018) Annex I EU Common Approach on the liability rules and insurance related to Connected and Autonomous Vehicles. In: European Added Value Assessment accompanying the European Parliament's legislative own initiative report, p 96

Fagnart J-L (1994) L'indemnisation des victimes d'accidents de la circulation après la réforme bâclée du 30 mars 1994. RGAR:12388

Fagnart J-L (2006) Petite navigation dans les méandres de la causalité. RGAR 2:14.080

Fagnart J-L (2017) La responsabilité du fait des choses. Wolters Kluwer, Waterloo

Fagnart J-L, Boularbah H (2000) La garantie et la responsabilité en matière de dommages causés par les produits. Bruylant, Brussels, pp 132–134

Fairgrieve D et al (2016) Part I. Product Liability Directive. In: Machnikowski P (ed) European product liability: an analysis of the state of the art in the era of new technologies. Intersentia, Cambridge, pp 46–47

Fruy G (2020) Attention à la fermeture automatique des portes! Les Pages: obligations, contrats et responsabilités 69:1

Glansdorff F (2013) Les conditions de la responsabilité extra-contractuelle. In: Van Den Haute E (ed) Le droit des obligations dans les jurisprudences française et belge. Bruylant, Brussels

Herrmann A, Brenner W, Stadler R (2018) Autonomous driving: how the driverless revolution will change the world. Emerald Publishing, Bingley, pp 18–19

Jacquemin H, Hubin J-B (2017a) Chapitre 2. La responsabilité extracontractuelle du faits des robots ou de l'intelligence artificielle. In: Jacquemin H, De Streel A (eds) L'intelligence artificielle et le droit. Larcier, Brussels, p 118

Jacquemin H, Hubin J-B (2017b) Titre 1. Aspects contractuels et de responsabilité civile en matière d'intelligence artificielle. In: Jacquemin H, De Streel A (eds) L'intelligence artificielle et le droit. Larcier, Brussels

Kalra N, Anderson J, Wachs M (2009) Liability and regulation of autonomous vehicle technologies. In: California PATH Research Report – RAND, p 22

Kemp R (2018) Autonomous vehicles – who will be liable for accidents? Digit Evid Electron Signat Law Rev 15:35

Kostiainen J, Tuominen A (2019) Mobility as a service—stakeholders' challenges and potential implications. In: Muller B, Meyer G (eds) Towards user-centric transport in Europe: challenges, solutions and collaborations. Springer, Cham, pp 239–254

Kruithof M (2015) Oorzaak of aanleiding? Geen causaal verband zonder causale bijdrage'. In: Vansweevelt T, Weyts B (eds) Actuele ontwikkelingen in het aansprakelijkheidsrecht en verzekeringsrecht. Iste Interuniversitair Congres over Aansprakelijkheids – en Verzekeringsrecht. Intersentia, Antwerp, pp 139–208

Kruithof M (2017) Chapter 11. Tort Law. In: Kruithof M, De Bondt W (eds) Introduction to Belgian law. Kluwer Law International, Alphen aan den Rijn, p 252

Kruithof M (2018) Tort law in Belgium. Kluwer Law International, Alphen aan den Rijn

Lambert C-E (2006) La responsabilité du fait des choses – rappel des conditions d'application et observations. RGDC 1:5–6

Lambert C-E (2016) La responsabilité du fait des choses – rappel des conditions d'application et observations. RGDC 1

Lohmann M-F (2017) Liability issues concerning self-driving vehicles. EJRR 7(2):337

Malengreau T (2019a) Automatisation de la conduite: quelles responsabilités en droit Belge? (Première partie). RGAR 5:15.578

Malengreau T (2019b) Automatisation de la conduite: quelles responsabilités en droit belge? (Deuxième partie). RGAR 6

Malengreau T (2020) Automatisation de la conduite: quelles responsabilités en droit Belge? In: Lazaro C, Strowel A (eds) Des Véhicules Autonomes à l'Intelligence Artificielle: Droit, politique et éthique. Larcier, Brussels

Marchetti R (2005) Quelques considérations à l'égard de la notion de vice et de l'exigence causale dans le cadre de la responsabilité du fait des choses vicieuses. RGDC 6

Martin-Casals M (2019) Causation and scope of liability in the internet of things. In: Lohsee S et al (eds) Liability for artificial intelligence and the internet of things. Hart Publishing, Oxford, pp 201–228

McCarthy M, Seidl M, Mohan S, Hopkin J, Stevens A, Ognissanto F, Kathuria N, Cuerden R (2017) Access to in vehicle data and resources. Publications Office of the European Union, p 259

Montero E (2006) Les produits défectueux dans un écheveau de responsabilités. RGDC 10:626

Narayanana S, Chaniotakisb E, Antoniou C (2020) Shared autonomous vehicle services: a comprehensive review. Transp Res Part C 11:255–293

Ninane Y, Van Zuylen J (2015) Le vice dont répond le gardien ou le propriétaire sur le fondement des articles 1384, alinéa 1er, et 1386 du Code civil. In: Delforge C, Van Zuylen J (coord) Les défauts de la chose. Responsabilités contractuelle et extracontractuelle. Anthemis, Limal

Ronneau V (2019) La garde de la chose: une figure, deux visages? Forum de l'assurance 195:97

Sauvage D-A (2020) Les véhicules autonomes et le droit de la circulation. RGDC 5:276

Schoubroeck C, Meurs T (2013) Vergoedingsregeling Zwakke Weggebruiker (art. 29bis WAM-wet). Kluwer, Mechelen, p 102

Sjafrie H (2020) Introduction to self-driving vehicle technology. CRC Press, Boca Raton, p 58

Smith B-W (2017) Automated driving and product liability. Mich State Law Rev 45

Stijns S, Samoy I (2020) Leerboek Verbintenissenrecht. Boek 1bis. Die Keure, Bruges

Swinnen K (2018) De inpassing van digitale producten in het Belgisch privaatrecht. TPR 3:1091

Tanghe J, De Bruyne J (2018) Software aan het stuur: aansprakelijkheid voor schade veroorzaakt door autonome motorrijtuigen. In: Vansweevelt T, Weyts B (eds) 3de conferentie ALLIC (Nieuwe risico's in het Aansprakelijkheids- en Verzekeringsrecht) Nieuwe risico's in het aansprakelijkheids- en verzekeringsrecht. Intersentia, Antwerp

Van Ommeslaghe P (2010) Droit des obligations. Tome II: Sources des Obligations. Bruylant, Brussels

Van Quickenborne M (2007) Oorzakelijk verband tussen onrechtmatige daad en schade. Kluwer, Mechelen, pp 31–46

Van Ryn J (1946) La responsabilité du fait des choses. JT:166

Van Schoubroeck C, Meurs T (2013) Vergoedingsregeling zwakke weggebruiker (art. 29bis WAM-wet). Kluwer, Mechelen, pp 20–23

Van Zuylen J (2013) Du fait justificatif à la force majeure: les visages contrastés de l'exonération de la responsabilité. In: Bockourt J, Callewaert V, Capart R et al (eds) Evaluation du dommage, responsabilité civile et assurances - Liber amicorum. Anthemis, Limal, p 273

Van Zuylen J (2015a) La responsabilité du gardien d'une chose affectée d'un vice (art. 1384, al. 1er, du Code civil). In: Cruysmans G (ed) Actualités en droit de la responsabilité. Bruylant, Brussels

Van Zuylen J (2015b) Le vice dont répond le gardien ou le propriétaire sur le fondement des articles 1384, alinéa 1er, et 1386 du Code civil, pp 20–21

Vandenberghe H, Van Quickenborne M (2011) Overzicht van rechtspraak. Aansprakelijkheid uit onrechtmatige daad 2000–2008. TPR 355

Vandenberghe H, Van Quickenborne M, Wynant L, Debaene M (2000) Aansprakelijkheid uit onrechtmatige daad (1994–1999) Overzicht van rechtspraak. TPR 1551:1564

Vansweevelt T, Weyts B (2009) Handboek buitencontractueel aansprakelijkheidsrecht. Intersentia, Antwerp

Wagner G (2018) Robot liability. In: Lohsse S, Schulze R, Staudenmayer D (eds) Liability for artificial intelligence and the internet of things. Hart Publishing, Oxford, p 39

Werbrouck J (2018) De productaansprakelijkheid voor zelfrijdende motorrijtuigen. TPR 55(1–2)

Liability for Automated and Autonomous Driving in Germany

Benjamin von Bodungen and Hans Steege

Abstract In 2017, regulations specifically addressing conditional vehicle automation were added to the Federal Road Traffic Act in Germany. In a further quantum leap, legal provisions for highly automated cars were introduced in 2021. The intention of the German Government of the time was to demonstrate the country's leadership in the development of self-driving vehicles and the surrounding legal ecosystem. This chapter examines the legal ramifications of vehicle automation in Germany specifically from a liability law perspective. The focus is first on SAE Level 0–2 vehicles, thereby carefully distinguishing between the driver, vehicle keeper and manufacturer as potential liability addressees and the preconditions for such liability to be available. In view of the rapid progress of vehicle automation, the chapter then places particular focus on the novel liability issues that the operation of motor vehicles at SAE Levels 3 and 4 gives rise to. As these automation levels have already been taken up by the 2017 and 2021 amendments to the Road Traffic Act respectively, it lends itself to elaborate on the merits and pitfalls of these novel statutory provisions, which may be of particular interest to readers from jurisdictions that are currently considering the introduction of a legislative framework for SAE Levels 3 and 4. What is more, the intricate allocation of liability in the context of emerging trends in the automotive sector such as artificial intelligence and software updates is discussed. Finally, liability for SAE Level 5 driving is dealt with. So far, no specific legal rules for the operation of fully automated vehicles exist in Germany. Hence, the adaptability of the existing civil liability system towards robotic vehicles will be analysed in the last part of this chapter.

B. von Bodungen
Bird & Bird LLP, Frankfurt/Main, Germany
e-mail: benjamin.von.bodungen@twobirds.com

H. Steege (✉)
Wolfsburg, Germany

H. Steege et al. (eds.), *Autonomous Vehicles and Civil Liability in a Global Perspective*, Data Science, Machine Intelligence, and Law 3,
https://doi.org/10.1007/978-3-031-41992-8_12

1 Introduction

Autonomous driving is currently a hot topic that is on everyone's lips and fires the imagination.[1] Automated and autonomous vehicles are being tested all over the world.[2] Numerous opportunities have been identified for how this disruptive technology is likely to affect us in the future, from urban and spatial planning to people's everyday lives.[3] The greatest promise is probably Vision Zero.[4] Compared to the numerous accidents, injuries and traffic fatalities, the realisation of accident-free road traffic would present a massive step forward with positive socio-economic side effects.[5] Likewise, the reduction of emissions and the increase of traffic efficiency must be referred to as opportunities.[6] Automation also promises a reduction in road traffic-related crimes.[7] With autonomous vehicles, driving under the influence of alcohol will no longer have negative consequences for the safety of road traffic, so that police checks for the purpose of verifying fitness to drive will become dispensable.[8] Last but not least, the assumption with automation is that it will be accompanied by an increase in mobility for physically impaired and elderly people, who will be empowered to participate in road traffic.[9]

1.1 Importance of Liability Law with Regard to the Regulation of New Technologies

It remains to be seen whether these numerous promises come true. From an epistemological point of view, there will only be sufficient empirical evidence to draw conclusions once tomorrow's robot vehicles have actually been tested in the field. By its very nature, this also holds true for the unintended side effects and risks

---

[1] See in this regard the statement of the German Federal Ministry of Digital Affairs and Transport dated 28 May 2021 that "Germany will be the world leader in autonomous driving", available at: https://www.bmvi.de/SharedDocs/EN/Articles/DG/act-on-autonomous-driving.html.

[2] A brief overview of autonomous driving developments in 30 countries around the world is available at: https://home.kpmg/xx/en/home/insights/2020/06/autonomous-vehicles-readiness-index.html.

[3] See Gruel and Stanford (2016), pp. 23–29.

[4] For an introduction to Vision Zero, see, e.g., Kim et al. (2017), pp. 1–2.

[5] See, e.g., the preamble to the Resolution on the Deployment of Highly and Fully Automated Vehicles in Road Traffic adopted by the Global Forum for Road Traffic Safety (WP.1) of the United Nations Economic Commission for Europe on 20 September 2018, available at: https://unece.org/DAM/trans/main/wp1/wp1doc/WP1_Resolution_Brochure_EN_web.pdf.

[6] Moriarty and Wang (2017), pp. 2111–2112.

[7] On automated driving, artificial traffic intelligence, and policing, see Müller (2022), pp. 24–27.

[8] On the potential impact of autonomous vehicles on alcohol consumption and drunk driving behavior, see Booth et al. (2020).

[9] Podszus et al. (2020), chapter 2.3, margin no. 90.

of automated and autonomous driving.[10] With a view to the future, it can be posited that many of the risks new technologies bring about only become apparent over time, when the technology is in use and adjustments to it can no longer be made easily. Hence, (parliamentary) technology assessment is faced with the challenge that it is never absolutely clear when the right time for regulation has come (the so-called Collingridge dilemma)[11] and that we are heading "with certainty into the uncertain"[12].

The regulation of new technologies and the associated risks provides a welcome starting point for addressing liability law. Its application in Germany with respect to automated and autonomous driving will be discussed in this chapter. It is thereby regarded as certain that technology is never one hundred percent free of errors. There will never be such an assurance, nor can it be reasonably expected.[13] In addition to the numerous legal implications associated with vehicle connectivity and automation—e.g., data protection law, vehicle homologation and registration law, telecommunications law, antitrust and competition law, administrative law, criminal law, and constitutional law—the question therefore arises as to who is liable for a manoeuvre that is—non-technically speaking—performed improperly by a robot vehicle and results in an accident causing personal injury or damage to property on the part of the occupants of the defective vehicle or other road users.

In Germany, the legal situation is particularly interesting as the legislator introduced regulations for "highly" and "fully" automated driving in the Federal Road Traffic Act (*Staßenverkehrsgesetz*—"StVG") as early as 2017. In addition, statutory provisions for autonomous driving recently entered into force in 2021 by way of amendment to the Federal Road Traffic Act. These legislative "updates" not only establish technical requirements for automated and autonomous vehicles but also address the operation of such vehicles and impose concomitant obligations on the persons involved, such as the driver, the vehicle keeper and the manufacturer. This allocation of duties has an immediate impact on the assessment of potential liability. In this respect, the German legal system provides for both contractual and non-contractual liability. We will limit ourselves to the latter in this chapter.

1.2 Triad of Non-contractual Liability with Regard to Road Traffic Accidents

Non-contractual liability, in turn, can be broken down into a well-established triad of liability in the context of road traffic accidents. First of all, Sect. 7 StVG subjects the vehicle keeper to strict liability for damage resulting from the death or injury of

[10] Chan (2017), pp. 214–215.

[11] See in detail Genus and Stirling (2018), pp. 63–64; Strand and Spaapen (2021), pp. 44–45.

[12] Renn (2014), p. 3.

[13] Federal Court of Justice, judgment of 16 June 2009—VI ZR 107/08, NJW 2009, 2952, 2954.

persons or the damage of property caused during the operation of a vehicle. Next, Sect. 18 StVG imposes fault-based liability on the driver. The driver is also liable under general tort law standards set out in Sect. 823 of the German Civil Code (*Bürgerliches Gesetzbuch*—"BGB"), which likewise require fault on the part of the driver. Lastly, the manufacturer of a defective product is liable for damage due to the deficiency of his product. German law does not provide for a special product liability law regime in the context of road traffic. Rather, product liability law rests on two generally applicable pillars. On the one hand, there is the European Product Liability Directive[14], which was implemented in Germany by way of the Product Liability Act (*Produkthaftungsgesetz*—"ProdHaftG"). On the other hand, manufacturers can be liable under the general tort law regime referred to above, which takes its starting point in the fault-based liability provided for in Sect. 823 BGB but has been refined and further developed by the courts over time. Today, tort law liability on the part of manufacturers is therefore based predominantly on precedent and well-established categories of cases that have emerged over time, which chiefly pertain to the manufacturer's violation of specific duties of care and organisational requirements during the various phases of production (i.e., construction, fabrication, user instruction, and monitoring of products in the field). This tort law liability is often also referred to as "producer" liability in contrast to "product" liability flowing from the Product Liability Act.

2 National Regulation and Comparison with the SAE J3016 Standard

Before the liability situation in Germany for the various levels of automation established by the international standard SAE J3016 will be dealt with, it must first be analysed which automation levels the newly enacted additions to the Federal Road Traffic Act mentioned above correspond to. This in turn is an indispensable prerequisite for the subsequent analysis of liability when operating systems at the different automation levels pursuant to SAE J3016.

2.1 *Road Traffic Act (***Straßenverkehrsgesetz***—StVG) as a Legal Source for Liability Law*

As set out above, the provisions of the Federal Road Traffic Act only concern the liability of vehicle keepers and drivers but do not address manufacturer liability. This

[14] Council Directive 85/374/EEC of 25 July 1985 on the approximation of the laws, regulations and administrative provisions of the Member States concerning liability for defective products, O.J. L 210/29 (1985).

immediately demonstrates that the statutory requirements established in the Federal Road Traffic Act can at most have an indirect effect on the liability of vehicle manufacturers. Conversely, changes to the Road Traffic Act are likely to have an immediate impact on the liability regime for keepers and drivers. In particular, any changes in the statutory requirements for driver conduct may have a direct bearing with respect to the fault-based rules for driver liability referred to above.

2.2 *Regulations for Highly and Fully Automated Driving, Sects 1a, 1b StVG*

The regulations on highly and fully automated driving were introduced by way of the Eighth Act Amending the Road Traffic Act (*Achtes Gesetz zur Änderung des Straßenverkehrsgesetzes*) of 16 June 2017[15]. They entered into force shortly thereafter on 21 June 2017. The most significant new provisions are to be found in Sects 1a, 1b and 63a StVG, with the latter devoted to data processing. Data processing dealt within Sect. 4.2.2.3.[16]

Sect. 1a StVG addresses the operation of motor vehicles that are equipped with "highly" or "fully" automated driving functions. With a view to the SAE J3016 classification, it could be assumed that this refers to SAE Levels 4 and 5. However, the explanatory memorandum to the new law reveals that the German legislator intended to rely on the definitions of the "Round Table on Automated Driving" instead.[17] In this context, the degree of automation and the legal expectations as regards the driver are defined as follows for highly automated driving functions: The system takes over both the lateral and longitudinal movement for a certain period of time in a specific situation. Hence, the driver does not have to monitor the system at all times. Only when necessary will he be requested with sufficient time reserve to take over the driving task. The automation system recognises all system limits, it is, however, not able to transfer the vehicle to a risk-minimising status in every situation.[18] The report of the "Round Table on Automated Driving" refers to a motorway chauffeur as an exemplary use case for highly automated driving.[19]

For the more advanced level of fully automated driving, the following description can be found in the report of the "Round Table on Automated Driving"[20]: The system completely takes over both lateral and longitudinal movement in a defined application. The human driver does not have to monitor the system. If the application is abandoned, the system prompts the driver to take over control of the vehicle with

[15] Federal Law Gazette 2017 Part I, p. 1648.

[16] In addition, several existing provisions of the Road Traffic Act were also amended.

[17] Steege (2021a), margin no. 2; Steege (2021b), p. 130.

[18] Federal Ministry of Transport and Digital Infrastructure (2015), p. 116.

[19] Federal Ministry of Transport and Digital Infrastructure (2015), p. 116.

[20] Federal Ministry of Transport and Digital Infrastructure (2015), p. 116.

sufficient advance warning. If no takeover occurs, the vehicle can, however, independently set itself to a system state that minimises risk. The fact that the system recognises all system limits and can transition to a risk-minimising state in every situation represents the main difference compared to highly automated vehicles. Conceptually, both levels aim at dual-use vehicles. With fully automated vehicles, however, it is possible that vehicles master a defined use case (e.g., a certain inner-city area) without driver intervention being necessary (or even technically possible) at any time. This is because, by definition, human take-over of the vehicle control may only become necessary when the vehicle operates outside the limits of its use case; furthermore, in the event of an incident or when the system limits are reached, the fully automated vehicle must be technically capable of transitioning to a risk-minimising state.

When cross-referencing the national law with the international standard SAE J3016, the following is to be noted: Highly and fully automated driving as per Sect. 1a StVG is typically grouped into Levels 3 and 4 of the SAE J3016 standard. More specifically, highly automated driving under German law is equated with SAE Level 3 (conditional automation), and fully automated driving under German law with SAE Level 4 (high automation). However, certain doubts about this classification already arise from the fact that the legislator did not incorporate the element of transfer to a risk-minimising state, which is essential for SAE Level 4, in Sects 1a, 1b StVG. It is also surprising that the legislative amendments to the Federal Road Traffic Act adopted in 2017 attach no differing legal requirements and consequences to the operation of vehicles equipped with highly automated compared to fully automated driving functions. Rather, both levels are treated equally with regard to the technical system requirements set out in Sect. 1a para. 2 StVG as well as the obligations imposed on the system user in Sect. 1b StVG. A meaningful difference between highly and fully automated vehicles is thus not apparent from the legislative text. Finally, there is no legislative statement in Sects 1a, 1b StVG that would fundamentally alter the traditional liability structure (see *supra* Sect. 1) with respect to highly and fully automated vehicles or even introduce special liability in this regard. Sect. 1a para. 4 StVG merely states that "the driver of a vehicle is also the person who activates a highly or fully automated driving function and employs it for the steering of the vehicle, even if he does not steer the vehicle himself within the scope of the intended use of this function".

2.3 *Regulations for Autonomous Driving, Sects 1d* **et seq.** *StVG*

In 2021, legal regulations for "autonomous" driving were introduced for the first time in Germany. This was accomplished by the Act Amending the Road Traffic Act and the Compulsory Insurance Act (*Pflichtversicherungsgesetz*—PflVG) of 12 July

2021, which is also known as the Act on Autonomous Driving.[21] It came into force on 28 July 2021 and added Sects 1d *et seq.* to the Federal Road Traffic Act.[22]

In a nutshell, Sect. 1d StVG defines autonomous vehicles as those that can perform the driving task independently of a person driving the vehicle. However, this only applies to previously specified and approved public roads, meaning that the vehicles with autonomous driving functions covered by the new law may only be operated within certain geographical restrictions. Furthermore, the vehicles concerned must be fitted with certain equipment that the law considers essential for the unassisted performance of the driving task within their respective operational design domain ("ODD"). The legal requirements with respect to this technical equipment are described in detail below (see *infra* Sect. 4.4.2.1).

Next, Sect. 1f StVG imposes obligations on the parties involved in the operation of vehicles with autonomous driving functions. Here you will find special legal duties placed on the vehicle keeper (see *infra* Sect. 4.3.2), the technical supervisor (see *infra* Sect. 4.2.1.2) and the manufacturer of such vehicles (see *infra* Sects. 4.4.2 and 4.4.3). In contrast, there are no requirements directed at the driver, which is a corollary of the fact that no driver is foreseen at this level of automation.

It is also worth mentioning that Sect. 1g StVG sets standards for the processing of data in the operation of autonomous vehicles and on the transmission of such data to public authorities or third parties in the event of, e.g., an accident. Next, Sect. 1h StVG deals with the subsequent activation of already installed (dormant) automated and autonomous driving functions. The testing of automated and autonomous driving functions is now addressed in Sect. 1i StVG, which establishes detailed requirements for the trialling of such functions. Finally, Sect. 1j StVG authorises the Federal Ministry for Digital and Transport to issue legal ordinances with the aim of further substantiating the provisions in Sects 1d *et seq.* StVG.[23]

As specified above, from their wording Sects 1a, 1b StVG refer to conditional and high driving automation and thus would seem to cover SAE Levels 3 and 4. Therefore, one would assume that the regulations for "autonomous" vehicles introduced in Sects 1d *et seq.* StVG would have to be attributed to the higher SAE Level 5. However, in view of the geographical restrictions placed on their operation, this is not the case. Rather, the most recent legislative amendments are intended to cover SAE Level 4 only.[24] Apart from the misleading reference to autonomous driving in the title of the law, it is then questionable what exactly the scope of the earlier Sects 1a, 1b StVG should be, which are commonly understood to also deal with SAE

[21] Federal Law Gazette 2021 Part I, p. 3108.

[22] For a detailed overview of the Act on Autonomous Driving, see Kleeman and Arzt (2021), pp. 99 *et seq.*; Steege (2021b), pp. 128 *et seq.*; von Bodungen and Mühlon (2022), pp. 5 *et seq.*

[23] On this basis, the Ministry has meanwhile issued the Ordinance on the approval and operation of motor vehicles with autonomous driving functions in specified operating areas (*Verordnung zur Genehmigung und zum Betrieb von Kraftfahrzeugen mit autonomer Fahrfunktion in festgelegten Betriebsbereichen*), which was published in Federal Law Gazette 2022 Part I, p. 986 and entered into force on 1 July 2022.

[24] Bundestag printed matter 19/27439, p. 16.

Level 4 automation in addition to SAE Level 3. A solution to this conflict is frequently seen in limiting the earlier amendments to SAE Level 3,[25] but this restrictive interpretation of the statutory text should best be confirmed by legislative clarification, which has not yet been forthcoming.

2.4 *Interim Result*

On the basis of a restrictive interpretation, the regulations on highly and fully automated driving (Sects 1a, 1b StVG) that entered into force in 2017 concern SAE J3016 Level 3 only. In contrast, the recently introduced regulations on autonomous driving (Sects 1d *et seq.* StVG) are aimed at SAE Level 4. It follows, therefore, that for vehicles of SAE Level 5, no regulations exist in Germany so far. Vehicles at this level of automation can be truly described as autonomous vehicles as their ODD is not subject to any restrictions and they thus satisfy the promise fuelled by the automotive industry of being operational "everywhere and anytime".

3 Liability for Vehicles of the SAE J3016 Standard Levels 0–2

In the following, the liability law regime in Germany for vehicles of levels 0–2 of the SAE J3016 standard will be considered.

3.1 *Description of Relevant Levels of Automation*

SAE Level 0 (No Driving Automation) means vehicles where the entire driving task (i.e., the dynamic Driving Task, "DDT") is carried out by the human driver, even if he is supported by active safety systems.[26]

SAE Level 1 (Driver Assistance) is described as "[t]he sustained and ODD-specific execution by a driving automation system of either the lateral or the longitudinal vehicle motion control subtask of the DDT (but not both simultaneously) with the expectation that the driver performs the remainder of the DDT".[27] This makes it more than clear that the vehicle only provides support to a limited extent and does not take over both the lateral and the longitudinal vehicle control. The driver must therefore not only monitor the system during the entire

[25] See, e.g., Deutscher Verkehrsgerichtstag (2018).

[26] Cf. the explanations in SAE International (2021), p. 30.

[27] SAE International (2021), p. 30.

driving process, but also perform the remaining tasks, such as steering (i.e., the lateral movement), while the system controls the vehicle distance and speed (i.e., the longitudinal movement). In addition, situations can easily arise in which the system cannot be relied upon to take over the tasks assigned to it, so that this has a direct impact on the human driver's duty of care.

SAE Level 2 (Partial Driving Automation) already represents a noticeable further development of vehicle automation. The SAE J3016 standard describes Level 2 as follows: "The sustained and ODD-specific execution by a driving automation system of both the lateral and longitudinal vehicle motion control subtasks of the DDT with the expectation that the driver completes the OEDR subtask and supervises the driving automation system".[28] OEDR refers in this context to object and event detection, recognition, classification, and response.[29]

In summary, the following should be noted: For SAE Level 0 vehicles, the driver must perform *all* driving tasks. In contrast, for SAE Levels 1 and 2, the driver is assisted by appropriate driving functions, but must still monitor the vehicle at all times and perform the driving tasks not assumed by the driving automation system being used.

3.2 *Driver Liability*

3.2.1 Driver Liability Under the Federal Road Traffic Act

3.2.1.1 General Overview

As mentioned above (see *supra* Sect. 1), the driver's liability is dealt with in Sect. 18 para. 1 StVG, which (in conjunction with Sect. 7 para. 1 StVG) provides that if a person is killed, the body or health of a person suffers harm or property is damaged in the operation of a motor vehicle, the driver is obligated to compensate the injured person for the resulting damage. However, in case of death or injury of one or more persons caused by the same event, the driver's liability is limited to an amount of five million euros in total; similarly, the driver's liability is capped at a total amount of one million euros in the case of damage to property, even if various items are damaged by the same event (Sect. 12 StVG). What is more, the driver's obligation to pay compensation is excluded if the damage was not caused by his fault (Sect. 18 para. 1 sentence 2 StVG). This demonstrates that driver liability under the Federal Road Traffic Act is designed as fault-based liability. Its rationale lies in sanctioning a culpable act on the part of the driver.

[28] SAE International (2021), p. 31.

[29] SAE International (2021), p. 40.

3.2.1.2 Capacity as Driver

The vehicle driver within the meaning of Sect. 18 para. 1 StVG has traditionally been the person who is driving the motor vehicle both laterally and longitudinally at the moment of the accident and has actual control of the steering wheel.[30] In vehicles of SAE Level 0, this is indisputably the human driver. After all, he is in control of the vehicle both in terms of its longitudinal and lateral movement and thus performs the entire driving task.

In contrast, SAE Level 1 vehicles are characterised by the fact that the automation system provides a certain degree of support by taking over either the lateral or the longitudinal vehicle control (see *supra* Sect. 3.1). The driver, however, must monitor the system during the entire driving process, as well as perform all other driving tasks. In addition, situations may occur that cannot be adequately dealt with by the system. Hence, also at this level of automation, the human being is to be regarded as the driver of the vehicle in accordance with Sect. 18 para. 1 StVG. It is the driver that has control over the operation of the vehicle at all times, since he performs either the lateral or the longitudinal driving task and can also override or deactivate the automation system at any time.

Finally, even with SAE Level 2 vehicles, where both the lateral and the longitudinal driving tasks are assumed by the driving system, the user of the automation system still legally qualifies as driver. Not only can he override and deactivate the automation system at any time, but he is also tasked with permanently monitoring the system.

3.2.1.3 Fault and Exculpatory Evidence of the Driver of the Vehicle

Liability pursuant to Sect. 18 para. 1 StVG is dependent on culpable conduct on the part of the driver. Such fault will be presumed in the case of a road traffic accident. It is therefore on the driver to rebut the statutory presumption of fault by providing evidence that he was driving in accordance with the road rules and did not negligently or even intentionally cause the personal injury or property damage at issue. The driver may succeed in exonerating himself, e.g., if a technical defect—such as a burst tyre or brake failure—is the root cause of an accident.[31] In addition, the driver must then also prove that he lost control of the vehicle due to this defect through no fault of his own.[32]

Since the fault of the driver is the decisive prerequisite for his liability, compliance with behavioural requirements embodied in the traffic rules as well as the observation of general duties of care is of particular significance for the exoneration

---

[30] Heß (2022), StVG Sect. 18, margin no. 3.

[31] See on this point, Higher Regional Court Saarbrücken, judgment of 3 August 2017—4 U 156/16, NJW 2018, 315.

[32] Heß (2022), StVG Sect. 18, margin no. 8.

of the driver. If the driver succeeds in demonstrating that he behaved in a roadworthy manner, then the presumption of fault is refuted.[33] In practice, therefore, the rebuttal of the presumption of fault is of utmost importance.

3.2.2 Driver Liability According to the German Civil Code

In addition to Sect. 18 StVG, driver liability can also be based on Sect. 823 para. 1 BGB. This general tort law liability is also contingent upon culpable behaviour on the part of the driver. However, in contrast to Sect. 18 StVG, the driver's fault is not presumed by law. Instead, the burden is on the injured party to prove culpable conduct on the part of the driver. Despite this burden of proof, which is disadvantageous for the victim of a traffic accident, Sect. 823 BGB also offers advantages to the injured party compared to Sect. 18 StVG. This is especially true in the case of accidents causing major damage because, unlike Sect. 18 StVG, the driver's general tort law liability as per Sect. 823 para. 1 BGB is *not* subject to any maximum limits. With regard to the duties of care, the breach of which may give rise to tort law liability, reference can be made to the preceding comments in relation to Sect. 18 StVG.

3.3 *Keeper Liability*

If a person is killed, the body or health of a person is injured or property is damaged in the operation of a motor vehicle, Sect. 7 para. 1 StVG requires the keeper of such vehicle to compensate the injured person for the resulting damage. We note that the keeper of a motor vehicle is not necessarily the owner but rather the person who uses it for his own account and has the power of disposal over it that such use requires.[34] Unlike the driver, the keeper is liable regardless of fault. This strict liability is the price to be paid by the keeper for operating a motor vehicle and thus exposing other road users to a source of danger. The liability of the vehicle keeper must be strictly distinguished from the driver's liability outlined above (see *supra* Sect. 3.2). In particular, the keeper does not need to have driven the motor vehicle or been inside it in order to be strictly liable. However, if the driver and the keeper are the same person, Sect. 7 StVG is *lex specialis* and takes precedence over Sect. 18 StVG.

In practice, the keeper's liability pursuant to Sect. 7 para. 1 StVG offers the injured party the advantage that, in contrast to the driver's liability as per Sect. 18 para. 1 StVG, it does not require any culpable behaviour on the part of the vehicle keeper. However, the keeper's liability is subject to the same caps as that of the

---

[33] Higher Regional Court Hamm, judgment of 27 May 1998—13 U 29/98, NZV 1998, 463; Higher Regional Court Saarbrücken, judgment of 3 August 2017—4 U 156/16, NJW 2018, 315.

[34] Federal Court of Justice, judgment of 3 December 1991—VI ZR 378/90, NJW 1992, 900, 902.

driver (see *supra* Sect. 3.2.1.1). Furthermore, it is excluded if the accident is caused by *force majeure* (Sect. 7 para. 2 StVG).

3.4 Manufacturer Liability

3.4.1 General Overview

In addition to the driver and the keeper of a motor vehicle at levels 0–2 of the SAE J3016 standard, its manufacturer may also be subject to liability under German law. As set out above (see *supra* Sect. 1.2), the liability of manufacturers (synonymous with producers) is thereby based on two separate but closely related legal regimes. On the one hand, there is the EU-wide harmonised liability regime that rests on the European Product Liability Directive (so-called "product" liability). On the other hand, manufacturer liability based on the general tort law provisions set out in the German Civil Code must also be considered (typically referred to as "producer" liability). The latter regime has been shaped by the courts in Germany over the last century. It largely overlaps with the liability standards under the national Product Liability Act, which implements the European Product Liability Directive in Germany. Where differences exist, this is explained below in the context of the producer's tort law liability.

Prior to putting a motor vehicle on the market, the respective car manufacturer is subject to a multitude of obligations.[35] The consideration of when these obligations arise is based on general principles of tort law duties of care, so that there are no automotive-specific features to be noted in this regard.[36] The decisive factors are, on the one hand, the product user's justified expectations in terms of product safety and, on the other hand, the effort the producer can reasonably be expected to put into the safe design, manufacture and distribution of its product.[37] The user's legitimate safety expectations are shaped in particular by the price of the product, the manufacturer's advertising statements and the intended purpose of the product.[38]

3.4.2 Scope of Manufacturer Tort Law Liability: Product

Producer liability under German tort law has a broader scope in terms of the products covered than harmonised product liability set out in the Product Liability Directive

---

[35] Cf. on the following obligations specifically in the context of automated and autonomous driving Steege (2021a), margin no. 34–39.

[36] Steege (2021c), p. 8.

[37] Federal Court of Justice, judgment of 7 June 1988—VI ZR 91/87, NJW 1988, 2611; Higher Regional Court Cologne, decision of 23 September 2015—5 U 189/14, BeckRS 2016, 06177, margin no. 13.

[38] Wagner (2020a), BGB Sect. 823, margin no. 822.

and its implementing act in Germany, the Product Liability Act. This is because tort law does not mandate that a *movable* object causes the damage in question, so that not only defective physical items but also deficient incorporeal products (e.g., software) that cause harm to their user are captured. Dogmatically, this follows from the fact that tortious producer liability under Sect. 823 para. 1 BGB does not hinge upon the defectiveness of a tangible product, but on the culpable breach of a duty of care or an organisational responsibility on the part of the manufacturer.[39] Application of the Product Liability Act, on the other hand, will only be triggered if the defect in a physical object causes damage.[40] Having said this, defective brakes or tyres are undoubtedly included in this narrower concept of product emanating from the Product Liability Directive, so that liability is conceivable under both liability regimes if the driver or other road users are injured as a result of a tyre blowout, for example.

3.4.3 Defects Up to the Time of Placing a Product on the Market

3.4.3.1 Design Defects

If the product design is not undertaken in such a way that an average user can operate the product safely within the scope of its intended purpose, then the producer is in violation of its specific duties of care regarding the design stage of production.[41] In this context, the manufacturer especially has the duty to identify the legitimate consumer expectations in terms of product safety. To this end, he must exhaust all available possibilities, such as analysing trade journals and conference reports as well as competitors' products for improved safety standards. Furthermore, the manufacturer must take into account the potential hazards both from an improper use as well as a foreseeable misuse of the respective product.[42]

Ultimately, the reasonable safety features to be integrated into the product design are to be assessed on a case-by-case basis in relation to the hazards posed by the product at hand,[43] thereby taking into account the legal interests potentially at stake.[44] Hence, in the case of significant dangers to life and limb, the manufacturer must take more extensive safety measures than in the case of potential property

---

[39] Steege (2021c), p. 7.

[40] Cf. von Westphalen (1990), p. 87.

[41] Federal Court of Justice, judgment of 5 February 2013—VI ZR 1/12, NJW 2013, 1302, 1303; Federal Court of Justice, judgment of 16 June 2009—VI ZR 107/08, NJW 2009, 2952, 2953; Kullmann (2002), p. 4.

[42] Federal Court of Justice, judgment of 5 February 2013—VI ZR 1/12, NJW 2013, 1302, 1303.; Federal Court of Justice, judgment of 7 June 1988—VI ZR 91/87, NJW 1988, 2611; Kullmann (2002), p. 4.

[43] Federal Court of Justice, judgment of 16 June 2009—VI ZR 107/08, NJW 2009, 2952, 2954; Federal Constitutional Court, decision of 16 October 1996—1 BvR 1179/95, NJW 1997, 249.

[44] Federal Court of Justice, judgment of 31 October 2006—VI ZR 223/05, VersR 2007, 72, 73.

violations or insignificant physical impairments only.[45] These manufacturer obligations that increase with the importance of the legal interests in question ultimately derive from the constitutional protection of life and physical integrity enshrined in Article 2 sentence 2 of the Basic Law of the Federal Republic of Germany (*Grundgesetz*—GG). If the product design falls short of these safety requirements, this constitutes a product defect at the design stage.[46]

3.4.3.2 Manufacturing Defects and Quality Control

Manufacturing defects concern the faulty manufacture of individual product components or products.[47] In this context, the manufacturer's organisational duties are of eminent importance. He must organise the production process in such a way that no machining, fabrication or manufacturing defects occur during production.[48] This structure of organisational obligations pertains to all of the manufacturer's operational processes and also impacts quality control. The interaction between production and quality control is of particular importance if it cannot be ruled out that certain manufacturing defects or human errors may occur during production.[49] In this case, quality control offers the advantage that production errors can be identified and corrected; conversely, when production has been optimised, the importance of quality control decreases.[50] However, quality management systems such as ISO 9000 *et seq.* that are frequently to be encountered in practice do not exempt the manufacturer from liability *per se*.

In the case of road traffic, defective transistors for the central locking,[51] unsafe braking systems,[52]defective push rods,[53] deficient valves,[54] and faulty airbags[55] have already been the subject of case law on manufacturing defects. The so-called push-rod case[56] illustrates the interconnection of production and quality control very well: On the one hand, the German Federal Constitutional Court insisted with

[45] Federal Court of Justice, judgment of 31 October 2006—VI ZR 223/05, VersR 2007, 72, 73; Federal Court of Justice, judgment of 16 June 2009—VI ZR 107/08, NJW 2009, 2952, 2954.

[46] Federal Court of Justice, judgment of 16 June 2009—VI ZR 107/08, NJW 2009, 2952, 2953.

[47] Federal Court of Justice, judgment of 19 April 1956—III ZR 227/54, VersR 1956, 410, 411; Michalski (1998), p. 962.

[48] Steege (2021c), p. 8; Kullmann (2002), p. 4.

[49] Steege (2021c), p. 8; Kullmann (2002), pp. 4 *et seq.*

[50] Wagner (2020a), BGB Sect. 823, margin no. 822.

[51] Federal Court of Justice, judgment of 31 March 1998—VI ZR 109-97, NJW 1998, 1942.

[52] Federal Court of Justice, judgment of 28 September 1970—VIII ZR 166/68, VersR 1971, 80.

[53] Federal Court of Justice, judgment of 17 October 1967—VI ZR 70/66, NJW 1968, 247, 248.

[54] Higher Regional Court Cologne, judgment of 16 November 1990—19 U 129/89, NJW-RR 1991, 740.

[55] District Court Munich, judgment of 19 April 1999—154 C 35416/98, NJW-RR 2001, 321.

[56] Federal Court of Justice, judgment of 17 October 1967—VI ZR 70/66, NJW 1968, 247, 248.

regard to the organisational obligations incumbent on the manufacturer that he must ensure that a suitable replacement is available if quality control personnel drops out. On the other hand, the court dealt in great detail with the specific requirements for quality control. This must be carried out by an employee selected on the basis of the following competences and qualifications: Training, reliability, ability and skill.

3.4.3.3 Suspicion of a Defect

The mere suspicion of a defect does not automatically justify the assumption that a product is indeed defective. Instead, the injured party must prove that a product defect exists.[57] Having said this, the ECJ's ruling of 5 March 2015[58] requires, at least in part, a reversal as far as the assessment of product defects in the case of a merely suspected defect is concerned. According to this ruling, a product is already considered to be defective if it has the significant potential to damage important legal interests such as the product user's life. The ECJ's position in relation to the Product Liability Directive can also be applied to producer liability under national tort law.[59] In this respect, too, as far as products belonging to the same group—such as cardiac pacemakers and implantable cardioverter defibrillators in the case of the ECJ's ruling—have a potential defect, the product user obviously cannot wait for the damage (i.e., potentially its death) to occur. Hence, such a product may be classified as defective without there being any need to establish that that the product actually exhibits such defect.

Against this background, the question would seem decisive whether it is reasonable for the person concerned not to use the product with a suspected defect in order to prevent possible damage. If the potential damage is significant, even low failure probabilities or a few defective products from the relevant production series may be sufficient to justify non-use of such product. With respect to automation products for road traffic use, there are typically high-ranking legal interests at stake. In case of the failure of such systems, considerable damage to the health and lives of accident victims must be feared. Hence, already a low failure probability may be sufficient to deem such product defective. On the other hand, the assessment may well be different for automation products if their use brings about "only" a risk of damage to property, but not of bodily injuries or even fatal impairments. This could be the case, e.g., for a parking assistant the failure of which is to be expected with a certain degree of probability. Overall, the lesser the damage to be feared, the higher the

[57] Established case law, see Federal Court of Justice, judgment of 26 November 1968—VI ZR 212/66, NJW 1969, 269, 274; Federal Court of Justice, judgment of 7 June 1988—VI ZR 91/87, NJW 1988, 2611, 2613; Regional Court Berlin, judgment of 10 February 2012—19 O 263/11, VersR 2013, 1454, 1455.

[58] ECJ, Judgment of 5 March 2015—C-503/13, C-504/13 (Boston Scientific Medizintechnik GmbH/AOK Sachsen-Anhalt et al.), NJW 2015, 1163.

[59] Spindler (2020), BGB Sect. 823, margin no. 649; Wagner (2020b), ProdHaftG Sect. 3, marginno. 54.

probability of its occurrence will have to be in order to assume a product defect. Ultimately, however, the circumstances of each individual case will be decisive in this regard.[60] For example, it might also be necessary to consider whether the user could have easily switched to products from other manufacturers. This question may have to be assessed differently in the case of automation products for road traffic than in the case of life-saving medical products.

3.4.3.4 Instructional Defects

Manufacturers are liable for damages that have occurred due to a lack of information of the user about which hazards are inherent to the product despite its defect-free manufacturing.[61] Such information enlightens the user about immanent dangers of the product[62] and enables him to avoid potential hazards during its use[63]. The manufacturer's duty to instruct also pertains to foreseeable product misuse by the user.[64] In determining when an obvious misuse is to be assumed, the legal interests at stake are of utmost importance, so that a stricter standard applies when the product user's life and physical integrity are in jeopardy.[65] The complete absence of a user manual constitutes an instructional error in any case.[66] On the other hand, too much information can overwhelm the user and also be harmful. Whether the manufacturer overshoots the mark or provides too little information can be a fine line, particularly with new technologies such as automated systems or artificial intelligence previously unknown to the user community. This will also very much depend on the presentation of the information in the individual case.

---

[60] Burckhardt (2015), p. 662.

[61] Federal Court of Justice, judgment of 13 February 1975—VI ZR 44/74, NJW 1975, 1457; Federal Court of Justice, judgment of 12 November 1991—VI ZR 7/91, NJW 1992, 560; Federal Court of Justice, decision of 10 June 1998—V ZR 324-97, NJW 1998, 2905; Federal Court of Justice, judgment of 7 October 1986—VI ZR 187/85, NJW 1987, 372.

[62] Higher Regional Court Frankfurt/Main, judgment of 2 March 1993—22 U 212/91, NJW-RR, 1994, 346, 347.

[63] Federal Court of Justice, judgment of 16 June 2009—VI ZR 107/08, NJW 2009, 2952, 2954; Federal Court of Justice, judgment of 11 July 1972—VI ZR 194/70, NJW 1972, 2217, 2220.

[64] Federal Court of Justice, judgment of 25 October 1988—VI ZR 344/87, NJW 1989, 707; Federal Court of Justice, judgment of 24 January 1989—VI ZR 112/88, NJW 1989, 1542; Federal Court of Justice, judgment of 12 November 1991—VI ZR 7/91, NJW 1992, 560; Federal Court of Justice, judgment of 27 September 1994—VI ZR 150/93, NJW 1994, 3349, 3350; Higher Regional Court Bamberg, judgment of 26 October 2009—4 U 250/08, VersR 2010, 403; Dahm-Loraing and Koyuncu (2010), p. 142.

[65] Federal Court of Justice, judgment of 5 February 2013—VI ZR 1/12, NJW 2013, 1302 margin no. 13; Federal Court of Justice, judgment of 12 November 1991—VI ZR 7/91, NJW 1992, 560, 561.

[66] Higher Regional Court Karlsruhe, judgment of 28 March 1996—7 U 61/94 VersR 1998, 63, 64.

3.4.4 Defects After Placing a Product on the Market

The manufacturer's product-related duties of care do not end at the point in time when the product is placed on the market.[67] Rather, they extend to the monitoring of its products in the field and also involve response obligations in the event of an identified hazard.

3.4.4.1 Product Monitoring

Product monitoring obligations arise from the general duties of care towards the legal interests of third parties and require the manufacturer to keep an eye on whether his product is still safe even after it has been placed on the market. At its core is the detection of previously unknown product hazards.[68] While manufacturer liability under the Product Liability Act only covers the phase until the product is placed on the market, tort law liability for inadequate product monitoring remains available and can be linked to the creation of a hazard by way of placing products on the market not known to be dangerous at that point in time. It is therefore irrelevant from which of the above-mentioned categories of defects (*supra* Sect. 3.4.3) such risks emanate.[69]

Active and Passive Product Monitoring

A distinction is to be made between so-called passive product monitoring obligations, where the manufacturer limits himself to taking complaints from its customers. Such exclusively random-based acquisition of knowledge of safety-relevant product deficiencies is by no means deemed sufficient.[70] Rather, active product surveillance is required on the part of the manufacturer. Hence, producers must on their own initiative evaluate technical literature, the results of academic conferences,[71] the experiences of competitor manufacturers and their product improvements with

[67] Staudinger (2019), BGB Sect. 823, margin no. 176.

[68] Federal Court of Justice, judgment of 16 December 2008—VI ZR 170/07, NJW 2009, 1080; Federal Court of Justice, judgment of 17 March 1981—VI ZR 286/78, NJW 1981, 1606; Federal Court of Justice, judgment of 27 September 1994—VI ZR 150/93, NJW 1994, 3349, 3350; Kullmann (2002), p. 6; Michalski (1998), p. 963.

[69] Michalski (1998), pp. 962 *et seq.*

[70] Federal Court of Justice, judgment of 7 December 1993—VI ZR 74/93, NJW 1994, 517, 519; Federal Court of Justice, judgment of 6 December 1994—VI ZR 229/93, NJW-RR 1995, 342, 343; Foerste (1994), p. 909; Michalski (1998), p. 963; Staudinger (2019), BGB Sect. 823, margin no.176.

[71] Federal Court of Justice, judgment of 12 November 1991—VI ZR 7/91, NJW 1992, 560; Federal Court of Justice, judgment of 17 March 1981—VI ZR 191/79, NJW 1981, 1603; Federal Court of Justice, judgment of 31 January 1995—VI ZR 27/94, NJW 1995, 1286, 1287 *et seq.*; Federal Court of Justice, judgment of 28 September 1970—VIII ZR 166/68, VersR 1971, 80, 81 *et seq.*; Brüggemeier (1982), pp. 1301 *et seq.*

regard to safety standards,[72] and utilise and record the knowledge gained in this way.[73]

Specifically in relation to motor vehicles, the evaluation of data available to the manufacturer due to vehicle connectivity is considered necessary in the context of active product monitoring.[74] The manufacturer's product monitoring obligations are also closely related to his design obligations (*supra* Sect. 3.4.3.1). If the producer becomes aware of safety-relevant product weaknesses in the context of his monitoring activities, he must eliminate these deficiencies from the product design. Otherwise, he is in violation of his tort law due diligence obligations. Accordingly, the design of products placed on the market after becoming aware of safety-related vulnerabilities is to be regarded as deficient.[75]

Finally, with regard to road traffic, there are high-ranking legal interests at stake when previously unidentified product risks materialise. Hence, the required intensity of the manufacturer's product monitoring is of significant importance. In this regard, the potential damage must be weighed against its likelihood of occurrence, whereby (economic) reasonableness represents a limitation here. In principle, the product monitoring obligation is of lesser intensity in the case of conventional products placed on the market, provided they have been sufficiently tested and exist in large numbers. The situation is different, however, in the case of newly developed products with considerable damage potential.[76] This can be assumed, for instance, in the case of novel and hitherto unknown automation products that are based on, or use, artificial intelligence. Producers should take this into account when placing such products on the market.

Interaction with Third-Party Products

It is debatable whether manufacturers also bear product monitoring obligation with regard to hazards arising from the combination of their products with accessories manufactured by third parties. The courts have indeed assumed such a duty.[77] In contrast, it has been posited that manufacturers are only responsible for the product risks they create themselves. Hence, if it is a third party product that increases the hazard or creates an independent risk, this is said to be outside the manufacturer's sphere of responsibility.[78] Accordingly, if the product manual warns against the use of third-party accessories or only permits the use of accessories that have been authorized or produced by the manufacturer, such a warning is generally regarded

[72] Federal Court of Justice, judgment of 17 October 1989—VI ZR 258/88, NJW 1990, 906, 907; Kunz (1994), p. 450; Michalski (1998), p. 963.

[73] Spindler (2020), BGB Sect. 823, margin no. 662.

[74] Chibanguza and Schubmann (2019), p. 316.

[75] Hofmann (2020), p. 286.

[76] Sack (1985), pp. 815 *et seq.*

[77] Federal Court of Justice, judgment of 9 December 1986—VI ZR 65/86, NJW 1987, 1009; Federal Court of Justice, judgment of 31 January 1995—VI ZR 27/94, NJW 1995, 1286, 1287 *et seq.*; Kunz (1994), p. 451.

[78] Ulmer (1988), p. 579; Spindler (2020), BGB Sect. 823, margin no. 665.

as sufficient and there is no legitimate basis for any further-reaching product monitoring obligations.[79]

A different assessment may by appropriate, however, if accessories supplied by third parties are commonly used in the market and the manufacturer must therefore foresee their combination with his own products.[80] Admittedly, it is not fully clear here when such obviousness is to be assumed, so that the boundaries become blurred. The legal analysis becomes even more uncertain with respect to the use of artificial intelligence for purposes of vehicle automation. In view of the connectivity and openness (i.e., multitude of interfaces) of artificial intelligence, safety-related risks are created by the manufacturer in this case that clearly reach beyond his own product. It is therefore not unreasonable to require the manufacturer in this context to direct its product monitoring measures also towards the combined use of its products with third-party accessories and services.

3.4.4.2 Product Warning

The manufacturer has a duty to warn of product hazards identified during product monitoring and to instruct the user accordingly.[81] Hence, manufacturers of products for the automotive industry are under an obligation to alert customers when product dangers become known to them. In the event of a particularly high potential for harm to life and limb, the manufacturer may even have to recall its products.[82]

Increasing Importance of Software Updates

The source code of a software is never completely error-free. The more lines of code a programme has, the more erroneous lines it will contain.[83] Software updates are therefore gaining more and more significance in the automotive industry. If security gaps are revealed to the manufacturer in the course of its product monitoring measures, he must modify the software architecture for all vehicles that are brought to the market from that point on in order to close the safety gaps identified. If the software is already used in vehicles that have been delivered to customers, there are two possibilities: Firstly, the manufacturer can warn product users of the respective dangers and secondly, the security gap detected can be closed by way of a software update. Particularly in the case of software used in road traffic—be it in assisted or partially automated vehicles, be it in highly and fully automated or even autonomous vehicles—security gaps and outdated software pose considerable risks to life and

[79] Federal Court of Justice, judgment of 9 December 1986—VI ZR 65/86, NJW 1987, 1009; Spindler (2020), BGB Sect. 823, margin no. 661.

[80] Spindler (2020), BGB Sect. 823, margin no. 665.

[81] Federal Court of Justice, judgment of 7 December 1993—VI ZR 74/93, NJW 1994, 517; Federal Court of Justice, judgment of 16 December 2008—VI ZR 170/07, NJW 2009, 1080.

[82] Thürmann (1999), p. 146.

[83] Raue (2017), p. 1841; Wiesemann et al. (2020), p. 139.

limb. In this context, software updates would appear to be an efficient way to effectively counter these risks.

Obligation on the Part of the Manufacturer to Update the Software?
Whether a tort law obligation on the part of manufacturers to provide software updates can be assumed is heavily disputed in current legal literature[84] but cannot be discussed here in greater detail. It should only be noted in this context that, especially in the case of automated vehicles, software updates would seem to offer a very suitable method of reacting quickly and effectively to identified hazards and potential damage scenarios. If one therefore advocates the tortious duty of car manufacturers to provide updates to rectify safety-relevant software glitches, it must also be clarified how long such duty exists after a vehicle has been brought to the market. Here, too, the criterion of reasonableness is likely to draw boundaries in terms of time. It is not reasonable to expect the manufacturer to provide software updates for all eternity even after a specific vehicle type has long since ceased to be produced and brought to the market. In practice, however, cases in which vehicle keepers or drivers omit to install software updates provided by manufacturers are likely to be just as relevant.

3.4.5 Causality Between Product Defect and Damage

For manufacturer liability to come into play, the product defect (or rather, in the context of producer liability, the manufacturer's breach of his duties of care under tort law) must be the cause of the violation of the aggrieved party's legal interests. The burden to provide this proof rests with the aggrieved party.[85] Admittedly, this hurdle may be difficult to overcome for the victim of a road accident, especially if complex or even AI-driven automation systems are involved. At first glance, this statement may seem contradictory because product liability law offers a clearly identifiable responsible party in the form of the manufacturer. Yet, particularly with highly complex AI-driven products, it can be a challenge for accident victims to identify the party responsible for a damage-causing product defect. Two aspects come into play here. First of all, the damage typically is hardly foreseeable and results from the autonomy of the self-learning artificial intelligence involved. Furthermore, in the case of AI-driven products employed in road traffic, the services and systems of various manufacturers will typically interact, which makes it extremely

[84] Opposing such obligation: Rockstroh and Kunkel (2017), p. 81; Schrader and Engstler (2018), p. 360; Spindler (2004), p. 3148; affirming the obligation to provide software updates: Raue (2017), p. 1844 and, with regard to the Product Safety Act, Wiebe (2019), p. 625.

[85] Federal Court of Justice, judgment of 24 January 1989—VI ZR 112/88, NJW 1989, 1542, 1545; Federal Court of Justice, judgment of 12 November 1991—VI ZR 7/91, NJW 1992, 560, 562 *et seq.*; Federal Court of Justice, judgment of 29 September 1981—VI ZR 147/80; VersR 1981, 1181, 1182; Higher Regional Court Karlsruhe, judgment of 10 October 2001—7 U 117/99, VersR 2003, 1584, 1586.

difficult to identify the manufacturer whose product was the root cause for the damage that occurred (phenomenon of multi-causality).[86]

Having said this, if artificial intelligence is not used in such a complex environment as described above and situations of multi-causality can be ruled out, it may indeed be promising for victims of road traffic accidents to avail themselves of the legal regime on product and producer liability. What is more, artificial intelligence does not necessarily even have to be the cause of a damage event. For instance, if a defective AI-product is used in the infotainment system and a traffic accident occurs because an engine control unit fails independently of this, the use of artificial intelligence is not causal for the damage from the accident. Given the unpredictability and uncontrollability of artificial intelligence, it seems unlikely in any case that it will be employed in the steering functions of vehicles in the near future.

3.4.6 Fault on the Part of the Manufacturer

Tort law liability hinges upon a culpable breach of the manufacturer's duties of care arising before or after placing a product on the market. Whether the manufacturer has acted negligently or even intentionally is not always easy to affirm. Pursuant to Section 276 para. 2 BGB, negligence is to be assumed if the manufacturer disregarded the ordinary standard of care that could be expected to be met by an average manufacturer. Such violation is to be denied if, based on the then current state of science and technology, it was not recognisable to the manufacturer that a product suffered from a defect that subsequently caused damage in road traffic. When assessing whether such a so-called developmental defect, which excludes fault of the part of the manufacturer, is to be assumed, the time of placing the product on the market is decisive. Even if a product defect was present at this point in time, no culpable breach of duty is to be assumed on the part of the manufacturer, as he cannot reasonably be expected to prevent defects that, according to the then up-to-date state of science and technology, were unforeseeable.

It remains to be seen whether and how the traditional principle of fault underlying tort law can adequately address the novel hazards associated with artificial intelligence. It goes without saying that in the absence of legal personality, artificial intelligence itself cannot be held liable, so that its "fault" is irrelevant.[87] Having said this, artificial intelligence does not act deterministically and its complexity and autonomy call into question whether the manufacturer can anticipate *ex ante* how it will develop and act in the future. In this situation, the fault principle would seem to reach its limits. The manufacturer is of course nevertheless obligated to compensate the victim according to the principles of strict liability under the Product Liability Act. However, also in the context of fault-based tort law liability, a tendency is

[86] Zech (2019), p. 207.

[87] On the liability of robots, see also Kluge and Müller (2017), pp. 24 *et seq.*

emerging in jurisprudence to impose particularly strict organisational and other duties of care on the producers of digital products, which blurs the boundaries to strict liability. Similarly, with respect to AI-driven products and the risks they pose, there is an ongoing discussion to extend strict manufacturer liability,[88] to set up liability funds,[89] or to introduce compulsory[90] insurance regimes.[91] Specifically with regard to road traffic, however, the existing liability system for motor vehicles has so far been considered sufficient.[92] This is all the more true if one considers that it is rather unlikely that artificial intelligence will be employed in the steering functions of vehicles of levels 0–2 of the SAE J3016 standard in the near future (see *supra* Sect. 3.4.5).

4 Liability for Vehicles of the SAE J3016 Standard Levels 3–4

In this part, automated driving systems (ADS) at levels 3 and 4 of the SAE J3016 will be considered.

4.1 Description of Relevant Levels of Automation

SAE Level 3 (Conditional Driving Automation) refers to "[t]he sustained and ODD-specific performance by ADS of the entire DDT under routine/normal operation with the expectation that the DDT fallback-ready user is receptive to ADS-issued request to intervene, as well as to DDT performance-relevant system failures in other vehicle systems, and will respond appropriately".[93] This description makes clear the increased degree of automation compared to SAE Level 2 driving functions, but at the same time illustrates that the human driver is assigned certain new kinds of duties when using an ADS at SAE Level 3. On the one hand, the driving function takes over longitudinal and lateral vehicle control and the driver can devote himself to non-driving related tasks, but on the other hand, he must remain

[88] Borges (2016), pp. 279 *et seq.*; Bräutigam and Klindt (2015), p. 1137; Schaub (2017), p. 348.

[89] European Parliament resolution of 16 February 2017 with recommendations to the Commission on civil regulation in the field of robotics (2015/2103 [INL])—P8_TA(2017)0051, p. 18, paras 58, 59.

[90] Spindler (2015), p. 775; opposing view: Horner and Kaulartz (2016), p. 14.

[91] Opposing view at least in relation to low-risk artificial intelligence: von Westphalen (2020), p. 254, who considers strict liability and shifting the burden of proof to be sufficient.

[92] von Bodungen and Hoffmann (2016b), p. 508; Grützmacher (2016), pp. 697 *et seq.*; von Westphalen (2020), p. 254.

[93] SAE International (2021), p. 31.

attentive for requests from the ADS to resume vehicle control. The exact calibration of this dividing line presents itself as a major challenge for the social and legal sciences.

SAE Level 4 (High Driving Automation) is characterised by "[t]he sustained and ODD-specific performance by an ADS of the entire DDT and DDT fallback".[94] This definition makes particularly evident that the presence of a human driver is no longer required at SAE Level 4 and that in-vehicle users are merely passengers with no obligation to respond to system failures. This applies, first and foremost, to vehicles that never leave the ODD of the SAE Level 4 feature (e.g., a shuttle bus on a university campus). However, also dual-use vehicles are conceivable that require a human driver to operate the vehicle outside the prescribed ODD of the ADS and allow the driver to switch to automated mode when the ODD (e.g., the motorway) is reached.[95]

As a result of the 2017 and 2021 amendments to the Road Traffic Act, regulations for motor vehicles with SAE Level 3 and 4 features now exist in Germany (see *supra* Sect. 2). In December 2021, the Federal Motor Transport Authority (*Kraftfahrt-Bundesamt*—KBA) granted the world's first type-approval for an SAE Level 3 ADS to the Mercedes-Benz "Drive Pilot", which is an automated lane keeping system (ALKS) for motorway-like roads up to a speed of 60 km/h.[96] In contrast, no major changes have so far been made to the liability law regime.

4.2 Driver Liability

4.2.1 Capacity as Driver

A majority of 88% of road accidents are currently due to human error.[97] In this respect, driver liability plays an important role from the victim's perspective. Having said this, vehicle automation raises the intriguing question as to whether driver liability will lose importance in the future if it is no longer the human being but rather the malfunctioning of an automated driving feature that causes an accident, thus leading to an increase in manufacturer liability (so-called liability shift).[98]

---

[94] SAE International (2021), p. 31.

[95] SAE J3016, version 4/2021, p. 32.

[96] Cf. KBA press release of 9 December 2021, available at: https://www.kba.de/DE/Presse/Pressemitteilungen/Allgemein/2021/pm49_2021_erste_Genehmigung_automatisiertes_Fahren.html.

[97] Federal Statistical Office (2018), p. 49.

[98] See in detail von Bodungen and Hoffmann (2016a, b); Hey (2019), pp. 138 *et seq.*

4.2.1.1 Vehicles of SAE Level 3

As explained above (see *supra* Sect. 3.2.1.1), the driver is the person who puts the vehicle into operation under his or her own responsibility, controls its lateral and longitudinal movement and thus exerts the actual control over the vehicle.[99] With automated driving at SAE Level 3, however, the vehicle software has the sole control over the lateral and longitudinal movement when the automation function is activated. This calls into question the driver status of the system user. For this reason, Sect. 1a para. 4 StVG now clarifies that the person who activates an automated driving function must be considered as the driver (see *supra* Sect. 2.2). As set out above, we thereby apply a strict interpretation, according to which Sects 1a, 1b StVG that entered into force in 2017 concern SAE J3016 Level 3 only (see *supra* Sects. 2.2–2.4).

Prior to the introduction of Sect. 1a para. 4 StVG, it was heavily disputed whether the person employing an SAE Level 3 system could continue to be considered the driver.[100] This question has now been answered in the affirmative by the German legislator. From the legislator's perspective, the mere possibility to override and deactivate the automation function at any time (cf. Sect. 1a para. 2 sent. 1 no. 3 StVG) grants the system user sufficient decision-making power in controlling the vehicle to justify his being qualified as the vehicle driver.[101] This rationale also applies when the system user turns away from the traffic situation around the vehicle[102] and even if he should decide to leave the driver's seat while the system is switched on[103].

4.2.1.2 Vehicles of SAE Level 4

In contrast, for "autonomous" vehicles within the meaning of Sect. 1d StVG, no driver liability is envisaged by the German legislator as these vehicles do not require a driver but rather only carry passengers (see *supra* Sect. 4.1). It is important to mention again that despite the misleading title of the law, the latest legislative changes refer to SAE Level 4 and *not* to SAE Level 5 (see *supra* Sect. 2.3). Due to the absence of a driver at this level of automation, driver liability according to Sect. 18 para. 1 StVG and Sect. 823 para. 1 BGB (see *supra* Sect. 3.2) is generally not available. This will only be different for dual-use vehicles that are being operated *outside* the ODD of the ADS (e.g., if the system user has resumed vehicle control at

[99] Federal Court of Justice, judgment of 22 March 1977—VI ZR 80/75, BeckRS 1977, 2968; Schrader (2016), p. 245; von Bodungen and Hoffmann (2016a), p. 452; Walter (2019), StVG Sect. 18, margin no. 6.

[100] Bewersdorf (2005), p. 106; Fleck and Thomas (2015), p. 1395; Schrader (2015), p. 3541.

[101] Schrader (2016), p. 245.

[102] Also Buck-Heeb and Dieckmann (2020), Chap. 3.1.1, margin no. 36.

[103] Greger (2018), p. 2.

the exit from the highway where the ODD of the ADS ended and he subsequently causes an accident).

It is noteworthy that the liability gap flowing from the removal of the driver is compensated, at least to some extent, by the fact that the German legislator introduced the requirement to provide for a technical supervisor, who will typically be located outside the driverless vehicle (see *supra* Sect. 2.3). Under certain conditions, the technical supervisor must then deactivate the vehicle remotely or clear certain driving manoeuvres (Sect. 1f para. 2 StVG). Unfortunately, the law is silent on the consequences in terms of liability of a breach of these duties. Extending the established driver liability for presumed fault pursuant to Sect. 18 StVG (see *supra* Sect. 3.2.1.3) to the technical supervisor was contemplated during the legislative process but did not find its way into the new law. This was due to the perceived lack of comparability with the much more extensive duties of a driver, because the technical supervisor is merely tasked with clearing driving manoeuvres suggested by the automated vehicle or alternatively deactivating the vehicle. In view of this exclusion from the more specific liability regime under the Road Traffic Act, in future the liability of the technical supervisor will have to be based on general tort law as set out in Sect. 823 BGB.[104]

4.2.2 Fault and Exculpatory Evidence on the Part of the Driver of the Vehicle

4.2.2.1 General Overview

As was explained above in relation to SAE Levels 1 and 2 (see *supra* Sect. 3.2), driver liability pursuant to Sect. 18 StVG and Sect. 823 para. 1 BGB requires culpable conduct on the part of the driver, whereby the driver's fault is presumed under the Federal Road Traffic Act but not when unlimited tort law liability under the German Civil Code is at stake. In any case, the driver does not have to be an ideal driver. It is sufficient if the driver lives up to the standard of care of an "averagely experienced driver in road traffic".[105] Directly connected with the question of whether the driver has satisfied this standard is the issue of which duties of care he has to comply with when operating an automated vehicle. Such duties will already apply prior to the start of a journey. This includes, for example, ensuring that the driver is in a fit state of health to drive the vehicle safely.[106] Such duty would seem highly relevant also at SAE Level 3, where the system user must resume vehicle control after being prompted by the system. Furthermore, Section 23 of the Road

[104] See in detail von Bodungen and Mühlon (2022), p. 10.

[105] Federal Court of Justice, judgment of 9 February 1982—VI ZR 59/80, VersR 1982, 442.

[106] Kaufmann (2020), Chap. 25, margin no. 321.

Traffic Ordinance (*Straßenverkehrsordnung—StVO*) places the onus on the driver to ensure that the vehicle is in a roadworthy condition and has no obvious defects.[107]

4.2.2.2 Recalibration of the Driver's Duties of Care in Sect. 1b StVG

With regard to conventional vehicles, the driver's duties of care chiefly relate to exercising lateral and longitudinal vehicle control at all times.[108] This obviously no longer applies in the case of automated driving. Accordingly, Sect. 1b para. 1 StVG now allows the user of an automated driving function at SAE Level 3 to veer away from the traffic and vehicle navigation, which would amount to a blatant violation of due care when piloting a conventional vehicle. However, the German legislator has not set out in detail which non-driving related activities (e.g., checking emails, watching a film or reading the newspaper) the system user may engage in after he has surrendered vehicle control to the automation system. This lack of clear guidance has been heavily criticised in German legal literature.[109]

Complicating matters further, the Federal Road Traffic Act requires the driver to remain perceptive to such an extent that he can resume control over the vehicle at any time with undue delay if the system prompts him to do so of if he recognises, or must recognise due to obvious circumstances, that the prerequisites for the intended use of the automated driving function no longer exist (Sect. 1b para. 2 StVG). With respect to system-side requests to resume vehicle control, the law also stipulates that they must not only be issued visually, acoustically, tactilely or otherwise perceptibly but also with sufficient lead time before control of the vehicle is transferred to the driver (Sect. 1a para. 2 sent. 1 no. 5 StVG). In this context, it is often assumed that a period of 7–10 s is sufficient for a person to take over the steering wheel again.[110] Yet, the findings of traffic psychology cast doubt on whether this rule of thumb is accurate. Depending on the physical and mental condition of the driver and the preceding type of distraction, the lead time required may be significantly longer.[111]

As set out above, the driver commits a breach of its duties of care not only if he fails to intervene despite timely warning, but also if he omits to do so although the conditions for the intended use of the automated driving function obviously no longer exist. When determining whether the driver is to be held responsible for such negligent ignorance, one must take into account that the law now permits him, in principle, to engage in non-driving activities. Conversely, this means that the relevant circumstances must be so evident that the driver can nevertheless become aware of them, which presumably would be the case when a tire blows out or when

---

[107] Federal Court of Justice, judgment of 9 May 1995—VI ZR 128/94 NZV 1995, 310.

[108] Armbrüster (2016), p. 208.

[109] von Kaler and Wieser (2018), p. 371; Balke (2018), p. 7; Greger (2018), p. 3; Schirmer (2017), p. 255.

[110] Radlmayr and Bengler (2015), p. 27.

[111] Krems and Beggiato (2020), Chap. 2.1.2, margin no. 22.

the vehicle starts driving in serpentine lines.[112] If, however, the term "obvious" is extended beyond such clear-cut cases by construing it more broadly,[113] the consequence may effectively be that the driver must constantly observe the traffic situation.[114] Otherwise, he risks being charged with negligent conduct and thus being held liable.

4.2.2.3 Data Storage as per Sect. 63a StVG

Sect. 63a para. 1 StVG entered into force together with Sects 1a, 1b StVG and stipulates that automated vehicles must store the position and time date determined by a satellite navigation system when there is a change in vehicle control between the driver and the automated system. Such storage shall also take place when the system prompts the driver to take over control of the vehicle or a technical malfunction occurs. As explained above with respect to Sects 1a, 1b StVG (see *supra* Sect. 2), the statutory reference in this context to SAE Level 3 *or* Level 4 is misleading and should be restricted to vehicles at SAE Level 3. This view is also supported by the fact that Sect. 63a para. 1 StVG would have no meaningful scope of application at SAE Level 4 because take-over scenarios within the ODD are not envisaged at this far-advanced level of automation (see *supra* Sect. 4.1).

The data stored on the basis of Sect. 63a StVG can assist the driver in rebutting the statutory presumption of fault in Sect. 18 StVG.[115] On the other hand, it will no longer be possible for drivers to make blanket assertions that the automation system was switched on when the vehicle was involved in an accident or that a technical defect caused the damage.[116] Rather, the data will show whether the automation system exercised vehicle control at the time of an accident or a technical defect had occurred. It should be noted, however, that Sect. 63a StVG confines data storage to the aforementioned—limited—set of data and does not intend to provide for a comprehensive recording of all data potentially relevant for accident investigation.[117]

---

[112] Cf. Bundestag printed matter 18/11776, p. 10.

[113] Such broad understanding is advocated by Berndt (2017), p. 125; Wolfers (2017), p. 12.

[114] In this vein, Kütük-Markendorf (2017), p. 353 advocate a narrow interpretation.

[115] Stender-Vorwachs and Steege (2020), Chap. 3.6.1, margin no. 201.

[116] Bundestag printed matter 18/11300, p. 24; Brockmeyer (2018), p. 260.

[117] Bundestag printed matter 18/11534, p. 26; König (2017), p. 252.

4.3 Keeper Liability

4.3.1 Strict Liability Under the Federal Road Traffic Act

It was already explained above that Sect. 7 para. 1 StVG imposes liability for personal injury and property damage on the keeper of a motor vehicle regardless of fault (see *supra* Sects. 1 and 3.3). This strict liability is the flip side of the operating risk of the vehicle to which the keeper exposes others. However, as culpable behaviour on the part of the keeper is not a prerequisite, his liability under the Road Traffic Act is limited. Interestingly, the legislator has doubled the liability cap for vehicles operating at SAE Levels 3 and 4 compared to conventional vehicles (see in this regard *supra* Sect. 3.2.1.1). Hence, in the case of death or injury inflicted upon a third party by way of the operation of an automated driving function at SAE Level 3 or 4, the keeper is strictly liable up to an amount of ten million euros. Maintaining strict keeper liability pursuant to Sect. 7 para. 1 StVG at this more advanced level of automated driving can be justified with a cost/benefit argument. The keeper enjoys the benefit of the use of the vehicle and must, as a trade-off, pay compensation to third-party victims harmed in the course of the operation.[118] A departure from the well-established liability regime for keepers specifically with respect to vehicles at SAE Levels 3 and 4 was therefore rejected.[119] For this very reason, the Compulsory Insurance Act (*Pflichtversicherungsgesetz*—PflVG) places an obligation on the keepers of such vehicles to take out liability insurance.

The liability of the keeper is only excluded if an accident is caused by *force majeure* (Sect. 7 para. 2 StVG) or in case of unauthorised use of a motor vehicle without the knowledge of the keeper (Sect. 7 para. 3 StVG). The *force majeure*-exception to the vehicle keeper's strict liability will only apply in the case of external, unforeseeable and unavoidable events.[120] A malfunction of the automation system does not satisfy these requirements, but rather it is part of the operation of the vehicle.[121] Therefore, in such a situation the keeper's liability will be available also in the case of Level 3 and 4 automated driving.[122] The assessment of external vehicle hacking is not quite as clear-cut, whereby the prevailing view is that this does *not* constitute *force majeure* due to its foreseeable nature.[123]

---

[118] Armbrüster (2017), p. 84; with regard to the opening of a source of danger also Federal Court of Justice, judgment of 26 April 2005—VI ZR 168/04, NJW 2005, 2081, 2082.

[119] See, e.g., Deutscher Verkehrsgerichtstag (2018); Buck-Heeb and Dieckmann (2020), Chap. 3.1.1, margin no. 17 *et seq.*

[120] von Bodungen and Hoffmann (2016a), p. 451; Ternig (2016), p. 306.

[121] Borges (2016), p. 274; Franke (2016), pp. 62 *et seq.*

[122] Greger (2018), p. 2; Borges (2016), p. 274; Fleck and Thomas (2015), p. 1394.

[123] Hammel (2016), p. 207; Ternig (2016), p. 306 takes a dissenting view in this context.

4.3.2 Fault-Based Liability Under the German Civil Code

In addition to strict vehicle keeper liability stipulated in the Federal Road Traffic Act, claims against the vehicle keeper can also be based on general tort law liability under Sect. 823 para. 1 BGB. In this case, the keeper must have engaged in intentional or negligent conduct, with the burden of proof resting on the injured party.[124] Particularly in the case of serious road accidents, a claim under tort law can be preferable to strict liability under the Federal Road Traffic Act because tortious liability does not provide for a maximum limit in terms of the amount of damages recoverable.

Importantly, the legal regulations introduced in 2021 for driving at SAE Level 4 place the obligation on the vehicle keeper to maintain the road safety and environmental compliance of its motor vehicle and to take all necessary precautions for this purpose (Sect. 1f para. 1 sent. 1 StVG). More specifically, the keeper must ensure regular maintenance of all vehicle systems required for operating the autonomous driving function (Sect. 1f para. 1 sent. 2 no. 1 StVG). At first glance, these obligations correspond to those that are imposed on the driver at SAE Level 3 (see *supra* Sect. 4.2.2.1). However, since the Federal Road Traffic Act no longer provides for a driver at SAE Level 4, these duties are assigned to the vehicle keeper instead to prevent a gap in liability that would otherwise arise.[125] What is more, the keeper's maintenance obligations are fleshed out in the "Ordinance on the approval and operation of motor vehicles with autonomous driving functions in specified operating areas" (see *supra* Sect. 2.3), which requires the keeper to (*inter alia*) perform an extended daily departure inspection prior to the start of operations as well as an overall vehicle inspection every 90 days (Section 13 paras 1, 7). Overall, the requirements placed on the keeper are so demanding that they can hardly be met by private persons, which is why it can be assumed that, at least for the time being, only entrepreneurs and commercial undertakings will be eligible as operators of autonomous vehicles for the purpose of transporting goods or passengers in designated areas.[126]

4.4 Manufacturer Liability

As the driving task becomes increasingly automated, the probability increases that a road accident is not due to human error in driving the vehicle, but to the failure or malfunction of an automated or autonomous driving function (see *supra* Sect. 4.2.1). This inevitably brings the manufacturer of the defective automation system into focus as the subject of liability, as well as—if not identical—the automaker (OEM)

[124] Federal Court of Justice, judgment of 12 February 1963—VI ZR 70/62, NJW 1963, 953.

[125] Bundestag printed matter 19/27439, p. 25.

[126] For further details in this regard, see Haupt (2022), pp. 167 *et seq.*; Steege (2022), pp. 165 *et seq.*

who brought the vehicle, including the defective automation function, onto the market as a complete product.

4.4.1 Wide-Ranging Convergence Between Manufacturer Liability Under Sect. 823 BGB and the Product Liability Act

It was set out in detail above (see *supra* Sects. 1.2 and 3.4.1) that manufacturer liability rests on two pillars, namely fault-based tortious liability pursuant to the German Civil Code (Sect. 823 BGB) on the one hand, and on the other hand the EU-wide standardised product liability regime pursuant to the European Product Liability Directive, which has been implemented in Germany by means of the national Product Liability Act and is typically construed as a strict liability system. Both strands of liability demonstrate far-reaching parallels, because the manufacturer's tortious duties of care in relation to the product design, manufacture and instruction are shaped according to the very categories that mold the legitimate safety expectations of the product user. As mentioned above, the overlap between both regimes does not encompass observation of products already placed on the market, which is exclusively addressed by tort law (see *supra* Sect. 3.4.4). However, the common ground increases even further if, contrary to the hitherto predominant legal opinion (see *supra* Sect. 3.4.2), one advocates a purposeful extension of the Product Liability Directive to incorporeal products such as software. This is particularly relevant in the case of automated driving systems, since these are essentially not tangible but software-driven products that interpret environmental data collected by sensors, deduce appropriate driving manoeuvres from this data, and trigger corresponding steering commands.

4.4.2 Design, Manufacturing and Instructional Defects

4.4.2.1 High Safety Expectations Regarding Automation Systems at SAE Levels 3 and 4

General Overview

According to traditional principles, manufacturers must take those precautions in the design and planning of a product that are both necessary and reasonable according to objective standards in view of the hazards posed by the respective product and the legal interests at stake (see *supra* Sect. 3.4.1). More specifically, it is reasonable for product users to expect manufacturers to implement those constructive safety measures that are feasible as practical solutions according to the state of the art in science and technology.[127] Irrespective of the novel risks of self-learning and collaborating

[127] Federal Court of Justice, judgment of 16 June 2009—VI ZR 107/08, NJW 2009, 2952, 2953; Spindler (2020), BGB Sect. 823, margin no. 638; von Bodungen (2022a), p. 4.

robot vehicles that will inevitably emerge in the future, already according to current standards these automobiles will have to be considered as bearing a significant physical risk potential for life and limb of their occupants and other road users due to their speed and mass alone, and consequently high safety expectations would seem justified as a starting premise for machine-piloted vehicles at SAE Levels 3 and 4.[128] This applies all the more to the implementation of information technology concepts for securing the automation system against cyber-attacks and hacking, since significant dangers to high-ranking legal interests such as the life and limb of road users are to be feared if essential safety functions are compromised or the vehicle steering system falls victim to an unauthorised external takeover.

The Human Driver as a Benchmark for Justified Safety Expectations
It is frequently posited that the legitimate user expectation also comprises that automated vehicle systems must be at least as safe as the human driver.[129] This is chiefly due to the prospectively higher performance of automated driving systems compared to the safety-related capabilities of human drivers. In this respect, the expectations of product users are also significantly shaped by the vision of accident-free automated road traffic ("vision zero") touted by the automotive industry. On closer inspection, however, a combined standard for determining the user's legitimate safety expectations would seem preferable.[130] In line with the aforementioned view, a software-based automation system is indeed to be considered defective from a product liability perspective if it causes an accident that would not have occurred to a human driver acting with average care in the identical traffic situation. From a system-related perspective, however, a product defect is also to be assumed if a carefully acting human would not have mastered the specific situation that caused the damage, but an alternative automation product would have. This follows from the fact that in the latter case the automation system causing the damage did not comply with the state of the art in science and technology as embodied by the alternatively available system that would have mastered the critical situation.

Use of Self-learning Adaptive Systems in Automated Vehicles
When manufacturers equip automated vehicles with self-learning adaptive software, the presence of a product defect suggests itself in view of the design-related indeterminacy of the programming. In this respect, a decoupling of adaptive software systems from all safety-critical vehicle functions may be indispensable, at least at present. In order to limit the risk of flawed code development and unpredictable decision-making inherent in self-learning systems, it would also seem advisable to "freeze" the software status of automation systems when they are put on the market and only enhance them subsequently by means of verified updates.[131] However, if

---

[128] Horner and Kaulartz (2016), p. 11; Steege (2021a), margin no. 56; von Bodungen (2022a), p. 3.
[129] Hey (2019), p. 52; Gomille (2016), p. 77.
[130] For a more detailed illustration of this combined standard, see von Bodungen (2022b), pp. 337 *et seq.*
[131] Hartmann (2017), pp. 7 *et seq.*; Zech (2019), p. 209; von Bodungen (2022a), p. 4.

the risk cannot be eliminated by way of software design that the programme code develops incorrectly and the automation system performs deficient driving manoeuvres, it is questionable whether the product may be marketed at all. In this respect, a comprehensive cost-benefit analysis is required, in which the damage to be feared and its probability of occurrence are to be included just as much as the product benefits.[132] The increase in road safety, mobility gains for persons unsuitable to drive motor vehicles as well as the improvement of traffic flow management (including reduced environmental pollution) argue in favour of the introduction of vehicles piloted by software that is continuously optimizing itself.[133] However, an instructional error is at issue if the product user is not adequately informed about the overall circumstances in order to enable him to make an informed decision as to whether he wants to expose himself to the unavoidable residual risks of the new technology (see *infra* Sect. 4.4.2.3).

4.4.2.2 Significance of Statutory Design Requirements

Statutory Design Requirements for Automation Systems at SAE Levels 3 and 4
It was already mentioned that the legal regulations introduced in Germany for automated vehicles at SAE Levels 3 and 4 stipulate very detailed technical requirements that the respective self-driving systems must satisfy (see *supra* Sects. 2.2 and 2.3). For instance, the earlier amendments that were introduced in 2017 still mandate the presence of a human driver, which is why they also stipulate various requirements for the technical design of the human-machine interface. Sect. 1a para. 2 sent. 1 StVG sets out in this context that SAE Level 3 systems must not only take over the vehicle control when activated, observe all traffic rules during phases of automation and be able to be overridden or switched off by the driver at any time. Rather, they must also recognize when vehicle control must be handed back to the driver and alert him thereof with adequate lead time. Such warning may be given optically, acoustically, tactilely or perceptibly in any other way and must also alert the driver in case he is using the system contrary to the manufacturer's system description.

In contrast, the most recent legislative additions to the Federal Road Traffic Act dispense with the requirement of the presence of a driver in the vehicle. Accordingly, Sect. 1d para. 2 StVG mandates that motor vehicles with a SAE Level 4 driving function must be fitted with technical equipment capable of independently performing the driving task within their pre-approved operational area without any driver intervention and to comply with all rules of the road applicable there. The complexity of this software programming task for vehicle manufacturers is exacerbated by the statutory requirement to provide for an accident prevention system that is not only designed to avoid and reduce damage but, in the event of unavoidable

[132] Federal Court of Justice, judgment of 16 June 2009—VI ZR 107/08, NJW 2009, 2952, 2953 *et seq.*; Wagner (2017), p. 733.

[133] For further details, see von Bodungen (2022b), p. 341.

alternative harm to different legal interests, must be able to take into account the importance of these legal interests at stake, with the protection of human life having the highest priority.[134] Finally, its design must be such that the motor vehicle automatically transitions to a minimal risk condition when a technical malfunction occurs that impairs the exercise of the automated driving function, when the boundaries of its specified operating area are reached, or when the continuation of the journey would require violating road traffic law (e.g., in the case of a traffic light showing permanent red due to a technical defect). In the latter case, the system must suggest possible driving manoeuvres to its technical supervisor, whom the Federal Road Traffic Act and the accompanying Ordinance assume to be located at a distance from the driverless vehicle (see *supra* Sect. 4.2.1.2). The automation system must then execute such manoeuvres upon approval from the technical supervisor, unless this would pose a danger to road users.

Implications for the Product User's Legitimate Safety Expectations

The German legislator has deemed the aforementioned technical requirements indispensable to ensure the safe operation of vehicles at SAE Level 3 and 4, respectively. At the same time, it is reasonable to assume that the product user's legitimate safety expectations will also be shaped by these legal design specifications.[135] This is because if a vehicle is brought to the market as an automated vehicle at SAE Level 3 or 4, then there is good reason for the vehicle user, as well as other road users, to assume that the vehicle complies with the requirements set out in Sect. 1a para. 2 sent. 1 StVG (for SAE Level 3) or Sect. 1e para. 2 StVG (for SAE Level 4). The justification for this assumption follows not least from the fact that the manufacturer is required by law to explicitly guarantee satisfaction of the respective technical requirements in the system description or operating manual (Sect. 1a para 2 sent. 2 StVG and Sect. 1f para. 3 no. 4 StVG). Hence, if the vehicle design falls short of the legally prescribed safety-relevant features, it will generally have to be regarded as deficient. For instance, if a vehicle classified by the manufacturer as SAE Level 4 does not satisfy the requirement to comply with all rules of the road applicable in its operational area, this will trigger product liability in the event of an accident due to the system overlooking or misinterpreting a particular traffic sign, for example.

Conversely, compliance with the technical requirements for SAE Level 3 and 4 vehicles set out in the Federal Road Traffic Act does not automatically exonerate the manufacturer from a liability perspective. It must be noted in this context that technical specifications set by vehicle registration law—such as those in Sect. 1a para. 2 sent. 1 StVG and Sect. 1e para. 2 StVG—only define minimum requirements for vehicle safety, whereas manufacturer liability law aims to allocate responsibility for those safety risks that materialize despite product registration.[136] By way of an

[134] Detailed requirements for the accident prevention system have now been set out in Sect. 3 para. 8 of the Ordinance on the approval and operation of motor vehicles with autonomous driving functions in specified operating areas; see in this regard also Steege (2022), p. 163.

[135] Greger (2018), p. 4; Ehring (2020), p. 1161; von Bodungen (2022a), p. 4.

[136] Gollrad (2020), p. 104; von Bodungen (2022b), p. 340.

example, in contrast to the requirements for SAE Level 4 vehicles, Sect. 1a para. 2 sent. 1 StVG does not mandate the ability of SAE Level 3 systems to transition the vehicle to a minimal risk condition if this should become necessary (e.g., by slowing down the vehicle and stopping it on the side of the road if the human driver does not resume vehicle control after having been prompted to do so with adequate lead time). Nevertheless, it seems justified that product users would expect such emergency system to be present. Otherwise, very significant safety risks for the driver and other road users could ensue if the driver was prevented from resuming control of the vehicle due to a medical emergency, for example, and the vehicle then participated in road traffic in an uncontrollable and driverless manner.[137]

4.4.2.3 Instructional Defects

Hazards associated with the use of automation systems, which the manufacturer is not already able to effectively address at the stage of the design of his products, must in any case be dealt with as part of the user information, provided that the automation product in question may be placed on the market despite existing residual risks (see *supra* Sect. 4.4.2.1). In this way, the user of the product is enabled to weigh up the advantages and hazards of using the product and to decide on his own responsibility whether he is prepared to expose himself to the unavoidable residual risks of the new technology or not.[138] An instructional defect, on the other hand, is to be assumed if the manufacturer does not adequately inform the product user about those operational risks of the automation system that cannot be eliminated by way of its design. In this case, liability of the manufacturer is conceivable if the insufficiently informed product user suffers damage. The content and scope of the manufacturer's instructions are likely to be subject to strict standards in view of the risk to life and limb that not inadequately informed product users and other third parties face in automotive road traffic.[139]

Manufacturers must provide clear, comprehensive and accurate information about the intended use, performance and any limitations of their automation products. With respect to vehicles at SAE Level 3, particular attention must thereby be paid to comprehensive information about the allocation of duties at the human-machine interface, which must clearly identify and delineate the responsibilities of the product user in transition scenarios.[140] The manufacturer best knows the safety-relevant features of his product and must fully inform the user thereof. In contrast, if the producer's system description remains vague, incomplete or even incorrect with

---

[137] von Bodungen and Hoffmann (2018), p. 102; Gollrad (2020), p. 104; von Bodungen (2022a), p. 4.

[138] Federal Court of Justice, judgment of 16 June 2009—VI ZR 107/08, NJW 2009, 2952, 2954; Wagner (2017), p. 748; von Bodungen (2022b), p. 342.

[139] Spindler (2015), p. 769; Hey (2019), p. 82; von Bodungen and Hoffmann (2016b), p. 505.

[140] Kessel and von Bodungen (2018), p. 234; Freise (2019), p. 70; von Bodungen (2022b), p. 342.

respect to the circumstances when the driver must resume the driving task, this may result in manufacturer liability due to an instructional error.[141] Against this background, not only explanations in the operating manual, but also situation-specific instructions on the appropriate use of automation products in the on-board vehicle system would seem advisable.[142] It also could be recommendable for manufacturers to offer a driver's training on the occasion of the purchase of an automated vehicle.[143] Such training should include warnings about foreseeable product misuse,[144] such as that the driver must not fall asleep after handing over the driving task to the automation system but has to remain vigilant at all times to resume vehicle control. The most recent legislative amendments in Germany require manufacturers to provide training for persons involved in the operation of SAE Level 4 vehicles, particularly in relation to the technical performance of the driving function (Sect. 1f para 3. no. 5 StVG). Yet, the amount of the information to be provided is likely to decline with increasing automation until the vehicle user is completely discharged from the driving task in autonomous driving at SAE Level 5.

4.4.3 Product Monitoring and Warning

As set out in detail above, even after products have been placed on the market, manufacturers retain an obligation under tort law to monitor them in the field for hitherto unknown hazards and to effectively counter thereby identified product risks (see *supra* Sect. 3.4.4). In view of the novelty and technical complexity of automated vehicles, manufacturers will have to keep themselves very closely informed about their safety-relevant performance in the field.[145] In this context, modern networking technologies allow real-time analysis of vehicle data. If safety-specific risks are identified in this process, response obligations on the part of the manufacturer ensue. More specifically, the assumption of a tortious duty to install corrective safety updates free of charge for digital products is debatable, particularly in view of the rather minor burden this places on the manufacturer (see *supra* Sect. 3.4.4.2). In this vein, the most recent legislative amendments concerning SAE Level 4 systems require the manufacturer to demonstrate throughout the entire operating period of the respective motor vehicle that its electronic and electrical architecture is secured against external attacks (Sect. 1f para. 3 no. 1 StVG). Given the rapid pace of technological development, it is unlikely that this obligation could be satisfied

[141] Kessel and von Bodungen (2018), p. 234.

[142] von Bodungen and Hoffmann (2016b), p. 505; Hey (2019), p. 84; von Bodungen (2022b), p. 342.

[143] Kessel and von Bodungen (2018), p. 235.

[144] Federal Court of Justice, judgment of 25 October 1988—VI ZR 344/87, NJW 1989, 707; Federal Court of Justice, judgment of 27 September 1994—VI ZR 150/93, NJW 1994, 3349, 3350.

[145] Gomille (2016), p. 80; Greger (2018), p. 4; von Bodungen (2020), p. 558; Steege (2021c), p. 6.

without designing the vehicle to allow for, and subsequently performing, software updates.

In contrast, manufacturer liability under the Product Liability Act has so far turned a blind eye on the phase after a product has been placed on the market (see *supra* Sect. 3.4.4.1). However, the focus on the preceding period until the product is brought to the market comes into question if the steering software installed in a vehicle is continuously updated and thus never completely released from the manufacturer's sphere of influence and organisation. In the case of such ongoing updating of the software, it is arguable, at any rate in the case of function-changing or function-improving updates, that each such update constitutes a renewed placing of the vehicle on the market, which in turn may trigger strict product liability (so-called dynamization or renewal of the time of placing a product on the market).[146] It goes without saying that the requirements of data protection law, including the EU General Data Protection Regulation[147], must also be observed in the context of each such update.[148]

5 Liability for Vehicles of the SAE J3016 Standard Level 5

5.1 Description of Relevant Level of Automation

In the following, SAE Level 5 vehicles will be considered. This level of automation, which is termed as full driving automation, pertains to "[t]he sustained and unconditional (i.e., not ODD-specific) performance by an ADS of the entire DDT and DDT fallback".[149] Vehicles at this automation level are characterised by the fact that they are no longer restricted to a territorial area, but can operate anywhere at any time without a driver. SAE Level 5 vehicles do not require a human driver and are "passenger only". A level of automation beyond this is not conceivable. At the present time, no legal regulations for SAE Level 5 driving exist in Germany. Nor have any legislative initiatives been announced in this regard. Therefore, questions of liability can only be assessed on the basis of the existing legal regime described above.

---

[146] Hacker (2020), p. 2145; Wagner (2020c), p. 729; von Bodungen (2022a), p. 5.

[147] Regulation (EU) 2016/679 of the European Parliament and of the Council of 27 April 2016 on the protection of natural persons with regard to the processing of personal data and on the free movement of such data, and repealing Directive 95/46/EC O.J. L 119/1 (2016).

[148] Hartmann (2017), p. 45; Steege (2019a), pp. 512 *et seq.*

[149] SAE International (2021), p. 32.

5.2 Driver Liability

In the case of fully automated driving at SAE Level 5, there will no longer be any need for human intervention in the ongoing driving process. Instead of a driver there will only be passengers in the vehicle that will not be tasked with any traffic-related duties of care.[150] Consequently, driver liability as per Sect. 18 para. 1 StVG (see *supra* Sect. 1.2) is not available at this level of automation. This is equally true for tortious driver liability based on Sect. 823 BGB and corresponds to the legal situation at SAE Level 4 (see *supra* Sect. 4.2.1.2).

5.3 Keeper Liability

As set out above, road traffic liability law traditionally rests on a triad of liability addressees and is further underpinned by compulsory liability insurance (see *supra* Sect. 1.2). Yet, if driver liability is not available at SAE Level 5 due to the full release of the human being from the driving task, this does not mean that keeper liability also falls away. This is because the keeper's liability is based on exposing others to a source of danger that is to be equated with the risks that the operation of his respective vehicle entails. In this respect, there are no differences in the starting point compared with conventional vehicles, even if the risks associated with vehicles at SAE Level 5 are expected to prove to be less in the future. For this reason, there are sound reasons to assume that keeper liability pursuant to Sect. 7 para. 1 StVG will persist with respect to fully automated vehicles.

5.4 Manufacturer Liability

First of all, it should be noted that vehicles operating completely independently nevertheless do not qualify as liability subjects themselves. Liability can only be assigned to natural persons or legal entities. Robots and vehicles driven by artificial intelligence lack the necessary legal capacity in this regard. So far, German law does not provide for the liability of such non-legal entities.[151] Instead, manufacturer liability as described above could conceptually be applied to fully automated vehicles without triggering any distortions of the legal system. In this respect, too, the decisive question will be what legitimate safety expectations users may harbour with regard to vehicles at SAE Level 5 (see *supra* Sects. 3.4.1 and 4.4.1).

---

[150] Balke (2018), p. 6; Lange (2017), p. 349; Stender-Vorwachs and Steege (2017), p. 115; von Kaler and Wieser (2018), p. 370; Walter (2019), StVG Sect. 18, margin no. 15.

[151] von Bodungen and Hoffmann (2015), p. 524; Kreutz (2020), Chap. 3.1.2, margin no. 41 *et seq.*

6 Conclusion

In Germany, regulations for SAE Level 3 motor vehicles have been in force since 2017. Detailed specifications for SAE Level 4 motor vehicles were added to the Federal Road Traffic Act in 2021 and have now been further fleshed out in the accompanying "Ordinance on the approval and operation of motor vehicles with autonomous driving functions in specified operating areas". Germany considers itself a global front-runner in this respect. In contrast, regulations for autonomous driving at SAE Level 5 do not yet play a role in the German legislator's consideration. Special regulations for SAE Level 1 and 2 vehicles do not exist either. Instead, traditional road traffic and liability regulations can easily be applied with respect to such vehicles. Similarly, despite the legislative changes with respect to SAE Level 3 and 4 vehicles, the existing liability system has remained largely unchanged. As has been shown above, the German legal system would seem to be able to cope with the liability implications of the rampant vehicle automation without the need for sweeping modifications.

References

Armbrüster C (2016) Verantwortungsverlagerungen und Versicherungsschutz – Das Beispiel des automatisierten Fahrens. In: Gless S, Seelmann K (eds) Intelligente Agenten und das Recht. Nomos, Baden-Baden, pp 205–224

Armbrüster C (2017) Automatisiertes Fahren – Paradigmenwechsel im Straßenverkehrsrecht? ZRP 50:83–86

Balke R (2018) Automatisiertes Fahren. SVR 18:5–8

Berndt S (2017) Der Gesetzentwurf zur Änderung des Straßenverkehrsgesetzes – ein Überblick. SVR 17:121–127

Bewersdorf C (2005) Zulassung und Haftung bei Fahrerassistenzsystemen im Straßenverkehr. Zur Verantwortlichkeit von Staat, Fahrer, Halter und Hersteller für die Sicherheit des Straßenverkehrs. Duncker & Humblot, Berlin

Booth L, Norman R, Pettigrew S (2020) The potential effects of autonomous vehicles on alcohol consumption and drink-driving behaviours. Drug Alcohol Rev 39:604–607

Borges G (2016) Haftung für selbstfahrende Autos. CR 32:272–280

Bräutigam P, Klindt T (2015) Industrie 4.0, das Internet der Dinge und das Recht. NJW 68:1137–1142

Brockmeyer H (2018) Treuhänder für Mobilitätsdaten – Zukunftsmodell für hoch- und vollautomatisierte Fahrzeuge? ZD 9:258–263

Brüggemeier G (1982) Produzentenhaftung nach § 823 Abs. 1 BGB. WM 36:1294–1309

Buck-Heeb P, Dieckmann A (2020) Zivilrechtliche Haftung von Halter und Fahrer bei Einsatz (teil-)automatisierter Fahrfunktionen. In: Oppermann BH, Stender-Vorwachs J (eds) Autonomes Fahren, 2nd edn. C.H. Beck, München. Chap. 3.1.1

Burckhardt M (2015) EuGH: Haftung für fehlerhafte Produkte (BB-Commentary). BB 25:661–662

Chan C-Y (2017) Advancements, prospects, and impacts of automated driving systems. IJTST 6: 208–216

Chibanguza K, Schubmann D (2019) Die Produktbeobachungspflicht im Wandel der Digitalisierung und die Haftung der handelnden Personen. GmbHR 110:313–321

Dahm-Loraing R, Koyuncu A (2010) Haftung für Medizinprodukte – Teil 2: Die deliktische Produzentenhaftung gem. § 823 BGB. PHI n/a:142–145

Deutscher Verkehrsgerichtstag (2018) Empfehlungen des 56. Deutschen Verkehrsgerichtstags 2018. NZV 31:69–71

Ehring P (2020) Produktverantwortung bei Konstruktion und Betrieb automatisierter Fahrzeuge. MDR 74:1156–1162

Federal Ministry of Transport and Digital Infrastructure (2015) Bericht zum Forschungsbedarf, Runder Tisch Automatisiertes Fahren – AG Forschung. https://www.bmvi.de/SharedDocs/DE/Anlage/DG/Digitales/bericht-zum-forschungsbedarf-runder-tisch-automatisiertes-fahren.pdf?__blob=publicationFile. Accessed 6 Apr 2022

Federal Statistical Office (2018) Verkehr, Verkehrsunfälle, Fachserie 8 Reihe 7. https://www.destatis.de/DE/Themen/Gesellschaft-Umwelt/Verkehrsunfaelle/Publikationen/Downloads-Verkehrsunfaelle/verkehrsunfaelle-jahr-2080700187004.pdf;jsessionid=A29C26B5E8BA44D1E0FD9496DC59463A.live711?__blob=publicationFile. Accessed 6 Apr 2022

Fleck J, Thomas A (2015) Automatisierung im Straßenverkehr. NJOZ 15:1393–1397

Foerste U (1994) Neues zur Produkthaftung – Passive Beobachtungspflicht und Äquivalenzinteresse. NJW 47:909–911

Franke U (2016) Rechtsprobleme beim automatisierten Fahren – ein Überblick. DAR 86:61–66

Freise R (2019) Rechtsfragen des automatisierten Fahrens. VersR 70:65–79

Genus A, Stirling A (2018) Collingridge and the dilemma of control: towards responsible and accountable innovation. RP 47:61–69

Gollrad M (2020) Baukastenmodell statt Automatisierungsstufen – Anknüpfungspunkte für die Regulierung automatisierter Fahrzeuge. RAW 8:103–108

Gomille C (2016) Herstellerhaftung für automatisierte Fahrzeuge. JZ 71:76–82

Greger R (2018) Haftungsfragen beim automatisierten Fahren. NZV 31:1–5

Gruel W, Stanford JM (2016) Assessing the long-term effects of autonomous vehicles: a speculative approach. Transp Res Proc 13:18–29

Grützmacher M (2016) Die deliktische Haftung für autonome Systeme – Industrie 4.0 als Herausforderung für das bestehende Recht? CR 31:695–698

Hacker P (2020) Europäische und nationale Regulierung von Künstlicher Intelligenz. NJW 73: 2142–2147

Hammel T (2016) Haftung und Versicherung bei Personenkraftwagen mit Fahrerassistenzsystemen, 1st edn. VVW GmbH, Karlsruhe

Hartmann V (2017) Here come the robots - Produkthaftung und Robotik am Beispiel des automatisierten und autonomen Fahrens, Teil 1: Anforderungen an die Konstruktion. PHI n/a:2–17

Haupt T (2022) Die Verordnung zum Gesetz zum autonomen Fahren. NZV 35:166–169

Heß R (2022) StVG Sect. 18 Ersatzpflicht des Fahrzeugführers. In: Burmann M, Heß R, Hühnermann K, Jahnke J (eds) Straßenverkehrsrecht, 27th edn. C.H. Beck, Munich

Hey T (2019) Die außervertragliche Haftung des Herstellers autonomer Fahrzeuge bei Unfällen im Straßenverkehr, 1st edn. Springer Gabler, Wiesbaden

Hofmann F (2020) Der Einfluss von Digitalisierung und künstlicher Intelligenz auf das Haftungsrecht. CR 35:282–288

Horner S, Kaulartz M (2016) Haftung 4.0. InTeR 4:22–27

Kaufmann M (2020) Haftung des Kraftfahrzeughalters und -führers. In: Haag K (ed) Geigel, Der Haftpflichtprozess mit Einschluss des materiellen Haftpflichtrechts, 28th edn. C.H. Beck, Munich

Kessel C, von Bodungen B (2018) Germany's new road traffic law – legal risks and ramifications for the design of human-machine interaction in automated vehicles. In: Zachäus C, Müller B, Meyer G (eds) Advanced microsystems for automotive applications 2017. Springer International Publishing, Cham, pp 227–236

Kim E, Muennig P, Rosen Z (2017) Vision zero: a toolkit for road safety in the modern era. Inj Epidemiol 4:1. https://doi.org/10.1186/s40621-016-0098-z

Kleeman S, Arzt C (2021) Das Gesetz zum "autonomen" Fahren in Deutschland. RAW 9:99–105
Kluge V, Müller A-K (2017) Autonome Systeme. InTeR 5:24–31
König C (2017) Gesetzgeber ebnet Weg für automatisiertes Fahren – weitgehend gelungen. NZV 6: 249–253
Krems JF, Beggiato M (2020) Verkehrspsychologische Aspekte im Kontext automatisierter und autonomer Fahrzeuge. In: Oppermann BH, Stender-Vorwachs J (eds) Autonomes Fahren, 2nd edn. C.H. Beck, München. Chap. 2.1.2
Kreutz P (2020) Autonomes Fahren: Produkt- und Produzentenhaftung. In: Oppermann BH, Stender-Vorwachs J (eds) Autonomes Fahren, 2nd edn. C.H. Beck, München. Chap. 3.1.2
Kullmann H (2002) Produkthaftung für Verkehrsmittel - Die Rechtsprechung des Bundesgerichtshofes. NZV 15:1–10
Kunz J (1994) Die Produktbeobachtungs- und die Befundsicherungspflicht als Verkehrssicherungspflichten des Warenherstellers. BB 49:450–455
Kütük-Markendorf M (2017) Die hoch- oder vollautomatisierte Fahrfunktion als Vorstufe zum autonomen Fahren. CR 32:349–355
Lange U (2017) Automatisiertes und autonomes Fahren – eine verkehrs-, wirtschafts- und rechtspolitische Einordnung. NZV 30:345–352
Michalski L (1998) Produktbeobachtung und Rückrufpflicht des Produzenten. BB 53:961–966
Moriarty P, Wang SJ (2017) Could automated vehicles reduce transport energy? Energy Proc 142: 2109–2113
Müller D (2022) Automatisiertes Fahren, künstliche Verkehrsintelligenz und polizeiliche Arbeit. ZdiW 2:22–27
Podszus S, Degirmenci K, Breitner B (2020) Wirtschaftliche Auswirkungen/Autonomous driving: framework development for a successful transition and adoption. In: Oppermann BH, Stender-Vorwachs J (eds) Autonomes Fahren, 2nd edn. C.H. Beck, München. Chap. 2.3
Radlmayr J, Bengler K (2015) Literaturanalyse und Methodenauswahl zur Gestaltung von Systemen zum hochautomatisierten Fahren. FAT Schriftenreihe 276. https://doi.org/10.13140/RG.2.1.2222.4401
Raue B (2017) Haftung für unsichere Software. NJW 70:1841–1846
Renn O (2014) Mit Sicherheit ins Ungewisse. Möglichkeiten und Grenzen der Technikfolgenabschätzung. APuZ 64(6–7):3–10
Rockstroh S, Kunkel H (2017) IT-Sicherheit in Produktionsumgebungen – Verantwortlichkeit von Herstellern für Schwachstellen in ihren Industriekomponenten. MMR 20:77–82
Sack R (1985) Produzentenhaftung und Produktbeobachtungspflicht. BB 40:813–821
SAE International (2021) Taxonomy and definitions for terms related to driving automation systems for on-road motor vehicles J3016™ APR202
Schaub R (2017) Interaktion von Mensch und Maschine. JZ 72:342–349
Schirmer J-E (2017) Augen auf beim automatisierten Fahren! Die StVG-Novelle ist ein Montagsstück. NZV 30:253–257
Schrader P (2015) Haftungsrechtlicher Begriff des Fahrzeugführers bei zunehmender Automatisierung von Kraftfahrzeugen. NJW 68:3537–3542
Schrader P (2016) Haftungsfragen für Schäden beim Einsatz automatisierter Fahrzeuge im Straßenverkehr. DAR 86:242–246
Schrader P, Engstler J (2018) Anspruch auf Bereitstellung von Software-Updates? Unklare Begründung eines eingeschränkt notwendigen Anspruchs. MMR 6:356–361
Spindler G (2004) IT-Sicherheit und Produkthaftung - Sicherheitslücken, Pflichten der Hersteller und der Softwarenutzer. NJW 57:3145–3150
Spindler G (2015) Roboter, Automation, künstliche Intelligenz, selbststeuernde Kfz- Braucht das Recht neue Haftungskategorien? CR 30:766–776
Spindler G (2020) BGB Sect. 823 Schadensersatzpflicht. In: Spickhoff A (ed) beck-online. GROSSKOMMENTAR, status: 1.3.2022. C.H. Beck, München
Staudinger A (2019) BGB Sect. 823 Schadensersatzpflicht. In: Schulze R (ed) Bürgerliches Gesetzbuch, Handkommentar, 10th edn. Nomos, Baden-Baden

Steege H (2019a) Autonomes Fahren und die staatliche Durchsetzung des Verbots der Rechtswidrigkeit. NZV 32:459–467
Steege H (2021a) Annex III to Sects 1a – 1c StVG. In: Haus K-L, Krumm C, Quarch M (eds) Gesamtes Verkehrsrecht, 3rd edn. Nomos, Baden-Baden
Steege H (2021b) Gesetzesentwurf zum autonomen Fahren (Level 4). SVR 21:128–137
Steege H (2021c) Auswirkungen von künstlicher Intelligenz auf die Produzentenhaftung in Verkehr und Mobilität. NZV 34:6–13
Steege H (2022) Das (vorerst) letzte Puzzlestück – das autonome Fahren nimmt Fahrt auf. SVR 22: 161–168
Stender-Vorwachs J, Steege H (2017) Facetten Automatisierten Fahrens – von Datenschutz über Fahrerflucht bis zur Verkehrskontrolle. BRJ 10:114–118
Stender-Vorwachs J, Steege H (2020) Grundrechtliche Implikationen autonomen Fahrens. In: Oppermann BH, Stender-Vorwachs J (eds) Autonomes Fahren, 2nd edn. C.H. Beck, München. Chap. 3.6.1
Strand R, Spaapen J (2021) Locomotive breath? *Post festum* reflections on the EC Expert Group on policy indicators for responsible research and innovation. In: Yaghmaei E, van de Poel I (eds) Assessment of responsible innovation. Routledge, London, pp 42–59
Ternig E (2016) Automatisiertes Fahren: Wer führt – Mensch oder Maschine? zfs 37:304–310
Thürmann D (1999) Rückruf und Haftpflichtversicherung nach AHB oder ProdHB. NVersZ 2:145–192
Ulmer P (1988) Produktbeobachtungs-, Prüfungs- und Warnpflichten eines Warenherstellers in bezug auf Fremdprodukte. ZHR 152:564–599
von Bodungen B (2020) Automatisiertes Fahren (Automotive). In: Sassenberg T, Faber T (eds) Rechtshandbuch Industrie 4.0 und Internet of Things, 2nd edn. C.H. Beck, Munich, pp 537–562
von Bodungen B (2022a) Zivilrechtliche Haftung beim Einsatz künstlicher Intelligenz im autonomen Straßenverkehr von morgen – Geltendes Recht und Reformbestrebungen auf EU-Ebene. SVR 22:1–6
von Bodungen B (2022b) Automatisiertes Fahren – Produkthafungsrecht. In: Chibanguza K, Kuß C, Steege H (eds) Künstliche Intelligenz, Recht und Praxis automatisierter und autonomer Systeme, 1st edn. Nomos, Baden-Baden, pp 329–346
von Bodungen B, Hoffmann M (2015) Belgien und Schweden schlagen vor: Das Fahrsystem soll Fahrer werden! NZV 28:521–526
von Bodungen B, Hoffmann M (2016a) Autonomes Fahren – Haftungsverschiebung entlang der Supply Chain? (1. Teil). NZV 29:449–454
von Bodungen B, Hoffmann M (2016b) Autonomes Fahren – Haftungsverschiebung entlang der Supply Chain? (2. Teil). NZV 29:503–509
von Bodungen B, Hoffmann M (2018) Hoch- und vollautomatisiertes Fahren *ante portas* – Auswirkungen des 8. StVG-Änderungsgesetzes auf die Herstellerhaftung. NZV 31:97–102
von Bodungen B, Mühlon S (2022) Roboter im Regelbetrieb – Überblick über das Gesetz zum autonomen Fahren. ZdiW 1:5–11
von Kaler M, Wieser S (2018) Weiterer Rechtsetzungsbedarf beim automatisierten Fahren. NVwZ 37:369–373
von Westphalen F (1990) Das neue Produkthaftungsgesetz. NJW 43:83–93
von Westphalen F (2020) Produkthaftungsrechtliche Erwägungen beim Versagen Künstlicher Intelligenz (KI) unter Beachtung der Mitteilung der Kommission COM(2020) 64 final. VuR 35:248–254
Wagner G (2017) Produkthaftung für autonome Systeme. AcP 217:707–765
Wagner G (2020a) BGB Sect. 823 Schadensersatzpflicht. In: Säcker F, Rixecker R, Oetker H, Limperg B (eds) Münchener Kommentar zum Bürgerlichen Gesetzbuch, 8th edn. C.H. Beck, Munich
Wagner G (2020b) ProdHaftG Sect. 3 Fehler. In: Säcker F, Rixecker R, Oetker H, Limperg B (eds) Münchener Kommentar zum Bürgerlichen Gesetzbuch, 8th edn. C.H. Beck, Munich
Wagner G (2020c) Verantwortlichkeit im Zeichen digitaler Techniken. VersR 71:717–741

Walter A (2019) StVG Sect. 18 Ersatzpflicht des Fahrzeugführers. In: Spickhoff A (ed) beck-online. GROSSKOMMENTAR, status: 01.09.2019. C.H. Beck, Munich
Wiebe G (2019) Produktsicherheitsrechtliche Pflicht zur Bereitstellung sicherheitsrelevanter Software-Updates. NJW 72:625–630
Wiesemann H, Mattheis C, Wense S (2020) Software-Updates bei vernetzten Geräten. MMR 23: 139-144
Wolfers B (2017) Selbstfahrende Autos: Ist das erlaubt? RAW 5:2–13
Zech H (2019) Künstliche Intelligenz und Haftungsfragen. ZfPW 5:198–219

The Law Applicable to Autonomous Cars Driving in France

Lionel Andreu

Abstract The law applicable to autonomous cars driving in France is recent. It was initially developed for the sole purpose of allowing "experiments" relating to these new vehicles. More recently, various laws and regulations have been passed to regulate them on a long-term basis. This paper presents the rules that apply in France, studying those that apply to the use of these vehicles and the liabilities.

The law applicable to autonomous cars driving in France is recent. It was initially developed for the sole purpose of allowing *"experiments"* relating to these new vehicles. This is the objective of an ordinance of the 3rd of August 2016 (hereinafter referred to as the *"Experimental Ordinance"*)[1] which regulates the movement for experimental purposes of *"vehicles with partial or total delegation of driving on a road open to public traffic"*, while making it subject to *"the issue of a* [ministerial] *authorisation intended to ensure the safety of the conduct of the experiment"*. This ordinance was supplemented by a decree on the 28th of March 2018[2] which specifies the conditions of the experiment (hereinafter referred to as the *"experimental decree"*). In particular, it provides for the compulsory presence of a *"driver"*[3] (having received prior appropriate training and able to take the control of the vehicle

[1] Ordinance No. 2016-1057, 3 August 2016 on the experimentation of self-driving vehicles on public roads.

[2] Decree No. 2018-211, 28 March 2018 on the experimentation of vehicles with delegated driving on public roads. See the order of 17 April 2018 on the experimentation of vehicles with driver delegation on public roads and the order of 26 May 2021 amending the order of 17 April 2018.

[3] On this text, see *infra,* n°12. The authorisation for experimentation may allow this *"driver"* to be physically outside the vehicle. He must then be able to take the control of the vehicle back at any time.

L. Andreu (✉)
Paris, France
e-mail: lionel.andreu@univ-poitiers.fr

H. Steege et al. (eds.), *Autonomous Vehicles and Civil Liability in a Global Perspective*, Data Science, Machine Intelligence, and Law 3,
https://doi.org/10.1007/978-3-031-41992-8_13

back at any time[4]), as well as the existence of a recording device making it possible to determine when the vehicle has been driven in partial or total delegation of driving mode. This decree was more recently amended by a decree on the 2nd of December 2020,[5] which provides a number of more or less important[6] clarifications.

However, these experimental texts are only ephemeral legislation. They are only a prerequisite—a sort of legal test laboratory—for more durable legislation under *"ordinary law"* intended to apply to automated vehicles moving on national territory out of any experimentation. Noting the shortcomings of the legislation in this area, we undertook, in 2017, to draft a report on the law applicable to autonomous cars.[7] Achieved in 2018, this report included an inventory of the situation as well as a set of texts proposals intended to *"facilitate a future reform"*.[8] This report was submitted in June 2018 to Mrs. Anne-Marie Idrac, High Representative for the autonomous vehicle development strategy. It was then published by Dalloz editor.[9]

This report was soon followed by a law.[10] On the 26th of November 2018, the Government submitted to the Senate a *"Mobility Orientation Bill"* proposing an *"in-depth reform* [of] the *general framework of mobility policies to offer our fellow citizens, throughout the country, travel solutions that meet their expectations and the challenges of today"*.[11] In particular, this bill included a section dedicated to

[4]In particular, in the event of an emergency or when the conditions of use of the vehicle defined for the experiment are no longer met.

[5]Decree No. 2020-1495, 2 December 2020, amending Decree No. 2018-211 of the 28th March 2018 on the experimentation of vehicles with delegated driving on public roads.

[6]See in particular *infra*, No. 20.

[7]V. L. Andreu et al., *Des voitures autonomes, Une offre de loi*, Dalloz 2018, No. 00.06: *"The legislation relating to experiments remains insufficient, however. It is not able to respond to the paradigm shift brought about by the introduction of autonomous cars. Until today, the cardinal principle in this area remains that of the prevalence of the human in the regulation of road behaviour. Humans are civilly liable for accidents involving the vehicle of which they are the driver or guardian. Humans are criminally responsible for offences committed with their vehicles. Humans are accountable to their insurers for accidents in which they have participated. In other words, the human being is master on board. In a world where the human being is increasingly relegated after the machine, these principles cannot continue to apply. As the European Parliament pointed out in a recent resolution, it must be considered that "the more autonomous a robot is, the less it can be considered as a simple tool controlled by other actors (such as the manufacturer, the operator, the owner, the user...); that in this respect the question arises whether the ordinary rules [...] are sufficient or whether new principles and rules are needed to clarify the legal responsibility of the various actors relating to the acts or inactions of a robot that cannot be attributed to a particular human actor, and to determine whether the robot's acts or inactions that caused damage could have been avoided. Applied to autonomous cars, this question suggests an important evolution in the rules applicable to vehicles to take account of their march towards autonomy.*

[8]L. Andreu et al., *op. cit.* p. VIII.

[9]For a summary of this report, see our interview in the Recueil Dalloz: L. Andreu, *"Des voitures autonomes - Une offre de loi"*, D. 2018, p. 2080.

[10]It is not clear whether there is a causal link between these texts and the report (it is more likely to be a simple correlation).

[11]Explanatory memorandum.

"Autonomous and connected vehicles", the first article of which empowered the Government to take measures by ordinance *"in order to adapt the legislation, in particular the highway code, to the case of the movement on the public highway of motorised land vehicles whose driving functions are, under conditions, specifically time, place and pre-defined movement, delegated in part or in full to an automated driving system, in particular by defining the liability rules"*.[12] A second article of the bill also provided for the Government to be empowered to issue an ordinance to *"make vehicle data accessible"* to certain persons. The text will finally be voted on the 19th of November 2019 and the law promulgated on the 24th of December 2019.[13]

Two years later, two drafts were written, which were soon adopted, after various amendments, by the Government, leading to Ordinance No. 2021-442 of the 14th of April 2021 on access to vehicle data (hereinafter referred to as the *"Data Ordinance"*) and Ordinance No. 2021-443 of the 14th of April 2021 on the system of criminal liability applicable in the event of the movement of an autonomous vehicle and its conditions of use (hereinafter referred to as the *"Criminal Liability Ordinance"*). The latter regulates especially [14] the criminal exemption of the user of the autonomous vehicle when an offence has been committed while the driving system was activated and, correlatively, the criminal incrimination of the manufacturer for a part of the offences attributable to the vehicle. [15] To this end, it gives the police, persons in charge of reporting traffic offences and the holder of the vehicle registration certificate access to the information stored by the data recorders. [16] It also provides for an information duty on professionals who sell or hire an automated vehicle to consumers. [17] This ordinance was completed by a Decree No. 2021-873 of 29 June 2021 (hereinafter referred to as the "decree on criminal liability").[18] This text includes especially a set of definitions—notably that of the *"delegated driving vehicle"*, [19] which, not unrelated to one of our proposals,[20] can be *"partially, highly*

[12] Article 31 of the law.

[13] Law No. 2019-1428 of the 24th of December 2019 on the orientation of mobility.

[14] The ordinance also regulates the *"Safety of automated road transport systems"* - which will not be dealt with in the limited scope of this contribution.

[15] See *infra*, n° 24.

[16] See *infra*, no. 20.

[17] See *infra*, no. 9.

[18] Decree No. 2021-873 of 29 June 2021 implementing Ordinance No. 2021-443 of 14 April 2021

[19] It is a *"motor vehicle of category M, N, L, T or C, as defined in 1, 2, 4 and 5.1, or urban shuttle as defined in 6.13, equipped with an automated driving system"*. From a terminological point of view, we note the shift in vocabulary between the law as passed - which does indeed mention the *"autonomous vehicle"*—and the ordinance and decree—which prefer the periphrase *"vehicle with delegated driving"* or the expression *"automated vehicle"*. These differences reflect an existing opposition between the authors. In favour of the term "autonomous car", see L. Andreu et al., *op. cit.*, No. 00.01, note 2. In favour of the periphrase, see I. Vingiano-Viricel, *op. cit.*, n° 2.

[20] V. L. Andreu et al., No. 01.42: *"We have already seen that, from the point of view of technology, a distinction is made between different kinds of vehicles, which are classified according to a*

or totally automated".[21] The Data Ordinance sets out in detail the persons authorised to access vehicle data, distinguishing between the various cases. [22] With these various texts, we were expecting to see an ordinance on the rules governing civil liability—particularly in the event of an accident.[23] The aforementioned Mobility Orientation Law gave the empowered government to issue a text *"defining the liability rules"*—without any specific restrictions. However, it is known that insurers weighed heavily against such a text, fearing that it would lead to a more general development of the rules governing traffic accidents, which could prove unfavourable to their interests. They seem to have won the support of the Government, which did not follow up on the empowerment given by Parliament and preferred to leave gaps rather than fill them.[24]

These different rules are, each in their own way, evidence of the same idea: the autonomous vehicle is a particular object whose use can be a source of damage. It is therefore necessary to regulate its use so as to reduce the risk of accidents and to define the liability—both civil and criminal—of those who contribute to this risk. In order to explain the rules that apply, we will distinguish between those relating to the movement of the vehicle (Sect. 1) and those relating to the liability of the persons involved (Sect. 2).

---

complex system. Although it is not necessary for the legislator to go into the details of this classification - which is difficult to apply in practice - it nevertheless seems useful to isolate two types of autonomous vehicles, since those with only limited autonomous driving devices raise fewer difficulties than those with advanced devices. The approach is that of the German legislator, who chose to define and regulate not simply autonomous cars, but more specifically "vehicles with highly or fully automated driving functions". The French legislator could draw inspiration from this analysis and, after defining the genus - autonomous cars, subject to a certain number of rules common to all its species - highlight and define highly autonomous cars - which would be the only ones concerned by more restrictive rules.

[21] Our proposal, however, was somewhat different and less complex. We proposed to distinguish between the genus—autonomous vehicle—and its species, among which was isolated the highly autonomous vehicle. Hence the proposal to add a 3° in Article L. 110-1 of the Highway Code stating: *"The term 'autonomous vehicle' means any motor vehicle equipped with autonomous driving devices. The vehicle is highly autonomous when the on-board technology allows the user not to constantly monitor the behaviour of the vehicle. Autonomous driving devices are those that govern the speed or trajectory of the vehicle"*. The definitions in the decree are different—and cryptic. For the delegated driving vehicle, see *supra,* note 20. For the partially automated vehicle: *"a vehicle equipped with an automated driving system exercising dynamic control of the vehicle in a particular functional design domain, which must make a request for recovery in order to respond to certain traffic hazards or failures during a manoeuvre in its functional design domain"*. For the highly automated vehicle: *"a vehicle equipped with an automated driving system exercising dynamic control of a vehicle in a particular functional design domain, capable of responding to any traffic hazard or failure, without exercising a handover request during a manoeuvre within its functional design domain"*. For the fully automated vehicle: *"a vehicle equipped with an automated driving system exercising dynamic control of a vehicle, capable of responding to any traffic hazard or failure, without exercising a demand for recovery during a manoeuvre"*.

[22] See *infra,* n° 19.

[23] On this point, see *infra,* n° 35.

[24] See *infra,* n° 35et seq.

1 Traffic

Autonomous vehicles cannot be allowed to move without restrictions: rules must govern the use of these dangerous objects. These restrictions concern both the user of the vehicle (Sect. 1.1) and the vehicle itself (Sect. 1.2).

1.1 The User

With regard to the user, it is necessary to ask under what conditions he is entitled to use the autonomous vehicle (Sect. 1.1.1) and how this user can be considered with regard to the texts that classically govern *"drivers"* (Sect. 1.1.2).

1.1.1 The Right to Use

Like any vehicle, the autonomous car is a complex object, the misuse of which can lead to damage.[25] The question therefore arises as to whether users should be allowed to use it freely or whether specific constraints should be laid down—such as compulsory training in the use of the autonomous vehicle. In the context of the experiments, the legislator has ruled that *"the driver must have received adequate prior training in the delegated driving functions implemented during the experiment"*.[26] Such a rule is not very troublesome in the case of experiments, which remain a limited phenomenon, for which the introduction of training is not particularly annoying. With the general deployment of autonomous vehicles, the question arises as to whether this system should be generalised, [27] bearing in mind that a requirement for prior training in autonomous driving is likely to slow down the development of the market, but that the safety of individuals has everything to gain from it, particularly in the first years of use of these vehicles.

---

[25] In particular, the vehicle user should be able to take control of the vehicle back if the driving delegation system is encountering difficulties or if it system requests the user to do so.

[26] Article 12, III, of the experimental decree.

[27] On the hesitation that may exist in this area, see L. Andreu et al., No. 01.122: *"On the one hand, one could consider that it is natural that at the stage of experimentation, which is made up of uncertainties about the exact risks associated to this type of driving, the driver should be specially trained to counter them, such precautions no longer being necessary once the test phase is completed. But on the other hand, it could equally be argued that if "adequate prior training in the functions of delegated driving" is required of automotive engineers, it should be required of users who do not have this expertise. A wise solution would probably be to decide in favour of safety standards, at least in the first years of movement of these vehicles, until it can be determined whether or not the difficulties in taming this type of driving are real.*

With a view to preserving the safety of individuals, we had expressed the wish that compulsory training for autonomous driving should be established. [28] However, it was not a question of providing for a specific driving licence authorising the use of autonomous vehicles—as may exist for passenger transport vehicles or most large two-wheeled vehicles—but only a compulsory training course of a few hours enabling a certificate to be obtained—as is already the case for the use of small two-wheeled vehicles. [29] However, the legislator has made a different choice. Following on from Mrs. Idrac's report, [30] the Criminal Liability Ordinance only provides for an information obligation for consumers who buy or hire an autonomous vehicle from a professional.[31] Thus, according to the new Article L. 224-68-1 of the Consumer Code, *"Prior to the conclusion of a contract of sale or hire of a vehicle with delegated driving as defined by the Highway Code, the professional shall provide the consumer with information relating to the conditions of use of the automated driving system with which the vehicle is equipped"*. The manufacturer *"shall also guarantee public access to information content through any medium"* (user manuals, explanatory videos, *etc.*). The aim is to ensure that users of autonomous vehicles are aware of the general operating principles—non-compliance with the text by the manufacturer being punishable by a fine.

However, such an obligation to provide information seems insufficient to us. There are risks of misuse of the vehicle by a user who has not fully understood the conditions of use of the vehicle and the obligation to supervise or take over the vehicle.[32] In particular, the obligation to provide information will not benefit persons who use the vehicle without having concluded a contract of sale or hire (e.g. a private individual who takes advantage of a vehicle loan from a professional or a relative) or those who are not consumers (e.g. a legal person; a professional individual who buys

[28] L. Andreu et al., *op. cit.* Hence the proposal to create an Article R. 221-8-1 in the Highway Code (inspired by Article R. 221-8) which states that *"The various categories of driving licence set out in the preceding articles allow the driving of an autonomous vehicle on condition that the holder has undergone practical training provided by an establishment or association approved under Article L. 213-1 or L. 213-7.*

[29] L. Andreu et al., *op. cit.* No. 01.124, which added that *"For future drivers, this certificate could be obtained at the same time as the standard licence. For drivers who already hold a licence, they should obtain this certificate by following the training course"*.

[30] A.-M. Idrac, Report, op. cit. p. 39, which provided for simple *"information for drivers on the type of technique and how it works"* in the form of *"written manuals"* or *"explanatory videos"* - a proposal which we considered insufficient: see L. Andreu et al., *op. et loc. cit.*

[31] It must be said that the Mobility Orientation Law was not sufficiently welcoming. Article 31 stated: *"It may be provided for in this respect to impose the provision of information or appropriate training, prior to the provision of vehicles with driver delegation, during the sale or hire of such vehicles"*. Although the text seems to leave room for manoeuvre by referring to *"appropriate information or training"*, the text seemed to limit either to cases where a sale or hire takes place. However, the logic of pre-driving training is to be a necessary precondition for the use of these vehicles, not their purchase or hire.

[32] In particular, the accident reported in 2016 by the press in which the driver of the Tesla vehicle involved was watching a movie at the time of accident.

the vehicle for his professional needs) or to those who would not contract with a professional (e.g. the buyer of a second-hand vehicle)—letting along the case where the vehicle has been bought abroad. As for the user manuals and explanatory videos, it is likely that they will rarely be consulted, as is the case for most complex products that people buy. The system will thus lead to many people who do not have sufficient knowledge to use such devices safely to move around the country without any particular restrictions—knowing that we will not even be able to verify that they know how to use them during a roadside check. It is therefore regrettable that the government has chosen to give priority to the economic development of autonomous vehicles over the safety of individuals.

1.1.2 The Status of the User

Beyond these technical aspects, the development of autonomous cars also raises a problem of status of the user. French legislation gives a central role to the *"driver" of* the vehicle. It is the driver (or the *"guardian" of* the vehicle) who is liable in the event of an accident [33]- it being specified that the driver, when he is the victim, is not compensated in the same way as the non-driver victim. [34] In the Highway Code, the driver is referred to in more than four hundred articles. Under these conditions, it is understandable that the development of autonomous vehicles is likely to give rise to difficulties in implementing the various texts concerned, insofar as it will be difficult to consider as a *'driver'* the person who has activated the *'delegation of driving'*—and who, by definition, is no longer driving. [35]

[33] See *infra*, n° 40.

[34] See *infra*, n° 41.

[35] There is no legal definition of a driver, but it seems in case law that the status of driver presupposes that the person concerned retains some control of the vehicle. Thus, a person who, in order to take a nap, "*leaves the controls of his vehicle, of which he has therefore not retained control*", is no longer a driver within the meaning of the Act of 5 July 1985: Cass. Crim., 31 May 2016, 15-83625. Similarly, a student of a driving school who is under the orders of his driving instructor does not have sufficient control of the vehicle to be considered as a driver: Cass. 2nd civ., 29 June 2000, No. 98-1.848. Conversely, a person who is behind the wheel of a vehicle towed "*with the aid of a short and rigid iron bar*", but who "*retains a certain amount of control in driving the towed vehicle*" (action on the steering wheel) remains a driver: Cass. 2nd civ., 14 Jan. 1987, No. 85-14,655. Comp. Cass. 2nd civ. 27 November 1991, 90-11,326. This led us to the following analysis: "*If we consider the case of a vehicle whose autonomy remains low, we can understand that the person who retains sufficient control of the vehicle remains the driver, even though he would have activated the autonomy device with which his car is equipped* [...] *Conversely, when the autonomy device is sufficiently strong to totally or partially relieve the user of driving the vehicle, the person who activates it ceases to be the driver. The person who activates an autonomy device may thus be in the same situation as a student in a driving school, when the instructor takes full control of the vehicle - holding the steering wheel and operating not only the pedals but also the gearbox - to show the student how to put the car in first gear and drive. While behind the wheel, the student is not a driver. A person who activates a driving device in an autonomous vehicle would no longer be a driver. Between these two extremes, the question arises as to when a driver who*

To resolve these issues, we proposed to take inspiration from German legislation by providing for the legal assimilation of the user of an autonomous vehicle to the driver. Accordingly, it would be possible to add to Article L. 110-1 of the Highway Code a sentence specifying that *"The person who activates the autonomous driving system of a land motor vehicle is deemed to be the driver"*.[36] Such a legal fiction would have many advantages. With regard to the traffic rules, it would make it possible to apply to the user of an autonomous vehicle the various rules relating to traffic which do not deserve to be set aside. [37] As for the texts governing civil liability and insurance, the proposal would make it possible to avoid many of the difficulties associated with the designation of a person liable in the event of a traffic accident. [38] This status would also make it possible to exclude the manufacturer of the vehicle to be considered as a driver. [39]

The legislator seems not to have seen the difficulty. Perhaps following an isolated analysis according to which the person using the vehicle would be considered as *"the driver in the broad sense"*, [40] the experiment decree states, in an embarrassing manner, that *"When the driving delegation functions are activated, a person drives the vehicle in the capacity of a driver"*[41]—it being specified that this person may be outside the vehicle.[42] However, when we examine the role of this person, we see that the status of *"driver"* is not real: this person does not in fact *"drive"* the vehicle, but only has, in certain situations, to *"take control of the vehicle, particularly in the event of an emergency to ensure the safety of the vehicle, its occupants and road users, or when the conditions of use of the vehicle for the experiment are not met*

activates an autonomous device ceases to be a driver. If one applies the criterion of control of the vehicle, it should probably be accepted that the loss of driver status implies a device that ensures a sufficiently high degree of autonomy of the car, to the point of allowing the user of the vehicle to cease to supervise it permanently. Below that point, the person concerned retains the status of driver within the meaning of the various texts governing motorised land vehicles". Comp, I. Vingiano-Viricel and D. Noguero, *Intelligence artificielle et véhicules autonomes*, in Droit de l'intelligence artificielle, Dir. A. Bensamoun and G. Loiseau, LGDJ 2019, n° 204: *"as long as the user has the duty to keep control of the vehicle, even if it is a power to be used occasionally (action on the pedals and/or the steering wheel), he also remains a driver within the meaning of the law"*. In the same article, these two authors refer to the user of a level 5 autonomous vehicle as a *"driver"*—while also referring to him as a *"passive passenger"* (see No. 166) and specifying that *"level 5 does not automatically eliminate the presence of a driver who could always take control of the vehicle back and drive it in the conventional way if he so wishes"*. See also *infra,* notes 41and 80.

[36] L. Andreu et al., *op. cit.* n° 01.72.

[37] See *infra,* n° 26.

[38] See *infra,* n° 40.

[39] See *infra,* note 129.

[40] I. Vingiano, *What legal future for the "driver" of an "intelligent car"?*, n° 15. Comp. I. Vingiano-Viricel and D. Noguero, mentioned *supra* note 36. See also *infra,* note 80.

[41] Art. 12, I, of the experimental decree.

[42] Art. 12, II, of the experimental decree.

anymore".[43] It is thus understood that the driver is not directly driving the vehicle (which is done by the machine: this is the whole meaning of "*delegation of driving*"), but only for more or less actively supervising it, with the duty to take the control of the vehicle back (and then, if necessary, to become a driver again) when certain conditions are met. It follows that the reference in the text to the person's status as a "*driver*" is indeed a legal assimilation—which is merely implicit here, rather than explicit, as we proposed.

The Criminal Liability Ordinance is no better. There is no article that expressly defines the user as the "*driver*". However, several articles can be understood as implicitly making such an assimilation, such as those which provide that taking back control of the vehicle is a duty incumbent on the "*driver*" (i.e. the person who, after activating the delegation of driving, no longer drives).[44] It could thus be understood, once again, that the user of the vehicle would implicitly be considered as a "*driver*" by the law—since the text designates as a "*driver*" a person who does not "*drive*". [45] However, one would have preferred this assimilation to be expressed, so as to dispel the doubts that the texts are bound to raise, particularly in criminal matters, where prevails the principle of strict interpretation of the texts that concern the "*drivers*"[46]—all the more so as the Data Ordinance does indeed make the distinction of the "driver" and the "user", [47] showing that one can use an autonomous vehicle without being a driver in the strict sense...

1.2 The Vehicle

With regard to the restrictions on the vehicle, the essential issue of data recording will be considered (Sect. 1.2.1), followed by the issue of signalling the autonomous nature of the vehicle to other road users (Sect. 1.2.2).

---

[43] Art. 12, II, of the experimental decree.

[44] Art. L. 123-1.

[45] See the Criminal Liability Ordinance, which states that *"The driver must be constantly in a position to respond to a request to take control of the automated driving system* "(Article L. 123-1 of the Highway Code): as long as this "*takeover*" does not happen, it is therefore the automated driving system that has "*control*" *of the* vehicle.

[46] See *infra,* n° 28 (for criminal liability) and 39 (for civil liability).

[47] See Article L. 1514-5 of the Transport Code, which considers the activation, deactivation and control of the automated driving system by the "*person concerned, driver or user of one of the vehicles involved*".

1.2.1 Data Recording

Among the main constraints relating to the vehicle, the requirement for a *"data recorder"* is in first place. This is of great interest since [48] the criminal liability law, and to a lesser extent civil liability law, [49] is likely to vary depending on whether the damage is caused by the driver, when he is in control of the vehicle, or by the machine, when the autonomous mode is activated. For this reason, all the texts and proposals relating to autonomous cars require the installation in the autonomous vehicle of data recording devices that make it possible to trace the conditions under which it has operated.[50]

However, the problem is not the principle of such a data logger. It is its conditions of use. What information should be recorded? Who can access it? When must it be deleted? The experimental decree initially provided that *"Vehicles shall be equipped with a recording device enabling to determine at any time whether the vehicle has been driven in partial or total delegation of driving mode. The data are automatically and regularly deleted. In the event of an accident, the data recorded during the last five minutes shall be kept by the authorisation holder for one year.*[51] However, this provision was still too restrictive in regard to the nature of the information collected and the length of time it was kept.[52]

It was only slightly expanded in 2020: the new version of the experimental decree provided for a recording device *"allowing to know at any time the status of delegation of driving. The data are automatically deleted after a period of four months. The driver of the vehicle has access to this data on request. In the event of an accident, the data recorded during the last five minutes are kept by the authorisation holder for one year.*[53] The information collected is thus more extensive. A longer keeping period is also set. It is also provided that the driver himself can have access to the information. This is a good improvement, but the text is still unsatisfactory: not only does it not specify precisely what information relates to *the "status of delegation of driving"*, nor the persons authorised to access it other than the user, but the period of data keeping provided for is in any case too short for the data recorder to be fully useful in practice.[54]

From this point of view, the Data Ordinance seems more satisfactory. It allows a longer keeping period and defines the information that may be transmitted and the

---

[48] See *infra*, n° 25.

[49] See *infra*, n° 39.

[50] L. Andreu et al., *op. cit.*, n° 01.113.

[51] Article 11.

[52] L. Andreu et al., *op. cit.* No. 01.113, proposing a duration of 6 years—corresponding to the duration of the statute of limitations for a criminal offence.

[53] Article 11.

[54] If an accident occurs, the action could be brought well after the one-year period provided for in the text, since the limitation period for liability actions is in principle 5 years (Article 2224 of the Civil Code) and can be 10 years in the case of personal injury (Article 2226 of the Civil Code).

persons entitled to receive it more carefully—although many uncertainties remain pending the implementing decree. In the event of an accident, the driving delegation data are transmitted to the services responsible for carrying out post-accident investigations *"for the sole purpose of studying and analysing accidents"*.[55] In the same event of an accident, the transmission of driving status data relating to the *"conditions of activation, deactivation and control of the automated driving system"* is provided for the benefit of *"insurance undertakings which insure the vehicles involved in the accident, for the purpose of determining the compensation necessary for the enforcement of the insurance contract concerned* [...] *exclusively when the processing of these data is necessary for the performance of the insurance contract concerned"* and for the benefit of the *"guarantee fund for compulsory damage insurance"* for *"the same purpose"* when *"no insurance undertaking is in a position to carry out compensation in the context of the performance of an insurance contract"*.[56] Only *"the strictly necessary data to determine whether or not the delegation of driving was activated, or whether or not control of the vehicle was taken, for the purpose of compensating victims in application of Law No. 85-677 of 5 July 1985 aiming to improve the situation of victims of traffic accidents and accelerating compensation procedures"*, is then transmitted. It is specified that the consent of the person concerned—*"driver or user of one of the vehicles involved"*—is not required. The ordinance also provides for the transmission of data *"anonymised by a process guaranteeing the irreversible removal of the link between the data and the serial number or any identifier of the vehicle, its driver, owner or hirer"* to various persons such as road infrastructure managers, police and gendarmerie forces, fire and rescue services, *etc.*[57] The purposes are specified in each case, whether it is accident prevention,[58] knowledge and mapping of the road infrastructure and its equipment[59] or knowledge of road traffic.[60]

In addition to this complex system, the Criminal Liability Ordinance also provides for access to the data of the driver delegation data recording system in three cases. Firstly, the police can access the vehicle data *"in the event of a traffic accident resulting in personal injury"*. Secondly, persons responsible for recording traffic offences may access the data *"when checking vehicles and their drivers"*. Finally, the holder of the vehicle registration certificate may have access to it in the event of *"the discovery of one of the offences mentioned in* [Article L. 121-3 of the Highway Code]". On this last point, the text gives rise to reservations, insofar as it only allows access to the data to the holder of the registration certificate, and only when the offence is covered by Article L. 121-3 of the Highway Code, whereas the user of the

[55] Article L. 1514-4 of the Transport Code.

[56] Article L. 1514-5 of the Transport Code.

[57] See also Articles L. 1514-6 et seq. of the Transport Code, which allow the manufacturer to access certain vehicle data.

[58] Article L. 1514-1 of the Transport Code.

[59] Article L. 1514-2 of the Transport Code.

[60] Article L. 1514-3 of the Transport Code.

vehicle could also have an interest in having access to the information, in particular when the offence committed is not covered by the aforementioned Article L. 121-3. [61] On this point, it is not clear why the ordinance was not inspired by the aforementioned experimental decree which allows the vehicle user to access the data. [62]

1.2.2 Vehicle Identification

The question of data recorders does not exhaust the subject of constraints relating to the vehicle. Another difficulty arises from the cohabitation of autonomous vehicles with conventional vehicles, which will be necessary, at least in the early stages of the deployment of these vehicles.[63] *Human* drivers adopt certain behaviours in view of the fact that they are confronted with other *human* drivers on the road. The introduction of autonomous vehicles will change this balance and may make it difficult for drivers to anticipate the behaviour of the machine.[64] Under these conditions, the question arises as to whether it would be useful to draw the attention of drivers to the fact that a vehicle driving near them is driven by the machine. In our proposal, a positive answer was advanced. It was suggested that there should be a *"device to alert other drivers on public roads that they are confronted with an autonomous vehicle so that they can adapt and not rely on a human reaction on their part"*.[65] Several ways of implementing such a system are obviously possible. The least restrictive would be to affix a sticker to the back of the vehicle representing the autonomous nature of the vehicle (*"VA"* for example), similar to the sticker on cars driven by learner drivers. [66]

---

[61] See *infra,* n° 24.

[62] See *supra,* n° 18.

[63] Rappr. I. Vingiano-Viricel and D. Noguero, *art. prev. mentioned,* n° 156.

[64] This was illustrated by an accident in the United States in May 2015. A Google Car was disrupted by a sandbag left on the road and had to deviate into another lane. However, the driver of an oncoming bus, believing that the vehicle would not attempt this movement, did not let the autonomous vehicle pass, thus causing the accident (P. Sirinelli, *"Grain de sable pour la voiture autonome"*, D IP/IT 2016. 161).

[65] L. Andreu et al., *op. cit.*, n° 01.112.

[66] Hence the proposal to enshrine in the Highway Code an article R. 412-18 (inspired by the current article R. 413-5, concerning the sticker for learner drivers): *"I. - Any driver of a highly autonomous vehicle must, while driving, visibly affix to the rear of his vehicle a distinctive sign, the conditions of use and model of which are set by joint order of the Minister of the Interior and the Minister of Transport. This signalling duty may be fulfilled by the use of a luminous device installed on the vehicle under the conditions laid down by joint order of the Minister for the Interior and the Minister for Transport.* (L. Andreu et al., *op. cit.*, No. 01.112). See I. Vingiano-Viricel and D. Noguero, *art. prev. mentioned,* note 81, who stress the risk of a *"destabilising surprise"* for other drivers.

Reading the experimental decree, in its 2020 version, this proposal seems to have been seen by the legislator. Whereas the previous text made no such provision,[67] the new text adds that the holder of the experimentation authorisation *"shall inform the public present in the field of experimentation, by any appropriate means, of the movement for experimental purposes of vehicles with partial or total delegation of driving"*.[68] We are more surprised to find nothing equivalent in the ordinances and decrees that are intended to govern the movement of autonomous vehicles outside of any experimentation. Should we see this as a sign that the public authorities have chosen to give precedence to the economic development of autonomous vehicles over the safety of individuals? The question leads us straight to the question of responsibilities.

2 Liabilities

As with any motor vehicle, the use of an autonomous vehicle may give rise to liability, both criminal (Sect. 2.1) and civil (Sect. 2.2).

2.1 Criminal Liability

The legislator regulates the use of land motor vehicles and, in order to reduce the risk of accidents, provides for a certain number of traffic offences, which are set out in the Highway Code: speed limits, penalties for driving after consuming alcohol or drugs, prohibition on using the telephone while driving, *etc.* In principle, these offences are charged to the *'driver' of* the vehicle (Art. L. 121-1 Highway Code) or, exceptionally, to the *'holder of the vehicle registration certificate'* (Art. L. 121-2 and L. 121-3 Highway Code).[69] In addition to these road traffic offences, there are, where applicable, offences under ordinary law contained in the Criminal Code: offences against life (Articles 221-1 and 221-6 of the Criminal Code), against the physical or psychological integrity of a person (Articles 222-7 and 222-19 of the Criminal Code), against the property of others (Article 322-1 of the Criminal Code), *etc.* It

[67] Article 13 of the experimental decree only provides that *"When the vehicle is intended for the public transport of persons, it shall bear a notice visible to all its occupants indicating that it is an experimental vehicle with delegated driving and the requirements that apply to the transport of minors"* (Art. 13, IV). The sign thus concerns only public passenger transport vehicles and benefits only the users of the vehicle, whereas it is third parties who, in our view, should benefit from the information.

[68] Article 10, paragraph 2.

[69] Targeting the holder of the registration certificate facilitates the enforcement of the offences concerned by attributing them to the holder of the registration certificate, without the need to establish the identity of the driver.

goes without saying that with the deployment of autonomous vehicles, these various offences cannot all be attributed to the vehicle user. This will be seen by distinguishing between road traffic offences (Sect. 2.1.1) and ordinary law offences (Sect. 2.1.2).

2.1.1 Road Traffic Offences

In a human-oriented system, the current enforcement system works quite well. Traffic offences are charged to the driver—or, in a number of cases, to the holder of the registration certificate. [70] Things are different in the case of an autonomous car. The unlawful behaviour of the vehicle cannot always be attributed to these persons, since it can be attributed to the machine—and therefore to some extent to the manufacturer of the vehicle. In order to take this into account, the texts delimit the boundaries of road traffic offences so as to exclude the criminal responsibility of the vehicle user when the vehicle's unlawful behaviour is characterised even though the machine was responsible for its movement.

This is the case in the Experimental Ordinance, which provides that "*The first paragraph of Article L. 121-1 of the Highway Code is not applicable to the driver during periods when the driver delegation system, which he has activated in accordance with its conditions of use, is in operation and informs him in real time that he is in a position to observe traffic conditions and to carry out any manoeuvre in his place without delay*". This text is broadly taken up by the Criminal Liability Ordinance, which makes some adjustments. The experimental text has the defect of largely excluding the criminal liability of the person concerned "*during periods when the system of delegation of driving* [...] *is in operation*". Such an exemption appears excessive, given that certain offences must remain attributable to the vehicle user even when the autonomous driving mode is activated (e.g. offences related to the consumption of alcohol or drugs). In our report, we proposed limiting the exclusion to cases where the user has activated an autonomous driving device "*for which the behaviour of the vehicle is solely responsible*".[71] From this perspective, it would not be sufficient for the driving device to be activated for the person concerned to escape the applicable criminal liability: only offences related to the behaviour of the vehicle would be covered and the exemption would presuppose that they were attributable solely to the autonomous driving system. The ordinance comes closer to this proposal by now referring to "*offences resulting from a manoeuvre of a vehicle*"—thus leaving open the possibility that the user of the vehicle may be punished for offences that are not related to the movement of the vehicle as such.[72]

---

[70] See *supra*, n° 24.

[71] L. Andreu et al., *op. cit.*, n° 03.65.

[72] It is regrettable that these texts do not more clearly exclude the application of Articles L. 121-2 and L. 121-3 of the Highway Code—which make the holder of the registration certificate liable in

Nevertheless, the person concerned may become criminally liable again in certain circumstances. The Experimental Ordinance thus provides that *"The same first paragraph is again applicable after a request of the driving system at the end of a period of time during which control of the vehicle may be taken back as specified in the experimental authorisation, of which the driver is informed. The same applies when the driver has ignored the obvious fact that the conditions for using the driving delegation system, as defined for the experiment, were not or were no longer fulfilled".*[73] A similar—but more refined—provision is also enshrined in the Criminal Liability Ordinance, which states that *"The provisions of the first paragraph of Article L. 121-1 are again applicable: 1° From the moment the driver exercises dynamic control of the vehicle after he has taken the control of the vehicle back; 2° In the driver do not resume driving at the end of the transition period following a request from the automated driving system [...]; 3° To the driver who does not respect the summonses, injunctions or indications given by the forces of law and order, or the rules of priority for the passage of priority vehicles of general interest".*[74] These various clarifications are explained by the legislator's desire to punish the user of the autonomous vehicle who resumes driving the vehicle[75] or who is in a situation in which he or she should have resumed driving.[76]

They will nevertheless give rise to a problem. Although the legislator probably intended to make it possible to admit the criminal liability of any user of the autonomous vehicle who finds himself in one of the situations referred to, the texts do not necessarily lead to this. First of all, it should be noted that the legislator has restored application of Article L121-1 for the *"driver"* of the vehicle. However, in a field where the principle of strict interpretation of criminal texts prevails, [77] it is far from certain that the judge will accept to consider as a *"driver"* someone who has not taken over the driving of a vehicle whose autonomy is so developed that the person concerned cannot be considered as a driver in the traditional sense of the term.[78] Some will claim that the text only governs the hypothesis of a vehicle whose

certain circumstances. In practice, this omission should not have any impact, since the manufacturer who should be considered as *the "true perpetrator of the offence"* within the meaning of Articles L. 121-2 and L. 121-3 mentioned above (these texts allow the holder of the registration certificate not to be liable for the financial consequences of the offence committed when he can designate this *"true perpetrator"*).

[73] Article 21-1.

[74] Article L. 123-1, paragraph 3, of the Highway Code.

[75] Moreover, there seems to be no need for a special text to govern this hypothesis, in which the person concerned becomes an ordinary driver again.

[76] The first paragraph of Article L. 123-1 of the Highway Code can be read *a contrario* to mean that in cases where Article L. 121-1 of the Highway Code does not cease to apply, it can be applied. The third paragraph makes it possible to make the text applicable again in cases where it is set aside. It should thus be understood that the cases in which criminal liability is not ousted (paragraph 1) are added to those in which criminal liability is restored (paragraph 3).

[77] Article 111-4 of the Penal Code: *"The penal law is of strict interpretation"*.

[78] See I. Vingiano-Viricel, *op. cit.* No. 2, which, in a cryptic footnote, seems to contradict the idea that highly or fully autonomous cars are driverless. However, the author does not consider in this

autonomy remains low so that the user's status as a driver can be accepted.[79] From this point of view, it seems to us that the legislator would have been better advised to take inspiration from the proposal to expressly assimilate the user of the vehicle to the driver so as not to leave any doubt on this subject. [80] Secondly, it should be noted that the text makes no provision for the person concerned to be criminally liable in the cases referred to. It only states that the provisions of Article L. 121-1 of the Highway Code *"are again applicable"*. However, for an offence to be established, it is not enough for this text to be applicable; its conditions of application must also be fulfilled—which is not a question of applicability, but of implementation of the text. From this point of view, for the offence to be established, Article L.121-1 requires not only that there be a *"driver"* on board the vehicle, which refers to the difficulty noted above, but also that there be one or more *"offences committed by the driver in driving* [the] *vehicle"*. Here again, if one understands the legislator's objective, the text may give rise to difficulties in criminal matters—for example, if the person concerned has left the driving position without taking over the driving of the vehicle, whose behaviour has consequently been contrary to the rules of the highway code. Will he be considered to have committed an offence *in driving the vehicle* [81]? The difficulty may be solved by considering that the implicit assimilation of the user to the driver of the vehicle would be coupled with an equally implicit assimilation of the use of the vehicle to its driving.

The question remains as to how to sanction road violations when the driver is not at fault: should it be assumed that no one would be responsible? A positive answer to this question would not be satisfactory. In order to protect individuals, the prohibitions of criminal law must persist and lead to the punishment of those who are at the origin of the illegal behaviour of the vehicle. The non-responsibility of the user then directs the analysis towards the vehicle manufacturer—at least the one who programmed the vehicle to drive in a way that enabled it to engage in the prohibited behaviour. To this end, the Experimental Ordinance provides that *"If the driving of the vehicle, whose driving delegation system has been activated and is functioning under the conditions provided for* [by the texts], *contravenes rules whose non-observance constitutes a contraventional offense, the holder of the authorisation shall be financially liable for the payment of the fines"*. Thus, the person concerned may be financially liable for offences related to the movement of the

No. the case where criminal law is involved. Comp. in the same work, nos. 118 and 132, where the author considers that the criminal law texts concerning *"drivers"* can be attributed to the person concerned. See also I. Vingiano-Viricel and D. Noguero, *art. Prev. mentioned*, no. 166, supra note 36.

[79] See *supra,* no. 11, in particular note 36.

[80] L. Andreu, *op. cit.* under No. 04.117, with the proposal for an Article L. 110-1 3° of the Highway Code providing: *"The person who activates the autonomous driving system of a land motor vehicle is deemed to be the driver"*.

[81] Comp. with Cass. Crim. 31 May 2016, 15-83,625, concerning a person who was taking a nap in his vehicle: *"the victim, who had left the controls of his vehicle, of which he therefore did not retain control, was not the driver at the time of the accident within the meaning of the law of 5 July 1985"*.

vehicle—note that the text only covers *"contraventional offenses"* and not *"delictual offenses"*.[82] This rule is broadly taken up in the Criminal Liability Ordinance, which however applies it to the *'manufacturer of the vehicle'*[83]- and which excludes *"the cases provided for in 3° of Article L. 123-1"*, *i.e.* the hypothesis of a failure to take control of the vehicle back in response to instructions from the forces of law and order or the passage of a priority vehicle.[84] It is thus the manufacturer who is made financially responsible for contraventional offenses committed *"during periods when the automated driving system exercises dynamic control of the vehicle in accordance with its conditions of use"*.[85] On the other hand, as soon as the user resumes driving the vehicle, the manufacturer ceases to be liable and it is the driver who becomes again the focus.

This system has its supporters.[86] However, it seems to be excessive to us. As we wrote a few months before, imputing criminal liability—or pecuniary liability—to the manufacturer for all contraventional offences that could be related to its vehicles *"would lead, for one and the same programming error, to making the manufacturer responsible for all offences committed by all vehicles it puts on the market. The constitutional principle of the necessity of penalties would not be satisfied"* (e.g. for a programming error leading to the vehicle not respecting perfectly the safety distances, the manufacturer could be pecuniarily liable for millions of offences and have to make a payment for an astronomical amount). [87] However, in order to avoid total immunity for the negligent manufacturer, we proposed to create a new offence designed to punish anyone who, through carelessness or negligence, has manufactured a driving system that does not allow road traffic to be respected: *"The fact that a professional manufactures, even through carelessness or negligence, a device that allows an autonomous vehicle to initiate an offence is punishable by one year's imprisonment and a fine of 15,000 euros.*[88] However, this

---

[82] In France, criminal law distinguishes the *"contravention"*, the *"délit"* and the *"crime"*. There is no prison sentence for a *"contravention"*. For a *"délit"*, there could be a prison sentence for a period which may not exceed 10 years. For the *"crime"*, the prison sentence could be longer.

[83] Art. L. 123-2 of the Highway Code.

[84] This exclusion is understandable: the manufacturer does not have to be liable if the user was obliged to take over the vehicle. In this respect, the second paragraph of Article L. 123-1 could also have been referred to (case of the driver's failure to take control of the vehicle back at the end of the transition period following a request from the driving system). The legislator may have taken the view that the end of the transition period meant that the vehicle was no longer in the *"periods when the automated driving system exercises dynamic control of the vehicle"*, so that the manufacturer's pecuniary liability during these periods did not have to be expressly excluded.

[85] See, with regard to ordinary offences, *infra*, n° 32.

[86] I. Vingiano-Viricel, *op. cit.*, n° 116.

[87] We therefore proposed the following text: *"The manufacturer of an autonomous vehicle cannot be held responsible for road offenses by the vehicle. He is, however, liable if he knowingly programmed the vehicle to produce such a breach"*: L. Andreu et al., *op. cit., n°* 03.122.

[88] L. Andreu et al., *op. cit.*, n° 03.123.

proposal was not followed by the legislator.[89] The lack of proportionality of these texts will obviously raise a debate on the constitutionality of them. The fact that the government has preferred to turn a blind eye to this problem does not mean that it will not be brought before the courts.

2.1.2 General Law Offences

The situation is different for ordinary offences (i.e. offences that are not specifically road traffic offences). They do not apply specifically to a person with the—contentious—status of *"driver"*, but to any person who meets the conditions of the offence referred to in a more general text—one who causes homicide, one who causes damage to the property of others, *etc.* No assimilation of the persons concerned to the driver is then necessary for the offence to be established against the person concerned—whether it be the user, the manufacturer or a third party. At most, one can wonder about the fate of natural persons who *"did not directly cause the damage, but who created or contributed to creating the situation that allowed the damage to occur or who did not take the measures to avoid it"*. According to Article 121-3 of the Criminal Code, they are only criminally liable *"if it is established that they have either deliberately violated a particular duty of prudence or safety provided for by the law or regulations, or committed a serious fault that exposed others to a particularly serious risk that they could not have ignored"*. To facilitate the implementation of the text, it is useful that a *"particular duty"* to take control back is provided for in the texts, the non-compliance of which may be criminally attributed to the vehicle user. This was the thrust of one of our proposals, [90] which is echoed in the Criminal Liability Ordinance, which provides that the person concerned *"must at all times be in a state and position to respond to a request to take control from the automated driving system"*.[91] This is a welcome clarification, which will make it possible to admit the criminal liability of the user of the autonomous vehicle when damage has resulted from his negligence.

However, the status of *"driver"* is sometimes necessary for the application of aggravating circumstances. For example, while manslaughter is traditionally punishable by three years' imprisonment and a fine of 45,000 euros,[92] the penalty is

[89] This is more easy to understand for the experimentation texts, which involve a small number of vehicles and require extensive monitoring by the licence holder. This is not the case for the Criminal Liability Ordinance, which will apply to a larger number of vehicles, used in normal traffic conditions.

[90] L. Andreu et al., *op. cit.*, 03.64, and the proposal for an Article R. 412-6 III of the Highway Code providing in particular that *"The user of an* [autonomous vehicle] *is required to exercise caution in the use of the autonomous driving device. In particular, he is required to take control of the vehicle back when he is or should have been aware of the machine's inadequacy to ensure compliance with traffic rules. This inadequacy may be the result of a warning from the vehicle itself.*

[91] Art. L. 123-1 of the Highway Code.

[92] Art. 221-6 of the Penal Code.

increased to five years' imprisonment and a fine of 75,000 euros when the criminal offence is committed *"by the driver of a motorised land vehicle"*.[93] Under these conditions, the question arises as to whether these aggravating circumstances can be applied to the user of an autonomous vehicle, who uses a land-based motor vehicle but does not drive it as such. Here again, we can see the advantage of expressly assimilating the vehicle user to the driver and, on the other hand, the limits of the Criminal Liability Ordinance, which did not explicitly enshrine this proposal.[94] This gap in the new text will thus also be felt in criminal matters. On the other hand, the ordinance is more comprehensive in its application of criminal law to the vehicle manufacturer. In the continuity of the criminal road texts previously examined, [95] it provides that *"During periods when the automated driving system exercises dynamic control of the vehicle in accordance with its conditions of use, the manufacturer of the vehicle* [...] *is criminally responsible for the offences of unintentional injury to life or limb provided for in Articles 221-6-1, 222-19-1 and 222-20-1 of the Criminal Code when it is established that there has been a fault, within the meaning of Article 121-3 of the same Code"*. [96] The interest of this text is not to provide for the criminal liability of the manufacturer for the ordinary offences: Article 121-3 of the Penal Code is naturally applicable to the manufacturer who has committed a *"fault of imprudence, negligence or failure to fulfil a duty of prudence or safety provided for by the law or the regulations"* within the meaning of this text, as long as the texts in question provide for the punishment of this type of fault.[97] The interest of the new text is in fact to specifically target the articles of the Criminal Code providing for the aggravating circumstance relating to the fact that the criminal fault was committed *"by the driver of a land motor vehicle"*—which does not naturally apply to the vehicle manufacturer.[98] The text thus makes it possible to

---

[93] Art. 221-6-1 of the Criminal Code.

[94] See *supra*, n° 14 and 28. *Contra*, I. Vingiano-Viricel, *op. cit.* No. 118, for whom *"the driver, a natural person, is criminally liable for damage caused to victims even in delegated driving mode"* (see also No. 132). The statement probably stems from the author's analysis according to which the user of an autonomous vehicle would always be considered as a driver, even when not driving: see *supra*, notes 36, 41 and 80.

[95] See *supra*, no. 29.

[96] The latter text refers to *"recklessness, negligence or failure to comply with a duty of care or safety provided for by law or regulation, if it is established that the perpetrator did not take the normal care, taking into account, where appropriate, the nature of his tasks or duties, his competences and the power and means at his disposal.*

[97] Moreover, the text should not be interpreted *a contrario* as meaning that the manufacturer would not be criminally liable outside the cases covered by it - for example in the case of endangering others (Article 223-1 of the Criminal Code).

[98] The clarification that the manufacturer's liability presupposes that the vehicle is used *"in accordance with its conditions of use"* may cause difficulties in cases where the failure to comply with the conditions of use turns out to have no particular impact in practice. For example, a person who has used the vehicle while intoxicated without this having any effect on the behaviour of the vehicle and, consequently, the damage caused by it. It is to be hoped that the case law will not interpret the text literally.

subject the manufacturer to the rules applying to drivers within the meaning of the aforementioned texts. If one considers that the aggravation of criminal liability is intended to reinforce the prudence of those who incur it, one will undoubtedly admit the interest of the extension made by the legislator.[99]

Another aspect of the issue is the risk of vehicle hacking. The central role played by the machine in this respect gives rise to the fear that an ill-intentioned person could interfere with the system and deliberately cause one or more accidents. From this point of view, the existing repressive texts seems satisfactory overall. Ordinary law offences could naturally be applied to a pirate who meets the conditions.[100] To this must be added the more specific offences governing *"attacks on automated data processing systems"*.[101] These include *"fraudulently accessing or remaining in all or part of an automated data processing system"*,[102] *"hindering or distorting the operation of an automated data processing system"*,[103] *"fraudulently introducing data into an automated processing system, extracting, holding, reproducing, transmitting, deleting or modifying the data it contains"*,[104] *etc*. This provision applies without much difficulty to autonomous vehicles.[105]

However, the Data Ordinance supplements this by providing for an information obligation on the part of the vehicle manufacturer, who must inform the national vehicle type-approval authority *"without delay after becoming aware of any electronic attacks which may affect the information systems contributing to the operation or safety of the vehicle"*.[106] Those who fail to comply with this obligation are liable to a substantial fine. This new obligation and criminal offence are certainly welcome. It should help to reduce the vulnerability of autonomous driving systems and thus the scope for criminal liability—as well as civil liability—for individuals.

[99] To avoid further confusing the concepts at work, the legislator could also have provided for an aggravating circumstance specific to the vehicle manufacturer rather than equating the manufacturer with the driver.

[100] Thus, when the hijacking of the vehicle results in death, one might naturally consider prosecuting the hijacker for homicide.

[101] They are provided for in Articles 323-1 et seq. of the Criminal Code.

[102] Article 323-1 of the Penal Code, with aggravated penalties when it results in the *"deletion or modification of data contained in the system"* or *"an alteration of the functioning of this system"*.

[103] Article 323-2 of the Criminal Code.

[104] Article 323-3 of the Criminal Code.

[105] The only difficulty relates to the case where the intrusion is part of a terrorist undertaking. Article 421-1 of the Penal Code lists the offences for which a terrorist motive may lead to an aggravated sentence, referring somewhat vaguely to *"computer-related offences defined by Book III of the present Code"*. As has been written, *"The legislator takes certain liberties in this area in relation to the titles that are supposed to group these offences. The Penal Code does not contain any division that falls under such a formula: there are no "computer-related offences", but "offences against automated data processing systems", the subject of Chapter III of Title II, which includes Articles 323-1 to 323-8. The reference is therefore not perfectly exhaustive, even if its scope, apparently well understood, does not allow one to speculate on a dead-end ambiguity"* (Y. Mayaud, *Terrorisme, Infractions, Comportements*, Répertoire de droit pénal et de procédure pénale, January 2020).

[106] Art. L. 1514-8 of the Transport Code.

2.2 *Civil Liability*

One of the main issues raised by the deployment of autonomous cars is the civil liability incurred in the event of an accident. [107] From this point of view, French law has, since a law of 5 July 1985,[108] organised a system of special liability that is particularly favourable to victims in the event of a traffic accident involving a land-based motor vehicle[109]: victims are intended to be compensated quickly and simply by an insurer—the subscription of an insurance contract being imposed by the Insurance Code [110] . However, this law presupposes that a traffic accident has first been identified. For this reason, the existence of a traffic accident will be studied first (Sect. 2.2.1) before looking at the conditions that victims must meet in order to receive compensation (Sect. 2.2.2).

2.2.1 The Existence of an Accident

The 1985 Act provides for compensation to victims on condition that *"a traffic accident involving a land motor vehicle and its trailers or semi-trailers, with the exception of railways and tramways running on their own tracks"*[111] can be identified. A traffic accident is defined as *"a fortuitous event, causing damage to persons or property, caused by the involvement of a motorised land vehicle moving on a public road or private place, or stationary in a public place or even a private place if it is used for parking"*.[112] From this point of view, there is no doubt that the autonomous vehicle is indeed a motorised land vehicle whose damage falls within the scope of the 1985 law whenever a traffic event can be characterised.[113] It is irrelevant in this respect whether the user of the vehicle is active or inactive.

---

[107] In this sense, L. Andreu et al. No. 02.01; M. Clement-Fontaine, *Les véhicules autonomes dans l'œil du cyclone des réformes de la robotique, en matière de données personnelles et de responsabilité civile*, in M. BEHAR-Touchais, *Les objets connectés*, IRJS, 2018, t. 96, p. 115-127; P. Pierre, *Quand E. Musk rencontrera R. Badinter, ou le pilotage automatique des véhicules automobiles à l'épreuve de l'indemnisation hexagonale des victimes d'accidents de la circulation*, RLDC 2016, n° 141, p. 33. The opposite analysis is regularly defended in colloquiums by representatives of the French insurance federation.

[108] Law No. 85-677 of 5 July 1985 aiming to improve the situation of victims of traffic accidents and accelerating compensation procedures.

[109] The producer is also liable for damage caused by a defect in his product on the basis of Art. 1245 et seq. of the Civil Code, which is not detailed here.

[110] Article L. 211-1 of the Insurance Code: *"Any natural person or any legal entity other than the State, whose civil liability may be incurred because of damage suffered by third parties resulting from harm to persons or property in which a vehicle is involved, must, in order to drive the vehicle, be covered by insurance guaranteeing this liability.*

[111] Article 1.

[112] Lexique des termes juridiques, Dir. S. Guinchard and T. Debard, Dalloz 2017.

[113] L. Andreu et al., *op. cit.*, n° 01.31.

However, the case law states that if the accident is a *"fortuitous"* event, no accident can be characterised in various situations where the damage is the result of a voluntary action by a person. Thus, there is no *"accident" in the* case of a deliberate burning of a vehicle which damages others,[114] or the deliberate collision of a vehicle which results in the injury of one of its occupants,[115] or when a pedestrian deliberately pushes the driver of a scooter who is then hit by a bus.[116] It does not matter to the case law whether the perpetrator of the voluntary action is the one whose liability is covered by the insurer or someone else: it is the very existence of a traffic accident that is ruled out and not the mere implementation of the insurance contract.

Applied to autonomous vehicles, this restrictive definition of accident could give rise to difficulties. Indeed, it leads to the 1985 law being set aside in the case where the damage results from the hacking of an autonomous vehicle.[117] In such a case, the damage is not the result of a *"fortuitous event"*, but the result of the voluntary act of a third party. The victim would not then be entitled to take action under the 1985 law against the pirate,[118] the drivers or guardians of the vehicles involved or their insurers. However, he would still be able to refer the matter to the "Commission d'indemnisation des victimes d'infractions" (CIVI), in order to be taken care of by the "Fonds d'indemnisation des victimes d'actes de terrorisme et autres infractions" (FGTI), except to note that the scope of the compensation would remain less than under the 1985 law. [119] Such an analysis is disturbing *"at a time when the risk of computer piracy appears to be a major source of concern and anxiety in the face of the development of autonomous cars"*.[120] It is therefore regrettable that the legislator, yielding to the insurers' lobby, did not dare to set the rules of civil liability law relating to traffic accidents involving an autonomous vehicle, which would have made it possible to include in the scope of protection the victims of an act emanating from a malicious third party. [121]

[114] Cass. 2nd civ., 5 March 2001, n° 99-16852.

[115] Cass. 2nd civ., 23 Jan. 2003, no. 00-21676; 22 Jan. 2004, no. 01-11665.

[116] Cass. 2nd civ., 11 Dec. 2003, n° 00-20,921.

[117] L. Andreu et al., *op. cit.* No. 02.81; I. Vingiano-Viricel, *op. cit.*, n° 50.

[118] He may of course act against him on the basis of Article 1240 of the Civil Code, which more generally provides for liability for fault.

[119] L. Andreu et al., *op. cit.*, No. 02.90: *"while damages resulting from serious personal injury are fully compensated by the FGTI, limits apply to minor bodily injuries and property damage: coverage is only provided if the victim's income is below a certain threshold and, in any event, there is a ceiling"*.

[120] L. Andreu et al., *op. and loc. cit.*

[121] *Contra*, I. Vingiano-Viricel, *op. cit.* Which observes, with regard to the risk of terrorist attacks by hacking into vehicles, that *"this situation is unfortunately not new and the treatment of its consequences is already taken into account by the law"*.

2.2.2 Compensation to Victims

The 1985 statute provides for a system of facilitated compensation for victims, who cannot have their compensation reduced or denied in case of *"force majeure or the act of a third party invoked by the driver or guardian of the vehicle"*,[122] just as they cannot have their compensation reduced or denied because of *"their own fault, with the exception of their inexcusable fault if it was the sole cause of the accident"*.[123] However, this statute gives rise to various difficulties in the presence of an autonomous vehicle.

The first relates to the very principle of compensation for the victim. Although the insurer has a duty to pay for the damages resulting from the accident, it is necessary, following the 1985 law, that the victim can establish a claim for compensation against a *"driver"* or a *"guardian"* of the vehicle.[124] The Court of Cassation has held that *"the liability insurer cannot be held liable to the victim when the latter cannot himself invoke a claim against the insured arising from the latter's liability"*.[125] Therefore, in the case where only one vehicle is involved in the accident, it is necessary to determine whether a *person other than the victim* can be considered as the driver or guardian of the vehicle in order to answer the question of whether he is entitled in principle to be compensated by an insurer under a motor vehicle insurance contract.[126] In particular, this could be the manufacturer or the person who designed all or part of the autonomous driving system. In the case where several vehicles are involved, the difficulties are fewer, if one accepts that the victim can obtain compensation regardless of whether or not it is an autonomous vehicle.[127] This analysis

---

[122] Article 2.

[123] Article 3, which provides some further clarification.

[124] Article 2. On the contrary, see the authors mentioned *above*, note 109.

[125] Cass. 1st Civ. 4 November 1986, 85-11972.

[126] On the attribution of driver status, see *supra*, n° 10. On the attribution of custody, see the long developments of I. Vingiano-Viricel, *op. cit.*, n° 58. Comp. L. Andreu et al., *op. cit.*, n° 02.101. All of this is obviously subject to the stipulations of the insurance contract, which may well provide for compensation of victims beyond the legal hypotheses.

[127] V. L. Andreu et al., *op. cit.* No. 02.101, who consider that *"the realisation of such a risk* [i.e. the risk of *"impossible compensation of the victim"*] *seems very unlikely in the light of the case law, which has a very flexible, not to say resolutely audacious, definition of custody where compensation issues arise"*. This means that, when the question will arises, the case law will be able to find a guardian of the vehicle to allow the victim to be compensated. See I. Vingiano, *art. prev. mentioned*, n° 14 and I. Vingiano-Viricel, *op. cit.* No. 58: the author proposes to apply the distinction between custody of the structure and custody of the behaviour to the autonomous vehicle. *Contra*, L. Andreu et al., *op. cit.*, n° 02.171. See also. D. Noguéro, *Loi Badinter, voiture autonome, robot, évolution du risque et information au regard de la protection des assurés*, in dossier spécial Comprendre et anticiper la révolution numérique en assurance (Special file on understanding and anticipating the digital revolution in insurance), Wolters Kluwer France, Actualités du droit, Tech&Droit, Oct. 13, 2017 (online) and Revue Lamy Droit de l'Immatériel (RLDI) Nov. 2017, No. 142: *"The obligation to pay the debt on the basis of the Badinter [1985] Act is hardly in doubt, one might think"*.

nevertheless gives rise to some difficulties which the legislator could have dispelled if he had not ignored the 1985 law problems.[128]

Another difficulty relates more specifically to the case where the victim of the accident is the user of the vehicle itself. The compensation of this particular victim is indeed hampered by the 1985 law which states that *"the fault committed by the driver of the land motor vehicle has the effect of limiting or excluding the compensation of the damage he has suffered"*.[129] Thus, a victim driver who, for example, has not fastened his seatbelt, may have his claim for compensation reduced or denied in view of his fault. On this point, however, it seems questionable to us to assimilate the vehicle user fully to a driver, insofar as the person concerned does not really exercise control of the vehicle. The legal assimilation of the user to the driver, which seems to govern the matter, [130] must not be turned against the victims when they are not, strictly speaking, driving the vehicle, and when they are closer to the *"passengers transported"* to whom the legislator wished to reserve special protection. [131] For this reason, we had proposed to specify in the 1985 law that if the fault committed by the driver of the land motor vehicle has the effect of limiting or excluding compensation for the damage he has suffered, the situation is different in the case of an autonomous car: *"The preceding paragraph does not apply to the driver of an autonomous car who, at the time of the accident, was authorised not to control the behaviour of the vehicle"*.[132] It is regrettable that this proposal was not retained by the ordinances that now govern the matter, which do not say anything about the applicable liability rules. This shows once again how regrettable it is that the Government has given in to pressure from insurers and renounced regulating the civil liability governing traffic accidents as Parliament had allowed it to do so. [133]

---

[128] In particular, one may wonder whether the user of the vehicle or the manufacturer or even whoever designed all or part of the autonomous driving system is likely to be liable as a *"driver"* or *"guardian" of* the vehicle within the meaning of the 1985 law - in particular in the case where the vehicle is not insured. For the possible analysis of the manufacturer as driver or guardian, see I. Vingiano-Viricel, *op. cit.* n° 58 and 69. A similar question arises with regard to *"professionals in the repair, sale and inspection of motor vehicles"*, whose liability is not covered by motor vehicle insurance (Article L. 211-1 of the Insurance Code).

[129] Article 4.

[130] See *supra*, no. 12.

[131] Cass. Com., 6 May 2021, 20-14551: *"the legislator, taking into consideration the risks associated with motor vehicle traffic, intended to reserve special protection for certain categories of road users, namely pedestrians, transported passengers, children, the elderly and the disabled"*.

[132] L. Andreu et al., *op. cit.*, n° 02.153. See I. Vingiano-Viricel, *op. cit.* n° 60, who questions the relevance of actual law.

[133] See *supra*, no. 4.

Civil Liability for Damage Caused by Autonomous Vehicles Under Spanish Law

Ángela Fernández Arévalo and Juan Pablo Murga Fernández

Abstract This study analyses the legal regime of civil liability for damages derived from accidents involving motor vehicles, whether autonomous or not. The irruption of such vehicles into legal traffic is a novelty that Spanish legislation must address. The legitimation for compensation will correspond to the owner of the damaged assets, the injured party and, if applicable, to the relatives, to the legally predetermined relatives with presumptive character in the event of death. Compensation for damages shall follow the general system of valuation for pecuniary damage or the legal system of valuation for personal injury, without the injured party having to suffer a loss in this respect depending on whether or not the vehicle is an autonomous one. The liability, in the absence of a driver, by virtue of the *pro damanato* principle, prevailing in Spanish law, must fall on the owner of the vehicle, and his liability must be interpreted as strict liability, and on the insurance company or the Insurance Compensation Consortium, as the case may be. If the loss was due to the defective nature of the vehicle, in any of its facets, the owner could claim against the manufacturer.

1 Introduction

1.1 Objectives

The aim of this paper focuses on the analysis of the legal regime of the obligation to pay compensation for material and personal damages resulting from a traffic accident involving an autonomous motor vehicle in Spanish law, as opposed to non-autonomous and not fully autonomous vehicles.

The interest of this topic lies in its actuality and the need for a legislative reform that contemplates the reality of autonomous vehicles. Civil liability arising from

Á. F. Arévalo (✉) · J. P. M. Fernández
Universidad de Sevilla, Sevilla, Spain
e-mail: arfernandez@us.es

H. Steege et al. (eds.), *Autonomous Vehicles and Civil Liability in a Global Perspective*, Data Science, Machine Intelligence, and Law 3,
https://doi.org/10.1007/978-3-031-41992-8_14

damage to property or personal injury resulting from a motor vehicle traffic accident is regulated in Spanish law in the Law on Civil Liability and Insurance for the circulation of motor vehicles (Ley de Responsabilidad Civil y Seguro por la circulación de vehículos de motor), in the Civil Code, in the Criminal Code and, where applicable, in the Organic Law on the Criminal Liability of Minors (Ley Orgánica de Responsabilidad Penal del Menor).

The chapter structure is the following. Firstly, the legal qualification and the determination of the applicable law of the conflicting interests will be analysed. Secondly, the legal regime of the legal obligation to compensate will be discussed. First, the question of legal standing, i.e. who can claim damages, is analysed. Secondly, the liability, in other words, who can be sued: therefore, determination of the liable parties, title of imputation, cause of exoneration, and organisation of the plurality of parties, if applicable. Thirdly, determination of the items making up the quantum of compensation, taking into account the principles governing the matter: the principle of rogation, the prohibition of unjustified enrichment, and the *pro damnato* principle. Fourth, the time-limit for the claim: duration, nature of the time-limit and the dies a quo for its calculation.

Finally, concerning the research method, in the chapter has been applied the analytical method, based on the critical analysis of the regulations applicable to the matter and the most recent specialised doctrine in this field.

2 Legal Qualification of the Issue and Determination of the Applicable Law

The cause of the damage referred to deserves the legal qualification of civil liability arising from tort or delict[1] resulting from the production of damage to property and/or bodily injury resulting from a motor vehicle accident. The legal consequence in Spanish law is the imposition of the legal obligation to compensate for the damage caused, which is regulated in Royal Legislative Decree 8/2004, of 29 October, which approves the revised text of the Law on Civil Liability and Insurance in the Circulation of Motor Vehicles (Real Decreto Legislativo 8/2004, de 29 de octubre, por el que se aprueba el texto refundido de la Ley sobre responsabilidad civil y seguro en la circulación de vehículos a motor).

The present analysis defends its applicability to the aforementioned case, highlighting the changes that would be necessary in view of the new reality of autonomous vehicles in their various levels (Levels 0–5).[2]

[1] On the criminal implications of the use of autonomous vehicles, Macías Espejo (2020), pp. 1–22.

[2] In this sense, Navarro-Michel (2020), pp. 941–961, and Álvarez Olalla (2019), pp. 145–170. Following a contrary position, Hernáez Esteban (2019), pp. 197–243.

The elements of the factual situation of this heteronomously constituted obligation are, therefore, damage, motor vehicle, and the fact of the circulation. The analysis of each of these elements is presented below.

2.1 Damage

In Spanish law there is no exhaustive list of compensable torts, nor is there a legal concept thereof. It is an undetermined legal concept, which our courts define as the damage to any property owned by the injured party. A distinction can thus be made between damage to property, whether movable or immovable; and damage to non-property assets: personal injury, by causing injury and/or death; and damage to honour, personal and family privacy and self-image.

The damage must be accredited and causally connected with the tortious conduct, without any interference in the causal relationship, such as force majeure and exclusive fault of the victim.

Specifically, the conceivable torts in these hypotheses analysed are damage to property and personal injuries: the right to life and physical integrity. The distinction is important in this matter in terms of the following two factors: first, the assessment of each; second, the title of imputation of liability in either case.[3]

2.1.1 Damage to Property

The economic assessment of damage to property is carried out in accordance with the provisions of article 1106 of the Civil Code: the damages compensation includes not only the value of the loss suffered, but also the value of the profits that have not been obtained, that is to say, the consequential damage and the loss of profit.

The imputation of this liability is a negligence-based liability system, as established in the Law on Civil Liability and Insurance in the Circulation of Motor Vehicles (Ley de Responsabilidad Civil y Seguro en la Circulación de Vehículos de Motor).

[3]Law on Civil Liability and Insurance in the Circulation of Motor Vehicles (Ley de Responsabilidad Civil y Seguro en la Circulación de Vehículo de Motor): Article 1. Civil liability.

1. El conductor de vehículos a motor es responsable, en virtud del riesgo creado por la conducción de estos, de los daños causados a las personas o en los bienes con motivo de la circulación.

En el caso de daños a las personas, de esta responsabilidad sólo quedará exonerado cuando pruebe que los daños fueron debidos a la culpa exclusiva del perjudicado o a fuerza mayor extraña a la conducción o al funcionamiento del vehículo; no se considerarán casos de fuerza mayor los defectos del vehículo ni la rotura o fallo de alguna de sus piezas o mecanismos.

En el caso de daños en los bienes, el conductor responderá frente a terceros cuando resulte civilmente responsable según lo establecido en los artículos 1.902 y siguientes del Código Civil, artículos 109 y siguientes del Código Penal, y según lo dispuesto en esta Ley.

2.1.2 Personal Injuries

Personal injuries are compensated in the field of motor vehicle traffic accidents in accordance with the legal system of assessment of personal injuries contained in the aforementioned law. The compensable elements are injuries and death.

Liability is based on a quasi-strict system, as it can only be excluded if force majeure or the exclusive fault of the victim can be proven.

2.1.2.1 Damages Arising from Personal Injuries

The law predetermines the legal standing to claim damages. Claims can only be made by the injured party. Relatives are only entitled to damages for non-pecuniary and pecuniary loss if the injured person is categorised as seriously injured, i.e. needs the assistance of third parties for essential daily activities such as washing, dressing or eating.

Injuries are economically assessed in two stages: injuries resulting from temporary damage and injuries resulting from permanent damage, i.e. sequelae. At each stage there is a legally structured system of tables which distinguishes between three main items: 1. basic compensation, identical for all individuals sharing the same established legal circumstances; 2. compensation for specific non-material damage; and 3. compensation for non-material damage.

2.1.2.2 Damages Resulting from Death

The law predetermines the right to damages for cause of death with presumptive character, due to the painful loss caused by the death of the victim of the accident. Legitimation is held *iure proprio*, never *iure hereditatis*. The law provides for the existence of five non-exclusive groups of relatives: widowed spouse and cohabitant *more uxorio* descendants, ascendants, siblings and next of kin.

The economic assessment of these damage is structured around three concepts: one, basic compensation, identical for any subject in this situation; two, compensation for particular damage, according to the specific circumstances of the injured party; and three, compensation for pecuniary damage, consequential damages and lost profits derived from the death of the injured party.

2.2 Motor Vehicle

The concept of motor vehicles is provided for in the Law on Civil Liability and Insurance in the Circulation of Motor Vehicles (*Ley de responsabilidad civil y seguro en la circulación de vehículos de motor*) and its Regulation. The Law and Regulation also establish the vehicles that are excluded from this category.

The Traffic Act (*Ley de Tráfico*) also defines the concept of a motor vehicle as being linked to a driver.[4] It is important to underline this circumstance because legally in Spain, a fully autonomous vehicle, i.e. without a driver, is not explicitly contemplated for civil liability purposes. This situation generates a clear need for reform of this law in order to include it and the relevant modifications in the legal system, which will be analysed below.

The concept of autonomous vehicle is defined in the Instruction 15/V-113, of 13 November 2015, of the Directorate General of Traffic, on the authorisation of tests or research trials carried out with automated driving vehicles on roads open to traffic in general (*Instrucción 15/V-113, de 13 de noviembre de 2015, de la Dirección General de Tráfico, sobre autorización de pruebas o ensayos de investigación realizados con vehículos de conducción automatizada en vías abiertas al tráfico en general*): "Any vehicle with motor capacity equipped with technology that allows it to be operated or driven without requiring the active form of control or supervision by a driver, whether such autonomous technology is activated or deactivated, permanently or temporarily. For this purpose, it is not considered as autonomous technology those active safety or driver assistance systems included as vehicle equipment which necessarily require active human control or supervision for their operation or driving. Vehicles incorporating technology with functions associated with automation levels 3, 4 and 5 listed in Table 1 are the subject of this instruction. AUTONOMOUS MODE: driving mode consisting of operating or driving the autonomous vehicle without the active control of a driver when its autonomous technology is activated. CONVENTIONAL MODE means a driving mode of an autonomous vehicle in which the autonomous technology is deactivated and the driving or operation of the autonomous vehicle must be performed under the active control of a driver.

---

[4]Royal Legislative Decree 6/2015, of 30 October, approving the revised text of the Law on Traffic, Circulation of Motor Vehicles and Road Safety (*Real Decreto Legislativo 6/2015, de 30 de octubre, por el que se aprueba el texto refundido de la Ley sobre Tráfico, Circulación de Vehículos a Motor y Seguridad Vial*). Annex I:

For the purposes of this Act and its complementary provisions, the following definitions shall apply:

1. Driver. Person who, with the exceptions of the second paragraph of point 4, controls the steering mechanism or is in command of a vehicle, or in charge of an animal or animals. In the case of vehicles operated on a learner-driver basis, the person in charge of the additional controls shall be regarded as the driver.

Table 1 Niveles de automatización

NIVEL	DENOMINACION	DEFINICION	TAREAS DE CONDUCCION CONDUCTOR	TAREAS DE CONDUCCION SISTEMA	CONDUCCION LONGITUDINAL (ACELERAR/FRENAR) Y LATERAL (DIRECCION)	CONTROL DEL ENTORNO	RECUPERACION DE LAS TAREAS DE CONDUCCION EN CASO DE CONTINGENCIA	TAREAS DE CONDUCCION REALIZADAS POR EL SISTEMA
0	SIN AUTOMATIZACION	El conductor realiza continuamente todas las tareas asociadas a la conducción, incluso cuando son mejoradas a través de algún aviso o la intervención de sistemas.	El conductor realiza continuamente la tarea de conducción dinámica lateral y longitudinal.	N/A	CONDUCTOR	CONDUCTOR	CONDUCTOR	N/A
1	CONDUCCION ASISTIDA	El sistema de ayuda a la conducción desarrolla una tarea específica, bien realiza la conducción dinámica lateral o longitudinal utilizando la información del entorno del vehículo, mientras que el conductor realiza el resto de tareas de conducción.	El conductor realiza continuamente la tarea de conducción dinámica lateral o longitudinal	El sistema realiza la conducción longitudinal o lateral que no esté realizando el conductor	CONDUCTOR Y SISTEMA	CONDUCTOR	CONDUCTOR	ALGUNAS
2	CONDUCCION PARCIALMENTE AUTOMATIZADA	El sistema de ayuda a la conducción desarrolla la conducción dinámica lateral y longitudinal utilizando la información del entorno del vehículo, mientras que el conductor realiza el resto de tareas de conducción.	Supervisión de las tareas de conducción dinámica y el entorno.	Conducción longitudinal y lateral en un caso de uso definido.	SISTEMA	CONDUCTOR	CONDUCTOR	ALGUNAS
3	CONDUCCION AUTOMATIZADA CONDICIONADA	El sistema de conducción automatizada desarrolla todas las tareas de la conducción con la expectativa de que el conductor responda adecuadamente a la petición de intervención por parte de éste.	No es necesaria la supervisión constante de la conducción automatizada pero siempre debe estar en una posición adecuada para reanudar el control.	Conducción longitudinal y lateral en un caso de uso definido. Reconoce sus límites de rendimiento y pide al conductor reanudar la tarea de conducción dinámica con margen de tiempo suficiente.	SISTEMA	SISTEMA	CONDUCTOR	ALGUNAS
4	CONDUCCION ALTAMENTE AUTOMATIZADA	El sistema de conducción automatizada desarrolla todas las tareas de la conducción, incluso si el conductor no responde adecuadamente a la petición de intervención por parte de éste.	El conductor no es requerido durante el caso de uso.	Conducción longitudinal y lateral en todas las situaciones de un caso de uso definido	SISTEMA	SISTEMA	SISTEMA	ALGUNAS
5	CONDUCCION PLENAMENTE AUTOMATIZADA	El sistema de conducción automatizada desarrolla todas las tareas de la conducción bajo todas las circunstancias de la vía y ambientales.	N/A	Conducción longitudinal y lateral en todas las situaciones encontradas durante toda la prueba. No se requiere conductor	SISTEMA	SISTEMA	SISTEMA	TODAS

2.2.1 Non Autonomous Motor Vehicle, Not Fully Autonomous and Compulsory Insurance. Levels 0–4

The motor vehicle is conceptualised in the Regulation of the Law on Liability and Insurance in the Circulation of Motor Vehicles as follows (*Reglamento de la Ley de Responsabilidad y Seguro en la Circulación de vehículos de Motor*): Article 1.1. "Motor vehicles, for the purposes of civil liability in the circulation of motor vehicles and the obligation to insure, are considered to be all vehicles suitable for circulation on the surface of the land and propelled by motor, including mopeds, special vehicles, trailers and semi-trailers, the putting into circulation of which requires administrative authorisation in accordance with the provisions of the legislation on traffic, circulation of motor vehicles and road safety. Exempt from the obligation to insure are trailers, semi-trailers and special towed machines whose maximum authorised mass does not exceed 750 kilograms, as well as those vehicles that have been temporarily or definitively removed from the Vehicle Register of the Directorate General of Traffic".

According to art. 1.2 of this Regulation, the following are excluded: "The following are not included in the material scope of this Regulation: a) Railways, trams and other vehicles that run on roads belonging to them. b) Electric motor vehicles which, due to their design, destination or purpose, are considered to be toys, in the terms defined and with the requirements established in article 1.1 of Royal Decree 880/1990, of 29 June, which approves the safety standards for toys, and its concordant and development regulations (*Real Decreto 880/1990, de 29 de junio, por el que se aprueban las normas de seguridad de los juguetes, y su normativa concordante y de desarrollo*). Wheelchairs shall also not be included in the material scope of this Regulation".

The legal configuration of a vehicle as a motor vehicle entails the immediate consequence of the obligation to subscribe an insurance contract.[5] In the event that such a contract does not exist or is not in force, a public law entity, the Insurance Compensation Consortium (*Consorcio de Compensación de Seguros*), assumes the obligation to compensate for the damage caused (article 11 of the Law on Civil Liability and Insurance in the Circulation of Motor Vehicles *-Ley de Responsabilidad Civil y Seguro en la Circulación de Vehículos de Motor-*).

Instruction 15/V-113, of 13 November 2015, of the Directorate General of Traffic, on authorisation of tests or research trials carried out with automated driving vehicles on roads open to traffic in general (*Instrucción 15/V-113, de 13 de noviembre de 2015, de la Dirección General de Tráfico, sobre autorización de pruebas o ensayos de investigación realizados con vehículos de conducción automatizada en vías abiertas al tráfico en general),* differentiates between 6 levels of autonomous vehicles, in Annex I Table 1 attached below. The only category that corresponds to a fully autonomous vehicle is category 5.[6] The remaining categories—from 0 to 4—comply with the figure of the motor vehicle of the civil liability regulated in the Law on Civil Liability and Insurance in the Circulation of Motor Vehicles.

2.2.2 Fully Autonomous Motor Vehicle and Compulsory Insurance. Level 5

The only category of autonomous or driverless vehicle is that set out in category 5.[7]

The obligation to subscribe a compulsory insurance is also provided for in Instruction 15/V-113 of 13 November 2015 of the Directorate General of Traffic on the authorisation of tests or research trials carried out with automated driving vehicles on roads open to traffic in general (*Instrucción 15/V-113, de 13 de noviembre de 2015, de la Dirección General de Tráfico, sobre autorización de pruebas o ensayos de investigación realizados con vehículos de conducción automatizada en vías abiertas al tráfico en general*): "The owner of the autonomous vehicle or any person having an interest in its insurance shall be obliged to conclude and maintain in force an insurance contract covering up to the amount of the limits of

[5] Article 2. The obligation to obtain insurance. 1. Every owner of motor vehicles normally based in Spain shall be obliged to take out and maintain in force a contract of insurance for each vehicle he owns, covering, up to the amount of the limits of compulsory insurance, the civil liability referred to in Article 1. However, the owner shall be relieved of this obligation when the insurance is concluded by any person who has an interest in the insurance, who must state the concept in which he is contracting.

[6] *Apud thema*, Navarro-Michel (2020), pp. 945–949; Vicandi Martínez (2019), pp. 102–103; Hernáez Esteban (2019), pp. 197–243.

[7] On terminological issues, Barrio Andrés (2019), p. 101.

compulsory motor vehicle insurance, as well as civil liability arising from possible damage caused to persons or property as a result of the circulation during the performance of the tests on roads open to traffic in general".

The particularity of the case, as can be easily deduced, is the absence of a driver, which must be taken into special consideration for the purposes of determining the liable parties.

2.3 Fact of Circulation

The Regulation on Civil Liability and Insurance in the Circulation of Motor Vehicles (*Reglamento de Responsabilidad civil y seguro en la circulación de vehículos de motor*) defines the fact of circulation in Article 2: "For the purposes of civil liability in the circulation of motor vehicles and of the compulsory insurance coverage regulated in this Regulation, the facts of circulation are understood to be those arising from the risk created by the driving of the motor vehicles referred to in the previous article, both in garages and car parks, and on public and private roads or land suitable for circulation, urban or interurban, as well as on roads or land which without having such suitability are in common use."

3 The Legal Regime of the Obligation to Compensate for Damage Caused by Autonomous Vehicles Under Spanish Law

3.1 Legitimation to Claim for Damages

The entitlement to damages varies according to the property damaged. A distinction must be made between property and bodily assets.

3.1.1 Damage to Property

In the case of damage to property, the owner of the damaged property has the right to damages.

3.1.2 Personal Injuries

3.1.2.1 Damages in Case of Death

In the event of death, the deceased does not have entitlement, i.e. the damages do not become part of his inheritance. It is a constant doctrine of the Supreme Court, and it is so stated in the Law, that the damages must be claimed by the person who suffers the loss, which the Law legally predetermines in certain relatives, and with a presumptive character, in order to avoid unjustified enrichment. These are, without an excluding character: the widowed spouse or the surviving cohabitant *more uxorio*; and certain relatives, whether biological, adoptive or functional: the descendants—preferably children; ascendants—preferably parents; siblings and next of kin.[8]

3.1.2.2 Damages in Case of Personal Injuries

In the case of personal injuries, only the injured person is entitled to damages, not his relatives nor other third parties. Only in the event that the injured person deserves to be classified as seriously injured (who needs the help of third parties for essential daily tasks such as washing, dressing or eating as a result of the severity of the consequences) would his relatives be entitled to claim damages for moral harm consisting of psychological treatment up to the amount stipulated by law. For the remaining moral damage—loss of quality of life as a consequence of the suffering of the injured relative—the right to compensation corresponds surprisingly to the great injured party, who should receive these amounts.[9]

3.2 Liability: Liable Parties and Organisation of the Plurality of Debtors/Tortfeasors

3.2.1 Non or Not Fully Autonomous Vehicles. Levels 0–4

Spanish law imposes the obligation to provide damages in the event of a motor vehicle accident, in Article 1: firstly to the driver, regardless of whether or not he is the owner of the vehicle, as the material agent of the damage; secondly, to the owner of the vehicle, regardless of whether or not he is the driver, provided that the relations of dependence established in the case of tort in Article 1903 Civil Code, and in the case of a crime in Article 120.5 of the Criminal Code; and thirdly to the insurance company or, where appropriate, to the Insurance Compensation Consortium (*Consorcio de Compensación de Seguros*).

[8]Royal Legislative Decree 8/2004, of 29 October, which approves the revised text of the Law on civil liability and insurance in the circulation of motor vehicles (*Real Decreto Legislativo 8/2004, de 29 de octubre, por el que se aprueba el texto refundido de la Ley sobre responsabilidad civil y seguro en la circulación de vehículos a motor*), Articles 36 and 61 et seq.

[9]Royal Legislative Decree 8/2004, of 29 October, approving the revised text of the Law on civil liability and insurance in the circulation of motor vehicles (R*eal Decreto Legislativo 8/2004, de 29 de octubre, por el que se aprueba el texto refundido de la Ley sobre responsabilidad civil y seguro en la circulación de vehículos a motor*), articles 36, 52, and 93 et seq.

3.2.1.1 Driver, Whether or Not Owner of the Motor Vehicle

The law establishes the liability of the driver, irrespective of whether or not he is the owner of the vehicle.

This system is applicable to level 0 vehicles, and to not fully autonomous vehicles, levels 1 to 4, since in these vehicles there must be a person who takes control of the vehicle if necessary, and can therefore be considered as a driver.[10]

3.2.1.2 Owner of the Motor Vehicle, Whether or Not Driver

In the event that the damage does not constitute a crime, but rather a tort, the persons who have any of the relationships of dependence with the driver provided for in article 1903 Civil Code are jointly and severally liable with the driver. These are the legal representatives of minors or disabled persons, dependents, whether they are employees—hired personnel, or in the broad sense, according to the Supreme Court, as persons to whom the use of the vehicle is assigned free of charge.

In the case of crimes against road safety, provided for in the Criminal Code (Articles 379 to 385 ter), Spanish law distinguishes whether the criminal is of legal age or under 18 but over 14 years of age.

If the driver is of legal age, the persons mentioned in Article 120.5 of the Criminal Code are vicariously liable: "natural or legal persons who are owners of vehicles likely to create risks for third parties, for crimes committed in the use of those vehicles by their dependents or representatives or authorised persons".

If the driver is under 18 but over 14 years of age, the persons provided for in Article 61.3 of the Organic Law on the Criminal Responsibility of Minors (*Ley Orgánica de Responsabilidad Penal del Menor*) are jointly and severally liable with him/her: "When the person responsible for the acts committed is a minor under eighteen years of age, his/her parents, guardians, foster carers and legal or de facto guardians, in that order, shall be jointly and severally liable with him/her for the damage caused. When these have not favoured the conduct of the minor with wilful intent or serious negligence, their liability may be moderated by the judge according to the case".

3.2.1.3 Insurance Company

The insurance company with whom the owner contracted the compulsory insurance is jointly and severally liable with the other responsible parties, unless the vehicle has been stolen.

---

[10] Regarding the driver in vehicles that are not fully autonomous, and his role in the ultimate control of the vehicle, see Navarro-Michel (2020), pp. 950–952.

3.2.1.4 Insurance Compensation Consortium (*Consorcio de Compensación de Seguros*)

The Insurance Compensation Consortium (*Consorcio de Compensación de Seguros*) is jointly and severally liable with the liable parties in the event of the absence of a compulsory insurance contract in force, or in the event of theft. As a public law entity, it is represented by the State Attorney (*Abogado del Estado*).

3.2.2 Fully Autonomous Vehicles (Level 5)

Damage caused to third parties in their property or body by fully autonomous vehicles must be compensated, but the problem that arises is to whom the obligation to compensate is imposed. There is no driver in such vehicles. The owner of the vehicle uses the vehicle, generating a risk, however minimal it may be, but his liability cannot be that of liability for the driver's fault, since the driver does not exist, it would have to be of a strict liability nature. The insurance company or, if applicable, the consortium would have to be liable for the owner. It would also be possible to apply the manufacturer's liability regulated in the General Law for the Defence of Consumers and Users (*Ley General para la Defensa de Consumidores y Usuarios*). Each of them is examined in detail below.

3.2.2.1 Absence of a Driver

The driver in fully autonomous vehicles becomes a mere passenger. In this situation, who is liable?

3.2.2.2 Owner of the Motor Vehicle: Manufacturer of the Defective Vehicle?

The figure of the owner of the motor vehicle as the liable party is provided for in the Law on Civil Liability and Insurance in the Circulation of Motor Vehicles (*Ley de responsabilidad civil y seguro en la circulación de vehículos de motor*). In the case of a fully autonomous vehicle, he is not liable for the behaviour of the driver, since the latter does not exist. But it is clear that its ownership is over an asset that is potentially generating a risk, even if minimised, from which he obviously obtains advantages.[11] The tendency towards the strict liability carried out in similar cases by our Supreme Court could determine his liability.[12]

[11] On this topic, Vicandi Martínez (2019), pp. 105–108.

[12] This is how our scientific doctrine has stated it: Badillo Arias (2016), p. 203; Navarro-Michel (2020), p. 954.

The necessary reform of Spanish law on this point should perhaps involve the strict attribution of liability for these vehicles.[13]

Would the vehicle owner be exonerated from liability for damage resulting from a malfunction of the vehicle[14]?

Article 1.1.2. of the Spanish law states that "in the case of damage to persons, he shall only be exonerated from this liability when he proves that the damage was due to the exclusive fault of the injured party or to force majeure unrelated to the driving or operation of the vehicle; defects of the vehicle or the breakage or failure of any of its parts or mechanisms shall not be considered cases of force majeure".

In this way, it would perhaps be more in line with the protection of possible injured parties to impose liability on the owner, and if the defect of the vehicle is proven, the corresponding claim by the owner against the manufacturer.[15,16] For this purpose, as in the case of air navigation, a black box in the vehicle could be required to prove what happened in the accident.[17]

3.2.2.3 Insurance Company and, Where Appropriate, the Insurance Compensation Consortium (*Consorcio de Compensación de Seguros*)

We have already seen in previous lines the obligation to subscribe civil liability insurance in all these cases, which is to increase the guarantee of the injured party in the accident to receive compensation.[18]

3.3 Petitum

Spanish law does regulate the economic evaluation of damages, again distinguishing between property damage and personal injuries.

[13] This view has also been expressed in Spanish scientific doctrine: Álvarez Olalla (2017), pp. 1–7; Navarro-Michel (2020), pp. 955–956.

[14] On the concept of defect and defective product applied to autonomous vehicles, Hernáez Esteban (2019), pp. 197–243.

[15] Navarro-Michel (2020), p. 956.

[16] On the disadvantages of the application of the General Consumer and User Protection Law to this case, see Navarro-Michel (2020), pp. 943–956. Also, Hernáez Esteban (2019), pp. 197–243.

[17] Ibid. On the economic repercussions of the existence of the black box, also Hernáez Esteban (2019), pp. 197–243.

[18] In this respect see Vicandi Martínez (2019), p. 105.

3.3.1 Property Damage

Damage to property is assessed in accordance with Article 1106 Civil Code, distinguishing between consequential damage and loss of profit. Their evaluation is not particularly controversial.

3.3.2 Personal Injuries

The assessment of damages for personal injuries or death is also regulated in Spanish law, in the Legal System of Evaluation of Bodily Injury, contained in the Law on Civil Liability and Insurance in the Circulation of Motor Vehicles (*Sistema Legal de Valoración del Daño Corporal, contenido en la Ley de Responsabilidad Civil y Seguro en la Circulación de vehículos de motor*). It would not make sense to distinguish whether the perpetrator of the damage is a non-autonomous or not fully autonomous vehicle or a fully autonomous vehicle. It is a principle governing Spanish law, the *pro damnato* principle, which implies the need for full reparation of the damage caused to the injured party by the accident.

3.3.2.1 Damages in the Event of Death

Damages for death resulting from a motor vehicle accident under Spanish law are structured around three groups of items, regulated in Table 1 of the mentioned Act. The first is a basic compensation, the same for all types of persons who are of the same age and family characteristics as those described therein. To this item must be added the particular pecuniary losses, depending on the specific circumstances of the deceased, and a third one, consisting of property damage, consequential damages and loss of earnings.

If the victim was at fault, the total amount shall be reduced by a percentage in which the judge shall specify the causal participation of the victim in his own damage.

3.3.2.2 Damages in Case of Personal Injuries

Damages for personal injuries resulting from a motor vehicle accident under Spanish law are structured in two phases. The first is that of transitory injuries, from the date of the occurrence of the accident until the complete healing or consolidation of the injuries, a date which is fixed with the health report, with or without after-effects. If there are after-effects, a second phase of permanent injuries is opened. Both phases are assessed for compensation purposes. Again, three items in each phase: basic compensation, particular moral injuries and pecuniary damage.

If the victim was at fault, the total amount will be reduced by a percentage in which the judge specifies the causal participation of the victim in his own damage.

3.4 Period of Validity of the Action: Duration, Nature and dies a quo

The action for damages in these cases is subject to a limitation period of one year, if the damages derive from a civil wrong, or five years, if they derive from a crime.

If the damages derive from a civil wrongful act, the period of limitation shall be one year, the dies a quo being that of the date of production of the damage, in the case of damage to property or death; or that of the date of the health discharge report in the case of injuries, Articles 1968 and 1969 Civil Code.

In the case of damages arising from a crime, the limitation period is 5 years, starting from the dates indicated in the previous point, articles 1964 and 1969 Civil Code, if the action is brought before the criminal court. If the civil action is brought before the civil courts, as permitted by article 109.2 of the Criminal Code, the period begins when the criminal proceedings end, as established in articles 111–114 of the Criminal Procedure Act (*Ley de Enjuiciamiento Criminal*).

Legal Sources

Real Decreto de 24 de julio de 1889 por el que se publica el Código Civil.

Ley Orgánica 10/1995, de 23 de noviembre, del Código Penal.

Instrucción 15/V-113, de 13 de noviembre de 2015, de la Dirección General de Tráfico, sobre autorización de pruebas o ensayos de investigación realizados con vehículos de conducción automatizada en vías abiertas al tráfico en general.

Ley Orgánica 5/2000, de 12 de enero, reguladora de la responsabilidad penal de los menores.

Real Decreto Legislativo 8/2004, de 29 de octubre, por el que se aprueba el texto refundido de la Ley sobre responsabilidad civil y seguro en la circulación de vehículos a motor

Real Decreto 1507/2008, de 12 de septiembre, por el que se aprueba el Reglamento del seguro obligatorio de responsabilidad civil en la circulación de vehículos a motor.

Real Decreto Legislativo 6/2015, de 30 de octubre, por el que se aprueba el texto refundido de la Ley sobre Tráfico, Circulación de Vehículos a Motor y Seguridad Vial

Real Decreto de 14 de septiembre de 1882 por el que se aprueba la Ley de Enjuiciamiento Criminal.

Real Decreto Legislativo 1/2007, de 16 de noviembre, por el que se aprueba el texto refundido de la Ley General para la Defensa de los Consumidores y Usuarios y otras leyes complementarias

Versión consolidada de la Convención sobre la circulación vial de Viena del 8 de noviembre de 1968 [2014] Report of the sixty-eighth session of the Working Party on Road Traffic Safety, http://www.unece.org/fileadmin/DAM/trans/doc/2014/wp1/ECE-TRANS-WP1-145e.pdf

References

Doctrinal Sources

Álvarez Olalla P (2017) Desafíos legales ante la circulación de los coches autónomos: implicaciones éticas, responsabilidad por accidente y ciberseguridad. Revista Doctrinal Aranzadi Civil-Mercantil 2:1–7

Álvarez Olalla P (2019) Responsabilidad civil en la circulación de vehículos autónomos. En: Monterroso Casado (Dir.)/A. Muñoz Villareal(Coord.), Inteligencia artificial y riesgos cibernéticos. Responsabilidades y aseguramiento, Tirant lo Blanch, Valencia, pp 145–170

Badillo Arias JA (2016) La responsabilidad civil automovilística. El hecho de la circulación, Thomson-Reuters Aranzadi, 2016

Barrio Andrés M (2019) Consideraciones sobre el coche autónomo. Actualidad Jurídica Uría Menéndez 14:101–108

Hernáez Esteban E (2019) Inteligencia artificial y vehículos autónomos: el régimen de la responsabilidad civil ante los nuevos retos tecnológicos. Revista Aranzadi de Derecho y Nuevas Tecnologías 49:197–243

Macías Espejo B (2020) Criminal compliance program y exención de responsabilidad penal ante la hipotética, si no inminente, industrialización del coche sin conductor. Revista Aranzadi Doctrinal 6:1–22

Navarro-Michel M (2020) La aplicación de la normativa sobre accidentes de tráfico a los causados por vehículos automatizados y autónomos. Cuadernos de Derecho Transnacional 12(1): 941–961. https://doi.org/10.20318/cdt.2020.5231

Vicandi Martínez A (2019) El contrato de seguro de automóvil frente a los coches autónomos: luces y sombras de los *smart cars*. Revista De Derecho, Empresa y Sociedad 14:101–109

Italian Civil Liability Applicable to Self-Driving Cars

Maria Cristina Gaeta

Abstract The development of vehicle automation systems has reached high levels and is projected, due to technological evolution, towards even faster progress. The introduction of autonomous vehicles on the market entails considerable advantages but it also poses a challenge to current regulation on civil liability.

The present contribution, focusing on Italian law, investigates problems concerning civil liability caused by traffic accidents between two vehicles, where at least one is self-driving. The chapter will offer a critical analysis of the current regulation on automobile accidents, with the view of evaluating whether Italian civil liability law offers adequate protection to the victim in case a self-driving car is involved. More precisely the chapter will verify whether the current Italian law provides a comprehensive regulation on autonomous vehicle accidents or, rather, if it is necessary to modify the existing regulation in order to identify more clearly the figure of the tortfeasor in such cases.

1 Introduction

The spread of new technologies requires a **rereading** of traditional **legal categories of civil** law, and, at times, **the creation of new ones**. Among the various legal categories, civil liability schemes can be altered by autonomous artificial agents in everyday relationships,[1] although the principles of civil liability remain currently unchanged in the Italian legal system. [2]

[1] Asaro (2016), p. 191.

[2] Irti (1990), p. 150, underlines that the flexibility of legal principles, which involves the construction of new civil law institutions, accompanies the interaction between civil law and technological society. Perlingieri (2015), pp. 1235–1246 and Stradella and Palmerini (2013), analyse the impact of technologies on legal relationships.

M. C. Gaeta (✉)
Department of Legal and Economic Sciences, Università degli Studi Suor Orsola Benincasa, Naples, Italy
e-mail: mariacristina.gaeta@unisob.na.it

H. Steege et al. (eds.), *Autonomous Vehicles and Civil Liability in a Global Perspective*, Data Science, Machine Intelligence, and Law 3,
https://doi.org/10.1007/978-3-031-41992-8_15

Within this perspective, **driverless cars pose a distinct challenge** as their nature might imply the **rethinking of the subject responsible for the damages caused by (semi)autonomous vehicles accidents. Traditionally, this subject** has always been the driver (and/or the owner) of the vehicle but, as it will be seen, in the case of self-driving cars also the system supervisor on board could be responsible and not the **common driver**. [3] Furthermore, the gradual **transition towards the owner liability system** is unavoidable, alternatively or jointly and severally with the **liability of the producer.**

The introduction of autonomous vehicles on the market entails considerable advantages but it also poses a challenge to the current regulation on civil liability. In Italy, the existing laws do not directly address the problem of **liability in the event of a car accident involving at least one (semi)autonomous vehicle**. Even if the civil liability for road traffic can still be applied to semi-autonomous vehicles (in whole or in part), there is **no comprehensive regulation** that allows to properly identify the relevant liabilities and/or the tortfeasor.

2 Civil Liability in Relation to Levels 0–2 of the SAE J3016 Standard for Driving Automation

2.1 Overview of the Italian Tort Law: From the Function of Liability to its Classification

In order to examine if and how the circulation of autonomous vehicles can affect the rules of tort law on road traffic, it is necessary to start from a **critical analysis of these rules concerning the development of self-driving cars**. In this way, *(i)* it will be possible to **ascertain whether the current rules on civil liability are adequate in offering effective redress** to the damaged party, and, if this is the case,

[3] Convention on Road Traffic, Vienna 8 November 1968. In 2014 the UN Working Party on Road Traffic Safety (WP1) proposed amendments, among others, to Article 8 and Article 39 of the Vienna Convention, with the objective of enhance and ensure that safety rules do not hamper the advancement of new technologies aimed at improving road safety. According to the amendment, 'systems which influence the way vehicles are driven', are deemed to be in accordance with Article 8 of the Vienna Convention. The amendment was adopted by the UN on 23 September 2015 and entered into force on 23 March 2016. It was submitted also by the Italian government, showing the extent to which, the new technology is important also for Italy. The amendment seems to be a step forward in the admission of semi-autonomous vehicles, but is still far from that of fully autonomous vehicles, *see* United Nation Economic and Social Council, Economic Commission for Europe, Inland Transport Committee: Working Party on Road Traffic Safety (WP1), Report of the sixty-eighth session of the Working Party on Road Traffic Safety, 17 April 2014, available at https://www.unece.org/fileadmin/DAM/trans/doc/2014/wp1/ECE-TRANS-WP1-145e.pdf;Working Party on Road Traffic Safety (WP1), Report of the Seventy–second session of the Working Party on Road Traffic Safety, 19 April 2016, available at https://www.unece.org/fileadmin/DAM/trans/doc/2016/wp1/ECE-TRANS-WP.1-153e.pdf.

(*ii*) *whether* **current civil liability rules can regulate civil liability for autonomous vehicles** (possibly through extensive or analogical reasoning, according to art. 12 of the Italian preliminary disposition to the Italian Civil Code), or it is necessary to modify the existing legislation by adopting an *ad hoc* framework of rules.

In the Italian Legal System, the current rules governing tort law are composed of statutory rules, general principles,[4] and case law.[5] The set of statutory rules include the Constitution (in particular articles 2, 3 and 32),[6] the Civil Code (articles 2043 and following),[7] and the special laws (first of all Consumer Code,[8] and Private insurance code[9]).[10] About the civil code, **article 2043** establishes the general rule for tort law

[4] As for the general principles, we are dealing with legal provisions of general content which express values to which other legal provisions, called rules are conformed. While the rules directly influence the behaviour or attribute regulatory powers, the general principles indirectly influence the behaviour, i.e. influencing the exercise of regulatory powers. Among the most important general principles in the field of civil liability, we must underline that of accountability, about the assumption of the consequences of one's own actions towards third parties and also towards oneself. In this field, another fundamental principle is the principle of '*neminem leadere*', which contains in itself the provisions of Article 2043 of the Italian Civil Code regarding the obligation to compensate damages, but also arises just as a general principle consisting in the prohibition of unlawful acts. *See* Alpa (2018), pp. 37 ff.

[5] In the Italian legal system, unlike what happens in the common law system (which is founded on the well-known '*stare decisis*' principle), the case law is not binding. However, judicial decisions are often taken into consideration by other judges who at a later date have to decide similar cases.

[6] Italian Constitution, in force since 1 January 1948. The article 2 of the Italian Constitution, on the fundamental rights, which is usually applied in connection with article 3 of the Constitution which regulates the principle of equality to decide on the compensation for the biological damage suffered (a type of non-material damage). Article 32 of the Constitution, on the other hand, protects physical integrity, that is, health intended as an asset and, consequently, any injury to it must be compensated. An authoritative doctrine, dealing with the overall picture that has been outlined in the matter of non-pecuniary damage, after the change of address of the Italian Supreme Court with the judgment no. 1361/2014 (which has established that, if the damage affects a person's constitutional right, it is a cause of compensation), believes that non-pecuniary damage returns to being, as in the past, the instrument to compensate for the infringement of non-pecuniary rights. *See* Franzoni (2010); Italian Supreme Court, civil section III, judgment 23 January 2014, no. 1361, that recognized for the first time, explicitly, the right to compensation for the damage to life or the damage from death of the victim that is transmissible to the heirs.

[7] Italian Civil Code, R.D. 16 March 1942, n 262, OJ 79.

[8] Italian Consumer Code, D.lgs. 6 September 2005, n 206, OJ 235.

[9] Italian Private Insurance Code, D.lgs. 7 September 2005, n 209, OJ 239.

[10] It is useful to highlight in this regard three important differences between the Civil Code and the special legislation of the Consumer Code, elaborated by Castronovo. Firstly, art 125 of the Consumer Code provides that the right to compensation for damages is prescribed in three years from the day on which the injured person was aware of the damage or the identity of the tortfeasor, or could have had it using the normal diligence; the civil code, on the other hand, provides for a five-year prescription period in the case of non-contractual civil liability (art 2947 Civil Code) and, in the case of contractual civil liability, the ordinary 10-year limitation period (art 2946 Civil Code). Secondly, art 126 of the Consumer Code provide a 10 year right to right to take legal action after the product has been put into circulation; this is not provided in Civil Code. Finally, Consumer Code, in art 121, introduces a novelty with respect to the civil code (in particular art 2055, civil code) regarding the right of claims based on joint and several liability, because Consumer Code provides that action against the others who are jointly and severally obliged has to be exercised only in proportion to the size of the risk attributable to each one. Castronovo (2006), p. 732.

in the Italian legal system, which recognises civil liability in '**any fault or negligent act, which causes unjust harm to others'.** Given the abstract nature of this formulation, the provision is considered by the academic literature as the general rule applicable to all cases of unlawful damage. In applying this rule, the judge will have to determine whether (given consideration to current societal values) a specific behaviour can be considered an unlawful act, after verifying the occurrence of all the structural elements of liability (objective and subjective requirements)[11] identified by article 2043.[12]

Civil liability, in the Italian legal system, has mainly a **compensative function** (i.e. prearranged to economically repair the damage suffered), in compliance with the duties of solidarity imposed by the Italian Constitution. At the same time, however, the injured party deserves appropriate compensation, which also considers the actual extent of the damage that he/she has suffered, in particular in relation to the development of new technologies, that place individuals in a position of 'technological vulnerability'. In this light, the current function of tort law is slowly changing. Currently, **the function of civil liability can have not only compensative but also, in some cases, a punitive function,** intended as punishment for the unlawful act of the tortfeasor and not mere restoration for the damage suffered by the damaged party.

Concerning the **types of civil liability**, Italian legislation contemplates different kinds of liability on the base of the fault. Indeed, alongside the traditional **liability based on the fault of the agent** (as for Article 2043 of the Italian Civil Code), there are cases of semi-objective liability (e.g. driver liability in compliance with art. 2054, para 1, Italian civil code), where the fault is presumed, as well as cases of **strict liability or objective liability or liability without fault** (e.g. art. 2054, para 4, Italian civil code on the liability of the driver and the other individuals listed in the same article, in case of a **maintenance defect,** or the producer liability in case of production defect according to Italian consumer code), where the subject who has no **fault for the harmful event has nonetheless to compensate the damage.**[13] In addition to the liability scheme classification on the base of the fault, the Italian legal system also provide a classification on the base of the direct or indirect action of the damaging party, that distinguish direct liability from indirect liability (e.g. art. 2054, para 3, Italian civil code on the liability of the owner of the vehicle, or art. 2049 of

---

[11] The objective ones are: unlawful act, unjust damage, causal link between the fact and the damage; instead, the subjectives are the guilt of the agent (for intentional fault or misconduct) and the imputability of the tortfeasor (that means the lack of natural incapacity under article 2046 Italian Civil Code).

[12] When resorting to any legal requirement, it will always be up to the judge to quantify the amount of the compensation, given that the article 2059 of the Italian Civil Code legitimates the injured to claim compensation for the negative consequences, including non-patrimonial ones.

[13] On product liability in the field of new technologies *see* Montinaro (2020), pp. 365 ff.; to in-depth product liability: Simonini (2017), pp. 557ff.; Recinto and Porcelli (2014), p. 449; Alpa et al. (1989), pp. 56 ff; Mazzon (2012), pp. 1063 ff.

the Italian Civil Code on the vicarious liability).[14] According to **indirect liability a specific subject is considered liable, even if he/she has not materially caused the damage.**

2.2 *Identification of the Tortfeasor in the Case of Human-Driven Vehicles up to Level 2 of Driving Automation of the SAE J3016 Standard*

The new scenarios of tort law testify to the need for a progressive transformation of the rules of liability also in Italy.[15] In this context is placed **article 2054 of the Italian Civil Code**, which is the basic provision of **Italian civil liability for damages caused by road accidents**. However, this disposition is not sufficient in regulating all of the new cases of civil liability that might be linked with the fast development of the technologies onboard the vehicles. For this reason, in this paragraph a de *iure condito* analysis of the Italian legislation on road traffic liability is carried out, starting from the driver's liability regulated under article 2054 of the Italian Civil Code. The aim is to evaluate how current law can be applied to semi-autonomous vehicle accidents.[16]

Article 2054, para 1, of the Italian Civil Code establishes that the driver is liable for the damage caused by the circulation of a vehicle without a rail guide, unless it is proven that he/she has done everything in his/her power to prevent the occurrence of the damage.[17] This is a particular type of civil liability, so-called **semi-objective**

[14]Vicarious liability is the responsibility of any third party that had the right, ability or duty to control the activities of someone else. A typical example is the liability of the employer for the acts of his employees. As he had achieved a return from employing individuals who contribute to the running of his business activity, he is also called to bear the costs of the damage that those persons produced to third parties when operating in his interest. As a matter of fact, the fault of selecting the given collaborator (i.e. the so-called '*culpa in eligendo*'), which implies a direct liability is a thesis outdated. Another example of vicarious liability could be the liability of parents or teachers for the acts of children or students, which came from the social risk of the role they play. However, another part of the literature believes that parents' liability or teachers' liability is a type of direct liability, as they have to supervise (not at any time) and educate their children/students and so influence their behaviour (i.e. the so-called '*culpa in vigilando*' or '*culpa in educando*'). It is undoubtedly to be included in the category of vicarious liability, instead, the liability of the supervisor for damage caused by the subject with lacking mental capacity, since it is not a legally responsible subject. Bianca (2021), pp. 662 ff.

[15]For a more detailed analysis on Italian liability regulation, *see* Gatt et al. (2020), pp. 239 ff.

[16]For an analysis of art. 2054 of the Italian Civil Code in relation with the principles of product liability in the field of autonomous vehicles, *see* Albanese (2019), pp. 996–1044; Pellegatta (2019), pp. 135–162; Losano (2019), pp. 423–441.

[17]Buffone (2016), pp. 95 ff.

liability, which provides for an **inversion of the burden of proof** established by Article 2697 of the Italian Civil Code.[18] Therefore, pursuant to the first paragraph of the article 2054, the driver's fault is presumed and it is its duty proving that he/she has done everything possible to avoid the damage.[19]

To understand the particular liability of the driver, it is first necessary to define his/her role. The **driver is the individual who has the control of the vehicle**. Therefore, the driver will answer for damages caused by the vehicle even if this has been, for example, parked without handbrake on a sloping road or has been parked in the middle of a roadway.[20] However, it is no longer considered as a driver the individual who has given the driving controls to others, even if temporarily and under his supervision. Between the case in which the **vehicle** is considered **under the control of the driver** and the one in which it is **not under the driver's control**, there is the case of **semi-autonomous cars where the driver is** inside the vehicle **but the driving is exercised to some extent by a driving system.** Depending on which driving tasks are exercised by the autonomous system the liability of the driver can be limited or excluded.[21] In the case of completely autonomous vehicles,[22] however, it seems logical to compare all the people in the vehicle to passengers of a train or an aeroplane: as for the passenger of the train or plane there is no liability in case of an accident, **so the passenger of the driverless car will not be held liable in the event of an accident.**[23]

In addition to the liability of the driver, it is necessary to consider that **article 2054, para 3, Italian Civil Code**, also provides for the **liability of the owner of the vehicle or of the subject with an equivalent legal status**. With the **progressive automation** of vehicles, we are moving further and further away from the liability of the driver while that of **the owner is still central, but there are cases in which both liabilities are predictable at the same time:** it is the case of **joint and several liability with the driver of the owner of the vehicle.** The '*ratio*' of owner

---

[18] Bianca (2021), pp. 550 ff.

[19] It should be noted that the burden of proof of the driver having acted diligently is valid only for damages caused to things or third parties unrelated to road traffic, instead, in the case of collision between two or more vehicles, the driver must also prove the fault of the other driver(s) (so-called presumption of equal co-liability, art. 2054, para. 2, Italian Civil Code). Moreover, in order to overcome the presumption of equal co-liability, it is also necessary to provide proof of the unpredictability of the conduct of the other driver. Italian Supreme Court, section III, Judgment, 14 October 2015, no 20618. For an in-depth analysis of the topic Buffone (2016), pp. 56 ss.

[20] Italian Supreme Court, section III, Judgment, 13 January 2015, no. 281.

[21] Bertolini (2020); Bertolini and Riccaboni (2021), pp. 243 ff.; Bertolini (2013), pp. 277 ff.

[22] Bertolini (2013), pp. 214 ff., specifies that even fully autonomous vehicles are object and not subject of law. Precisely these are products created by the work of man in order to satisfy certain human needs and, therefore, in the event of a defect, the liability would be attributed to the producer of the same. The EU approach concerning thee personhood is well analysed by Bertolini (2020), pp. 35 ff.

[23] Palmerini and Bertolini (2014), pp. 112 ff., they clarify that the problem of liability in road traffic is perhaps the most relevant issue in relation to driverless cars, as it could have a strong technology-chilling effect, delaying their placing on the market.

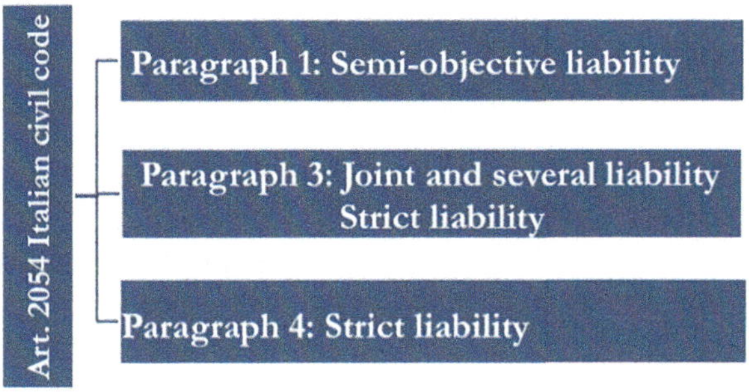

Fig. 1 Scheme of article 2054 of the Italian Civil Code

liability lies in the fact that the damage falls within the sphere of freedom of action of the owner and of the other mentioned subjects in the paragraph 3 who, in addition, are those who can **bear the economic loss of paying the damage occurred by road accident (i.e. deep pocket concept).**[24] Anyways, **the driver, the owner**, and the other subjects similar to them, have a **right of recourse against the producer if the vehicle accident depended on a vehicle defect**.

Another case of strict liability is that described in **paragraph 4 of article 2054** Civil Code (see Fig. 1) which establishes the objective responsibility of the driver and of the other persons listed in the same article, in case of a **production defect or maintenance defect**.[25] The Italian Supreme Court has stated that the strict liability of the driver and of the owner (as well as of the other mentioned subjects) is joint and several to the producer's liability (regulated under Italian Consumer Code and the

[24] Deep pocket is a concept used in the law and economics of tort law and refers to the idea that the risk of an activity (the compensation of damage) should be borne by the individual who is in a good position to handle it. This principle is applicable to owner's liability, as well as to insurer's liability and is independent for the fault criterion.

[25] The Italian Supreme Court, not distinguishing between maintenance defect and construction defect, qualifies the liability that derives from it in terms of strict liability, while admitting that the causal link between the defect of construction or the lack of maintenance, on the one hand, and the unjust damage, on the other hand, can be interrupted if an external factor occurs that involves the occurrence of the damage (Italian Supreme Court, Judgment, 9 March 2004, no. 4754). Moreover, With regard to the burden of proof, the Supreme Court has made it clear that it is task of the damaged party, the burden of proving the existence of the construction defect or maintenance defect and the related causal link with the damaging event, while the driver and the owner to get free must prove that the damage was due to a different cause from those listed in the last paragraph of the article 2054. Italian Supreme Court, section III, Judgment, 19 February 1981, no. 1019.

PLD[26]), according to article 2055 of Italian Civil Code (even if the sources of liability are different).[27]

In this regard, currently, the **damages caused by the defects of the vehicle intelligent systems** are regulated under paragraph 4, article 2054, Italian Civil Code only in a general way. In such cases, **precise rules for the liability of the producer** should be established[28]

At the end of the analysis of the **existing Italian legislation on the liability** of the driver, the owner and the subject with equivalent legal status, and the producer, it appears clear that the exsistent regulation can be applied up **to level 2 and, as it will be seen, also to the intermediate automation level**. For level 5 of automation, instead, **a modification of the existing legislation** or **even the introduction of an *ad hoc* framework of rules** should be provided.

In order to be able to assess which is the right way of regulating autonomous vehicles, it is necessary to test them on the road, as it is not possible to regulate a legal case that is not known. For this reason, **in Italy in 2017, Article 1, para 72 of law no. 205/2017** entered into force authorising **the testing** of **connected driving solutions on 'Smart Roads'**.[29] The law called for the introduction of a decree of the Minister of Infrastructure and Transport for the implementation methods and operational tools of the experimentation.

In **2018** (with a little delay compared to what was required at the legislative level) the **Ministerial Decree on Smart Road** was published.[30] The decree had a dual purpose. On the one hand, it aimed at promoting the technological adaptation of the infrastructures according to the 'Smart Roads' rules, in line with European and International requirements. On the other hand, the decree had the objective of

---

[26]The Product Liability Directive (PLD) was implemented in Italy by the d.P.R., 24 May 1988, n. 224, Implementation of the EEC directive n. 85/374 relating to the approximation of the laws, regulations and administrative provisions of the Member States regarding liability for damage caused by defective products, pursuant to art. 15 of the law of 16 April 1987, n. 183, and then repealed by d.lgs. 6 September 2005, n 206, OJ 235 (Italian Consumer Code), which at articles 114 and following, incorporates the provisions of the PLD. For the analysis of PDL *see* Machnikowski (2016). Currently the PLD is under revision. *See* European Parliament and Council, Proposal for a Directive on liability for defective products, 28 September 2022, COM(2022) 495 final.

[27]Italian Supreme Court, section III, Judgment, 9 March 2004, no. 4754; Italian Supreme Court, section III, 6 August 2004, Judgment, no. 15179; Carnevali (1974), p. 42, specifies that, although the prediction of the liability of the driver, the owner, and of the other subject mentioned in the third paragraph of the article 2054 has the purpose of facilitating the injured in obtaining compensation, it is not possible to exclude any liability of the producer for vehicle defects.

[28]Al Mureden (2012), pp. 1506 ff. Most recently: Al Mureden (2019), pp. 895 ff; Al Mureden and Ruffolo (2019), pp. 1704 ff.

[29]Italian Law of 27 December 2017, no. 205, OJ 302. The Italian regulation in this field started with the adoption of the National Action Plan on Intelligent Transport Systems – ITS (D.M. 12 February 2014, n 44).

[30]Italian Ministry of Infrastructure and Transport, Decree of 28 February 2018, n 70, OJ 90, on the implementation methods and operational tools of road testing of Smart Road solutions and connected and automatic driving. On Smart Roads decree, *see* Losano (2019), p. 423.

guaranteeing and promoting autonomous and connected vehicle testing, in consideration of the possibility that autonomous vehicles will soon be introduced on the market. In particular, the Decree establishes **the role of the supervisor**, starting from the assumption that in the highest automation level, there is no real driver but only a supervisor who must be able to resume control of the vehicle if needed. Given the impossibility to apply the rules of the Italian Traffic Code, which does not allow self-driving cars' road circulation, the decree constitutes the legal reference for driverless car testing on specific Italian roads and public areas, which can only take place within the established conditions.

After this first step forward, another significant Italian initiative was the signature of the **Memorandum of Understanding** between the Ministry for Technological Innovation and Digitization, and the Ministry for Infrastructure and Transport, on **May 2020**.[31] The aim of the memorandum of understanding was the development of innovative mobility through research and testing of autonomous and connected vehicles. The memorandum has the objective of developing and supporting applied research, experimentation and prototyping, production and training of new professionals in the field of innovative vehicles, as well as creating interest in services with a social impact on the Country. The agreement also supports and encourages collaboration and partnership between public institutions, companies, universities, and research institutions on smart vehicle testing.

3 Civil Liability in Relation to Levels 3 and 4 of the SAE J3016 Standard for Driving Automation. The Application of the Existing Legislation to (Semi)autonomous Vehicles

The question of the identification of the tortfeasor in case of a road accident, as has been shown, is not easy to solve at all. As demonstrated in several studies, starting from 2011,[32] usually traditionally road accidents are due to incorrect driver behaviour, while rarely they are a consequence of vehicle defects. On the other side, in the

[31] Italian Ministry for Technological Innovation and Digitization, and Italian Ministry of Infrastructure and Transport, Memorandum of understanding on innovation for autonomous and connected driving in urban and suburban areas, available at https://www.mit.gov.it/sites/default/files/media/notizia/2020-05/Guida_Autonoma_MID_MIT_.pdf, accessed on 15 June 2022. Still on the topic of innovation, but not strictly on autonomous vehicles, we report: Working Group on Artificial Intelligence and Blockchain, appointed by the Italian Ministry of Economic Development, on Proposals for an Italian strategy for artificial intelligence, 2019, available at: https://www.mise.gov.it/images/stories/documenti/Proposte_per_una_Strategia_italiana_AI.pdf, accessed on 15 June 2022; Italian Ministry for Technological Innovation and Digitization, Strategy for technological innovation and digitization of the country 2025, available at: https://docs.italia.it/italia/mid/piano-nazionale-innovazione-2025-docs/it/stabile/index.html, accessed on 15 June 2022, 2020.

[32] DEKRA Study (2011) La sicurezza stradale di uomo e tecnologia. Strategie politiche di prevenzione, available at: https://www.dekra-roadsafety.com/media/dekra-vsr-2011-it.pdf, accessed on 15 June 2022; Kalra N, Anderson J, Wachs M (2009) Liability and Regulation of Autonomous Vehicle Technologies, Study report of Berkeley University of California, available at: https://www.rand.org/pubs/external_publications/EP20090427.html, accessed on 15 June 2022.

Fig. 2 Types of product defects

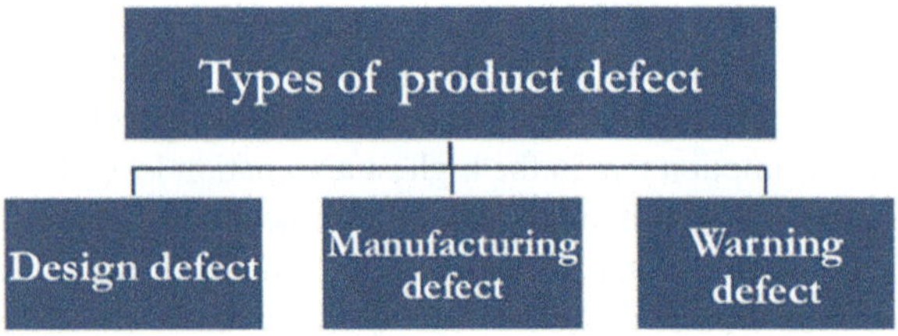

case of (semi)autonomous vehicles, it is logical that the producer liability is more frequent than that of the driver or the owner, even if their joint and several liability with the producer can be applicable in the event of a vehicle defect (cases that increase proportionally with the level of automation). In any case, the producer and the owners of a completely autonomous vehicle appear to be the main ones responsible for the accidents.

Concerning the producer liability, as already indicated, the Italian regulation is now included in the consumer code. More in detail, Article 114 cons. code (art, 1 of the PLD) provides that 'The producer shall be liable for damages caused by a defect in his product'.

The production is composed of three different phases: the design phase, the manufacturing phase, and the information phase. For each of these phases, different rules are set and have to be followed by the producer. Failure to comply with these rules entails the general duty to pay for any damage caused as a result of a defective product—in our case an ADAS (Advance Driver Assistance Systems)[33] or the onboard intelligent system of the autonomous vehicle (see Fig. 2).[34]

---

[33]To an in-depth analysis of the ADAS definition and functioning, please see the technical introduction to this book.

[34]In this regard, the so-called Machinery Directive 2006/42/EC intervened to harmonize the health and safety requirements that machinery must have and to allow the free circulation of products conforming to it within the European market. Autonomous vehicles would fall under the category of 'machines' provided by article 1, letter a) of the Directive and explained in its art 2. It regulates the obligations that producers must respect in order to be able to place the product on the market (Article 5, Directive 2006/42/EC). In particular, the producer must ensure that the machinery meets the essential health and safety requirements and, in addition, must provide the consumer with all the necessary warnings. Before being placed on the market, the product must positively pass the procedures for assessing the conformity of machinery provided in article 12 of the Directive. At the end of these procedures, according to Article 16 of the Directive, the manufacturer can issue the EC marking as declaration of conformity. Regulation of the European Parliament and of the Council 765/2008/EC of 9 July 2008 setting out the requirements for accreditation and market surveillance relating to the marketing of products and repealing Regulation (EEC) No 339/93 [2008] OJ L 218/30, regulates the marking procedure, by which the producer assumes responsibility for the conformity of the product with the EU rules. Further obligations on producers are provided by the Directive of the European Parliament and of the Council 1999/44/EC of 25 May 1999 on certain aspects of the sale of consumer goods and associated guarantees [1999] OJ L 171/12.

During the design phase, the producer-designer should evaluate the actual amount of information that the ADAS should provide to the driver, inserting only those necessary to make the driver understand what to do and additional/superfluous information that would only create confusion.[35] Secondly, the designer should conduct an ergonomic analysis of the intelligent devices, i.e. studies on the ADASs and the functioning for which they are designed, in order to improve the overall performance of the system and driver's satisfaction. The analysis of the acceptance by the driver of the automatic systems, based on the human-robot interaction (HRI) analysis, is also extremely important, with regard to the behavioural changes of the driver who interacts with these new technologies. Concerning the production phase, the assistance systems must be included in vehicle approvals, implying that, if approved, the producer has been compliant with the safety standards. Finally, concerning the information phase, instructions must be clear and easy to be understood by the category of the average consumer.

In level 4 of driving automation, it is therefore evident that, as a general rule, the liability rarely falls on the driver, and it is limited to the cases of a driving error, and more frequently falls on the owner for the only fact to have bought the vehicle and may or may not be inside it at the time of the accident; in any case, the manufacturer will be liable in the event of a fault of the driving system. As will be demonstrated, according to an economic analysis of non-contractual civil liability, the risk of the accident and the consequent compensation should fall on the subject who can economically bear it. [36] In a logic of the correct allocation of risk, it is evident that the owner of the vehicle, that is the person who purchased it, is economically stronger than the driver of the vehicle (when the two subjects are different) and is subject to held liable, but the driver's liability (if existent) remains in the case of accident due to his/her negligent behaviour. Then, with the achievement of the highest level of automation, the liability progressively passes on the producer.[37] As a matter of fact, the producer can bear the economic risk of damage, even better than the owner of the vehicle, also thank to adequate professional insurance. In this case, to not distort the market and frustrate any car manufacturer which wishes to produce automated vehicles, the cost of the manufacturer insurance should be spread between the manufacturer who gets the insurance coverage and the owner of the vehicle who pays the price of the vehicle, which is proportionally augmented covering a part of the insurance policy cost, as well as a more performing vehicle.

[35] Regarding the driving of an autonomous vehicle, requirements for driving license should also be changed, including the understanding and interaction with the ADAS and the onboard computer, characterised by a more or less high level of AI.

[36] Calabresi and Melamed (1972), pp. 1096 f., state that the ability of the tortfeasor to prevent a given damage is taken into account, together with other conditions such as the ability of the same party to lower transaction costs.

[37] The choice of producer's liability could be justified based on the ability of the producer to (better) insure himself and therefore handle costs associated with the production and distribution of the product (including transaction costs). Castronovo (2006), p. 687.

4 Civil Liability in Relation to Level 5 of the SAE J3016 Standard for Driving Automation: Self-Driving Cars as a Product

4.1 The Extensive Application of Existing Legislation versus the Need for a Specific Regulation to Identify the Tortfeasor in Case of Self-Driving Cars Accidents

Concerning level 5 of driving automation, it is of fundamental importance to identify the hypotheses in which the owner's liability is replaced with those of producer liability, **considering the self-driving cars as a product** and **the producer as the stronger party of the non-contractual relationship**.

In relation to the application of producer liability to fully autonomous vehicles, doubts also arise regarding the consumer code, because it still does not properly take into account the impact that new technologies had in terms of new damages and the new party involve. Indeed, the consumer code does not keep up with the times. For this reason, it appears necessary to carry out a *de iure condendo* analysis to verify if it appears necessary to modify the existing regulation or introduce an *ad hoc* one for self-driving cars regulation in Italy.

At the European level, there was initially a conservative attitude towards product liability legislation (PLD), which was also confirmed in Italy. In particular, in 2018, the EU Commission considered the PLD an adequate regulatory tool, even though today's products are much more complex.[38] However, the same Commission observed that it is necessary to clarify the legal understanding of certain concepts applied to new technologies (such as product, producer, defect, and damage) and look closely at certain products such as autonomous vehicles, which may pose a challenge to the performance of the Directive.

Autonomous vehicles are robots equipped with artificial intelligence.[39] For this reason, they fall within the product category, but the issue concerns the relationship between the product and the autonomous vehicle, as the latter is a product that has considerable peculiarities, first and foremost that of the AI.

In this regard, it should be verified if the producer can be held liable for an accident brought about by a self-driving car. As a matter of fact, even the ability to intelligently interact with the surrounding environment and, at the higher automation level, the ability to learn, is not enough to identify the robot (and so the self-driving car) as a subject whose actions could be considered the consequence of self-determination and awareness. Even if the robot can, in some cases, be autonomous

---

[38] European Commission Report and Staff Working Document, Liability for damage from defective products (Directive 85/374/EEC), 7 May 2018, COM 2018/246 final and SWD/2018/157.

[39] On the legal aspects of robotics and Artificial Intelligence in Italy and Europe please refer to: Alpa (2020); Amidei (2019), pp. 1715 ff.; Capilli (2019), pp. 621–631; Costanza (2019), pp. 1686–1689; Pagallo (2017), pp. 615 ff.; Procida Mirabelli di Lauro and Feola (2020), pp. 508–558; Ratti (2020), pp. 1175 ff.; Ruffolo (2020a, b).

and not predetermined in its interaction via the software algorithm, from a legal point of view it is still an object and, more precisely, a product.[40]

This clarification helps to evaluate the producer liability regulation applicable to self-driving cars, but it remains to be determined whether to apply the consumer code (PLD at EU Level) by legal analogy or extensively.

With regard to the application of producer liability to self-driving cars, the issue concerns the adequacy of the product liability rules to smart vehicles.[41] The producer liability, under some concepts and certain terms already individuated, is ineffective to regulate autonomous vehicles organically. In this context, the possible solution could be the extensive interpretation[42] of the existing legislation on product liability in order to fully re-enter in the scope also the driverless car. *Analogia legis* is not the solution as a legal regulation already exists, even though should be interpreted extensively to be applied also to autonomous vehicles.[43]

At the same time, in a scenario characterized by legislation that does not properly keep up with the technological era, it seems extremely important the recent intervention of the European legislator, for classifying AI (AI Act proposal)[44] and regulating AI liability (AILD proposal),[45] as well as for revising the PLD,[46] in order to identify the new liability profiles deriving from accidents caused by AI, in some cases, such as that of autonomous vehicle, intended as examples of AI embedded in robotics. However, in specific sectors, such as automotive, we need sector-specific laws for self-driving cars: the differences between the different

---

[40] Bertolini (2013), pp. 235 ff., who also affirm that the issues of liability involving robotic applications fall within the product liability rules; Salanitro (2020), pp. 1246–1276; Santosuosso (2020).

[41] Bertolini (2013), pp. 235 ff.

[42] The interpretation according to the intention of the legislator can give rise to the so-called extensive interpretation of the law, with which the words of the law are given a broader meaning than the literal one (art 12 the preliminary provision to the Italian Civil Code).

[43] Every national legal system is endowed with the character of completeness: it can not admit having a regulatory gap, but it must be able to give a solution to every possible conflict that is generated among those who are subjected to that legal system. However, the legislator can not foresee all possible conflicts, establishing as a rule for every possible case. Therefore, to fill the inevitable gaps, the art 12,para 2, of the preliminary provision to the Italian Civil Code, requires providing for the analogical application of the law (*analogia*). If a dispute can not be decided by a specific provision, the provisions regulating similar cases or similar matters must be applied (so-called '*analogia legis*'). The *extrema ratio* is to decide a legal case according to the general principles of the legal system (so-called '*analogia iuris*', provided by same article 12, section 2, when there are no laws to regulate a specific phenomenon.

[44] European Parliament and Council, Proposal for a regulation on laying down harmonised rules on artificial intelligence (artificial intelligence act) and amending certain union legislative acts, 21 April 2021, COM(2021)206 final (revised 3 November 2022).

[45] European Parliament and Council, Proposal for a Directive on adapting non-contractual civil liability rules to artificial intelligence (AI Liability Directive), 28. September 2022, COM(2022) 496 final.

[46] European Parliament and Council, Proposal for a Directive on liability for defective products, 28 September 2022, COM(2022) 495 final.

robotics applications are too significant to allow for a single 'law of robotics'.[47] In this regard, although in general terms, autonomous vehicles seem to be considered as high-risk AI systems, they will not be directly regulated by the AI Act proposal.[48]

4.2 *Third-Party Insurance, as an Effective Insurance Policy also for (Semi)autonomous Vehicles*

Concerning the civil liability for damages deriving from the circulation of vehicles, the so-called compulsory insurance[49] is provided at both European and national levels. Compulsory insurance means that vehicles can only circulate if insured. At the European level, the reference legislation is the Directive 2009/103/EC,[50] which, in article 3, expressly provides that each Member State shall take all appropriate measures to ensure that civil liability relating to the circulation of vehicles normally based in its territory is covered by car insurance. In particular, the insurance policy must cover the damage caused in the territory of a Member State according to the legislation in force in that State and the damage, which may be suffered by the citizens of a Member State. Therefore, the European legislation defers to the national legislation for the regulation of the concrete national case, even if it contains general principles that must be transposed in the Member States' law.[51] First of all, it highlights the principle that the injured party must claim compensation to the insurance company undertaking of the person who caused the accident or to its

[47] Pagallo (2013) inaugurates the field of robot law, from the point of view of legal philosophers and positive law experts, exploring the three legal fields of law: criminal law, contractual liability and non-contractual liability.

[48] Indeed, the proposed AI Act, Article 2, para. 2, provides that high-risk AI systems covered by EU harmonisation legislation listed in Annex II, section B, are generally not subject to the AI Act. This Annex lists Regulations 2018/858 and 2019/2144, which, in turn, now also address automated driving systems.

[49] Bertolini (2016), pp. 291 ff.; Bertolini et al. (2016), pp. 381 ff., explains that normally, the parties are left free to decide whether to enter an insurance contract or not, in order to manage a risk, they are exposed to. However, in some cases, the legislator deeming that the risk associated to a certain activity is too high, or that moral behaviour may negatively affect the spreading of the damages that arise, and for this impose a duty to purchase insurance (normally third-party insurance). The traffic insurance, as well as the professional liability insurance, are examples of compulsory insurance.

[50] Directive of the European Parliament and of the Council 2009/103/EC of 16 September 2009 relating to insurance against civil liability in respect of the use of motor vehicles, and the enforcement of the obligation to insure against such liability [2009] OJ L 263/11.

[51] To this general principle of compulsory insurance there is a derogation provided for in Article 5 Dir 2009/103/EC in favor of certain natural or legal persons, public or private, included in a list that the Member State must notify to other States and to the EU Commission. However, the Member State must take the appropriate measures to indemnify the persons for the damage caused in their own state or in that of the other Member States by vehicles belonging to the aforementioned persons exempt from compulsory insurance. To this end, it shall also draw up a list of the authorities or bodies responsible for the indemnification to be communicated to the Commission.

claims representative.[52] This principle is important because implies that the insurance scheme adopted is that of third-party insurance, with the provision also of direct compensation in specific circumstances.

In compliance with EU regulation, indeed, also in Italy, compulsory insurance follows the rules of third-party insurance. The current insurance legislation is the Italian Private Insurance Code, which has applied the principle of compulsory insurance,[53] introducing also a series of innovations, among which one of the most important is direct compensation (article 149).[54] It provides that, in specific cases, the injured person can go directly to his/her insurance company to get compensation for damages in case of an accident caused by other parties, without having to request the payment of damages to the company of the damaging party.[55] In addition, only ones who are insured with an insurance company belonging to the Direct Compensation Agreement (DCA, in Italian CARD), pursuant to art. 13 of the D.P.R. n. 254/2006,[56] can take advantage of the direct compensation.

For damages caused by the insured to third parties, the insurance company of the insurer, that collect the annual insurance premium,[57] replaces the insured by paying the damages. The damages for which the driver (and also the other subjects expressly provided for in article 2054 of the Italian civil code) is liable, are those caused 'to people or things'. In the formula adopted, no distinction is made between people and things transported or not, but contemporary case-law also includes people and things transported.[58] Moreover, if the damage derives from the collision with another vehicle, the person transported may act against the other driver assuming an equal responsibility of the parties under article 2054, paragraph 2, Italian Civil Code. In

[52] Art 22, Dir 2009/103/EC.

[53] Italian Private Insurance Code, D.lgs. 7 September 2005, n 209, OJ 239.

[54] The same prevision is provided at EU level by Art. 18, Dir 2009/103/EC.

[55] Buffone (2016), pp. 157 ff. The direct compensation procedure operates only if: (1) the motor vehicle accident occurred in Italy, in the Republic of San Marino or in Vatican City; (2) it involved only two vehicles, both identified, insured and registered in Italy; (3) and if there has been damage to property and/or persons, but minor value. It seems clear, therefore, that this is an exception to the third-party scheme, and that, when applicable, operates according to the scheme of the first-party insurance, even if the requirements are different and far more specific for the direct compensation than the first-party insurance.

[56] D.P.R.,18 July 2006, n. 254 regulations governing the direct compensation for damages arising from road traffic, OJ 199. In other words, these companies must establish a specific office, Company Service Reference Conventions Reference (SARC), for the management of the reports relating to the CARD between the member companies. The SARC also manages the relationships that derive from other agreements related to motor vehicle dealings, such as the Third-Party Transports Convention (CTT). Foreign insurance companies, on the other hand, must apply for membership of ANIA (Italian National Association of Insurance Companies) pursuant to Article 2 CARD.

[57] The insurance premium is the amount that the insured party corresponds to the insurance company to obtain insurance coverage.

[58] In the past, on the contrary, a restrictive interpretation of the rule was preferred.

addition, the people transported are entitled, in the presence of a transport contract, to act against the driver for damages according to contractual civil liability.[59]

Concerning the threshold of the insurance premium, i.e. a minimum that insurance companies must necessarily guarantee as compensation, is provided a basic threshold with the possibility of increasing the insurance premium.[60]

Briefly stated the principles of Italian insurance regulation in road traffic, the open issue concerns how autonomous vehicles affect compulsory insurance: who should be the insured? It should be an increase or a decrease in the insurance premium?[61]

Nonetheless, the evaluation of risk management, placed on the basis of the definition of the insurance premium, is based on the definition of its frequency and its severity. These aspects are difficult to analyse due to (a) the technical complexity of robotic devices (including driverless cars), (b) the lack of sufficient data concerning the potential risks and the accidents they may cause, (c) the uncertainties with respect to liabilities that producers and users may face (as already explained).[62]

However, despite the difficulties of the case and considering the strong impact that autonomous vehicles will have on the insurance system, it is necessary trying to focus on the principles that can ensure a high level of protection for the parties involved.

In the case of autonomous vehicles, the aspect that in practice really affects the injured party is to obtain compensation for damage. Therefore, even more than identifying the tortfeasor, it is important to establish who should compensate for the damage, given that the tortfeasor and the other subjects called to compensate for the damage may not coincide. This happens in all cases of a car accident in which compensation is paid not by the person held liable (driver and/or owner of the vehicle), but by the insurance company of the injured party, if the direct compensation scheme is applied, otherwise from the insurance company of the tortfeasor.

Moreover, it is necessary that the injured person can always obtain compensation for the damage suffered. For this reason, it is essential to maintain the current system of compulsory insurance, in the sense that, in the case in which the duty of mandatory insurance bears on the owner, the vehicle cannot be put into circulation if the owner is not insured or, on the contrary, the case in which the duty of

[59] In Italy, the transport contract is governed by articles 1678 ff., Italian Civil code.

[60] Historically, insurance companies have always been inclined to recognize the benefits of road safety from automotive technologies and have encouraged their policyholders to invest in them (e.g. Black box). *See* art. 132 ter, lett b) Italian Private Insurance Code.

[61] Perlingieri (2015), pp. 1241 ss., wonders about new forms of compulsory insurance required by the development of robotics, assuming the use of a criterion based on the level of learning and adaptation of the robot: a greater ability to learn of the robot would result in less responsibility for the producer (taking account of its power to give instructions to the robot), while it would entail less responsibility for the user. The author starts from the assumption that the robot can not be recognized civil liability as it does not seem necessary to completely overcome the concept of a robot as a legal good (in Latin '*res*'). This reconstruction reflects the legislation currently in force in the different legal system and in the European one, but there is a strong impulse of innovation that could also lead to the recognition of the legal personality to the robot.

[62] Bertolini et al. (2016), p. 387.

mandatory insurance bears on the producer, the vehicle cannot be placed on the market if the producer is not insured. At the same time, an *ad hoc* fund should be provided for cases in which the injured party does not have an insurance company to claim. It happens, for example, when the owner or producer of the vehicle is not insured, the owner of the vehicle that caused the accident is unknown or, in general, it is impossible to identify the tortfeasor.

Finally, an assessment should be made on the amount of the insurance premium, evaluating its variations based on the development of artificial intelligence in vehicles. A possible solution could be to evaluate the increase or decrease of the insurance premium in relation to the percentage of car accidents involving vehicles equipped with driving technologies and, in an evolutionary way, autonomous vehicles. It seems plausible to believe that the new on-board technologies imply a reduction in accidents and therefore a consequent reduction in the vehicle's owner insurance premium, towards the increase in the manufacturer's insurance premium, in consideration of the inevitable increase in the assumptions of the latter's liability.[63] This solution would seem to be consistent with the two cornerstones on compulsory insurance. The first point[64] concerns the exceeding of the limit of fault as the sole criterion for imputation of liability, both in the case of the liability of the owner and in that of the producer.[65] The second point, instead, concerns the development of a formula that always guarantees compensation for the damage that occurred.

5 Conclusions: The Need for an *ad hoc* Framework of Rules for Vehicles of Level 5 of Automation

As has been attempted to demonstrate, **the current civil liability regulation** in road traffic perfectly regulates the case of accidents involving traditional vehicles and can be applied extensively to semi-autonomous vehicles, but with the achievement of

[63] In this context, in Italy the art 132 ter, letter b), Italian Private Insurance Code, provides the duty for the insurer to apply a compulsory discount if on-board the vehicle are installed or are already preexisting electronic mechanisms that record the activity of the vehicle, called 'black box' or equivalent. After this first phase, as the vehicles become more and more autonomous and, at the same time, the driver will be denied the possibility of intervention during the driving phase, the responsibility will pass entirely to the producer, since there is no longer any subject qualifying as a driver. This will correspond to a new allocation of risk, which will probably result in an insurance system parametrised no longer on the driver but on the characteristics of the vehicle itself.

[64] Franzoni (2010), pp. 1 ff.

[65] In fact, as already explained, there is the semi-objective liability of the designer or the informer, and the strict liability of the manufacturer. Likewise, to the semi-objective liability of the driver is added the strict liability in certain cases expressly provided by the law.

full automation, it **does not appear a functional** regulatory tool for to the protection of the injured party.[66]

More precisely, the liability rules in road traffic regulate the liability of the driver or the owner (and the subject similar to him) in the circulation of traditional vehicles, and could be applied only to human-driven vehicles and, at most, to semi-autonomous vehicles until the most important driving commands are in the power of the human driver, who have to monitor the environment (i.e. until level 3–4 of automation of the SAE J3016 standard). Undoubtedly, also in these cases, if the accident was caused by a product defect, the driver/owner of the vehicle, condemned to the compensation for damages, has the right of recourse against the producer of the vehicle or its defective component part. Alternatively, in the litigation phase (and therefore before the condemnation) the examination of the specific case will allow to verify whether it is a hypothesis of joint and several liability of the driver/owner and the producer or whether it can instead be considered exclusively liable one of them. In this light, the importance, supported by many parties, of the introduction of Data Sharing or Data Record systems, which make it possible to check whether at the time of the accident the vehicle was in autonomous or manual driving mode, emerges.

With higher automation (i.e. levels 4 and 5 of automation of the SAE J3016 standard), instead, the role of the driver is progressively lost because the vehicle drives itself and all the onboard users are considered passengers, to whom the liability in road traffic is not applicable. In this case, the owner of the vehicle, for the mere fact of having purchased the vehicle could be held liable, as he/she bears the risk of a possible accident and the consequent compensation for the damage. At the same time, however, the (semi)autonomous vehicle is a product and, as a product, falls within the scope of the consumer code (PLD at EU level). This interpretation of the law seems to be the most functional to protect the interests of the parties.[67]

In the case of total automation, instead, an *ad hoc* reference legislation is needed. It should not concern robotics in general but should be specifically focused on autonomous vehicles.

For this reason, **it is needed a regulation for Robotics and AI, plus sector-specific laws for specific fields of Robotics which are more complex** and which presents more risks, such as that of self-driving cars, with particular regard to liability issue, related insurance issues, as well as data protection,[68] cybersecurity and ethical issues. As there is a specific normative discipline for traditional vehicles,

[66]The case of the civil liability for self driving cars is also explored by Calabresi and Al Mureden (2020), pp. 7 ff.; Davola and Pardolesi (2017), pp. 616 ff.; Ruffolo and Al Mureden (2019), pp. 1704 ff.

[67]Reed et al. (2016), p. 5 explain that the product liability could be a good option for regulating machine learning damage (included that coming from self-driving cars) but, because it is a new technology no laws have yet in force.

[68]For an in-depth analysis of the damage caused by semi-autonomous vehicles, please refer to Gaeta (2019). Concerning the European analysis of the data protection issues, *see* Gatt et al. (2018), pp. 1 ff. With particular regard to data protection and self-driving cars, may I also refer to Gaeta (2018), pp. 147 ff.; Gaeta (2017), pp. 1 ff.

so is needed one for autonomous vehicles, which will favour the development and diffusion of fully autonomous vehicles. This desirable legislative reform should be composed of hard law, i.e. a specific binding legislation such as a regulation (which is immediately applicable in all the European Member States), soft law (as, for example, Ethical code,) standardisation (i.e. technical rules) and techno-regulation, which is the ability of the technology to make law and not only to be the object of law but also a regulation tool.[69]

In conclusion, carrying out a *de iure condendo* analysis, in case of a **semi-autonomous** accident, the possible solution should be to apply, as appropriate, the civil code for **driver's liability** or **owner's liability** and the **consumer code in case of product defect**. Lastly, in an evolutionary way, **for the fully autonomous vehicle** (level 5 of the SAE J3016) it could be **necessary to introduce an *ad hoc* framework of rules.** The framework of rules should be based on the respect by the producers of the ***ad hoc* standards** that should be introduced and that, however, **represent a minimum threshold of the producer's liability**. More in detail, the producer should be held liable mainly in case of a malfunction of the vehicle or absence of adequate warnings provided to the driver, if he did not comply with the standards, that the producer must respect to be put the self-driving cars into circulation. If, instead, the producer had complied with the standards, then a **case-by-case study** will have to be carried out, taking into account certain parameters which include those actually indicated by consumer code, in order to determine the reason for the malfunctions or failure to warn and establish the possible liability. In this scheme, the extension of the producer's civil liability would not undergo significant alterations, compared to the current context, and the cost of accidents caused by autonomous vehicles conforming to the standards would **usually continue to be of the owner of the vehicle.**[70] According to this thesis, civil liability for fully autonomous vehicle accidents will be shared between the owner of the vehicle and the producer of the same.

Furthermore, during this transition period that will lead us to fully autonomous driving, the roads will be travelled with both **traditional vehicles and increasingly autonomous cars**, and the coexistence between the two types of vehicles will need a regulation based on technical rules to discipline their interaction.[71]

Anyhow, the regulation of self-driving cars cannot be national, considering the large capacity of the vehicles to move from State to State. At least **we need European legislation**, something that the European Authorities have been aware.

In this scenario, a balance between tradition and innovation it will be fundamental, **using existing legal categories where possible** (e.g. driver's liability, owner's liability producer's liability).

Table 1 shows the levels of automation of the SAE standard, identifying for each of them the tortfeasor in case of an accident produced by the vehicle with a more or

[69] A typical example of techno-regulation is represented by the data protection by design, expressly provided by the art 25 of the GDPR, which expressly provide that the product itself, thanks to the way it has been designed, has to guarantee the protection of the personal data of the data subject.

[70] Al Mureden and Ruffolo (2019), pp. 1704 ff.

[71] Abraham and Rabin (2019), pp. 127 ff.

Table 1 Liability and insurance policy in relation with the automation level

	SAE automation levels	Liability	Applicable law	Insurance policy
0	No automation	**Driver/Owner of the vehicle** (subject with an equivalent legal status), with the right of recourse against the producer of the vehicle or of its defective component part	**National rules of car accident liability** (if necessary, identified according to the rules of Private International Law) and eventually European and national Product liability regulation	Car Insurance policy of the vehicle owner
1	Driver assistance	**Driver/Owner of the vehicle** (subject with an equivalent legal status), with the right of recourse against the producer of the vehicle or of its defective component part	**National rules of car accident liability** (if necessary, identified according to the rules of Private International Law) and eventually European and national Product liability regulation	Car Insurance policy of the vehicle owner
2	Partial automation	**Driver/Owner of the vehicle** (subject with an equivalent legal status), with the right of recourse against the producer of the vehicle or of its defective component part	**National rules of car accident liability** (if necessary, identified according to the rules of Private International Law) and eventually European and national Product liability regulation	Car Insurance policy of the vehicle owner
3	Conditional automation	**Owner of the vehicle** (subject with an equivalent legal status), and **the driver of the vehicle** if he had a behaviour that does not comply with his driving tasks	**National rules of car accident liability** (if necessary, identified according to the rules of Private International Law) and eventually European and national Product liability regulation	Car Insurance policy of the vehicle owner
4	High automation	**Owner** (subject with an equivalent legal status) **and/or producer** of the vehicle	**National rules of car accident liability** (if necessary, identified according to the rules of Private International Law) and/or European and national Product liability regulation	Car Insurance policy of the vehicle owner and/or ad hoc Professional insurance of the producer and a compensation fund for autonomous vehicles

(continued)

Table 1 (continued)

	SAE automation levels	Liability	Applicable law	Insurance policy
5	Full automation	**Producer** of the vehicle and or **owner** of the vehicle (subject with an equivalent legal status)	**New framework or rules** for autonomous vehicles	Car Insurance policy of the vehicle owner and/or Professional insurance of the producer and a compensation fund for autonomous vehicles

less high level of automation and, consequently, the subject who bears the duty of purchasing the mandatory insurance contract.

Law

Italian Private Insurance Code, D.lgs. 7 September 2005, n 209, OJ 239.
Italian Consumer Code, D.lgs. 6 September 2005, n 206, OJ 235.
Italian Data Protection Code, D. lgs. 30 June 2003 n 196, OJ 174.
Italian Electronic Communication Code, D.lgs. 1 August 2003, n 259, OJ 214.
Reform of the Italian system of Private International Law, L. 31 May 1995, n 218, OJ 128.
Italian Traffic Code, D.lgs. 30 April 1992, n 285, OJ 114.
Italian Civil Code, R.D. 16 March 1942, n 262, OJ 79.
Italian Code of Civil Procedure, R.D. 28 October 1940, n 1443, OJ 253.
D.M. 28 February 2018, n 70, implementation methods and operational tools of road testing of Smart Road solutions and connected and automatic driving, OJ 90
D.M. 12 February 2014, n 44, adoption of the National Action Plan on Intelligent Transport Systems (ITS).
D.P.R. of 18 July 2006, n 254, Regulation governing the direct compensation for damages deriving from road traffic, pursuant to article 150 of the legislative decree September 7, 2005, n. 209 (Italian Private Insurance Code), OJ 199.

References

Abraham KS, Rabin RL (2019) Automated vehicles and manufacturer responsibility for accidents: a new legal regime for a new era. Va Law Rev 105:127–171
Al Mureden E (2012) Sicurezza "ragionevole" degli autoveicoli e responsabilità del produttore nell'ordinamento giuridico italiano e negli Stati Uniti. Contr impr 6:1506 ff
Al Mureden E (2019) Autonomous cars e responsabilità civile tra disciplina vigente e prospettive de iure condendo. Contr impr 3:895–924

Al Mureden E, Ruffolo U (2019) Autonomous vehicles e responsabilità nel nostro sistema ed in quello statunitense. Giur it 7:1704–1715
Albanese A (2019) Danni da veicoli ad elevata automazione. Europa dir priv 4:996–1044
Alpa G (2018) La responsabilità civile. Principi. Utet, Turin, pp 37 ff
Alpa G (2020) Diritto e intelligenza artificiale. Pacini, Pisa
Alpa G, Bin M, Cendon P (1989) La responsabilità del produttore. In: Galgano F (ed) Trattato di diritto commerciale e diritto pubblico dell'economia, XIII. Cedam, Milan, pp 56 ff
Amidei A (2019) Intelligenza Artificiale e product liability: sviluppi del diritto dell'Unione Europea. Giur it 7:1715–1730
Asaro PM (2016) The liability problem for autonomous artificial agents. In: AAAI Spring Symposia. Stanford University, Stanford, pp 190–194
Bertolini A (2013) Robots as products: the case for a realistic analysis of robotic applications and liability rules. LIT 5(2):277 ff. https://doi.org/10.5235/17579961.5.2.214
Bertolini A (2016) Insurance and risk management for robotic devices: identifying the problems. GJ 16(3):291–314. https://doi.org/10.1515/gj-2015-0021
Bertolini A (2020) Artificial Intelligence and Civil Liability. Study for the Policy Department for Citizens' Rights and Constitutional Affairs Directorate-General for Internal Policies. European Union, PE 621.926
Bertolini A, Riccaboni M (2021) Grounding the case for a European approach to the regulation of automated driving: the technology-selection effect of liability rules. Eur J Law Econ 51:243–284. https://doi.org/10.1007/s10657-020-09671-5
Bertolini A, Salvini P, Pagliai T et al (2016) On robots and insurance. Int J Soc Robot 8:381–391. https://doi.org/10.1007/s12369-016-0345-z
Bianca CM (2021) La responsabilità. In: Trattato di diritto civile, vol 5. Giuffrè Francis Lefebvre, Milan, pp 662 ff
Buffone G (ed) (2016) Responsabilità civile automobilistica. Cedam, Padua, pp 95 ff
Calabresi G, Al Mureden E (2020) Driverless Cars e responsabilità civile. Rivista di diritto bancario, Suppl Issue I:7–21
Calabresi G, Melamed DA (1972) Property rules, liability rules, and inalienability: one view of the Cathedral. Harv Law Rev 85:1096 f
Capilli G (2019) Responsabilità e robot. NGCC 3:621–631
Carnevali U (1974) La responsabilità del produttore. Giuffrè, Milan, p 42
Castronovo C (2006) La nuova responsabilità civile, 3rd edn. Giuffrè, p 732
Costanza M (2019) L'intelligenza artificiale e gli stilemi della responsabilità civile. Giur it 7:1686–1689
Davola A, Pardolesi R (2017) In viaggio col robot: verso nuovi orizzonti della r.c. auto ("driverless")? Danno e resp 5:616–629
Franzoni M (2010) Il danno risarcibile. In: Trattato della responsabilità civile, vol 2, 2nd edn. Giuffrè, Milano
Gaeta MC (2017) The issue of data protection in the Internet of Things with particular regard to self-driving cars. DIMT, pp 1 ff
Gaeta MC (2018) La protezione dei dati personali nell'Internet of Things: l'esempio dei veicoli autonomi. Dir inf 1:147 ff
Gaeta MC (2019) Liability rules and self-driving cars: the evolution of tort law in the light of new technologies. Editoriale Scientifica, Naples
Gatt L, Montanari R, Caggiano IA (2018) Consent to the processing of personal data: a legal and behavioural analysis. Some insights into the effectiveness of data protection law. EJPLT 1:1 ff
Gatt L, Caggiano IA, Gaeta MC (2020) Italian tort law and self-driving cars: state of the art and open issues. In: Oppermann BH, Stender-Vorwachs J (eds) Autonomes Fahren. Technische Grundlagen, Rechtsprobleme, Rechtsfolgen. C.H. Beck, Munich, pp 239–255
Irti N (1990) Digesto civ., sez. civ., 4th edn. UTET, Turin, p 150
Losano M (2019) Verso l'auto a guida autonoma in Italia. Dir inf e inform 2:423–441

Machnikowski P (ed) (2016) European product liability. An analysis of the state of the art in the era of new technologies. Intersentia, Cambridge
Mazzon R (2012) La responsabilità oggettiva e semioggettiva. UTET, pp 1063 ff
Montinaro R (2020) Responsabilità civile da prodotto difettoso e tecnologie digitali tra soft law e hard law. Persona e Mercato 2:365 ff
Pagallo U (2013) The law of robots: crimes, contracts, and torts. Springer, Berlin
Pagallo U (2017) Intelligenza Artificiale e diritto. Linee guida per un oculato intervento normativo. Sistemi intelligenti 3:615–636. https://doi.org/10.1422/88512
Palmerini E, Bertolini A (2014) Regulating robotics: a challenge for Europe. Upcoming issues of EU law for the IURI Committee, EU Parliament Workshop 2:94–129
Pellegatta S (2019) Autonomous driving and civil liability: the Italian perspective. Rivista di Diritto dell'Economia, dei Trasporti e dell'Ambiente 17:135–162
Perlingieri C (2015) L'incidenza dell'utilizzazione della tecnologia robotica nei rapporti civilistici. Rass dir civ 4:1235–1246
Procida Mirabelli di Lauro A, Feola M (2020) Diritto delle obbligazioni. ESI, Napoli, pp 508–558
Ratti M (2020) Riflessioni in materia di responsabilità civile e danno cagionato da dispositivo intelligente alla luce dell'attuale scenario normativo. Contr Impr 3:1175 ff
Recinto G, Porcelli M (2014) La responsabilità per danno da prodotti difettosi. In: Recinto G, Mezzasoma L, Cherti S (eds) Diritti e tutele dei consumatori. ESI, Naples, p 449
Reed C, Kennedy E, Silva SN (2016) Responsibility, autonomy and accountability: legal liability for machine learning', Legal Studies Research Paper no. 243. Queen Mary University of London, p 5
Ruffolo U (2020a) Intelligenza artificiale. Il diritto, i diritti, l'etica. Giuffrè, Milan
Ruffolo U (2020b) Intelligenza artificiale e responsabilità civile. Giuffrè, Milan
Ruffolo U, Al Mureden E (2019) Autonomous vehicles e responsabilità nel nostro sistema ed in quello statunitense. Giur it 7:1704–1715
Salanitro U (2020) Intelligenza artificiale e responsabilità: la strategia della Commissione Europea. Riv dir civ 6:1246–1276
Santosuosso A (2020) Intelligenza artificiale e diritto. Mondadori, Milan
Simonini GF (2017) La responsabilità del produttore: presunta o oggettiva? Recenti prospettive di indagine. Il dir. dell'economia 30(93):557ff
Stradella E, Palmerini E (2013) Law and technology. The challenge of regulating technological development. Pisa University Press, Pisa

Civil Liability for Self-Driving Cars in Dutch Law

Eric Tjong Tjin Tai

Abstract This chapter discusses the Dutch rules on civil liability for accidents involving traditional cars and self-driving cars. In Dutch law, various parties may be liable in the case of accidents with a self-driving car: the owner or keeper, the driver, the possessor, the manufacturer, the car mechanic. In practice, the owner or keeper will primarily be held liable as art. 185 WVW imposes strict liability for all accidents, even for self-driving cars. As insurance companies will cover the damages for such accidents, they may take recourse on car manufacturers on the basis of product liability or negligence. The driver (who is usually also covered by the car insurance) may also be liable on the basis of negligence if he could and should have intervened to prevent an accident. In the case of level 4 or 5 cars, the main issue is whether the courts will allow a driver to not pay attention continuously.

1 Introduction

In a codified system like Dutch law, autonomous vehicles do not pose a fundamental problem. The general clauses on fault-liability and strict liability provide sufficiently open norms to give a victim of an accident by an autonomous vehicle various angles for obtaining compensation. Nonetheless there are several issues and areas for improvement. Meanwhile the Dutch legislator has already allowed tests with autonomous vehicles. In this chapter I will discuss the current state of the law, focusing on the different grounds of liability and the liable persons, as well as the way in which the various rules interact. I will also discuss the role of insurance.

E. Tjong Tjin Tai (✉)
Tilburg University, Tilburg, Netherlands
e-mail: t.f.e.tjongtjintai@tilburguniversity.edu

H. Steege et al. (eds.), *Autonomous Vehicles and Civil Liability in a Global Perspective*, Data Science, Machine Intelligence, and Law 3,
https://doi.org/10.1007/978-3-031-41992-8_16

2 Liability for Level 0–2 Cars

Dutch liability law regarding level 0–2 cars consists of a combination of fault liability and strict liability rules. In order to describe who is liable and what the criteria for liability are, we need to examine the various applicable grounds of liability. These grounds generally apply concurrently: if the criteria for two different grounds are fulfilled, both grounds may be applied. A few exceptions are mentioned in Sect. 2.6.

2.1 Specific Liability for Cars

First of all, the owner or the keeper of a car is liable for accidents involving a car driving on a public road. This liability is based on art. 185(1) WVW (*Wegenverkeerswet*, Road Traffic Act), which states (in translation):[1]

> If a motorized vehicle that drives on the road is involved in a traffic accident which caused damage to persons or goods not transported in that vehicle, the owner or keeper - if there is a keeper - of the vehicle are liable for the damage, unless it is likely that the accident is due to force majeure, including the case that the accident was caused by someone for whom neither the owner nor the keeper are accountable.

The owner is the person who according to the rules of property law is owner,[2] which in principle is the person to whose name the car is registered.[3] If there is a possessor (*bezitter*) who is not the owner, the possessor will be considered as the owner for the purpose of the WVW.[4] For the sake of simplicity we will hereafter only speak of the owner.

The keeper (*houder*) is the person who has the car under his control[5] on the basis of a lease-purchase agreement (*huurkoop*), on the basis of a right of usufruct, or an another basis has the car under his control for long-term use.[6]

[1] See generally on this provision Sieburgh (2019), nrs. 273–294; van Dam (2020), nrs. 602–610. On the relation to self-driving cars and the entire framework of liability law see, besides the literature refered to later, van Wees (2004), Van Wees (2010), Engelhard (2017), Engelhard and De Bruin (2018), Vellinga (2014), and extensively Vellinga (2020a).

[2] Art. 5:1 BW (Burgerlijk Wetboek, the Dutch Civil Code).

[3] Art. 1(3) WVW. See art. 36 WVW on the license plate registration.

[4] This follows from art. 1(2) WVW, see also Sieburgh (2019), nr. 281. See Sect. 2.3 for the meaning of 'possessor'.

[5] In Dutch this reads 'onder zich heeft', which means that the person has direct physical control over the object, such as when the car is parked somewhere but you have the car key. The 'keeper' in this context is more restricted than the general concept of 'keeper' in the Dutch civil code (cf. art. 3:108 BW), where a person in short-term control also may qualify as a keeper.

[6] Art. 1(1)(o) WVW. An example of the last category is a parent lending his car to his child indefinitely.

The liability of keeper prevails; only if there is no keeper will the owner be liable.[7] When I speak of 'owner or keeper' hereafter I refer to this alternative liability where the keeper takes precedence.

Art. 185(2) WVW extends the liability of owner or keeper to acts by other persons who have driven the car by permission of the owner or the keeper of the car. This amounts to a specific form of strict liability. This provision may also apply if the car key has been stolen,[8] although in such a case the owner or keeper could also be liable on the basis of negligence.[9]

Art. 185(3) WVW excludes the application of this article for cases of damage to unleashed animals, and to other motorized vehicles (including persons and goods in these vehicles).

The liability on the basis of art. 185(1) WVW is quite extensive and is only limited by force majeure (*overmacht*). If the owner or keeper does not make force majeure plausible,[10] the owner or keeper is liable even if no fault on his side has been proven.[11] A defect of the car does *not* constitute force majeure.[12] Neither does a sudden bodily incontinence (seizure, fainting), nor bad weather, nor complicated situations on the road.[13]

In practice courts have only recognized force majeure when the accident is *solely* caused by someone for whom the owner or keeper is not liable.[14] This primarily means a fault of the victim that was so unlikely that the driver did not have to take it into account when driving. Acts of third parties for who the owner or keeper is not accountable can also constitute force majeure, but often such acts can be foreseen and do not amount to force majeure.[15] If there is no force majeure, liability may still be reduced by contributory negligence.

There is an important restriction to the possibility of contributory negligence. In case the victim is a child below the age of 14, the Dutch Supreme Court has declared the rule that there is always 100% liability on the side of the car owner, regardless of the actual contribution by the victim.[16] In case of other categories of vulnerable traffic participants such as elderly people, pedestrians, cyclists, the rule is that there

---

[7] Sieburgh (2019), nr. 281.

[8] HR 31 January 1992, NJ 1992/319.

[9] HR 2 December 1966, NJ 1967/42 (Herring stand).

[10] 'aannemelijk' in Dutch. In practice it may be hard to make force majeure 'plausible', even though it seems to imply a lower standard than proving force majeure.

[11] For instance if it has not become clear what caused the accident: HR 17 November 2000, NJ 2001/260.

[12] HR 16 april 1942, NJ 1942/394, see Sieburgh (2019), nr. 282.

[13] Sieburgh (2019), nr. 283.

[14] Sieburgh (2019), nr. 284, Rutten and Oskam (2016), Lavrijsen and Weitering (2019).

[15] Cf. art. 185(2) WVW.

[16] Sieburgh (2019), nr. 285. E.g. HR 1 June 1990, NJ 1991/720, HR 31 May 1991, NJ 1991/721, HR 2 June 1995, NJ 1997/700, HR 17 November 2000, NJ 2001/260. These cases have to do with contributory negligence and force majeure in turn, but the end result is the 100%-rule described in the main text.

is at least 50% liability, while the remainder of the damage is divided on the basis of contributory negligence and/or force majeure.[17] The only exception to this liability is if the accident is due to intention or gross negligence by the victim.[18]

Damages are assessed and awarded according to the general rules (see Sect. 2.2).

2.2 *General Fault Liability: Negligence*

The general fault liability provision of art. 6:162 BW (Burgerlijk Wetboek, Dutch Civil Code) applies as well to car accidents.[19] This provision is mainly used in case art. 185 WVW is not applicable, that is to say in case of accidents of a car with other motorized vehicles, or for the damage to passengers of (and goods transported in) the car itself.[20] The provision will mainly be used for a fault committed by the driver of the car, although negligence can also be used to hold other parties liable, such as an automotive mechanic.

According to art. 6:162 BW a person is liable if he acts wrongfully, thereby causing harm. He acts wrongful if he violates a legal obligation, infringes a right, or acts negligently (in violation to the standard of care expected in society). For accident with cars, the third category, often called 'breach of a duty of care' (*schending van een zorgplicht*), is the most relevant. In order to determine negligence or lack of care, courts use the so-called Cellar hatchway test,[21] which involves four factors: the probability of carelessness by the victim, probability of accidents happening, seriousness of the consequences, and burden of precautionary measures. The courts may also apply the test of how a reasonable person would have acted. For car accidents this norm is influenced by traffic rules: wrongfulness is determined mainly on the basis of whether the driver obeyed traffic rules, in particular the duty to drive safely (art. 5 WVW). Negligence is readily accepted if there is even a slight error.[22]

Besides the presence of a wrongful act, liability requires culpability (art. 6:163 BW), causality, and harm/damage. A fifth requirement is relativity (*relativiteit*)[23] of the rule that is violated, however, that is presumed to exist in case of wrongfulness, and relativity will in particular be present in case of traffic accidents leading to

[17] Sieburgh (2019), nr. 286. E.g. HR 28 February 1992, NJ 1993/566, HR 14 July 2000, NJ 2001/417.

[18] An example is HR 30 March 2007, NJ 2008/64: a dark-skinned person crossing an unlit road at night in dark clothes.

[19] Art 185(4) WVW explicitly allows concurrent actions on other grounds.

[20] About the liability towards passengers see van Dam (2020), nr. 612.

[21] Named after the leading case HR 5 November 1965, NJ 1966/136, about a visitor to a pub falling down a cellar hatch.

[22] HR 26 March 1982, NJ 1982/292 (Wenmakers/Smeets), van Dam (1985), p. 261; van Dam (2020), nr. 611-1.

[23] This is similar to the notion of scope of the rule, *Schutzzweck*.

property damage or personal injury. Culpability is presumed as well, and the threshold of being culpable is very low.[24] Even in the absence of culpa, the accident may still be ground for liability if the cause of the accident is to be attributed to the defendant on the basis of a statute or generally accepted social norms (*in het verkeer geldende opvattingen*).[25] As regards causality, both factual and legal causality are required (art. 6:162 BW and 6:98 BW). The Dutch courts have accepted a so-called 'reversal rule' in case the driver violated a clear safety norm,[26] such as the speed limit or the prohibition on drunk driving. In such cases causality between the violation and the harm (accident) is presumed and needs to be refuted. This amounts to a rule of *res ipsa loquitur*.

The rules on assessment of damages can be found in art. 6:95-110 BW. All kinds of loss may be compensated: material loss as well as immaterial damage. Material loss may consist of property damage, personal injury and pure economic loss. There is no general exclusion of compensation for pure economic loss, although courts may be more prone to decide that pure economic loss is too remote (lacks legal causality) in comparison to property damage and personal injury.

The rules on contributory negligence (art. 6:101 BW) apply to liability on the basis of art. 6:162 BW as well. The 100% and 50% rules described in Sect. 2.1 are not applicable.

2.3 Strict Liability of the Possessor of the Car

Another ground of liability is the strict liability of the possessor of the car. Art. 6:173 BW states that the possessor of a movable object of which it is known that it will create a particular danger for persons or objects if it does not meet the demands that can be required in the given circumstances, is liable for the damage when this danger materializes. This liability does not apply if there would not be any fault in the sense of art. 6:162 BW if the possessor had known of the risk. That exception applies for instance to cases of legitimately caused harm.

The possessor (*bezitter*) is the person who has control over the car and claims to be owner.[27] Hence if the actual owner is not the possessor, only the possessor (having actual control) is liable on this basis. Conversely, a driver who only borrows the car is not possessor and is not liable on this ground.

[24] HR 11 November 1983, NJ 1984/331 (deer crossing road).

[25] Art. 6:162(3) BW.

[26] For example HR 19 December 2008, NJ 2009/28 (Smeets/Heerlen).

[27] A possessor is a person in a position of a keeper (having control of the car) and also claiming to be owner (art. 3:107 Burgerlijk Wetboek, Dutch Civil Code). This distinguished it from a keeper who acknowledges that someone else is owner (such as a driver who has rented the car).

If the object was used in the course of business (for instance a company truck used to make deliveries), the business is held liable instead of the keeper (art. 6:179 BW).[28]

If the object is also a product in the sense of product liability (on which see Sect. 2.4), the keeper is not liable (art. 6:173(2) BW),[29] unless

- it is plausible (*aannemelijk*), given all circumstances of the case, that the defect already existed when the object was brought into circulation, or
- insofar the damage is material damage that is not compensated due to the limits of compensation of such damage for product liability.[30]

The key question for applying this form of strict liability is when an object does not meet de demands that can be required in the given circumstances (art. 6:173 BW). Dutch lawyers use 'defective' as shorthand for the longer phrase of this provision. The legislative history shows that this provision particularly aimed at objects that are intrinsically dangerous (such as explosives), and moving objects.[31]

As the rule derives from French tort law, in particular what is now art. 1242 Code civil 2016,[32] it makes sense to look at French case law for a further interpretation. In France this form of liability has received generous application, whereby an object is easily found to be deficient if it is found wanting in any sense. The only requirement is that the object was in some sense active, if there is an 'act of the thing' (*fait de la chose*).[33] Although there is very little case law to go by, on the basis of an analysis of the parliamentary history of the article and the case law on the corresponding provision in French law, it seems fair to presume that a partly or wholly autonomous vehicle would be deficient when it would exhibit behaviour that would not be accepted from a human driver.[34] I.e. the car would be treated as if it were a diligent human driver. For instance, a level 0–2 car that automatically parks would be considered defective if it crashed into the car behind it. The car is supposed to be able to park properly like a diligent human driver would.

[28] See further Bauw (2015).

[29] Since the manufacturer is liable in these cases on the basis of product liability.

[30] € 500, see Sect. 2.4.

[31] Tjong Tjin Tai and Boesten (2016), pp. 657–664, s. 4.5.

[32] Formerly Art. 1384 Cc 1804.

[33] Tjong Tjin Tai (2017), pp. 123–132, s. 7; Tjong Tjin Tai (2018), pp. 55–82; Viney et al. (2013).

[34] Tjong Tjin Tai and Boesten (2016), pp. 657–664, s. 4.5; Tjong Tjin Tai (2017), pp. 123–132, s. 7.

2.4 Product Liability

Finally, the *car manufacturer* may be liable on the basis of product liability.[35] The rules for product liability are an implementation of the Product Liability Directive 85/374, and are therefore identical to the rules in other European Union member states. The main requirement is that the product is defective, has a defect (*gebrek*). Defects are usually categorized in

- product defects (such as a defective sensor),
- information defects (such as incorrect operating instructions), and
- design defects.

For autonomous vehicles the last category seems particularly important, as it can be argued that a vehicle that does not respond correctly to complicated traffic situations has been developed and tested incorrectly or does not have sufficient safety measures implemented to cope with such conditions. Liability is restricted to the state of the art: the manufacturer is not liable for risks that would have been undetectable at the time of putting it into circulation, given the state of the art (art. 6:185(1)(e) BW).

A complication is that errors may be introduced because of defects in a software update of the car. Software as such appears to lie outside the application of product liability, although there are different opinions.[36] In the present context the issue is moot, as the software will be made available and integrated by the producer of the car, and can therefore be considered as simply a part of the complete car.[37]

Only two kinds of damage are compensated: personal injury, and damage to property in the private sphere. The property damage is only compensated when it exceeds the amount of € 500.[38] Pure economic loss is outside the scope of product liability.

There is hardly any case law about application of product liability to cars, and none of it concerns accidents. While one might be inclined to conclude that this means that cars never malfunction, it may also indicate that drivers find it hard to prove that the sole cause of the accident was a malfunction of the car. On a more practical note, a likely explanation for the lack of case law is the role of insurance companies in handling car accident claims (Sect. 2.6).

[35] Lavrijsen and Weitering (2019), pp. 167–171; van Wees (2015b), pp. 170–180, s. 3.4; van Wees (2015a), pp. 313–321, s. 4.3; Van Wees (2018); Schreuder (2014), s. 2.1.2; de Vey Mestdagh and Lubbers (2015), pp. 267–280, s. 3.2.3; Vellinga (2020b), pp. 66–71.

[36] Vellinga (2020b), pp. 66–71 mentions different opinions.

[37] Schreuder (2014), s. 2.1.1; van Wees (2015b), pp. 170–180, s. 3.3; de Vey Mestdagh and Lubbers (2015), pp. 275–276; Vellinga and Vellinga (2015); Tjong Tjin Tai and Boesten (2016), pp. 657–664; Vellinga (2020b), pp. 66–71.

[38] Sieburgh (2019), nr. 267.

2.5 Other Persons Liable for Negligence

The general rule of negligence (Sect. 2.2) can be applied to other persons besides the driver. For example, the *possessor* may be negligent if he did not keep the car in good condition and as a result the car crashed.[39] Of course the strict liability of art. 6: 173 BW would also apply in such a case.

Another example is if the car has been repaired incorrectly and subsequently crashes because of the faulty repair. The mechanic who repaired the car might be held liable for negligence.

The manufacturer may also be liable on the basis of negligence to third parties.[40] This provides an alternative ground besides product liability and avoids the restrictions in application of product liability, in particular the restriction to tangible goods and the exclusion of pure economic loss.

Finally there is discussion as to whether supervisory agencies may be liable, in particular the Rijksdienst wegverkeer (RDW) that needs to approve cars before they are allowed on Dutch public roads.[41]

2.6 Insurance

Some additional comments need to be made regarding insurance.[42] Liability on the basis of the WVW in practice does not directly bear on the owner/keeper, because there is mandatory car insurance (art. 2 WAM[43]). Somewhat simplified, the duty to have car insurance rests on the owner and the possessor.[44] If there is a keeper the duty rests on the keeper instead of on the possessor.[45] The insurance must cover the liability of the keeper, driver and passengers of the car for damage to other persons and goods, including damage to passengers and goods within the car itself.[46]

In such cases the insurance company will take over the claims process and pay damages; the victim has a direct action on the insurance company for his damages

[39] In earlier case law this was not too quickly assumed: van Dam (2020), nr. 609.

[40] HR 2 December 2011, NJ 2012/197, HR 25 November 2005, NJ 2009/103, about facts occurring before product liability was established in the code.

[41] Vellinga (2020b), pp. 66–71.

[42] See also van Wees (2016), pp. 29–34; van Dam (2020), nr. 602-2.

[43] Wet Aansprakelijkheidsverzekering Motorrijtuigen, Car Liability Insurance Act.

[44] Art. 2(1) WAM refers to the possessor (*bezitter*) and the person in whose name the license plate has been registered (which will usually be the owner). These are the same persons who may be liable for accidents under art. 1(2) WVW. The difference here is that both have a duty to arrange insurance for the car, while art. 1(2) WVW provides for an alternative liability.

[45] Art. 2(2) WAM. Keeper (houder) is defined in art. 2(2) WAM in the same way as in art. 1(1) (o) WVW. The text suggests that the owner remains obliged to arrange car insurance even when there is a keeper.

[46] Art. 3 WAM.

(art. 6 WAM, also art. 7:954 BW). If the possessor/owner, in spite of his legal obligation, did not have insurance, a government-appointed fund[47] will compensate the victim.

Because of the ubiquity of insurance, insurers tend to settle claims out of court amongst each other if possible (particularly in cases where the accident involved two cars crashing into each other).

2.7 *The Complete Overview of Liability for Level 0–2 Cars*

The liability rules described above apply not only for level 0, but also for level 1 and 2 automation.[48] These rules lead to a distinction between two categories:

- accidents with victims that are not driving or passenger in a motorized vehicle, and
- accidents with victims in another motorized vehicle (including passengers in the car itself).

For the first category the owner/keeper is the primary responsible party on the basis of art. 185 WVW, which is a form of strict liability. The owner/keeper of a car is liable also if the accident was primarily caused by a defect of the car, such as mistakes in appreciation by the AI. In practice insurance will pay the damages; the insurance company can take recourse on the manufacturer in case the accident was caused by a defect in the sense of product liability.

For the second category the driver is the primary responsible party, based primarily on negligence (art. 6:162 BW), where a high standard of care is required. The driver is responsible even for automated processes of level 1 and 2, as he is able to take over driving any time and decides whether to invoke or interrupt an automated process or not. The fact that he can prevent accidents imposes an obligation to do so.[49] The driver will have to take into account possible errors and mistakes by the AI. Hence it is nearly impossible for the driver to escape liability in case the direct cause of the accident was a malfunction of the AI: the driver is supposed to watch out for such defects. The driver could, if the car is defective, take recourse on the person liable for the defect, either the owner/keeper (and possible the car mechanic or the manufacturer).

The net effect is that for cars of level 0–2 victims are protected for any behaviour from the car-driver combination which deviates from what could be expected from highly diligent human drivers.

[47] The Waarborgfonds Motorverkeer, see art. 23–26 WAM. The fund obtains its means by mandatory contributions by car insurance companies operating in The Netherlands, art. 24 WAM.

[48] Lavrijsen and Weitering (2019), pp. 167–171, s. 5.

[49] See also 185(2) WVW which could be applied by analogy.

It is possible that there are other parties who are also liable on another ground, such as product liability, strict liability for defective objects, or negligence. In such cases those parties are jointly and severally liable:[50] the victim can choose whom to sue for the entire damage. The defendant may subsequently sue the other parties for proportional contribution to the award of damages,[51] which is determined by the proportionate contribution of each to the risk (which may also be corrected by considerations of fairness, such as the seriousness of the fault).[52] However, in practice, due to mandatory insurance, the victim simply negotiates with the insurance company of the car owner (as the car owner is liable for any accident with the car). Hence the complications that could arise from the presence of multiple liable parties do not materialize in practice. Art. 185 WVW is the main ground for liability against non-motorized victims.

The victim could choose to sue the manufacturer instead of the owner/keeper or driver, but that would not normally happen given that the latter parties are usually insured and would be easier to hold liable. To sue the manufacturer, the victim would have to prove the presence of a defect in the car.

Assuming that it is proven that there is an accident and the car owner is liable, the insurance company may take the position that damages should be reduced because of lack of causality or because of contributory negligence. The insurance company may at its discretion choose to take recourse on other parties who may have contributed to the accident, but may also decide not to do so.

As regards the proof by the victim: it will usually suffice to prove that there was an accident that seems likely to be caused by the car. For art. 185 WVW this would suffice. If the case is based on negligence, the victim will have to prove negligence or a defect, but the court may find sufficient evidence in supporting circumstances (i.e. indications that the car drove too fast for the specific traffic conditions). I.e. the court would not need to investigate the specific state of the AI in the car, but could suffice with looking at the external behavior in the given case. To determine this behavior it may be useful to have access to the data inside the car[53] (for example to determine the speed, whether it detected obstacles) but that is not always necessary.

Finally, it is also possible that the car owner wishes to sue the manufacturer for the damage to his own car. This will follow the normal rules of product liability and negligence. The owner/keeper may find it difficult to prove that the accident was caused by a product defect. Dutch procedural law has several means by which to obtain relevant evidentiary materials, but it may be costly and time-consuming to litigate such a case. For insurance companies these restrictions are not an obstacle if

[50] Art. 6:99 and 102 BW.

[51] Art. 6:102 and 101(2) BW.

[52] Cf. art. 6:101(2) BW.

[53] The Dutch law of civil procedure provides the victim with an action that may allow him access to such data (the 'exhibitieplicht' of art. 843a Wetboek van Burgerlijke rechtsvordering, Dutch Code of Civil Procedure). This is somewhat similar to the American instrument of 'discovery'.

the case is sufficiently important. Furthermore if it concerns a general defect, a mass claim action could be instigated.

The Dutch system provides grounds to hold several different persons liable. A practical complication is that some grounds of liability are not cumulative but alternative; they do not lead to joint and several liability but only to alternative liability. Examples are the liability of the possessor on the basis of art. 6:173 BW versus the liability of the manufacturer on the basis of product liability, and the liability of the possessor versus the liability of the business user on the basis of art. 6: 173/179 BW. The victim has to ensure that he sued the correct person, or if he wants to play it safe, should sue both possible defendants in the same procedure.

3 Liability for Level 3 and Level 4 Cars

For level 3 (conditional automation) and 4 (high automation) cars, the general legal framework described in the previous section applies as well, as will be discussed below (Sect. 3.1). However, there is also specific legislation for cars that do not have a driver in the car: this is aimed specifically at cars of level 3 and 4. This is discussed in Sect. 3.2.

3.1 Applicable Rules

Liability for cars of level 3 and 4 would principally follow the general rules outlined in the previous section. The same liability addressees can be distinguished. If the accident is solely due to a normal driving error, the general rules for liability of the driver and owner/keeper of art. 6:162 BW and art. 185 WVW apply (Sects. 2.1 and 2.2). These do not need further consideration here. Matters become more complicated in case the accident is wholly or partly due to incorrect behavior of the car. For those situations, the following remarks hold.

The *owner/keeper* of the car will primarily be liable for accidents caused by a defect. Defects of the car are no excuse under art. 185 WVW. The owner/keeper can take recourse on the manufacturer in case the conditions of product liability are fulfilled (see below). In practice this would be at the discretion of the insurer of the owner/keeper, as the insurance company would compensate the victim and would take over a possible claim on the manufacturer or other third parties.

For the *driver* the liability would also remain as stringent as before, as the driver can intervene and should intervene in case of danger. The driver could be liable on the basis of negligence even if the primary cause of the accident is a defect of the car, provided that the driver was actually able to prevent the defect from causing an accident (such as interrupting the automation and take over full control). To be able to do so, the driver will have to remain as diligent as with normal cars, as he has a duty of care to watch out for errors by the car's ADAS.[54] In a 2019 report on partly autonomous cars,[55] the Dutch Safety Board described several cases of accidents occurring because the driver relied overly on the car's AI, highlighting the necessity of continuous attention when driving. Examples are a car on cruise control (at 180 km/hour, far over the speed limit) not recognizing a truck moving into his lane in order to overtake another truck, and therefore not braking; a truck not recognizing a low loader carrying a bulldozer and therefore not braking.

Even if the car manufacturer advertised that the driver need not pay attention unless warned by the car, such a proclamation cannot be invoked against the victim; at best it can give rise to a claim by the driver on the manufacturer for an information defect. The mere fact that the car is significantly automated would be no excuse for a driver. Only if the defect was of such a nature that the driver could not intervene (for example, the brake system did not work) would the driver not be liable. This could be difficult to prove as it would usually be presumed that the driver was at fault when the car under his control caused an accident.

However, if automation at level 4 would advance so far that a driver would not actually need to pay attention all the time, arguably the legal position would amount to level 5. If it becomes socially and morally acceptable that the driver of a level 4 car would not pay attention, courts might treat drivers of level 4 cars similar to drivers of level 5 cars (on which see Sect. 4). In other words, it may be doubted whether the law would make the technical difference between level 4 and 5 decisive for the behavior that is to be expected of the 'driver'.

The *technical* distinction between level 4 and level 5 is therefore from a legal point of view irrelevant; what is decisive is the *legal* distinction between a car where the driver will always have to be alert to dangers, and a car where the driver is allowed not to watch out unless he receives specific indications of malfunctioning or danger (Sect. 4). It is of course possible that the legislator intervenes and sets a specific standard in a statutory provision, but in the absence of such intervention it is up to the courts to decide on this matter. If experiences with level 4 cars are good, social perception might change about what kind of behavior is to be expected.

The liability of the *manufacturer* and/or *car mechanic* on the basis of negligence could be heightened in the case of a level 3 or 4 car, as such cars would take over much more duties. It would therefore be more likely that an accident would be caused by a defect in the car, which may require the persons who can control the risk of defects to take more extensive precautionary measures. The liability of the manufacturer on the ground of product liability (Sect. 2.4) will not change.

For the victim the increased possibility that the accident is due to a defect of the car normally wouldn't matter, as the car insurance of the owner/keeper or driver would cover his damage anyway. The position of the owner/keeper of the car may be different: it is possible that the insurance does not cover the full material damage to

[54] A duty to pay attention at all times is also laid down in art. 8(6) Vienna Convention on Road Traffic, ratified for The Netherlands, see Tractatenblad 1974, 35.

[55] Dutch Safety Board (2019).

the car (dependent on his insurance policy), and the consequence of an insurance claim is usually that the owner/keeper has to pay a higher insurance premium. The owner/keeper may therefore wish to obtain damages from the manufacturer.

3.2 Specific Rules for Self-Driving Cars (Level 1–4)

Recently the Dutch legislator adopted a system of licensed admission for self-driving cars. The rules do not follow the distinction between levels 0–5, but rather use as criterion whether the driver is physically in the vehicle.

As of 1 July 2019,[56] art. 149aa and 149ab WVW allow such cars, where the driver is not present in the vehicle, on public roads for experimental purposes if a license has been given by the Ministry of Traffic (presently under the Ministry of Infrastructure and Water Management). As the rules refer to a driver, they only apply to level 1–4 cars; level 5 cars would not be allowed unless there is still a person (driver) who monitors the car and can intervene if necessary. The only change is that the driver need not be inside the car.

The license may remove the applicability of several of the usual rules for cars. The license can be refused in particular (art. 149aa(5) WVW) if the experiment is not in the interests of persons in traffic and passengers, road safety, keeping the road in good condition, freedom of traffic (art 2(1) WVW), does not contribute to innovation of road safety, sustainability, or preventing traffic congestion, or the lifting of rules cannot be combined with the purpose of the to be lifted rule.

The license will be provided for no more than three years (art 149ab(1) WVW), and will specify inter alia on which roads the experiment may be conducted, during which period, in which weather conditions and at which times, which safety measures are being taken, where the driver is located and how many vehicles the driver may control simultaneously. The license can be revoked if the rules are not followed.

These rules therefore merely allow experimental self-driving vehicles on public roads, with possibly specific restrictions for each license. They do not change the liability rules.[57] The legislator explicitly stated:

> Furthermore, during the experiments there must at all times be a driver who permanently monitors, regardless whether he is inside or outside the vehicle. The driver can always intervene whenever necessary.

and

---

[56] Staatsblad 2019/240, entry into force of Law of 28 September 2018, Stb. 2018/347. See also Stb. 2018/479 on changes in the rules for admission of licensed cars to private roads, and Vellinga, Vellinga and Van Wees 2016 on earlier developments.

[57] See the parliamentary history, Kamerstukken II 2017-2018, 34 838, nr. 3, p. 3.

> there should at all times be a human driver, who can respond to signals or situations that the vehicle can encounter on the road and thereby can influence the behaviour of the vehicle.[58]

The driver is therefore not allowed to watch movies or otherwise let his attention waver.

The victim should not be worse off if a self-driving vehicle is involved.[59] To ensure this, the person requesting a license must at the time of application show that he has adequate insurance.[60] The rules on mandatory insurance for cars also apply to self-driving vehicles.[61]

These considerations therefore confirm the existing level of liability.

4 Liability for Level 5

As yet there are no specific regulations for level 5 cars, which is not surprising as such vehicles cars not allowed on public roads. None have been admitted yet by the Rijksdienst voor het Wegverkeer (the Road Traffic Government Service). Furthermore the current rules presume the presence of a driver.[62] Assuming that a level 5 car is allowed on the road, such a vehicle would be considered an unmanned vehicle. The person in the vehicle would simply be a passenger, as it is the vehicle that does the driving. It seems likely that new rules will be drawn up before level 5 cars are admitted. The existing rules would apply as follows to level 5 cars.

The *owner/keeper* of the car will be liable for defects in level 5 cars, on the basis of art. 185 WVW, as this provision does not require the presence of a driver and can be applied directly to level 5 cars. As damages for this form of liability are covered by insurance companies, it is not unlikely that insurance companies would change the cost of insurance to reflect how safe the car actually drives.[63] As insurance companies collect data about accident-rates of cars and drivers, they should be able to do this. Through this mechanism there is an economic incentive on owners to buy cars that have been observed to be relatively safe, while this market preference provides car manufacturers in their turn with an incentive to improve the safe driving capabilities of their cars.

Besides liability on the basis of art. 185 WVW, the *possessor* may be liable on the basis of art. 6:173 BW: a defective object. As explained in Sect. 2.3, an autonomous

---

[58] Kamerstukken II 2017-2018, 34 838, nr. 3, pp. 7 and 10.

[59] Kamerstukken II 2017-2018, 34 838, nr. 3, pp. 15–16.

[60] Kamerstukken II 2017-2018, 34 838, nr. 3, pp. 16, 21.

[61] Kamerstukken II 2017-2018, 34 838, nr. 3, p. 16.

[62] See also art. 8(1) Vienna Convention on Road Traffic that states that moving vehicles must have a driver. Vellinga (2020b), pp. 66–67 argues for adaptation of the Convention definition.

[63] For human drivers this is already standing practice: a owner/driver who causes an accident has to pay higher insurance premiums than other drivers, which premium is reduced according to the number of accident-free years.

car can be considered defective if it makes a mistake that a highly diligent driver would not have made. The liability of possessor is therefore rather extensive. In practice this ground of liability will rarely find application. Not only are the owner, keeper and possessor often the same person, but art. 185 WVW also provides a more secure basis for a liability claim as it doesn't require the victim to prove the existence of a defect. However, in specific instances there might be reason to use art. 6:173 BW. In case art. 6:173 BW applies and the car is used in the course of business, the liability rests not on the possessor but on the *business* that operates the car (art. 6:179 BW). Therefore a platform that offers driverless cars is liable for defects on the basis of art. 6:173 and 179 BW, even if the cars themselves are not owned by the platform but are leased. Incidentally, in that case the platform would be keeper in the meaning of the WVW.

In the case of a fully autonomous vehicle, level 5, it seems incorrect to speak of a *driver*. For the present analysis, however, we can refer to the driver as the person who effectively can intervene in the operation of the car. This is to be distinguished from a mere passenger who (as in a taxi) can only provide instructions to destination, but does not have access to the controls. A level 5 car could be operated as a taxi if it completely separates the passenger from the actual controls, or could still have a driver if there is a seat with steering wheel or other forms of control. There are boundary cases: presumably cars might have an emergency stop (like trains and subway cars) that can also be operated by passengers. We should distinguish between three cases: mere passengers without any possibility of intervention, passengers who can press an emergency button, and passengers (such as owners/ keepers) who can take over the complete controls whenever they choose. The first group will generally not be liable on the basis of negligence as they lack any control over the way the car drives.

As to the liability of passengers with the possibility to intervene (by using an emergency button or by taking over controls), matters are not entirely clear. Liability would have to be based on negligence (art. 6:162 BW).[64] The first question that needs to be answered is whether the passenger would have an obligation to continually pay attention (and to intervene if necessary). If so, the passenger would effectively be treated like a driver, and the analysis for level 3 and 4 cars would apply.

If the passenger did not have a duty to continual attention, the passenger would have to be held liable for an omission, having neglected to intervene. This is possible in Dutch law: the general liability rule of 6:162 BW also covers omissions. The Dutch Supreme Court has elaborated this by distinguishing three kinds of cases:

[64] It should be pointed out that the mere fact that a level 5 car would be admitted on Dutch roads is not a defence against liability on the basis of negligence. A license does not block a claim on the basis of negligence: HR 10 March 1972, ECLI:NL:HR:1972:AC1311, NJ 1972/278 (Vermeulen/ Lekkerkerker), HR 21 October 2005, ECLI:NL:HR:2005:AT8823, NJ 2006/418 (Ludlage/Van Paradijs).

omissions that are part of negligence, omissions to fulfill a special obligation to supervise a certain danger, and pure omissions.[65]

It is unclear in what category the passenger should be placed. It could be argued that someone who chooses to be transported in a level 5 vehicle would take upon himself a special duty of care (*bijzondere zorgplicht*) to protect other persons in traffic against the danger, and would thereby be required to pay more attention than normal passengers in normal cars would. If so, the question would be when the passenger would have sufficient awareness or knowledge of the imminent risk to be obligated to intervene. Conversely, it is also possible that a court would assume that a passenger in such a case does not need to exhibit such extreme care, and would only have to pay attention and intervene if there are relevant indications of a higher risk of malfunction of the software (because of messages in the media, error messages about updates or so, signs that the sensors would be damaged, odd behaviour).[66]

Alternatively, it is possible that the 'driver', as being a mere passenger, would not have a special duty of care regarding the behavior of the car. The case would then have to be decided as a case of pure omission. This requires at least actual knowledge with the passenger about the danger in question,[67] and a possibility of intervention. For instance: the car drives straight towards a child crossing the street, while the car gives no indication that it would brake, and the passenger sees this happening, but does not push the big red button 'Emergency brake' even though he is clearly aware of this possibility.

Finally, in any case the passenger might be negligent by choosing to set the car in motion in violation of warnings by the manufacturer. An example may be driving the car in fully autonomous mode in weather conditions where it has not been approved for. In such cases there is clear negligence, while the car may not be considered defective.[68]

To sum up: the passenger/'driver' will have at least a residual duty of care, but it is uncertain what this duty will involve.

The liability of *manufacturer* and *car mechanic* would be the same as for level 3–4 cars (Sect. 3.1). Arguably the concrete obligations of the manufacturer would be more stringent, as more safety tests and precautions would be required of a level 5 car. Furthermore it is theoretically possible that the victim chooses to sue the manufacturer instead of the owner/keeper. As far as damages are concerned, there is no advantage in doing so for Dutch cars (which are covered by insurance), but a victim might do so for other reasons, such as exposing defective cars.

---

[65] HR 22 November 1974, NJ 1975/149, Tjong Tjin Tai (2007), pp. 2540–2546.

[66] This might be deduced as a residual duty of care, as has been assumed in the case of giving a dangerous object in custody, HR 29 April 2011, NJ 2011/406 (Melchemie/Delbanco).

[67] Tjong Tjin Tai (2007), pp. 2540–2546 cf. HR 27 November 2015, NJ 2016/245.

[68] It can be argued that the car is defective as it should have detected (and be able to detect) the adverse weather conditions and refused to drive in such circumstances.

The main burden of liability for a level 5 vehicle would therefore probably still rest on the owner/keeper of the car on the basis of art. 185 WVW, with a residual role for art. 6:173 BW for the possessor/business user in cases where art. 185 WVW is not applicable. The burden of paying damages would in such cases be borne by insurance companies. It is possible that insurance companies would as a matter of course then reclaim the damages they paid from the car manufacturers on the basis of product liability and/or negligence. Although victims might also sue car manufacturers directly, this seems less likely as it is harder for a victim to sue on the basis of product liability than to claim under art. 185 WVW or 6:173 BW. In certain cases the 'driver' of the car could be held liable on the basis of negligence as well.

5 Conclusion

The liability for self-driving vehicles under Dutch law is a complicated network of overlapping and interacting rules, with a few clear principles. In Dutch law the focal points of liability are the car owner and driver, who are liable even if the direct cause of the accident is a defect of the car. The actual financial burden of paying damages is shouldered by mandatory insurance. Even for higher levels of automation Dutch liability law tends to hold the owner and driver liable and requires them to pay a certain amount of attention to indications that the car is malfunctioning. In the case of a passenger of a self-driving taxi, liability would rest on the owner, but it is still conceivable that a passenger is liable on the ground of negligence. While the victim will usually be compensated by the car insurance company, it is possible that insurance companies will take a more active role in taking recourse on car manufacturers where the accident is at least partly caused by a defective car.

Laws

Law of 28 September 2018, Stb. 2018/347
Code civil 2016 (France)
Wetboek van Burgerlijke rechtsvordering (Rv), 2002, Dutch Code of Civil Procedure
Wegenverkeerswet (WVW), 1994, Road Traffic Act
Burgerlijk Wetboek (BW), 1992, Dutch Civil Code
Vienna Convention on Road Traffic, 1968
Wet Aansprakelijkheidsverzekering Motorrijtuigen (WAM), 1963, Car Liability Insurance Act

References

Bauw E (2015) Onrechtmatige daad: aansprakelijkheid voor zaken, 2nd edn. Monografiën BW B47. Wolters Kluwer, Deventer

de Vey Mestdagh K, Lubbers J (2015) 'Nee hoor, u wilt helemaal niet naar Den Haag...' Over de techniek, het recht en de toekomst van de zelfrijdende auto. Ars Aequi 64(4):267–280

Dutch Safety Board (2019) Who is in control? Road safety and automation in road traffic, research report November 2019. Available via https://www.onderzoeksraad.nl/en/page/4729/who-is-in-control-road-safety-and-automation-in-road-traffic. Accessed 31 Mar 2020

Engelhard E (2017) Wetgever, pas op! De (vrijwel) autonome auto komt eraan. Ars Aequi 66(3): 230–236

Engelhard EFD, de Bruin RW (2018) Liability for damage caused by autonomous vehicles. Eleven Publishing, Den Haag

Lavrijsen N, Weitering M (2019) De zelfrijdende auto en het overmachtsverweer van art. 185 WVW. Verkeersrecht 67(5):167–171

Rutten R, Oskam V (2016) Overmacht in het verkeer; een zeldzaamheid of toch niet? Verkeersrecht 64(9):322–329

Schreuder AI (2014) Aansprakelijkheid voor 'zelfdenkende' apparatuur. Aansprakelijkheid, Verzekering & Schade 14(5/6): nr 20

Sieburgh CH (2019) Mr. Asser's Handleiding tot de beoefening van het Nederlands Burgerlijk Recht, vol 6-IV: Verbintenissenrecht: de verbintenis uit de wet, 15th edn. Wolters Kluwer, Deventer

Tjong Tjin Tai TFE (2007) Nalaten als onrechtmatige daad. Nederlands Juristenblad 82(40): 2540–2546

Tjong Tjin Tai E (2017) Aansprakelijkheid voor robots en algoritmes. Nederlands Tijdschrift voor Handelsrecht 14(3):123–132

Tjong Tjin Tai E (2018) Liability for (semi-)autonomous systems. In: Mak V, Tjong Tjin Tai TFE, Berlee A (eds) Research handbook in data science and law. Elgar Publishing, Cheltenham, pp 55–82

Tjong Tjin Tai E, Boesten S (2016) Aansprakelijkheid, zelfrijdende auto's en andere zelfbesturende objecten. Nederlands Juristenblad 91(10):657–664

van Dam CC (1985) De Hoge Raad op zoek naar de perfekte automobilist. Verkeersrecht 33:261

van Dam CC (2020) Aansprakelijkheidsrecht, 3rd edn. Boom juridisch, Den Haag

van Wees KAPC (2004) Aansprakelijkheidsaspecten van elektronische rijtaakondersteunde systemen in het wegverkeer. Verzekeringsrecht 2004/5

van Wees KAPC (2010) Over intelligente voertuigen, slimme wegen en aansprakelijkheid. Verkeersrecht 58(2):33–44

van Wees KAPC (2015a) Enkele juridische aspecten van de (deels) zelfrijdende auto. Computerrecht 32(6):313–321

van Wees KAPC (2015b) Aansprakelijkheidsaspecten van (deels) zelfrijdende auto's. Aansprakelijkheid, Verzekering & Schade 15(5):170–180

van Wees KAPC (2016) Zelfrijdende auto's, aansprakelijkheid en verzekering; over nieuwe technologie en oude discussies. Tijdschrift voor Vergoeding Personenschade 19(2):29–34. https://doi.org/10.5553/TVP/138820662016019002001

van Wees KAPC (2018) Voertuigautomatisering en productaansprakelijkheid. Maandblad voor Vermogensrecht 26(4):112–122. https://doi.org/10.5553/MvV/157457672018016004002

Vellinga NE (2014) De civielrechtelijke aansprakelijkheid voor schade veroorzaakt door een autonome auto. Verkeersrecht 62: nr 151

Vellinga NE (2020a) Legal aspects of automated driving: on drivers, producers, and public authorities. Dissertation, University of Groningen

Vellinga NE (2020b) Juridische aspecten van het gebruik van autonome auto's. Verkeersrecht 68(3):66–71

Vellinga NE, Vellinga WH (2015) Enkele verkeersrechtelijke aspecten van toelating van (deels) zelfrijdende of autonome auto's tot het wegverkeer. Verkeersrecht 63: nr 35

Vellinga NE, Vellinga WH, van Wees KAPC (2016) Testen van autonome of zelfrijdende auto's op de openbare weg. Verkeersrecht 64(6):218–227

Viney G, Jourdain P, Carval S (2013) Traité de Droit Civil: Les conditions de la responsabilité, 4th edn. LGDJ, Paris

The Civil Liability for Self-Driving Cars in Austria

Jörg Zehetner, Merve Cetin, and Angelika Holzer

Abstract Austrian law provides a sufficient legal basis for assessing liability issues in connection with self-driving cars. A driver's liability according to the rules of the Austrian Civil Code (ABGB) may be difficult to argue and especially to prove as an unlawful and culpable behaviour by a legal subject may not be established. However, the holder of a motor vehicle could generally be held liable according to the rules of the Austrian Railway and Motor Vehicle Liability Act (EKHG), which provides for a strict (yet limited) liability. In addition, the manufacturer of the autonomous vehicle as well as the software producer are liable for any damages caused by a malfunction of a vehicle's control software according to the rules of the Austrian Product Liability Act (PHG).

1 Self-Driving Cars

1.1 Definition of Autonomous Driving

In the Austrian legal system, there is no legal definition of autonomous driving or autonomous vehicles to date. For this reason, at this point reference is made to the definition of the Association of Automotive Engineers (SAE International) in the so-called "Standard J3016", which divides the automation of vehicles into six levels. The levels range from level 0 (no automation) to level 5 (full automation).

In **level 0**, there is no automation, which is why the driver's full attention is required while driving. In automation **levels 1 and 2**, the monitoring and intervention of the driver is necessary. In level 1, there are only certain assistance systems (e.g. distance control system) and level 2 is partially automated (e.g. lane keeping function). According to SAE, drivers of vehicles at automation **level 3** (Conditional

We thank our colleague at KWR Armin Zarghami for his valuable inputs.

J. Zehetner (✉) · M. Cetin · A. Holzer
KWR Karasek Wietrzyk Rechtsanwälte GmbH, Wien, Austria
e-mail: Joerg.Zehetner@kwr.at

H. Steege et al. (eds.), *Autonomous Vehicles and Civil Liability in a Global Perspective*, Data Science, Machine Intelligence, and Law 3,
https://doi.org/10.1007/978-3-031-41992-8_17

Automation) must take control of the vehicle at all times in critical situations. At **level 4** (High Automation), control by the driver is no longer required for predefined applications (e.g. motorway driving), but the driver must also take over the wheel when requested by the system. At **level 5**, the computer system takes over the complete control of all vehicle systems of the car and a driver is no longer needed.[1] In this sense, "autonomous driving" without a driver describes level five automation. The term autonomous driving should be distinguished from the general term of "automated driving":[2] Automated driving refers to self-driving cars in which a human must be present at all times and able to take control; i.e. essentially level three of the J3016 standard (Fig. 1).[3]

1.2 Traffic Law in Austria

According to § 102 para 3a KFG (Kraftfahrgesetz = Motor Vehicles Act),[4] the driver may transfer certain driving tasks to assistance systems or automated or networked driving systems present in the vehicle if provided by regulation. The precondition for this is that these systems are approved or that these systems meet the requirements for test purposes specified in the regulation.

On the basis of § 102 para 3a KFG the **Automated Driving Regulation (Automatisiertes Fahren-Verordnung, AutomatFahrV)** was adopted.[5] According to this regulation during **test drives** by research institutions and vehicle manufacturers, certain driving tasks, such as keeping distance, accelerating, braking or lane keeping, may be assigned to an assistance system or automated or networked driving system present in the vehicle, provided that compliance with the provisions of the Motor Vehicles Act (KFG), the Railway Crossing Ordinance (Eisenbahnkreuzungsverordnung)[6] and the Immission Control Act-Air (Immissionsschutzgesetz-Luft)[7] is guaranteed in any case.

Test drives can be authorised for example in the following cases: for **autonomous minibuses** (§ 7 AutomatFahrV), **vehicles with motorway pilots for automatic lane changing** (§ 8 AutomatFahrV) **and self-driving army vehicles** (§ 9 AutomatFahrV).[8] The precondition for steering such an automated test vehicle is that the test driver must be able to resume driving duties at any time. According to §

[1] Eustacchio (2017), p. 509; Klar (2017), p. 434; Romaniewicz-Wenk and Jirak (2018), p. 466.

[2] Amlacher and Andréewitch (2017a), p. 168.

[3] standards.sae.org/j3016_201609/.

[4] BGBl 1967/267.

[5] BGBl II 2016/402, Automatisiertes Fahren Verordnung – AutomatFahrV.

[6] BGBl II 2012/216.

[7] BGBl I 1997/115.

[8] §§ 7, 8 and 9 AutomatFahrV.

Fig. 1 SAE International Standard J3016 levels of automation (https://www.sae.org/news/2019/01/sae-updates-j3016-automated-driving-graphic (7.09.2021))

3 AutomatFahrV the driver has to take his designated place in the vehicle in an intended manner and also remains responsible for resuming his driving tasks.[9]

In March 2019, in addition to test drives, a basis in section 3 of the AutomatFahrV was created for **approved driver assistance systems**, specifically parking assistance and motorway assistance, to be used on roads with public transport in Austria without the driver having to continue to have one hand on the steering wheel at all times which is, in principle, mandatory pursuant to § 102 para 3 KFG.[10] The driver can transfer certain driving tasks to the assisting systems, but he remains responsible for taking over the driving tasks again at any time. There is therefore no shift in liability and the general civil liability regulations of the ABGB (Allgemeines bürgerliches Gesetzbuch, Austrian Civil Code) and EKHG (Eisenbahn- und Kraftfahrzeughaftpflichtgesetz) still apply (see Sects. 2 and 3).[11]

In 2022 the testing of vehicles was regulated in more detail.[12] While the AutomatFahrV created the legal framework for testing automated vehicles on roads with public traffic, the amendment of the AutomatFahrV has the following purpose: This amendment is intended to additionally[13] enable the testing of automated vehicles for passenger and freight transportation (§ 7a AutomatFahrV, § 7b AutomatFahrV), automated work machines (§ 9b AutomatFahrV), automated driving on and off highways (§ 8a AutomatFahrV), and automated parking services (§ 9a AutomatFahrV). This is based on stricter requirements for test drivers, a mandatory route analysis and the associated risk management.[14]

2 Fault-Based Liability According to the Austrian Civil Code (ABGB)

2.1 Liability of the Driver

According to the Austrian Civil Code (ABGB), the conditions for the existence of a duty to compensate damages are

a. the existence of **damage**,
b. the **causation** of the infliction of damage,
c. the **unlawfulness** and
d. the **culpability** of the tortfeasor.[15]

[9] Eisenberger et al. (2016), p. 388.

[10] First amendment to the Automated Driving Ordinance, BGBl. II Nr. 66/2019.

[11] Roubik (2019), p. 154.

[12] BGBl II 2022/143.

[13] In addition to the already mentioned § 7 AutomatFahrV, § 8 AutomatFahrV and § 9 AutomatFahrV.

[14] Nikowitz (2022), p. 201.

[15] Krejci and Zehetner (2020), pp. 187 et seq; Welser and Zöchling-Jud (2015), pp. 356 et seq.

2.1.1 Liability for Level 0–2 cars

There are no specific characteristics for vehicles of level 0–2 with regard to liability according to the Austrian Civil Code in terms of the usual liability regime. For this reason, the liability for these cars will not be further discussed.

2.1.2 Liability for Level 3 and 4 Cars

The use of vehicles of levels 3 and 4 does not change the **liability** of the **driver** for culpably caused damage.[16] However, the duties of care may be set lower: It is suggested that the higher the level of automation, the lower the driver's duties.[17]

2.1.3 Liability for Level 5 Cars

The requirements for the liability of a driver of a level 5 car are, in principle, the same as for all the other levels.

However, *Amlacher N, Andréewitch M*[18] see problems on the level of **unlawfulness** with regard to autonomous vehicles because the judgment of unlawfulness always refers to a conduct and not to a success. Conduct is unlawful if it violates commandments or prohibitions of the legal order or offends against morality. However, only legal subjects (= natural or legal persons) can engage in such behaviour, whereas a self-driving car makes autonomous decisions and in this respect cannot set any unlawful human actions.

Further, another prerequisite of unlawfulness is that the damage caused was in some way foreseeable. Since the developed software learns independently, for example through updates or deep learning, it is precisely not foreseeable in which specific direction the AI will continue to develop.[19]

Difficulties also arise at the level of **culpability**:[20] The autonomous driving system could not be considered culpable by definition because it does not have a legal personality. Culpability must also be traced back to the behaviour of a human being.[21]

In conclusion, unlawfulness and culpability cannot be directly attributed to the autonomous driving system itself. A damage claim requires the unlawful and culpable behaviour of a legal subject which[22] may be difficult to establish. Some suggest that a driver of a level 5 car would only be obliged to check the technical

---

[16] Roubik (2018), p. 9.

[17] Eustacchio (2017), p. 510; Harnoncourt (2016), p. 547; Graul (2020), p. 25.

[18] Amlacher and Andréewitch (2017b), p. 209.

[19] Bruckmüller and Schumann (2017), p. 139; Graul (2020), p. 26.

[20] Amlacher and Andréewitch (2017b), p. 210.

[21] Graul (2020), p. 26; Reinisch (2019), p. 303; Rabl (2018), p. 223.

[22] See also Rabl (2018), p. 223.

devices before driving, such as obvious soiling of the sensors, and to intervene in emergencies. Only by breaching these obligations, a damage claim could arise.[23]

It should be borne in mind that **evidence** for a damage claim must be provided by the injured party. *Amlacher N, Andréewitch M*[24] therefore propose to resort to the simplification of proof by prima facie evidence (*Anscheinsbeweis*). The prerequisite for this, however, is the existence of a typical sequence of events. To do this, it would be necessary to find out which errors occurred in the autonomous driving system, which is why access to the vehicle's data, or possibly even to the underlying source code, would be required. *Amlacher N., Andréewitch M's* idea of equipping autonomous vehicles with a kind of "black box"—similar to what is already used in aircraft today—is a possible solution to detect errors more quickly in the future.

2.2 Liability of Programmers, Technicians and Producers

According to fault liability, in addition to the driver, the individual programmer, technician, engineer and the producer who acted in breach of duty of care can be considered as opponents of a damage claim:

A **programmer** who creates software that is reprehensibly faulty, or a **technician** who arranges for the installation of non-functional environmental sensors, are likewise possible liability opponents. However, in most cases it will be difficult to find the specific person who is responsible for the error that caused the accident.[25]

The **producer** is liable if the fault of the individual programmer or technician can be attributed to the producer. In Austria, the well-known rules on vicarious liability (*Gehilfenhaftung*) and the contract with protective effect for the benefit of third parties are used (*Vertrag mit Schutzwirkung zugunsten Dritter*) for this purpose. While the conduct of vicarious agents towards third parties is only attributed to the producer according to § 1315 ABGB (i.e. almost never),[26] the producer must accept the attribution of his vicarious agents in relation to contractual partners according to § 1313a ABGB (*Erfüllungsgehilfenhaftung*).[27] Persons towards whom a contract provides protective effects are treated equivalent to contractual partners within the meaning of § 1313a ABGB. These persons must be foreseeably affected in order to benefit from the protective effect; *Harnoncourt*[28] cites as an apt example that the final purchaser of a motor vehicle is foreseeably affected by the contract between the producer and the intermediary.

[23] Eustacchio (2017), p. 510; Harnoncourt (2016), p. 547; Graul (2020), p. 25.

[24] Amlacher and Andréewitch (2017b), p. 210.

[25] Amlacher and Andréewitch (2017b), p. 210; Harnoncourt (2017), p. 111; Harnoncourt (2016), p. 547.

[26] Amlacher and Andréewitch (2017b), p. 210; Templ (2016), p. 12.

[27] Graul (2020), p. 25; Harnoncourt (2017), p. 111; Harnoncourt (2016), p. 547.

[28] Harnoncourt (2017), p. 111; Harnoncourt (2016), p. 547.

3 Strict Liability According to the Austrian Railway and Motor Vehicle Liability Act (EKHG—Eisenbahn- und Kraftfahrzeughaftpflichtgesetz)

3.1 Strict Liability in Austria

The strict liability according to the EKHG is a special liability regime. In contrast to the general liability for fault according to the ABGB (Austrian Civil Code), neither unlawful nor culpable conduct is required for the liability to arise. Liability applies even in the absence of unlawful and culpable conduct.[29] Due to the high risks associated with traffic, the legislator wanted to create a stricter liability regime than fault-based liability.[30]

The option to choose between EKHG and ABGB can become significant in practice, especially because of the upper liability limit in § 15 EKHG. Apart from that, the same person is not always the subject of liability. If, for example, the holder of a vehicle lends it to his wife and the wife culpably causes an accident, the injured party has the choice whether to claim against the husband as the vehicle holder under the EKHG or against the wife under the Austrian Civil Code (ABGB).[31] In practice, both the holder of the vehicle and the driver are held liable, usually together with the motor vehicle insurer (see also Sect. 3.4).

3.2 Scope of the EKHG

The basic case[32] of the EKHG is

- an accident
- during the operation
- of a motor vehicle or a railway (also cable cars, chair lifts,. . .).

An **accident** is a sudden external event, e.g. a collision with a car or hitting a pedestrian.[33] The term of an accident in the sense of the EKHG can be transferred to autonomous driving.

A **motor vehicle** is a vehicle used in accordance with the Austrian Motor Vehicle Act (KFG) as for use on the road, which is driven by technically released energy (§ 2 para 1 KFG) and has a design speed of at least 10 km/h.[34] The law intends to

[29] Krejci and Zehetner (2020), p. 196; Perner et al. (2019), p. 363.

[30] Perner et al. (2019), p. 371; Welser and Zöchling-Jud (2015), p. 441.

[31] Perner et al. (2019), p. 363.

[32] Perner et al. (2019), p. 371; Welser and Zöchling-Jud (2015), p. 441.

[33] Perner et al. (2019), p. 372.

[34] Perner et al. (2019), p. 372; Welser and Zöchling-Jud (2015), p. 442.

exempt slow vehicles (below 10 km/h design speed) from the strict liability of the EKHG, as the legislator assumes that high speed is one of the main dangers of operation.[35] Since according to § 2 para 1 subsec 1 KFG, the legal definition of a motor vehicle depends on its propulsion by technically released energy, it does not matter whether it is automated, autonomous or controlled by a human being—provided that a speed of 10 km/h is exceeded.[36]

In general, a vehicle is **in operation** when it moves by means of its own engine power. Correctly, the vehicle is already in operation when it is involved in traffic.[37] According to case law[38] (since there is no legal definition), the decisive factor for operation is not whether the motor vehicle was still in operation at the time of the accident, but whether the accident is related to one of the dangers of the motor vehicle.[39]

3.3 Compensable Damages

According to the EKHG there is a liability for personal and property damage. This liability is limited by a maximum limit in § 15 (in principle: EUR 2,080,000, for personal damage) and § 16 EKHG (in principle: EUR 1,300,000, for property damage).

The liability for **personal damage** is pursuant to §§ 1325 et seq ABGB (e.g. treatment costs, loss of earnings etc).[40] For persons who were transported without the will of the keeper ("stowaways"; *Schwarzfahrer*), § 3 para 1, 2 EKHG) there is an exclusion of liability according to § 3 EKHG. Persons active during operation (e.g. the driver, § 3 para 3 EKHG) are likewise not protected. Driven persons, e.g. the co-driver, are well within the scope of protection of the EKHG, since they are not active during operation.[41]

In case of **property damage**, compensation is based on the general rules of the **ABGB**. Lost profit (*entgangener Gewinn*) is not compensable under the EKHG, only in case of gross negligence it can be claimed through competing claims from culpability.[42] In the event of **damage to objects** carried by the motor vehicle, they shall only be compensated under the EKHG if they are carried by a passenger as

[35] Danzl (2018), p. 111.

[36] Danzl (2018), p. 112; Eustacchio (2017), p. 510; Harnoncourt (2017), p. 112.

[37] Perner et al. (2019), p. 372.

[38] OGH 22.8.1996, 1 Ob 9/96.

[39] Eustacchio (2017), p. 511.

[40] Perner et al. (2019), p. 372.

[41] Perner et al. (2019), p. 372; Welser and Zöchling-Jud (2015), p. 444.

[42] Perner et al. (2019), p. 372.

hand luggage. Otherwise, the provisions of the contract of carriage shall apply (§ 4 EKHG).[43]

3.4 *Persons Liable*

In general, the **holder of a motor vehicle** is the one who is being held liable (§ 5 EKHG). The holder is the one who has the power of disposal over the motor vehicle and operates it on his own account. It does not depend on the ownership of the vehicle or the registration (tenants, lessees, etc. can also be holders). If a vehicle has multiple holders, they can be held jointly and severally liable (§ 5 EKHG) and can take recourse internally (§ 869 ABGB).[44]

Pursuant to § 19 para 2 EKHG, the holder is to be attributed with the fault of others who were operating the motor vehicle with his will. § 19 para 2 EKHG establishes a vicarious liability that goes beyond §§ 1313a and 1315 ABGB.[45]

If damage has been caused by several motor vehicles or railways, § 8 EKHG imposes **joint and several liability towards the injured party**. § 11 EKHG resolves the question of recourse according to whether there was negligence (which is not necessary for external liability) and the extent of the (extraordinary) operational risk.[46]

Insurance is compulsory for motor vehicles in Austria. This means that every holder of a motor vehicle must take out liability insurance. The insurance is jointly and severally liable with the holder (§ 26 KHVG) (see also Sect. 3.1). In this way, the injured party has a **sufficient liability fund**.[47]

3.5 *Relief from Liability*

Although the EKHG standardises a strict liability, an exemption is possible according to the rule of § 9 EKHG. Thus, the EKHG does not stipulate a pure liability for success.

As a first step, it is necessary to check whether an **accident** has occurred at all during the operation of a motor vehicle or railway (step 1: liability). If there has been an accident, liability generally applies, unless the accident was caused by an unavoidable event. **Unavoidable events** are events that would be described as force majeure as well as special natural events such as an earthquake, rockfall or a

[43] Welser and Zöchling-Jud (2015), p. 442.

[44] Welser and Zöchling-Jud (2015), p. 443.

[45] Danzl (2018), p. 577.

[46] Perner et al. (2019), p. 373.

[47] Perner et al. (2019), p. 373.

lightning strike that kills or blinds the driver (step 2: exemption from liability).[48] Despite being caused by an unavoidable event, liability nevertheless applies if the accident was caused by an **exceptional operational risk**. An exceptional operating risk is everything that is more dangerous than the normal operating risk of the vehicle, i.e. every dangerous situation that is not necessarily connected with operation (skidding, heart attack of the driver, driver fainting) (step 3: liability).[49]

In brief, there is no liability if the sphere of the operator is completely free of deficiency, except in the case of an exceptional operational risk.

3.6 Strict Liability According to the Austrian Railway and Motor Vehicle Liability Act (EKHG) for Level 0–2 Cars

The strict liability according to the EKHG for vehicles of level 0–2 follows the regime just explained under Sects. 3.2–3.5 and has no special features.

3.7 Strict Liability According to the Austrian Railway and Motor Vehicle Liability Act (EKHG) for Level 3 and Level 4 Cars

The strict liability according to the EKHG for vehicles of level 3 and level 4 follows the regime just explained under Sects. 3.2–3.5 and has no special features.

3.8 Strict Liability According to the Austrian Railway and Motor Vehicle Liability Act (EKHG) for Level 5 Cars

Self-driving cars also fall under § 2 para 2 EKHG. Therefore, if an accident occurs during the operation of an autonomous vehicle, the EKHG applies.[50] The holder of a motor vehicle could generally be held liable even if the accident is caused by an autonomous vehicle.[51]

Liability under EKHG is excluded according to § 9 EKHG if the accident was caused by an unavoidable event. It could be argued that errors in the programming of

---

[48] Danzl (2018), p. 274.

[49] Perner et al. (2019), p. 374; Welser and Zöchling-Jud (2015), pp. 444 et seq.

[50] Danzl (2018), p. 109; Harnoncourt (2017), p. 112; Eustacchio (2017), p. 510.

[51] Harnoncourt (2017), p. 112; Harnoncourt (2016), p. 548.

the autonomous driving system constitute an unavoidable event that excludes liability under the EKHG. However, such an argument would probably have little chance of success: The exemption from liability does not apply if the unavoidable event is caused by a defect in the condition or failure of the devices of the motor vehicle. According to *Amlacher N, Andréewitch M*,[52] an error in the programming of an autonomous vehicle or another error in the control system could be qualified as defect in the condition or a failure of the devices, which is why the exemption from liability would not apply.

Therefore the exceptions of § 9 EKHG do not apply to the holder of a self-driving car. The vehicle holder does not benefit from an exception to liability even if the self-driving car does not recognise other vehicles and causes an accident due to the defective software or sensors or a hacker attack, since according to *Harnoncourt*[53] these are not unavoidable events within the meaning of § 9 EKHG.

As mentioned before, pursuant to **§ 19 para 2 EKHG**, the holder is to be attributed with the fault of others who were operating the motor vehicle with his will.[54] However, it has not yet been fully clarified whether this vicarious liability of the holder should also apply to the use of new technological aids, such as assistance systems and computer programs of an autonomous motor vehicle.[55]

4 Strict Liability According to the Austrian Federal Act Governing the Liability for a Defective Product (Product Liability Act; PHG)

4.1 Regulatory Purpose of the Austrian Product Liability Act (Produkthaftungsgesetz—PHG)

The object of product liability is not the liability of the seller for the defectiveness of an object, but the liability of the **producer** for the **dangerousness of the object** produced by him. The Product Liability Act (PHG), which is based on the Product Liability Directive,[56] covers tortiously protected goods such as life, health and property.[57]

The liability of the producer is to be distinguished from the liability of the retailer. This culpability-based liability according to the Austrian General Civil Code (ABGB) rarely applies, since the retailer did not cause the defectiveness of the

[52] Amlacher and Andréewitch (2017b), p. 211.

[53] Harnoncourt (2017), p. 113; Harnoncourt (2016), p. 549.

[54] Danzl (2018), p. 577.

[55] Harnoncourt (2016), p. 549; Harnoncourt (2017), p. 114; Eustacchio (2017), p. 511.

[56] 85/374/ECC, Council directive of 25 July 1985 on the approximation of the laws, regulations and administrative provisions of the Member States concerning liability for defective products.

[57] Krejci and Zehetner (2020), p. 197; Welser and Zöchling-Jud (2015), p. 451.

product and usually cannot recognise it.[58] Nevertheless, warranty claims against the retailer are possible, as they do not require culpable behaviour.

4.2 Compensable Damages

The strict liability according to the PHG can be claimed not only by the **purchaser** of the goods, but also by **third parties**. Compensation is paid for personal injury and property damage caused by defects that the product showed when being put on the market by the liable party.[59]

Product (§ 4 PHG) is any movable tangible property, even if it is part of another movable property or combined with an immovable property, including energy.

Liability applies to personal injury as well as damage to property (§ 1 PHG). The extent of compensation for personal injury is determined by §§ 1325 et seq ABGB (e.g. treatment costs, loss of earnings etc). Damage to property is only compensated if it occurs to physical objects that differ from the product. For this reason, damage caused to the product itself as a result of the product defect is not compensated.[60]

§ 2 para 1 PHG significantly restricts compensation for property damage for entrepreneurs. Damage to property used for business purposes is not compensated, i.e. damage suffered by an entrepreneur who uses the property predominantly in his business is excluded from compensation.[61] This exclusion of liability seems to have a more limited scope than Art 9 Product Liability Directive which is the basis for § 2 para 1 PHG. However, the European Court of Justice ruled that the national regulation is compliant with Art 9 Product Liability Directive.[62]

Furthermore, § 2 para 2 PHG stipulates a deductible for the injured party. It amounts to EUR 500, per damaging event.[63] The deductible can be claimed based on the ABGB, which does not stipulate a deductible.

4.3 Persons Liable

The liable persons (§ 1 para 1 PHG) are (i) the entrepreneur by whom it was produced and put into circulation, (ii) the entrepreneur by whom it was imported into and put into circulation in the European Economic Area ("importer").[64] The

[58] Rabl (2016), p. 35; Welser and Zöchling-Jud (2015), p. 452.

[59] Welser and Zöchling-Jud (2015), p. 452.

[60] Rabl (2016), p. 63; Welser and Zöchling-Jud (2015), p. 453.

[61] Krejci and Zehetner (2020), p. 197; Rabl (2016), p. 87; Welser and Zöchling-Jud (2015), p. 454.

[62] ECJ 4.6.2009, C-285/08; Rabl (2016), p. 87.

[63] Krejci and Zehetner (2020), p. 197; Rabl (2016), p. 90; Welser and Zöchling-Jud (2015), p. 454.

[64] Welser and Zöchling-Jud (2015), p. 454.

producer in the sense of § 1 para 1 PHG is the person who manufactured the finished product, a raw material or a component part as well as anybody who presents himself as producer by affixing to the product his name, trade mark or other distinguishing feature (§ 3 PHG).

4.4 Causation by Product Defect (§ 5 PHG)

The damage to be compensated under the PHG must be caused by a product defect. A product shall be deemed defective if it does not provide the safety which, taking all circumstances into account, may be reasonably expected. The expectations depend on (i) the presentation of the product, (ii) the use to which it can reasonably be expected that the product would be put, (iii) the time when the product was put into circulation (§ 5 PHG). The definition of § 5 PHG covers all defects assigned to the producer. Design errors, i.e. errors in planning, production errors, i.e. errors in execution including "runaways" (*Ausreißerschäden*), and instruction errors, which are understood to be the omission of information on hazardous properties, fall under this definition.[65]

4.5 Putting into Circulation (§ 6 PHG)

A product shall be deemed put into circulation as soon as the entrepreneur—irrespective of the title—has transferred it to another person into the latter's power of disposition or for the latter's use. Dispatching the product to the customer shall be deemed sufficient.

4.6 Exclusion of Liability (§§ 8 et seq PHG)

The obligation to pay compensation under the PHG is excluded primarily by proving that the characteristics of the product could not have been recognised as a defect according to the state of science and technology at the time it was put into circulation (§ 8 PHG). In a development process as fast as that of AI, this will be relatively often the case.[66] There is therefore **no liability for the development risk**. This means that research in the sense of technical progress is able to develop new techniques and new

[65] Rabl (2016), pp. 134 et seq; Welser and Zöchling-Jud (2015), p. 456.
[66] Reinisch (2019), p. 303.

products that may be associated with dangers that are not yet known and also not recognisable, for which the producer does not bear the risk.[67]

4.7 Liability of the (Software) Producer

Since accidents involving autonomous vehicles are likely to be mostly due to errors in the control **software**,[68] this could provide a basis for arguing a claim against the **producer**. However, complex products such as cars usually have different producers. As mentioned above the producer in the sense of § 1 para 1 PHG is the person who manufactured the finished product, a raw material or a component part (§ 3 PHG). Therefore, the manufacturer of the car can be subject to liability as well as the software producer.[69]

When it comes to the **manufacturer of the car**, the prevailing view is that the manufacturer is liable for any damages caused by a malfunction of a vehicle's control software.[70]

Whether the **software producer** is liable for such damages as well, depends on whether the software implemented in autonomous vehicles can be qualified as a movable tangible product within the meaning of § 4 PHG. In general a **distinction** is made between **operating system software, computer-related software, standard software and individual software**:[71]

Mass-produced operating system software and **computer-related software** distributed by means of (physical) **data carriers** is qualified as a **product** within the meaning of § 4 PHG. In the case of **standard software** it is still considered quite **controversial** whether it falls under the concept of a product of the PHG.[72] An **individual programme** stored on a data carrier is mostly not qualified as a product within the meaning of the PHG, as the intellectual performance is the main focus here.[73]

According to *Rabl C*[74] it is not necessary to differentiate between **standard** and **individual software**, since according to the conception of the PHG, the way in which a product is produced is basically irrelevant. Thus, the software producer is liable if the software controls devices by means of electronic impulses and the stored information directly leads to destruction or damage. In case the software was

[67] Rabl (2016), p. 211; Welser and Zöchling-Jud (2015), p. 458.

[68] Amlacher and Andréewitch (2017b), p. 212.

[69] Amlacher and Andréewitch (2017b), p. 212.

[70] Posch and Terlitza (2016), § 4 PHG Rz 10; Rabl (2016), p. 127; Amlacher and Andréewitch (2017b), p. 212; Eustacchio (2017), p. 512; Harnoncourt (2017), pp. 115 et seq.

[71] Amlacher and Andréewitch (2017b), p. 212.

[72] Posch and Terlitza (2016), § 4 PHG Rz 10.

[73] Posch and Terlitza (2016), § 4 PHG Rz 10.

[74] Rabl (2016), p. 127.

implemented by way of data transmission, the software shall not be considered a tangible product.

According to *Amlacher, Andréewitch*,[75] the control software of autonomous vehicles is usually not individual but standard software that is delivered in a large number of cars. They argue that the control software of an autonomous vehicle can certainly be seen as a product within the meaning of the PHG: On the one hand, it is not typical standard software that is merely executed on a PC, but a thing that comes quite close to a physical product. On the other hand, the software becomes a component of the end product through its implementation in the car. In conclusion, *Amlacher, Andréewitch* consider it admissible to raise damage claims against a software producer.

Koziol H, Apathy P, Koch B[76] argue that it cannot make a difference whether software is stored on a data carrier or transferred digitally in another way. Accordingly, they see software as a product within the meaning of § 4 PHG and assume the liability of the software producer.

Templ[77] takes into account the German literature, where software also falls under the regime of the German PHG, and assumes that in the future the PHG will also be applied in Austria in the case of mere software defects.

We agree with the prevailing view that the manufacturer of the end product, thus the car, is liable for any damages caused by a malfunction of a vehicle's control software. Besides the manufacturer the software producer is liable as well under the PHG irrespective of whether the software was stored on a data carrier or transferred digitally. Otherwise software producers who transfer the software digitally would be better off compared to software producers who provide their software on a data carrier. This would be an unobjective differentiation.

As a result, the manufacturer of the end product as well as the software producer could be subject to damage claims. In case a damage was caused by a software defect and the end producer had to compensate for damages, he could seek recourse from the software producer according to § 12 PHG.[78]

This being said, according to the current case law a manufacturer of a component part cannot be held liable for damages of the end product (*Weiterfresserschaden*).[79] Thus, the software producer could not be held liable for damages of the car.[80] However, the opinion of the Austrian Supreme Court is viewed critically in literature and some support the liability of the manufacturer of a component part.[81]

[75] Amlacher and Andréewitch (2017b), p. 212.

[76] Koziol et al. (2014). Pt B Rz 28, Pt B Rz 131 et seq; dissenting Eustacchio (2017), p. 512.

[77] Templ (2016), p. 13.

[78] For further details see Harnoncourt (2017), pp. 120 et seq.

[79] OGH 3.2.1994, 8 Ob 536/93.

[80] Eustacchio (2017), p. 512; Harnoncourt (2017), pp. 118 et seq.

[81] Koziol et al. (2014). Pt B Rz 28, Pt B Rz 96 et seq; Rabl (2016), pp. 67 et seq.

5 Conclusion

The current legal situation provides a sufficient basis for assessing liability issues in connection with self-driving cars. A liability under ABGB may be more difficult to argue and especially to prove as an unlawful and culpable behaviour by a legal subject may not be established. However, the holder of a motor vehicle could generally be held liable if the accident is caused by an autonomous vehicle under the EKHG. In addition, the manufacturer of the autonomous vehicle as well as the software producer are liable for any damages caused by a malfunction of a vehicle's control software under the PHG.

Law

OGH 13. 9. 2012, 6 Ob 215/11b

References

Amlacher N, Andréewitch M (2017a) Rechtliche Fragen des autonomen Fahrens – Verkehrsrecht (Teil I). jusIT 2017:167–171
Amlacher N, Andréewitch M (2017b) Rechtliche Fragen des autonomen Fahrens – Haftung (Teil II). jusIT 2017:209–214
Bruckmüller K, Schumann S (2017) Automatisiertes und autonomes Fahren: Strafrechtliche Rahmenbedingungen in Österreich. In: Eisenberger I, Lachmayer K, Eisenberger G (eds) Autonomes Fahren und Recht. Manz, Wien
Danzl K-H (2018) EKHG Eisenbahn- und Kraftfahrzeughaftpflichtgesetz, 10th edn. Manz, Wien
Eisenberger I, Gruber CJ, Huber A, Lachmayer K (2016) Automatisiertes Fahren. ZVR 2016:383–392
Eustacchio A (2017) Automatisiert zum Recht. ZVR 2017:509–515
Graul S (2020) Haftung autonomer Fahrzeuge – eine zivilrechtliche Perspektive. ELSA Austria Law Rev V:24–30. https://doi.org/10.33196/ealr202001002401
Harnoncourt M (2016) Haftungsrechtliche Aspekte des autonomen Fahrens. ZVR 2016:546–553
Harnoncourt M (2017) Haftungsrechtliche Aspekte des autonomen Fahrens. In: Eisenberger I, Lachmayer K, Eisenberger G (eds) Autonomes Fahren und Recht. Manz, Wien
Klar W (2017) Das vernetzte Fahrzeug als erster Schritt der Automatisierung. ZVR 2017:434–438
Koziol H, Apathy P, Koch B (2014) Haftpflichtrecht III, 3rd edn. Jan Sramek Verlag, Wien
Krejci H, Zehetner J (2020) Privatrecht, 9th edn. Manz, Wien
Nikowitz M (2022) Verordnung für das Testen automatisierter Fahrzeuge: zweite Novellierung. ZVR 2022:196–201
Perner S, Spitzer M, Kodek G (2019) Bürgerliches Recht, 6th edn. Manz, Wien
Posch W, Terlitza U (2016) In: Schwimann M, Kodek G (eds) ABGB Praxiskommentar, 4th edn. LexisNexis, Wien
Rabl C (2016) Produkthaftungsgesetz: Kommentar, 1st edn. LexisNexis, Wien
Rabl T (2018) Künstliche Intelligenz oder künstliche Aufregung: drei Thesen zur Digitalisierung, ecolex 2018:223
Reinisch F (2019) Künstliche Intelligenz – Haftungsfragen 4.0. ÖJZ 2019:298–305

Romaniewicz-Wenk M, Jirak C (2018) Autonomes Fahren in Österreich – eine Utopie? ecolex 2018:466–470

Roubik M (2018) Automatisiertes Fahren auf Straßen mit öffentlichem Verkehr – Rechtliche Rahmenbedingungen im Vergleich. Bundesministerium für Verkehr, Innovation und Technologie, Wien

Roubik M (2019) Novellierung der Verordnung über die Rahmenbedingungen für automatisiertes Fahren. ZVR 2019:154–157

Templ H (2016) Über "die Haftungsfrage" von selbsttätig am Straßenverkehr teilnehmenden Kfz. ZVR 2016:10–14

Welser R, Zöchling-Jud B (2015) Bürgerliches Recht II, 14th edn. Manz, Wien

Liability in the Case of Autonomous Cars in Romania

Vasile Luha, Bogdan Florea, and Silvia Maican

Abstract The liability is based in the Romanian contemporary civil law on the idea of culpa; still, there are cases when the liability has an objective nature, without culpa, founded on the idea of guarantee, based on the activity risk, the authority risk or a combination of the previous two types of risk, in order to establish the support of the objective guarantee obligation.

A type of objective tort liability is the liability for goods, where the central subject is the legal guardian.

The liability for the traffic accidents caused by vehicles equipped with driver assistance systems is involved only under the conditions of the liability for one's own actions; for the autonomous vehicles, the legal guardian of the accident causing vehicle during the automatic use, is held liable.

1 The Civil Liability for Traffic Accidents in Romania in the Context of EU Legal Provisions

We analyse in this chapter the Romanian legislation of the civil liability regarding the civil liability resulting from the involvement in road accidents of vehicles equipped with various levels of automation. We do not take in consideration the civil liability for damages caused by the respective vehicle in situations where the damage is not a consequence of the traffic, e.g. an explosion during the stationing of the vehicle.

The situation of the autonomous vehicles was discussed by the European Committee on Crime Problems, which analysed the criminal liability in the accidents caused by the completely autonomous vehicles (no driver involved) and depicted that the common feature of the automatized equipment types (including the autonomous vehicles) is that of the ability and necessity to collect information, react and learn from an "experience" without any human interference, so the producers,

V. Luha · B. Florea · S. Maican (✉)
University "1 Decembrie 1918" of Alba Iulia, Alba Iulia, Romania

H. Steege et al. (eds.), *Autonomous Vehicles and Civil Liability in a Global Perspective*, Data Science, Machine Intelligence, and Law 3,
https://doi.org/10.1007/978-3-031-41992-8_18

programmers, and users cannot predict the possible actions of an AI robot. Therefore, the possibility of damage causing cannot be reduced to zero. The European Committee on Crime Problems states that this aspect suggest two conclusions, excluding each other, on the liability for negligence: no one is responsible because the machine acted alone and, alternatively, a producer can foresee a prejudice and, therefore, should face a strict, de facto, liability, for the results of the actions of a robot.[1]

In addition, on 17 May 2018, the Commission released a communication setting out a European strategy on connected and automated mobility. This communication contains a common vision and identifies supporting actions to develop and deploy key technologies, services and infrastructure.

The European Commission proposes for the autonomous vehicles to be equipped with data recording devices for establishing who operated the vehicle (the autonomous system of the vehicle or the human driver) during an accident. Related to the compensations for the victims, the Commission shows that the Directive 2009/103/CE[2] relating to insurance against civil liability in respect of the use of motor vehicles, and the enforcement of the obligation to insure against such liability already establishes a rapid compensation of the victims, including the case of an autonomous vehicle involved in an accident. It is also underlined that, after the payment of the compensations, the insurer can open a legal action against the vehicle producer according to the Directive 85/374/CEE on the approximation of the laws, regulations and administrative provisions of the Member States concerning liability for defective products.[3]

The CEE communication takes in consideration, in our opinion, the ideal situations when the vehicle has a liability insurance and the accidents results due to an abnormal functioning of the intelligent machine destined for private use, a case when the liability belongs to the producer.

The communication does not take in consideration the situation when the vehicle does not have a liability insurance (independent of the existence of an organism compensating the victims of accidents caused by uninsured vehicles), when the prejudice is higher than the insured sum, when the insurance company cannot be identified, and when an abnormal functioning of the system or the victim's culpa cannot be proved.

Further, the CEE Directive 85/374/CEE on the approximation of the laws, regulations and administrative provisions of the Member States concerning liability for defective products takes in consideration only the goods destined for private use. In the situation of a professional use of the vehicle, the insurer paying the

[1] https://rm.coe.int/cdpc-2019-17-draft-assessment-of-the-answers-to-the-questionnaire-on-a/168098e24c, p. 17, accessed on October 28, 2020.

[2] https://eur-lex.europa.eu/legal-content/EN/ALL/?uri=celex%3A32009L0103, accessed on September 10, 2020.

[3] https://eur-lex.europa.eu/legal-content/EN/TXT/?uri=CELEX:52018DC0283, accessed on October 28, 2020.

compensations might face obstacles in the legal actions against the producer. In these conditions, an unjustified difference of treatment would appear between the insurance company intending to open a legal action against the producer of the good used in private actions and the insurance company intending to open a legal action against the producer of the good used with a professional purpose. The Law 240/2004 on the producers' liability for the damages generated by defected[4] brings in the Romanian legislation the provisions of the Directive 85/374/CEE only considers defective products intended for private consumption and used by consumers.

By analysing the provisions of article 3 from the Directive 2009/103/CE, we deduce that the indemnity insurance covers the damages produced by vehicles, with no reference to the person driving the vehicle or to the owner or regular user of the vehicle.

In the same time, article 12 of the same Directive shows that the insurance must cover the liability in the case of personal injury of all the passengers, others than the driver. There is no legal effect for any legal provision or clause in the insurance agreements, which try to exclude the third parties affected by the accidents when the vehicles are operated by: (i) unauthorised persons (excepting the third parties voluntarily embarked in the vehicle, if the insurer can prove their knowledge of the fact that the vehicle was a stolen good); (ii) persons who do not have a valid driving licence; or (iii) persons who do not respect the technical legal obligations relating to the state and safety of the respective vehicle. Also, the family members of the insured person, the driver and any other liable person is involved in an accident and are covered by the insurance agreement are not excluded, based on their relationship, from the benefits of the insurance for personal injury. The insurance covers the personal injuries and the material damage suffered by pedestrians, cyclists and other unmonitored users of the roads, who, after a traffic accident involving a vehicle, have the right to compensations according to the national civil law.

Relating to the compensation procedure regulated in art. 22 from the Directive for vehicles insurance, we observe that this text mentions the request for compensation to be addressed to the insurance company in agreement with the driver or to the representative for the resolution of the compensation requests.

Thus, even if the text of the Directive shows that the insurance is for the vehicle and not for the driver, the procedure for the compensations indicates in article 22 the insurance company in agreement with the person causing the accident.

The orientation toward the insurance of the vehicle and not toward the insurance of the person responsible for the accident has consequences on the victim's compensation in the case of an accident caused by a vehicle of SAE 4 and 5 level with a complete autonomous driving system, because, if the vehicle is the subject of the insurance, the victim will expeditiously benefit of compensations in all situations.

On the other hand, if the subject of the insurance is the person responsible for the accident, then, in the case of complete autonomous vehicles, we face a dilemma. Who is the person responsible for the accident if there was no driver? The electronic

[4]Republished in Monitorul Oficial al României, Part I, no. 313/22 April 2008.

person (e-personhood) is not recognised in any country, as stated by the European Committee on Crime Problems.[5] In this context, the orientation toward the insurance of the vehicle and not of the person seems to be the one insuring an expeditiously compensation to the victim.

The Romanian traffic laws do not regulate at this moment the special cases of the semi- or completely autonomous vehicles. Thus, for the accidents produced in Romania by vehicles with any level of automation, the problem of the liability and compensations must be discussed in relation with the present legislation whose vision is oriented toward the identification of the person's liability for covering the resulted prejudice.

2 The Liability for Level 0–2 Vehicles

The liability is based in the contemporary Romanian civil law on the idea of culpa. Next to it, there are situation when the liability is objective, not based on culpa, and founded on the idea of guarantee.

Culpa maintains its fundamental position as a principle value for tort liability. Art. 1357 from the Romanian Civil Code (further called Civil Code) stipulates: "*(1) The person who causes damage to another person by an unlawful act committed guilty of it shall be obliged to repair the damage (1) (2) The offender is liable for the lightest culpa*".

From provisions in art. 1349 (1), (2), art. 1357–1371, art. 1381–1395, regulating the tort liability for the prejudice caused by a person, results the fact that the following conditions must be cumulatively met to engage the tort liability: (i) the existence of the prejudice; (ii) the existence of tort; (iii) the existence of a relation cause-effect between the tort and the prejudice; (iv) the existence of culpability of the person causing the prejudice, meaning the intention, negligence or imprudence of that person.

The person invoking the tort liability must prove the cumulative existence of all four conditions using all the legal means of evidence, including witnesses and simple presumptions.

The prejudice is represented by the damaging result or result, patrimonial or moral, which is the consequence of violating or harming the legitimate rights and interests of a person.[6]

The tort is represented by the behaviour violating the imperative legal norms or the good morals, without the intention of producing legal effects against their author, which are still legally produced. In other words, by tort we understand the human

[5] https://rm.coe.int/cdpc-2019-17-draft-assessment-of-the-answers-to-the-questionnaire-on-a/168098e24c, p. 5, accessed on October 28, 2020.

[6] Pop et al. (2015), pp. 329–330.

action or inaction against the law or good morals, which produce the violation or breach of the subjective rights or legitimate interests of another person.[7]

The causal relation between tort and prejudice relates to the fact that the tort caused the prejudice and it represents the third essential condition of the tort liability. Thus, for the right to compensations to be born, there must be established a relation cause- effect between the tort and the prejudice. Both for the tort liability and for the contractual liability, the civil liability does not exist outside the causal relation between the tort and the prejudice.

The following aspects must be considered to the establishment of the causal relation specific to the tort liability:[8] (i) the cause-effect relation specific between the actions or inactions of the people, with an illicit character, and the prejudice must be established. The context explaining the appearance of a prejudicial effect, under the relation of natural causality, cannot be ignored; (ii) the illicit act can be represented by actions or inactions, therefore it must be established if the distinction is important for the establishment of the prejudice; (iii) the causality relation is not in direct connection with the relation between act and prejudice; it can also be a mediate relation, when the act produces a situation that allows other factors to cause the prejudice.

The Romanian doctrine established the culpability as representing the authors' psychic attitude by the time of the illicit act or before it, in relation with the act and its consequences.[9] The definition formulated in the doctrine underlines the essence of the problem by referring to the intellectual and will foundation of the human action, and, in the same time, to the connection between the man's actions with its consequences (the objective element). We highlight that issue is the attitude imputable to its author, the individual's subjective attitude, requiring the realization of the social significance of the act, the anti-social character and the will to act and to assume the legal consequences.[10]

Disregarding the form, the culpability requires the existence of the following factors: (i) the intellectual factor (the voluntary human act contains a complex cognitive process, requiring the knowledge of the object, reality, the realization of the needs and desires, the mediation and the regulation of the means and their correlation); (ii) the volitional factor (the psychic activity aiming to reach goals, requiring the deliberation and the decision for a specific behaviour to be manifested).

Distinct from the tort liability for one's act, the Romanian Civil Code regulates the objective tort liability, founded on the idea or the legal obligation of guarantee, based on the activity risk, the authority risk or a combination of the two, in order to establish the obligation of objective guarantee. The cases of liability regulated by the connected legislation are based on the obligation of guarantee, sometimes associated with the principle of precaution.

---

[7] Pop et al. (2015), p. 337.

[8] Costin and Luha (2007), pp. 123–125.

[9] Eliescu (1972), p. 176; Stătescu and Bîrsan (2008), p. 176.

[10] Romoşan (1999), p. 13.

A type of objective tort liability is the liability for objects where the legal guardian is the subject, a notion explained further in our work, for the prejudice caused by the SAE 3–4 level vehicles.

For the collision of vehicles, if there is possible to establish the culpa of those involved, they will be liable in the conditions of the tort liability for one's act. In relation with the other drivers, that are also legal guardians of their vehicles, this act represents a case of force majeure (particularly in relation to the natural and anonymous event produced independently from the human desire. E.g.: the sudden appearance of a man, object, or animal on the road, the incomplete or inexistent signalling of the road), possible to abolish the liability if they did not contribute to the creation of the prejudice.

In the situation when the culpa belonging to one or several drivers cannot be proved, the problem of engaging the liability of the legal guardians of the vehicles involved in the accident appears; in this case, the provisions of art.1376 (2) Civil Code are applicable: "*The disposition of paragraph 1 are to be applied also in the case of vehicles collision or in similar cases. Still, in such cases, the obligation to recover all the prejudice belongs to the person whose culpable act meets the conditions of force majeure in relation with the others*".

The act of a third party of breaching from culpa the traffic law can be considered, for the other participants to the traffic, a case of force majeure if it has an objective character, unpredictable and irresistible. Therefore, it should be discussed the problem of applying the principles of tort liability for one's act for the drivers, founded on the culpability of the behaviour while driving or the liability of the legal guardian of the vehicle, who guarantees an objective liability, independent of any culpa. In Romania, the traffic on public roads of vehicles, pedestrians and other categories of road users, the rights, obligations and responsibilities of individuals and legal entities, as well as the duties of public administration authorities, institutions and organizations are subject to the provisions of the Government Emergency Ordinance (Ordonanța de urgență a Guvernului) no. 195/2002 regarding the traffic on public roads,[11] as amended and supplemented, further named OUG no. 195/2002.

The special norm regulating traffic on public roads aims to ensure a fluent and safe circulation and to protect the life, corporal integrity and health of all traffic participants or in the nearby area of the traffic, also to protect their legitimate rights and interest, the public and private property and the environment.

In this respect, OUG no. 195/2002 defines the public road as any path for terrestrial communication open to the public and especially organised for the pedestrian and car traffic, except for the railways. The legal provisions expressly mention that the roads that are closed to the public circulation are signalled at their entrance by visible inscriptions.

According to art. 75 in OUG no. 195/2002, the traffic accident is the event cumulating the following conditions:

[11] Republished in Monitorul Oficial al României, Part I, 670/ 3 August 2006.

(a) it occurred on a road open to public traffic or originated in such a place;
(b) caused the death, the injury of one or more persons, or damaged at least a vehicle or produced other material prejudice;[12]
(c) the event involved at least a moving vehicle.

By corroborating the text of the special norm, OUG no. 195/2002, with the provision of the Civil Code, it results that, for the accidents produced on the public road, the liability will be determined according to the special norm, and for the events involving vehicles outside the public roads, the liability is engaged according to the common law.

The circulation of all participants to the traffic on the public roads must take place in strict accordance with the rules established by the special norm. Therefore, the behaviour to be adopted by the participants to the traffic is that of conforming to the legal provisions regulating the traffic on public roads.

When a participant breaks the law and causes an accident, his behaviour is not the expected and accepted behaviour, so it represents a case of force majeure for the others, as external, unpredictable, invincible and inevitable event, as provisioned in art. 1376 (2) of the Civil Code.

Levels SAE 0–2 of automation of the vehicles mean that the human drives even if the assistance systems are active. Even if the driver's feet are off the pedals and he does not have the control of the wheel, he is responsible for the breaking and acceleration procedures, so, in case of an accident, the human driver is fully responsible.

Example given, the assistance systems for the SAE 0–2 level vehicles are limited to emit alerts and momentary assistance. The equipment of the vehicle with assistance systems as the urgent automatic breaks, the alert on a dead point, etc. do not diminish the human responsibility. These systems are created to assist the human driver, and not to replace any of the mandatory maneuvers for the safety of driving.

Also, this type of assistance systems for brakes and acceleration installed on SAE 1 and 2 vehicles do not exonerate the liability of the human driver. Even if the vehicle has systems for its on-road maintenance or an adaptive radar system, they still need to be operated under the supervision of the human driver.

The Romanian legislation does not regulate the conditions to exonerate the liability for the driver who causes a traffic accident by exclusively counting on the assistance systems of the vehicle, because he has the obligation to permanently supervise the systems and take the control when necessary.

In Romania, the person driving the vehicle that breaches with culpability a rule of the public traffic is responsible for a traffic accident, meaning that we are in the

[12]The High Court of Cassation and Justice of Romania, by the Decision 8/2019, published in Monitorul Oficial al României, Part I, 424/ 30 May 2019 decided that the injury of one or several persons, as mentioned by art. 75b), contained in the definition of the traffic accident, as premises for leaving the scene of an accident, regulated by de art. 338 (1) in the Romanian Penal Code, also refers to self-injury, when teh only injured person is the driver of the vehicle involved in the accident.

presence of tort liability for one's act, if the following conditions are cumulatively met: (i) the existence of a prejudice; (ii) the existence of an illicit act; (iii) the existence of a causal relation between the illicit act and the prejudice; and (iv) the existence of culpability for the one causing the prejudice.

Thus, the driver of a SAE 0–2 level vehicle is the subject of a tort liability for his act and cannot invoke the lack of liability for breaking the traffic law as a result of using the assistance systems. E.g., a driver cannot invoke the fact that he did not respect the priority of another vehicle because his adaptive radar system did not slow by itself in an intersection and caused an accident, because the system requires the driver to set a specific speed that will be maintained until an obstacle is perceived (i.e. the stop lights of another vehicle in front of it). This system is not destined to recognise the traffic signs and the human driver must deactivate it outside the highway or any other road allowing a constant speed.

Still, if a driver in the above presented situation can prove that the assistance system did not function in its normal parameters and did not allow him to take the control of the vehicle, when he tried to do it, we are in a situation of the producer's liability for a defective product, if the prejudiced person is a consumer. In this situation, the producers' liability is objective, based on the idea of guarantee and not on the culpability of the person causing the accident.

If the injured person is not a consumer, the driver of the vehicle legally summoned to cover the prejudice caused by the ineffective functioning of the assistance system could call in his guarantee the producer or the seller of the vehicle that, due to its defects, stopped the driver from controlling it and preventing the accident.

Another case of liability engaged for other person than the person responsible for the accident is when the driver is underage, prohibited by court or is an agent of another person.

The liability for the prejudice caused by a minor or a person in the above-mentioned cases is regulated by art. 1372 in the Civil Code. The common element of tort liability is the lack or the decrease of the exercise capacity and, in these cases, the liability of the legal tutor is engaged.

At the base of the relation between the doer and the responsible person is the obligation to supervise and control of the living manner of the minor's or of a person under guardianship. The provisions refer to the permanent supervision as the only modality to have authority on the supervised persons, justified by their unpredictable behaviour, lacking discernment and able to cause prejudice to other people. The individuals accepting the supervising obligation (next to the immediate supervision, they accept the guidance and the control of the acts of those supervised) expressed their agreement to prevent the prejudice of other people and to be liable for the consequences.

Related to the tort liability of the principal for damages caused by illicit acts of his agents, this type of liability appears only in the situation where the offenders cause unjust damages to third parties through an illegal extracontractual activity, and it is regulated by art. 1373 in the civil Code. The argument is represented by an activity of directing, supervising, and controlling another person.

Between the doer (the agent—who caused a prejudice by illicitly acting) and the responsible person (the principle—the person performing the activity of directing, supervising, and controlling the agent) a principal—agent relation is established, allowing the functioning of the liability mechanism for the prejudice produced in relation with the delegated responsibility or function. The engagement of the tort liability of the principal is possible only if specific general conditions (prejudice, illicit act, and the causal relation between the illicit act and the prejudice) and special conditions related to the doer are met (the existence of the agent-principal relation between the doer of the illicit act leading to prejudice, as agent, and the liable person, as principal; the illicit and prejudice causing act of the doer must be in connection with its agent responsibilities or function[13]).

The victim of the agent's illicit act has the right to choose to claim compensations from the agent, from the principle, or from both; in the last case, we have a joint liability of the agent and of the principle.

Therefore, when a minor, a person under judicial interdiction or under the direction, supervision and control of another person drives a vehicle on a public road and causes an accident with prejudices for a third party, others' liability can be engaged: parents, tutors, and principals.

3 The Liability for the SAE 3–4 Level Vehicles

If, in the case of SAE 0–2 vehicles, we speak of systems for the driver's assistance, for the SAE 3 and 4 level, the concept changes and we must take in consideration the systems with automatic driving, which can autonomously drive a vehicle in certain conditions and will not function outside these conditions. If, the conditions for the autonomous driving of the vehicle are met and the autonomous systems are active, the human does not drive, not even when he/she occupies the driver seat.

The difference between level 3 and level 4 resides in the fact that, for the level 3, the system may require the human driver to take the control, and in this case, he/she must drive.

Therefore, for the SAE 3 level, although the automatic system controls the functioning of the car when it is in operation, the human driver must be prepared to take over when required to do so by the system.

The equipment on the automatized vehicles for data recording to establish who was driving (the autonomous system or the human driver) during an accident should also provide information on a possible request of the system for the human driver to take control of the vehicle prior to the accident.

Identifying the driver of the vehicle (human driver or automated system) at the time of the accident or, as the case may be, proving that the automated system made a

[13] Eliescu (1972), pp. 287–293; Stătescu and Bîrsan (2008), pp. 146–154; Boilă (2008), pp. 279–366.

request for the human driver to take control of the vehicle prior to the accident, is extremely important in determining liability for damage caused by the vehicle involved in the accident. Thus, if the human driver was driving the vehicle when the accident occurred, the aspects exposed for the SAE 0–2 level vehicles are applicable. If the vehicle was driven by the automatic system, which did not request the human driver to take the command, we are in a situation in which the liability of the human driver cannot be engaged because he/she did not have the direction, control and supervision of the vehicle. Therefore, the tort liability for the human driver's act is excluded.

There is no Romanian regulation for the vehicles with automatic driving systems, but the Civil Code regulates, as presented above, the liability for objects.

The foundation for the liability for objects as understood from the express provisions of art. 1376 (1) in the Civil code is objective, independent of any culpa and implies object legal guardian's liability. The central idea for the engagement of legal guardian's liability is related to his legal obligation of guaranteeing the risk of an abnormal functioning of the object in the eventuality of a prejudice. According to the Romanian traditional doctrine, the liability of the legal guardian is direct and objective, derogatory from the liability for one's act—a technical form that expresses a guaranteeing obligation.[14] Now, the establishment of the liability for objects is founded on the idea of guarantee for the activity risk and it is dominant in the contemporary Romanian doctrine, highlighting the necessity of a complex insurance system for the civil liability and the possibility to transfer completely the prejudice from the insured person to the insurance company. The necessity of a new guaranteeing fund was also underlined, in order to strengthen the sources of damage reparation and to pay more attention to the victims, next to the objective establishment of the tort liability for the legal guardian and the expansion of the insurance system for goods.[15]

According to art. 1377 in the Civil Code, the guarding of the object belongs to the owner or to the person who, based on a legal provision or contract, or even de facto, has an independent control and supervision of the object and use the object for private interests.

The notion of "object guarding" led to the following question in the Romanian doctrine: what does this security represent?

It has been shown that the "legal guardianship" of an object represents the power of direction, supervision and control that a person can exercise *de jure* or *de facto,* independently over an object that he uses directly or indirectly in his own interest. The legal guardianship is different from the material guardianship; the material guardianship is also a power to direct, supervise and control an object, but but the person exercising the material guardship does not exercise this power independently

[14] Eliescu (1972), pp. 371–374; Boilă (2008), pp. 161–162.

[15] Pop (2000), p. 302.

and in his own interest, but in the interest and under the authority of the legal guardian of the object, whose orders and instructions are followed.[16]

In the same time, the legal guardship can be split between the owner or the possessor and the temporary holder (tenant, borrower, depositary, etc.): the legal guardship of the structure remains to the owner or to the possessor and the legal guardship of the use and maintenance of the object passes to the temporary holder.[17]

For the existence and the engagement of the legal guardian's liability for the damage caused by objects, several general conditions and some specific conditions must be met.

The general conditions that must be fulfilled for the engagement of the legal guardian's liability are the existence of the prejudice and the existence of the causal relation between the action of the object and the prejudice.

The special conditions refer to the quality of legal guardian of the object that caused the prejudice and to the effects of the liability.

The quality of legal guardian from whom the compensation is claimed is not necessary to be proved; it is presumed to the contrary that the quality belongs to the owner, the holder of the real right or to the possessor. The legal guardian of the objective can decline the liability by proving that the prejudice is caused by force majeure, an act of the victim, or an act of a third party with no accountability for the legal guardian. Art. 1380 in the Civil Code stipulates, "For the enumerated cases (...) there is no obligation to repair the prejudice, if this was exclusively caused by an act of the victim, the act of a third party or a case of force majeure".

Related to the effects of the liability, we observe that, if the conditions are met, the victim is entitled to pretend the compensations from the legal guardian of the object.

If the object causes the prejudice during the time that the material guardship belongs to a different person than the legal guardian, the victim can claim compensations from the material guardian, but only in the conditions and based on the material guardian's liability (articles 1357–1371 Civil Code). If the victim claims the compensations from the legal guardian of the object, whose liability is objective and more aggravated, the legal guardian pays the compensations and is entitled to a back down action against the material guardian, if he can prove that the conditions of the liability for material guardian's own deed are met, as provided by article 1357 Civil Code.[18]

In case of vehicles collision, some notions need to be clarified. Firstly, we observe that the text of the Romanian Civil Code does not distinct between the collision of vehicles on the public roads and outside them.

The collision of the vehicles was established as a special hypothesis of liability for the prejudices caused by vehicles only when there is no possibility to establish a specific person's culpa. Each guardian of the vehicle can claim compensation from the legal guardians of other vehicles involved in the accident and is bound, at his

---

[16] Pop et al. (2015), p. 377.

[17] Pop et al. (2015), p. 379.

[18] Stoica and Baias (1987), pp. 48–53.

turn, to compensate the others. The legal disposition consecrates the idea of reciprocity of risks, meaning that all legal guardians are jointly liable to ensure the recovery of all prejudice.

If an accident is produced when the automatic system drives, the following hypotheses may appear: (a) the engagement of the legal guardian's liability; (b) the engagement of the direct liability of the producer for the defective products in the case of the consumers; or (c) the lack of the legal guardian's or producer's liability if they can prove that the prejudice is caused by: a case of force majeure; the victim's act; or, a third party's act.

(a) In the case of SAE 3 level vehicle, driven by the automatic system which did not require the driver to take control, and in the case of the SAE 4 level vehicle driven by the automatic system, in the absence of liability insurance or for the part exceeding the insured amount, the legal guardianship will be held responsible for the damage caused by his objects, which is an objective liability based on the idea of guarantee.

Thus, the legal guardian of the vehicle will be responsible independently of any fault for the damage caused by the vehicle under his guard. This objective liability, based on the idea of a guarantee, may also be considered to cover for the situation in which a defect of the product cannot be identified, since the automated system has learned from its own experience without human interference and has developed a behaviour which could not be predicted neither by its manufacturers, neither by its programmers nor by its users.

The objective liability of the legal guardian for things does not impinge in any way on those laws laid down by the European Committee on Crime Problems concerning liability for negligence. Thus, although no one is responsible for the machine acting on its own, the liability of the legal guardian is bound by the legal guarantee obligation.

(b) If the damaged person is a consumer and it has been proven to be a vehicle malfunction, the manufacturer or seller of the goods whose failure to function led to the accident and thus to the occurrence of a damage could be held directly under the responsibility of the manufacturer or seller of the goods.

If the accident was due to a defect of the used good which, however, is not intended for private consumption, the producer is not directly liable, so that the legal guardian of the vehicle sued to cover the damage caused by a malfunction of the good, must introduce legal proceedings for the warranty against the manufacturer or the seller of the good whose defect in operation caused the accident and, consequently, the occurrence of a damage.

(c) In the third hypothesis, Article 1380 Civil Code states, "there is no obligation to cover the damage where it is solely caused by the act of the victim itself or by a third party or is the result of a case of force majeure".

The legal guardian or the manufacturer is exempted of liability if he proves that the cause of the defect in a product which caused the damage is force majeure, the victim's act or the act of a third person, if the conditions of force majeure are met.[19]

Therefore, to the extent that the victim's behaviour was such as to create a malfunction of the automatic system, the liability of the legal guardian is exempted if the latter can prove that the victim caused the accident. Thus, a pedestrian crossing the road through a place where pedestrian traffic is usually prohibited that caused the vehicle's automatic driving system to be unable to reduce speed until the impact and pedestrian injury to be avoided, shall not be attributed to the automatic system.

In addition, the act of a third party that caused the malfunction of the automatic driving system is exempting from liability for both the legal guardian and the manufacturer. For instance, the acts of covering or destruction of a road indicator by a person may prevent the automatic system from recognizing that indicator and may cause an accident. It should be noted that, as compared to the legal guardian, the manufacturer may be a third party if it is proved that the accident occurred because of a defect in production or the designing of the product.

As for force majeure, we state that Article 1351(2) Civil Code stipulates that: "Force majeure is any external event, unforeseeable, absolutely invincible and inevitable".

Force majeure is an external circumstance, not related to the person or to the good, which caused the damage, or to his qualities. This requirement is met by the followings: extraordinary events not related to the will and activity of a human (natural cataclysms); human-made events (acts of war, revolutions). The unforeseeable event is based on a social-objective unpredictability caused by conditions that are not specific to the person held to be responsible, and are determined on the extraordinary nature of the event. For there to be force majeure it is enough for the unpredictability of the event to be normal, there is no need for absolute, extraordinary or exceptional unpredictability. Invincibility and inevitability must be absolute, with the assessment being based on the condition and possibilities of the person who takes the utmost due care and diligence of a man. If the impossibility is relative and the person held to be responsible was unable to make every effort to avoid that event, he will be forced to cover the damage.[20]

Where force majeure is the sole cause of injury, there shall be no causal relationship between the unlawful act and injury, so it is not possible to impose the civil liability. If force majeure is not the only cause of the damage, but is caused by a plurality of causes of which only one or some of them meet the character of force majeure, the person to be held responsible shall be partially exempted from civil liability.

[19] Chirică (2017), pp. 461–464.

[20] Vasilescu (2012), p. 374.

4 The Liability for the SAE 5 Level Vehicles

As presented *supra,* there are no legal regulations in Romania for the autonomous vehicles so the liability for the prejudices resulting from an accident caused by a SAE 5 level vehicle will be engaged according to the provisions of the common law, which is the Civil Code and the subsequent legislation (i.e. the legislation regulating the prejudice caused by the defects of the products), similar to the situation of the SAE 4 level vehicles with an automatic driving system that totally controlled the vehicle.

The automatic driving systems of SAE 5 level will not require the person to take over driving, these systems being able to drive in any situation. Similar with the SAE 4 level, the pedals and wheel are optionally installed.

When there are no pedals or wheel, the individual in the vehicles does not have any direct control on the vehicle, but he/she can eventually give or modify a command by using an electronic system connected to the vehicle. E.g., the person in a SAE 5 level car introduces a destination in a mobile application, which is transmitted to the electronic system of the car and the command will be executed. If, during the travel, the person in the vehicle wants to modify the command, he/she must give a new command, following the same steps as for the prior command and using the same electronic means.

In these conditions, it is irrelevant if the person has or has not a licence for using on public roads the vehicle he/she is traveling with, because the human has no direct control on the vehicle and is not able to react and take control over the vehicle in a possible dangerous situation undetected by the automatic system.

Therefore, in the case of a vehicle with a SAE 4 or 5 automatic system, no culpa can be identified for the person in the car. Thus, all the persons in this car are, in case of accident, entitled to demand compensations for injuries, even if one of them occupies the driver's seat if the automatic system of the car does not allow, or did not require, the human driver to take the control.

5 Conclusions

Concluding, in case of an accident provoked by a vehicle with a SAE 4 or 5 level of automation, whose automatic system drives without requiring or allowing the person to drive, the victims will receive compensations from the insurance company of the vehicle; the liability of the legal guardian of an autonomous SAE 4 or 5 is engaged when: (i) the vehicle does not have a liability insurance; (ii) the prejudice is larger than the insured sum; (iii) the insurance company cannot be identified; (iv) an accident took place without being possible to prove a defective functioning of the system, the victim's culpa or a third party's culpa and neither a case of force majeure.

In case the legal guardian is a consumer, the good is used for private interests, and it is possible to prove a defect of the product, the liability will belong to the producer. If the injured person is not a consumer, the legal guardian of the car summoned to court to cover the prejudice caused by the vehicle under his guard, will have to call in his guarantee the manufacturer or the seller of the vehicle and will have to prove that a defect of the product was the cause of the accident.

The objective liability of the legal guardian for goods does not interfere with the provisions of the European Committee on Crime Problems on the liability for negligence. Therefore, although no one is responsible for a vehicle equipped with Artificial Intelligence acting on its own, the liability of the legal guardian is engaged based on the legal obligation of guarantee.

In Romania, at this point, it is not possible to discuss a situation of transfer of liability from the legal guardian to the Artificial Intelligence vehicle who acted independently having the direction and total control of the vehicle, because the electronic person (e-personhood) is not recognised as legal person. Such a conception will contradict the principle of the legal guardian's liability, because, even if the AI vehicle is able to autonomously drive, it is under the direction, control and supervision of a human who introduce the car in traffic and use it for various purposes (own use, rental, etc.). Therefore, the legal guardian has the obligation to insure the car. If the liability of the legal guardian is not engaged and he/she does not fulfil the legal obligation of ensuring the vehicle, the victims of an accident provoked by the vehicle do not receive compensations.

The legal orientation and the jurisprudence in Romania do not consider the liability *de facto* of the producer for all the actions of the AI robot. Still, since law is a living domain and the concept of autonomous vehicle is a new concept in full evolution, we cannot exclude a possible future orientation toward this type of liability.

Laws

Directive 2009/103/EC of the European Parliament and of the Council of 16 September 2009 relating to insurance against civil liability in respect of the use of motor vehicles, and the enforcement of the obligation to insure against such liability, available online at: https://eur-lex.europa.eu/legal-content/EN/ALL/?uri=celex%3A32009L0103. Accessed 10 September 2020

Romanian Civil Code from July 17, 2009 (or Law nr. 287/2009)

LAW no. 240 from June 7, 2004, regarding the producers' liability for the damages generated by the defective products

Government Emergency Ordinance (OUG.) no. 195/2002 regarding the traffic on public roads

Council Directive 85/374/EEC of 25 July 1985 on the approximation of the laws, regulations and administrative provisions of the Member States concerning liability for defective products https://eur-lex.europa.eu/legal-content/EN/TXT/?uri=CELEX:52018DC0283. Accessed 28 October 2020

References

Books

Boilă LR (2008) Răspunderea civilă delictuală obiectivă. Ed. C.H.Beck, Bucureşti

Chirică D (2017) Tratat de drept civil. Contracte speciale. In: Vânzarea şi schimbul, vol I, 2nd revised edn. Ed. Hamangiu, Bucureşti

Costin MN, Luha V (2007) Teoria generală a obligaţiilor, vol II. Ed. Risoprint, Cluj Napoca

Eliescu M (1972) Răspunderea civilă delictuală. Ed. Academiei, Bucureşti

Pop L (2000) Teoria generală a obligaţiilor. Ed. Lumina Lex, Bucureşti

Pop L, Popa IF, Vidu SI (2015) Tratat elementar de drept civil. Obligaţiile. Ed. Universul Juridic, Bucureşti

Romoşan ID (1999) Vinovăţia în dreptul civil român. Ed. All Beck, Bucureşti

Stătescu C, Bîrsan C (2008) Drept civil. Teoria generală a obligaţiilor. ed. IX, revised, Ed. Hamangiu, Bucureşti

Stoica V, Baias FA (1987) Acţiunea în regres a paznicului juridic împotriva paznicului material. Studii şi cercetări juridice. Serie veche, no. 1/1987, Bucureşti

Vasilescu P (2012) Drept civil. Obligaţii. Ed. Hamangiu, Bucureşti

Online Publications

European Committee on Crime Problems (CDPC) (2019) Assessment of the Answers to the Questionnaire on Artificial Intelligence and Criminal Justice (using the example of Automated Driving). https://rm.coe.int/cdpc-2019-17-draft-assessment-of-the-answers-to-the-questionnaire-on-a/168098e24c. Accessed 28 Oct 2020

Autonomous Vehicle and Liability in Sweden

Anton Olsson and Karl Montelius

Abstract This chapter is the Swedish part of a research project that intends to edit a comparative description and analysis of civil liability in the context of automated vehicles for selected countries.

The chapter shows that Swedish liability legislation regarding vehicles focuses mainly on three liability subjects: the manufacturer, the owner, and the driver.

When the vehicle is used in traffic, the driver is the main liability subject. This poses legal difficulties in the context of automated vehicles. Swedish law does not provide a legal definition of a "driver"; however, it is clear from case law that a driver is a physical person and not software or a computer.

Our conclusions are that the Swedish liability legislation is adapted for driving on levels 0–3 according to the SAE J3016 standard. The current legislation regarding traffic insurance is such that it could be applied for all levels of automated vehicles. Traffic insurance will remain an important part of liability law for automated vehicles on levels 4 and 5.

For vehicles on automation levels 4 and 5, the existing legislation within Swedish liability law requires adaptations, especially with regards to the driver's responsibilities during automated travel. Currently, Swedish law depends on the courts to develop the concept of "driver" and the driver's responsibilities to function in a context with automated vehicles. We therefore invite the Swedish legislature to promptly continue to develop the Swedish liability legislation in light of the technical advances within the field of automated vehicles.

A. Olsson (✉) · K. Montelius
MAQS Advokatbyrå, Gothenburg, Sweden
e-mail: anton.olsson@maqs.com

H. Steege et al. (eds.), *Autonomous Vehicles and Civil Liability in a Global Perspective*, Data Science, Machine Intelligence, and Law 3,
https://doi.org/10.1007/978-3-031-41992-8_19

1 Introduction

Automated vehicles are about to be increasingly common on our roads as technological progressions and development continue, and this will cause a major change in the way we view mobility. A shift towards automated vehicles will also entail challenges to the current Swedish legal system regarding vehicles and certainly give rise to new liability issues, which is the subject matter of this chapter.

The legal landscape in Sweden relating to vehicles, drivers and traffic is complicated and consists of many different laws and regulations that have been in continuous development over the last 130 years. As the technical development has been extensive within both the automotive industry and society in general, legislation has been developed over the years to adapt to these changes and will be forced to continue to do so in the future.

The current legislation presupposes that vehicles are designed and manufactured for a physical driver, who is obligated to remain in control of the vehicle at all times. As the implementation of more automated and self-driving vehicles becomes a reality, there is an obvious need for new and updated legislation in Sweden. The legislation process to adapt to the future of self-driving vehicles has been ongoing for several years. For example, a public report (SOU 2018:16) which was published in 2018, included several proposals for new legislation with adaptations to automated driving. Except for legislation regarding trial operations, the legislator has however not yet adopted any changes or new laws.

The aim of this chapter is to provide an overview of the Swedish legal system as regards liability relating to accidents where vehicles are involved. This chapter will cover topics that are relevant in the context of liability, such as liability addressees, prerequisites for liability, and insurance coverage. Section 2 of this chapter introduces the general principles for liability under Swedish law and explains when someone may be held liable for damages that are caused by his or her actions or omissions. Furthermore, Sect. 2 accounts for the different liability addressees and who may be held liable for damages caused in an accident. Section 3 elaborates on the legal concept of "driver", which becomes more difficult to define as the degree of a vehicle's autonomy increases. In Sects. 4–5, we will discuss the implications of Swedish liability law in relation to the higher degrees (levels 3–5) of automated vehicles according to the SAE J3016 standard. Section 6 provides a brief description of trial operations, which currently is the only area where the Swedish legislator has provided specific regulations relating to autonomous vehicles. Finally, in Sect. 7, we will include a summary of the chapter.

2 Liability for 0–2 Level Cars: General Principles for Liability and Liability Addressees

2.1 The Functions of Liability Law

Driving a vehicle is an activity clearly associated with risk. Accidents involving vehicles cause property damage, personal injuries, and death every year. Even so, the use of vehicles is a necessary part of our everyday lives and we must likely accept that accidents and injuries will continue to occur, even with self-driving and autonomous vehicles.

The primary function of liability law in this context is to direct responsibility for the costs and consequences of an accident. Will the costs be borne by the injured party, or can the costs be transferred onto someone else involved who bears a greater responsibility for the accident? In some situations, for example where the vehicle as such is defective and not safe, the responsibility and costs for an accident should not be borne by either of the parties directly involved in the accident, but instead with the party responsible for putting the vehicle into circulation on the market. Thus, liability law also contains incentives for manufacturers, importers, and distributors to ensure that vehicles meet certain safety standards. Therefore, another important function of liability law is to prevent accidents from occurring in the first place by requiring that vehicles are safe enough for their intended use.

2.2 General Principles of Liability and Compensation

If a vehicle related accident occurs, the injured party may consider someone else responsible and ask the question whether it is possible to claim compensation for the damages incurred from the party liable for the accident. This situation is governed in general by the Tort Liability Act (*Sw: Skadeståndslagen 1972:207*).

According to the Tort Liability Act, the injured party may hold another party accountable for personal injury and property damage incurred due to an accident, if the other party caused the accident with intent or through negligence.[1] The Tort Liability Act is subsidiary in nature and will not be applied if there is more specific legislation applicable in the situation, or if the parties are in a contractual relationship that governs the issue of liability between them. The principle of intent or negligence as a prerequisite for liability for personal injury or property damage is a general principle under Swedish liability law, and will be applied by default unless specific regulations or contractual provisions clearly state otherwise.

[1] Chapter 2 Section 1 Tort Liability Act.

Liability for pure financial loss (i.e. not damage to person or property), generally requires that that the liable party has committed a criminal offence, or acted in breach of an agreement.

If the injured party is able to hold another person liable for the damages incurred due to an accident, the fundamental principle according to chapter 5 of the Tort Liability Act is that the injured party should be compensated as if the accident never occurred. This assessment should be as precise as possible. In Sweden, compensation for damages thus intends to have a repairing effect for the injured party and is not directly intended to function as a penalty for the liable party or to otherwise have a deterring effect in relation to the general public.

In order to claim compensation for damages, the injured party is required to prove (i) that the damage is a result of the accident caused by the other party, (ii) that the other party caused the accident with intent or through negligence, and (iii) that the damage is an adequate result of the accident. The injured party always has a general obligation to reasonably mitigate its costs and losses due to an accident.

In case of a vehicle related accident, the property damage (for example costs for repair work or purchasing a replacement car) is often possible to calculate without greater difficulty. Damages for personal injuries are however often more complicated to calculate. According to the Tort Liability Act, damages due to personal injuries includes *inter alia* medical costs, loss of income as well as compensation for physical and emotional suffering.[2] As physical suffering may last for a longer period of time, potentially for the duration of the injured party's lifetime, it is often hard to determine the compensation to the injured party in order to completely achieve the repairing effect "as if the accident never occurred". Instead, the court may award the injured party compensation as a lump sum based on the Traffic Accident Board's guidelines that include templates and standard amounts to be paid depending on the degree of disability incurred.[3]

2.3 The Concept of Negligence

As mentioned above, the general prerequisite for liability according to Swedish law is negligence. Liability mainly arises due to negligent acts, but may also arise due to omissions in situations where action is prescribed by law.

Establishing negligence is a rather abstract assignment. A person who causes damage or injury is negligent if he or she has not acted in accordance with how a reasonable and considerate person would have acted under similar circumstances (the principles of "bonus pater familias"). In a sense, the concept of negligence thus sets out a societal code of conduct against which people's actions may be judged.

[2]Chapter 5 Section 1 Tort Liability Act.

[3]The Traffic Accident Board (2021) Guidelines regarding compensation for claims due to traffic injuries. https://www.trafikskadenamnden.se/skadereglering/#Link12056. Accessed 8 March 2021.

However, whether a person has acted negligently is up to the court to determine in the individual case.

Sometimes, laws and other regulations provide guidance as to the reasonable and considerate course of action in certain situations. There is for example an abundance of traffic rules and regulations, which include obligations and guidance regarding how people should act in traffic situations. As such, acting in breach of traffic rules while participating in traffic will most likely be considered as negligent. If there are no specific laws or regulations applicable to the situation, the assessment will be made based on the risks that arise as a consequence of the action. If the risks for damages are high, a person is required to undertake more deliberate considerations before acting. This assessment is stricter when there is a risk for personal injury.

2.4 Liability Addressees

2.4.1 The Manufacturer

One way to prevent accidents is to ensure that vehicles in circulation on the market are safe, meaning that vehicles as products reach a certain degree of safety and reliability. As such, according to the Product Safety Act (*Sw: Produktsäkerhetslagen 2004:451*, the manufacturer of a vehicle has a general responsibility to ensure that vehicles released on the market are safe and free from security flaws.[4]

According to the Product Liability Act (*Sw: Produktansvarslagen 1992:18*), which complements the Product Safety Act, the manufacturer of a product has a strict liability for security flaws in the product.[5] For this purpose, a "product" means all movables[6] and the Product Liability Act does hence not directly apply to intellectual creations such as software. The responsibility also applies *inter alia* to a party who has imported the product to the EEA, and for a party who has marketed the product as its own by for example placing its trademark on the vehicle. In the following, we refer to these parties collectively as "the manufacturer". The Product Liability Act is a Swedish implementation of EU directive 85/374, which aims to form a common minimum level of product safety within the EU.

The strict liability for manufacturers according to the Product Liability Act includes a liability for personal injuries, and damages to consumer property caused due to a security flaw in a product. Hence, this is a deviation from the general principle of negligence as a prerequisite for liability. A product has a security flaw if it does not provide the safety that can reasonably be expected, taking into account the expected use and marketing of the product, instructions for use, the time it was put into circulation, and any other relevant circumstances. Due to the heavy

[4] Sections 7–8 Product Safety Act.

[5] Section 1 Product Liability Act.

[6] Section 2 Product Liability Act.

harmonisation, these general criteria are fundamentally similar within the EU member countries. The liability for the manufacturer applies for a period of ten years from when the product was put into circulation on the market by the manufacturer.[7]

If a product does not meet the security requirements, the product may suffer from a security flaw and the manufacturer may be obligated to recall the product. The manufacturer may also be liable to compensate personal injury and property damage caused by the security flaw.

In one case, the Swedish Supreme Court established strict liability for personal injuries due to security flaws in food.[8] The case regards a teacher who was infected with salmonella from food that was served in the school cafeteria. Considering the risk for serious injuries connected to food consumption, the manufacturer's possibilities to exercise caution and supervision during manufacture, as well as possibilities for insurance, the Supreme Court ruled that the manufacturer (the school that served the food) was strictly liable for the personal injuries incurred by the teacher.

In one case regarding property damage, a gasoline provider was found liable towards a consumer for damage to the consumer's car due to impurities in the gasoline.[9] The Supreme Court stated that the consumer had no practical opportunity to examine the product and thus had to rely on the supplier that the product could be used without causing damage.

In a commercial relationship, the injured party was awarded compensation for property damage when the counterparty's product was defect in relation to the agreed quality even though the Product Liability Act and the Purchase Act (*Sw: Köplagen 1990:931*) normally do not award such compensation.[10]

2.4.2 The Owner/Traffic Insurance

According to the current legislation, the owner's main responsibility is to hold and maintain a traffic insurance. In the public reports SOU 2018:16 and Ds 2021:28, an increased responsibility for the owner has, however, been proposed. In addition to the current responsibility, the owner is proposed to be liable for breaches of traffic rules and should be imposed with a fine unless the breach is due to a defect in the driving system that does not fall within the owner's control.

Below we will elaborate on the traffic insurance system as it applies under current legislation.

[7] Section 12 Product Liability Act.

[8] NJA 1989 s. 389.

[9] NJA 1985 s. 641.

[10] NJA 2001 s. 309. The case concerned a delivery of motor oil that contained zinc, contrary to the parties' agreement regarding the quality of the oil. The motor oil caused damages to the buyer's diesel engines. Because the supplier had undertaken to deliver a certain quality, the supplier was found liable also in relation to the damages to the buyer's diesel engines. References are made by the Supreme Court inter alia to the Product Liability Act.

Under the Traffic Damages Act (*Sw: Trafikskadelagen 1975:1410*), it is mandatory to maintain a basic liability insurance for every registered vehicle in use in Sweden.[11] The liability to maintain this insurance primarily lies with the vehicle's owner. The exception regards leased vehicles where the lessee instead should maintain the insurance. The mandatory insurance covers personal injuries and property damage that the vehicle causes on someone else's property in the event of an accident. A prerequisite is that the damage is caused as a result of the vehicle being "in traffic".

If the owner wants to include coverage for damages to its own property (e.g. on the vehicle), the owner can purchase additional voluntary insurance. Voluntary insurance comes as semi-insurance or complete insurance with a varying degree of coverage. If the owner wants the insurance to cover damages to the vehicle in case of an accident, a complete insurance coverage is required.

The system with mandatory liability insurance is incorporated to guarantee and procure the payment of compensation even in cases where the vehicle owner has not paid the insurance premiums or does not hold a traffic insurance at all.

The insurance will take over the owner's or the driver's personal liability to compensate any damage occurred. Therefore, it is not primarily necessary to investigate who or which vehicle caused the damage. An injured party may turn directly to the insurance company where the other vehicle is insured.

If a vehicle owner fails to hold the mandatory insurance, and the vehicle is driven by the owner or a third party with the owner's approval, the owner will be committing a crime and may be sentenced with a fine. The owner may also be subject to an injunction to fulfil his obligations to procure insurance. If a vehicle without insurance is involved in an accident, all insurance companies that are approved to provide traffic insurance will be jointly liable to pay compensation for personal injuries incurred by the uninsured vehicle.[12] The insurance companies have a right to reclaim parts of the compensation paid from the person who was liable to maintain insurance.[13]

If the injured party chooses to claim damages directly from the liable party under the Tort Liability Act (see Sect. 2.2 above), instead of through the traffic insurance system, the liable party will be entitled to assume the injured party's claim as an injured party under the relevant traffic insurance (up to the amount actually paid to the injured party).[14]

If an insurance company has paid compensation under the Traffic Damages Act for a damage that is covered/payable under the Product Liability Act, the insurance company will be entitled to assume the injured party's right to claim damages under the Product Liability Act (up to the amount actually paid to the injured party).[15] The

[11] Section 2 Traffic Damages Act.

[12] Section 16 paragraph 2 Traffic Damages Act.

[13] Section 24 paragraph 2 Traffic Damages Act.

[14] Section 19 Traffic Damages Act.

[15] Section 20 Traffic Damages Act.

manufacturer will thus not be put in a better position as a consequence of that the injured party is compensated by the insurance company, it will only have an insurance company as counterparty instead.

If damage has been caused by intent or gross negligence, the insurance company that has paid compensation to the injured party under the traffic insurance, will be entitled to assume the injured party's right to claim damages from the liable party (up to the amount actually paid to the injured party).[16]

There are several cases from the Supreme Court regarding whether or not the injured party is entitled to claim compensation from the traffic insurance, and below we will refer to a few examples:

NJA 2019 s. 89—Forest Fire

In July 2014, a very extensive forest fire broke out in Västmanland, Sweden. The fire was caused by a forest vehicle that carried out soil preparation work on a clear-cut located on a property. The vehicle's tires had metal straps fastened around them. The speed was estimated to be 2–3 km/h. Some part of the vehicle or its connected equipment caused ignition of vegetation or other organic material while driving. The fire spread and developed into one of the largest Swedish forest fires in modern times. The owner of the property claimed compensation from the insurance company under the traffic insurance in the amount of 8,500,000 SEK (~850,000 EUR). The Supreme Court came to the conclusion that the injured party was entitled to compensation *inter alia as* the vehicle had been used in its normal and intended way, and that there is a risk for this kind of damage (fire) when a forest vehicle is used in traffic, i.e. the damage was foreseeable.

NJA 2016 s. 142—Mobile Crane

A mobile crane was used for construction work in a harbour. The mobile crane was equipped with two engines; one for transportation and one for lifts. When used for lifts, the engine for transportation could not be used. The mobile crane was insured as a motor vehicle. In connection with the construction work, the mobile crane overturned/fell on a warehouse, which was damaged as well as goods in the warehouse. The claimant claimed compensation in the amount of 436,656 SEK (~43,000 EUR). The Supreme Court came to the conclusion that the damage was not caused whilst the vehicle "was in traffic" as the accident occurred when the vehicle was used solely as a mobile crane. The injured party was therefore not entitled to any compensation from the traffic insurance.

NJA 2017 s. 622—School Fire

In 2009, a few kids intentionally caused a fire in a vehicle that was parked close to a school building. The fire caused a short circuit in the vehicle, which caused the vehicle to jump/move forward towards the school building. The fire thereafter spread to the school building, which burned down to the ground. The injured party claimed compensation in the amount of 95,777,330 SEK (~9,600,000 EUR) from the

[16] Section 20 paragraph 2 Traffic Damages Act.

insurance company. The Supreme Court came to the conclusion that the damage was not caused whilst the vehicle "was in traffic", as the fact that the vehicle had moved forward was not a result of someone using the vehicle or otherwise had caused the vehicle to be in a traffic situation. The injured party was therefore not entitled to any compensation from the traffic insurance.

2.4.3 The Driver

At present, the main responsibility for a vehicle, while the vehicle is used in traffic, rests with the driver. The driver's general liability for damages caused while driving is set out in the Tort Liability Act, which, as described above under Sect. 2.2, sets out the basic requirements for when and who is responsible for damage, or if the damage should instead be borne by the injured party. The driver can be held liable for damages caused due to intent or negligence, for example if the driver caused the accident by acting in breach of traffic rules. In practice, the Tort Liability Act is rarely used in connection with traffic accidents, since the injured party can claim compensation through the traffic insurance without having to prove who caused the accident and if the accident was caused negligently.

3 The Driver and Driver's License

3.1 The Concept of "Driver"

According to article 8 in the Geneva Convention on Road Traffic from 1949, there is a requirement that each vehicle should have a driver and that the driver at each time should be able to control his or her vehicle. The Swedish Traffic Ordinance (*Sw: Trafikförordningen 1998:1276*) includes similar provisions and assumes that there is a physical person present who controls the vehicle. The Traffic Ordinance requires that all drivers take all necessary actions that are required to avoid traffic accidents. It is also stated that a vehicle is not to be driven by a person who, due to illness, sleepiness, alcohol, drugs, or similar, is not fit to drive the vehicle in a safe manner.[17] Thus, the Traffic Ordinance does lay down requirements on how to drive the vehicle, which implies that the law requires that *someone* drives a vehicle and that it is that person who should take precautions and make sure that the vehicle is driven in a safe manner. It is mainly also the driver who can be held liable for crimes relating to traffic accidents. However, the legislator assumes that someone other than the actual driver can be considered being the driver from a legal perspective, as well as that the "legal driver" does not have to be in the vehicle itself. The first situation is for example the case in practice driving under the Driver's Licence Act (*Sw:*

[17] Chapter 3 Section 1 Traffic Ordinance.

Körkortslagen 1998:488) where the "teacher" is considered being the driver during the practice session.

The number of cases regarding the definition of "driver" and the extent of the driver's responsibilities are few and relatively old, but some conclusions can be made that are relevant in this context:

- It is possible for a driver to be responsible for at least two vehicles at the same time. An example of this is towing, where the driver of the vehicle towing a second vehicle has been found liable for both vehicles simultaneously.[18]
- It is possible for a person to face criminal liability for several vehicles at the same time without being the driver. An example of this is a case where a military commander faced criminal charges in his role as a participant in traffic, and not as a driver, and was convicted for acting negligently in traffic due to orders he gave to subordinates who were driving in a military convoy. [19]

3.2 Driver's License

In Sweden, a person needs to obtain a driver's license for driving a motor vehicle. There are different authorizations depending on the vehicle, and for passenger cars, the driver can obtain the driver's license from the age of 18. Today, a driver's license with authorization B (without a condition code/restriction) gives the driver a right to drive cars and light trucks regardless of what different technical support systems and aids the vehicle has (for example parking assistance, cruise control or automatic transmission). However, different technical support systems can affect the assessment of the driver's competence in the driving test for the driver's license.

The requirements imposed on vehicles used in driving tests for driver's licenses are regulated in EU directives and in the Swedish Transport Agency's regulations. The Swedish Transport Agency's regulations on driving tests state that the driving examiner may decide that the use of technical support systems should be limited, which for example has resulted in that that parking assistance normally is not used when the driving license candidate's manoeuvrability is to be assessed.

In the future, it might be possible to expand the area of conditional driver's licenses. Today it is for example a requirement to take a driving test in a vehicle with a manual gearbox. However, it is permitted to take the test with an automatic transmission, but then the driver's license will be limited to apply only to vehicles with an automatic transmission. If new condition codes are introduced in the future, it could in theory be possible to use different levels of automated driving in driving tests. After such a test, the driver's license could be restricted with the condition that it is only permitted to drive with the same aids/technology. In the future, this could

[18]NJA 1934 s. 616, NJA 1972 s. 177.

[19]NJA 1969 s. 220.

open up for the elderly or disabled to keep their driver's license longer with conditions on using aids. The different conditional codes for driving license are currently regulated by EU directives, even though it is permitted for member states to apply national codes.

4 Liability for Level 3 and 4 Cars

The liability for level 0–2 cars are described above under Sect. 2. Below, we describe how Swedish liability law may be applied on levels 3 and 4.

There are no specific legal regulations in place yet which establish how to determine liability in case of an accident involving cars on level 3 and 4. For the time being, the same general liability rules, which apply at levels 0–2, will apply to level 3 and 4 cars as well. However, there is nothing in the Swedish road traffic legislation that prevents or prohibits the use of self-driving or autonomous vehicles in the road transport system.

The current Swedish road traffic legislation is based on a driver being responsible for driving the vehicle. As long as there is someone who can be considered to be a driver in or in connection with the vehicle, automated vehicles from level 3 and upward can be operated. As accounted for above in Sect. 3, Swedish law does not include a general definition of "driver" or "to drive" a vehicle. These fundamental concepts have been left to the courts to define depending on the circumstances in the individual case. So far, the concept of "driver" does not extend beyond physical drivers.

Considering that the existing case law concerning the concept of "driver" refers to a physical person, cars on level 3 and 4 pose certain challenges to the current Swedish liability system. From level 3 and upward, the car's automated driving features do the actual driving (in most situations). If there is a physical person in the "driver's seat", he or she would primarily take the role as passenger. Thus, Swedish law will need to provide an answer to the question who or what will have the responsibilities as driver when the vehicle's autonomous driving features are activated.

The public report SOU 2018:16, "The Road to Autonomous Vehicles", includes several proposals for new legislation, and *inter alia* proposes that an autonomous vehicle must have a physical driver during autonomous driving.[20] According to the proposal, it should be the owner's responsibility to ensure that there is a physical driver present inside or outside the vehicle during autonomous driving. The owner should also be responsible for the tasks performed by the autonomous driving features and that the vehicle follows traffic rules. The driver should be primarily responsible for maintaining a driver's licence and for taking control of the vehicle

[20] SOU 2018:16, *The Road to Autonomous Vehicles*, p. 93.

upon the vehicle's request. If the driver negligently would fail to take manual control of the vehicle, the driver would commit a criminal offence.

The idea to maintain a responsible driver during automated driving[21] has been further elaborated upon in the public report Ds 2021:28, which suggests introducing a new role named "driver at the ready". This person should be the one who activates the automated driving or takes control over the vehicle when required. The driver at the ready needs to be prepared to immediately take control of the vehicle if requested by the system and to otherwise take the measures required to ensure adequate traffic safety. Criminal charges are the suggested consequence for failing to do so.[22]

These proposals may provide a solution to the issues with the "driver concept", but will also entail a rather dramatic shift in liability where the vehicle's owner will bear the primary liability for the vehicle's traffic behaviour. Time will tell if the legislator will adopt these proposals into law.

In conclusion, the liability for a vehicle while in traffic currently rests with the physical driver. By introducing automated vehicles this will change. One solution to the liability issue may be that the owner is appointed as the primary liability addressee in case of accidents, at least when automated driving is activated, which would be consistent with the mandatory traffic insurance system. Suggestions in this direction have recently been presented in the above mentioned public report Ds 2021:28.

5 Liability for Level 5 Cars

As for level 3 and 4, there is no specific legislation relating to level 5 cars in Sweden. Unless new legislation comes into force, any question of liability involving a level 5 car would have to be assessed in accordance with the current liability legislation as described in Sect. 2. The system with mandatory traffic insurance for the car owner would entail that all cars will be insured even if they qualify as level 5 cars. The same applies for liability, where each subject involved in an accident may be held liable according to the general principle of negligence as well as under product liability law for the manufacturer.

For level 5 cars, where the vehicle's autonomous driving features will perform all of the actual driving, it becomes increasingly clear that the existing division of liability and concept of liability need to be developed. With technical developments in respect of vehicles, traffic and roads, it is not unlikely that the courts' assessment of driver liability in traffic will change. A higher degree of automation of the vehicles will probably mean that fewer people will be found guilty of crimes and offences

[21] The Swedish legislator has in its public reports changed from using "autonomous driving" to more consistently using "automated driving" (year 2018 compared to 2021).

[22] Ds 2021:28, *The Question of Liability during Autonomous Driving and new Legislation to Promote an Increased Use of Geo-Fencing*, pp. 20–22.

relating to accidents and infringements of the law. In the absence of legislation, the courts may demand in line with current case law that a physical person must have had, in some form, the opportunity to influence or intervene in the motion of the vehicle to be deemed to be driving. Cases involving people being regarded as driving negligently will probably become less common when the physical driver's actual control over a vehicle will be reduced as automation increases.[23] It is difficult to argue that someone was negligent if he or she had received trustworthy information that the vehicle can handle all situations in a safe manner (under certain specified conditions such as limits based on road type) and that the vehicle does not need any assistance or supervision by a human driver while in self-driving mode. Under such circumstances, it is likely that the liability for the vehicle's behaviour in traffic will be increasingly directed at the party in control of the software and hardware controlling the vehicle, i.e. the manufacturer. If the owner is required to assume liability for the vehicle's performance, as the public report SOU 2018:16 suggests, this would place a greater risk on the owner who may not have any real influence over the vehicle's automated driving features. Additional measures may need to be considered in order to achieve a fair division of liability between the owner and the manufacturer.

In a level 5 car, it is likely that liability issues will be more closely related to the design of the vehicles' systems (both hardware and software) and the vehicles' interplay with other vehicles, pedestrians and other road users. It could for example relate to that the vehicle has not showed enough attention and consideration to other vehicles or participants in traffic by conducting a sharp or unexpected turn. Further, it could relate to the persons involved having used the technique incorrectly, where the information provided by the car manufacturer to the user will be of importance.

However, if a human driver is abusing the technology in any way, he or she can be found negligent. It will therefore become important to clearly inform users/drivers about what the technology can do, any limitations of the technology, what the human driver is expected to do, and what the risks are if the instructions are not complied with. It will also be important that it is clear to a human driver whether the vehicle is in a self-driving mode or whether it is expected that the vehicle will be operated manually. Furthermore, the driver education that physical drivers complete to obtain their driver's license will need to be adapted to include education about how the systems work, what can be expected of them and what the drivers' role and responsibilities are.

In order to determine liability for traffic accidents, it will be important that there is a "black box" that records data from the vehicle's sensors and systems. The recording of such data will be extremely important in order to find out if someone has been negligent or whether or not the vehicle was in self-driving mode. It will also be important to adopt regulations regarding who should be entitled to obtain the data

[23] See for example The Swedish Transport Agency (2014) Pilot study regarding Autonomous Driving, Dnr TSG 2014-1316, p. 46 (available at https://www.transportstyrelsen.se/globalassets/global/press/autonom_korning_forstudie.pdf).

in the "black box". Additional important future issues in this regard are, in addition to liability, related to data privacy and the fair and proper use of the large amounts of data that the vehicle will collect during transport.

6 Trial Operations

In 2017 the Trial Operation Ordinance (*Förordning (2017:309) om försöksverksamhet med självkörande fordon*) entered into force that gives entities a possibility to apply for a permission to conduct trial operations on public roads in Sweden. Anyone who seeks permission must be able to prove that the trial operation is conducted in a safe manner. The permission is granted by the Swedish Transport Agency and a permit can be granted with conditions. The ordinance clearly states that when test operations are carried out with an automated vehicle, there must at all times be a physical driver inside or outside the vehicle. Furthermore, it is stated that a person who conducts trial operations without a permit will commit a crime and be fined.

A basic condition for obtaining a permit is that the applicant can prove that it can conduct the trial operation in a traffic-safe manner. This is done by showing that all risks with the trial operation are taken care of, including a description of the fully or partially automated functions of the trial. Automatic functions are evaluated based on the vehicle's ability to function independently, both lateral and longitudinal control included.

As of today, only a few companies have obtained permits by the Swedish Transport Agency. In August 2020 the Swedish government announced that the ordinance will become permanent (it was initially only effective until July 1, 2022). The name of the ordinance will also be changed from "Ordinance regarding trial operations with self-driving vehicles" to "Ordinance regarding trial operations with automated vehicles", which is well in line with the development during the last years where more focus has been on *automated* vehicles and not only *self-driving* vehicles.

7 Summary and Conclusions

7.1 General Conclusions

The technical development regarding automated vehicles requires that the Swedish legislation is updated to better suit and address the new reality where the driver's role as the main liable person for accidents gradually will change. The legislator has initiated its work of adapting the legislation to the new technical landscape, but there is still a lot to do and it is crucial that the legislation work continues.

The system with mandatory traffic insurance is well established in Sweden and functions properly. It is therefore likely that the owner in the future will continue to have an obligation to hold the car insured, so that compensation for property damage and personal injuries caused by a traffic accident also in the future first can be paid by the relevant insurance company. The insurance company will thereafter try to reclaim the cost from someone else, for example the manufacturer under the product liability legislation or the driver who has caused the accident.

One of the main questions relating to automated vehicles relates to how the product liability legislation will develop and handle automated vehicles with levels 4–5. Will the legislator and the courts accept a certain degree of accidents even if caused by automated vehicles, or will most of such accidents be deemed being caused by a safety flaw in the vehicle? A manufacturer cannot reasonably anticipate all possible scenarios and circumstances that an automated vehicle could encounter in reality. At the same time, it is difficult to expect that the general public will accept accidents with automated vehicles without the manufacturers assuming any liability.

Software is already an important component in today's vehicles, and will be a fundamental component by automation at level 4–5. However, as product liability is designed today, consumers may not hold software developers directly liable in the event that an accident is caused due to a security flaw in the software. In such case, the product liability will be borne by the vehicle or component manufacturer who then has to reclaim damages from the software developer on a contractual basis. As the software in automated vehicles will have been developed over a long period of time, and in many cases by several companies, it will also be interesting to see how the vehicle manufacturers may prove that the software has been developed in a safe manner and what requirements that will be put on them to fulfil the burden of proof.

It is a difficult task to find a fair balance of liability for vehicle accidents at level 4 and 5 between the liability subjects, as autonomous machines are not recognised as subjects under Swedish law. Even so, it is possible that accidents will occur because of what the vehicle has taught itself through its autonomous systems. In such situations, it may be unsatisfying to hold a driver or manufacturer accountable if there are no security flaws. Perhaps responsibility for the owner through the owner's mandatory insurance policy is the most efficient way to close the liability gap.

7.2 Liability While Driving Within the Levels of Automated Vehicles

7.2.1 Levels 0–3

Swedish liability legislation regarding vehicles focuses on mainly three liability subjects: the manufacturer, the owner, and the driver. When the vehicle is used in traffic, the driver is the main liability subject. This poses legal difficulties in the context of automated vehicles.

Swedish law does not provide a legal definition of a "driver"; however, it is clear from case law that a driver is a physical person and not software or a computer.

Our conclusions are that the Swedish liability legislation is adapted for driving on levels 0–3 according to the SAE J3016 standard. The current legislation regarding traffic insurance is such that it could be used for all levels of automated vehicles.

7.2.2 Levels 4 and 5

As regards levels 4 and 5, the existing legislation within Swedish liability law is not adequate and does currently depend on the courts to develop the concept of "driver" to also include automated vehicles. As stated above, the current legislation regarding traffic insurance could however be applied also for levels 4–5 cars as the legislation only requires that the vehicle should be "in traffic".

The manufacturers will have the same liability under the product liability legislation regardless of level of automation, but there will probably be more focus on the vehicle itself in connection with an accident. Implementation of "black boxes" in vehicles as a standard will also make it easier, to some extent at least, to determine why an accident occurred and who (if someone) caused it.

References

Swedish Laws

Product Safety Act (*Sw: Produktsäkerhetslagen 2004:451*)
Driver's Licence Act (*Sw: Körkortslagen 1998:488*)
Product Liability Act (*Sw: Produktansvarslagen 1992:18*)
Traffic Damages Act (*Sw: Trafikskadelagen 1975:1410*)
Tort Liability Act (*Sw: Skadeståndslagen 1972:207*)

Swedish Governmental Ordinances

Trial Operation Ordinance (*Förordning (2017:309) om försöksverksamhet med självkörande fordon*)
Traffic Ordinance (*Sw: Trafikförordningen 1998:1276*)

Swedish Case Law

NJA 2019 s. 89
NJA 2017 s. 622
NJA 2016 s. 142

NJA 2001 s. 309
NJA 1989 s. 389
NJA 1985 s. 641
NJA 1972 s. 177
NJA 1969 s. 220
NJA 1934 s. 616

Public Investigations and Reports

Ds 2021:28 *The Question of Liability during Autonomous Driving and new Legislation to Promote an Increased Use of Geo-Fencing*
SOU 2018:16 *The Road to Autonomous Vehicles*
The Swedish Transport Agency (2014) Pilot study regarding Autonomous Driving, Dnr TSG 2014-1316 available at: https://www.transportstyrelsen.se/globalassets/global/press/autonom_korning_forstudie.pdf

Online Sources

The Traffic Accident Board (2021) Guidelines regarding compensation for claims due to traffic injuries. https://www.trafikskadenamnden.se/skadereglering/#Link12056. Accessed 8 March 2021

Part VI
The United Kingdom

Automated Vehicle Liability in Great Britain

Felix Boon

Abstract The UK Government has sought to position the UK as a world-leading development hub for automated vehicles. It recognised that an effective regulatory model for automated vehicles was necessary to achieve this. After an extensive review of UK road traffic law by the Department for Transport, the Law Commission of England and Wales, and the Scottish Law Commission, were tasked with 'reviewing the UK's regulatory framework to enable the safe and effective deployment of automated vehicles on Great Britain's roads.' Meanwhile, the world's first insurer liability scheme for automated vehicles was enacted in the Automated and Electric Vehicles Act 2018.

This chapter provides a detailed overview of the existing law and insurance requirements for vehicles at SAE Levels 0–2 in Great Britain. Thereafter, it carries out an extensive analysis of the new insurer liability regime for 'automated vehicles' (SAE Levels 3–5) under the Automated and Electric Vehicles Act 2018. It also sets out a detailed review of the various recommendations of the Law Commissions in their three Consultation Papers and their Joint Report in 2022. Finally, it examines how automated vehicle insurers might then recover their losses under the Consumer Protection Act 1987, along with the law of tort/delict and contract.

1 Overview

1.1 Introduction

This chapter is made up of four sections. Section One outlines Connected and Automated Vehicle ("CAV") development in the UK and introduces the main sources of law on this subject. Section Two examines the traditional liability system for road traffic accidents, covering vehicles with no automation features up to and including vehicles with Advanced Driver Assistance Technology (ADAS).

F. Boon (✉)
2 Tintock Place, Dullatur, North Lanarkshire, Glasgow, UK

H. Steege et al. (eds.), *Autonomous Vehicles and Civil Liability in a Global Perspective*, Data Science, Machine Intelligence, and Law 3,
https://doi.org/10.1007/978-3-031-41992-8_20

Section Three discusses insurer liability for road traffic accidents involving vehicles with Automated Driving Systems (ADS) of SAE Level 3–5. Section Four addresses claims by insurers against automated vehicle manufacturers, software developers and service providers.

1.2 Connected and Automated Vehicles

As discussed in Chapter ‘Introduction’, the taxonomy of automation most commonly applied in this space is that developed by the Society of Automotive Engineers (SAE) in their technical document J3016 (Society of Automotive Engineers International April 2021). The taxonomy centres around performance of the dynamic driving task (DDT). Whereas driver assistance features such as cruise control, adaptive cruise control and lane assist are designed to help a driver perform the dynamic driving task, automated driving systems (ADS) are the “. . .hardware and software that are collectively capable of performing the entire DDT on a sustained basis, regardless of whether it is limited to a specific operational design domain (ODD). . .” (SAE 2021, 3.2). An ADS feature is part of an ADS and designed to operate in a specific ODD (SAE 2021, 3.7). At SAE Levels 0–2, it is usually considered that human drivers must remain ‘in the loop,’ actively monitoring and controlling the vehicle (SAE 2021, 3.8). While the level of human oversight necessary at SAE Level 3 is not uncontroversial, vehicles with automation levels at SAE Level 3 or greater will be subject to a separate regulatory regime for ‘automated vehicles’ in Great Britain.

1.3 UK Road Traffic Law: Regulatory Reviews

The UK Government recognised that the development of automated vehicles in the UK could only flourish if the UK’s regulatory environment allowed it. It conducted an initial regulatory review of UK Road Traffic Law in 2015 (Department for Transport 2015). While no significant barriers to CAV use were identified, it was nonetheless accepted that regulatory reform would be needed before CAVs could be put onto UK roads at scale. Consequently, the first CAV-specific piece of legislation anywhere in the word was enacted. Namely, the Automated and Electric Vehicles Act 2018.[1] The Department for Transport and Centre for Connected and Autonomous Vehicles (CCAV) also asked the Law Commission of England and Wales, and the Scottish Law Commission to “. . .review the UK’s regulatory framework to enable the safe and effective deployment of automated vehicles on Britain’s

[1]The Act gained Royal Assent on the 19th of July 2018 and came into force on 21 April 2021.

roads…" (Law Commissions 2018, 1.16-19).[2] In particular, their focus was on 'identifying gaps and uncertainty in the law, and suggesting reforms that ensure the regulatory framework is fit for purpose' (Law Commissions 2022, 1.8). This considerable project began in 2018 and concluded in 2022.[3] The Law Commissions' final report was published on the 25th of January 2022 (Law Commissions 2022) and contains 75 recommendations. Most notably, this included new legislation, as well as the creation of a number of new legal entities, agencies, criminal offences and regulatory duties.

When the review began, the view was reached that, absent any clear prohibition on using an automated driving system (Department for Transport 2015; Law Commissions 2018, 4.78-85) and provided the vehicle conformed with the relevant UK laws, CAVs of any level of automation would be legal to use on UK roads and other public places (CCAV et al. 2019). In their first consultation paper, the Law Commissions proposed a "…new system to authorise automated driving systems [under which], unauthorised systems should be prohibited." (Law Commissions 2018, 4.105.) The Law Commissions concluded that vehicles with automated driving system features should be subject to an authorisation process before their use is permitted on public roads and spaces in Great Britain (Law Commissions 2022, 3.67). This is given effect in Part 1 of the new Automated Vehicles Bill, which is discussed further below.

1.4 UK Law: Sources and Structure

The law in the UK governing civil liability for road traffic accidents has grown slowly, and developed out of a variety of domestic, EU and international sources (Merkin and Hemsworth 2015). This has been supplemented by a variety of national and international standards and guidance. Currently, the primary legislation is contained within the Road Traffic Act 1988, the Road Vehicles (Construction and Use) Regulations 1986, and the Automated and Electric Vehicles Act 2018. The law in this area will be expanded further once the Automated Vehicles Bill completes its passage through Parliament. Aiming to 'put safety and the protection of the user at the heart of its new regime,' the King's Speech expressly stated that the Bill was drafted to implement the Law Commissions' recommendations following their 4-year review of self-driving regulation (King's Speech Briefing Notes 2023). As such, notwithstanding their limited value as an aid to construction of the relevant legislation in this area, the Law Commissions' Consultation Papers and Final Report

[2]The scope of the Joint Consultation is relatively narrow. For example, it does not consider matters such as data protection, theft or cyber security. Similarly, it focusses on passenger transport and not goods deliveries.

[3]Rather than speculating on what might be, this chapter attempts as far as possible to describe the law as it is. Nonetheless, the Law Commissions' recommendations are likely to be adopted and so, where appropriate, draws on their analysis and references their recent proposals.

nonetheless provide useful insights into Government policy and potential future legislation. Particularly so while a number legislative changes are still being ironed out in Parliament. Consequently, in addition to detailed analysis of the 2018 Act, the Law Commissions' work is referenced extensively throughout this chapter.

The United Kingdom of Great Britain and Northern Ireland is a unitary state, comprising England and Wales, Scotland and Northern Ireland. They represent three discreet legal jurisdictions. Further, Scotland, Wales and Northern Ireland all have devolved legislatures. In each case, the UK Parliament retained legislative competence in respect of various general and specific matters. In Scotland[4] and Wales,[5] legislative competence for certain transport matters, insofar as they were subject to retained EU law, was specifically reserved to the UK Parliament in Westminster. Product standards, safety, and liability are also specifically reserved matters.[6] In Northern Ireland, the UK Parliament has not excepted or reserved matters relating to road vehicles under the Northern Ireland Act 1998.[7] Likewise, while product safety is a reserved matter in Northern Ireland,[8] product liability is not. Consequently, many of the main provisions discussed below concerning liability and insurance contained within the Road Traffic Act 1988, the Consumer Protection Act 1987 and the Automated and Electric Vehicles Act 2018, apply to Great Britain but not Northern Ireland. As such, this chapter concerns itself with the law governing liability for automated vehicle accidents in Great Britain.

1.5 EU Law

The United Kingdom joined the European Communities in 1973 by way of the 1972 Treaty of Accession. EU law was given force in the UK's domestic law by the European Communities Act 1972. In the intervening period, a substantial swathe of EU law was transposed, directly and indirectly, into UK law. Following a referendum on the 23rd of June 2016, the UK voted to leave the EU. It did so on the 31st of January 2020. Nonetheless, EU law continued to apply in the UK during an 11-month transition period provided for by the Withdrawal Agreement and which expired at 11 pm (UK Time) on the 31st of December 2020. Thereafter, a significant body of EU law (retained law) was effectively directly converted into UK law via the European Union (Withdrawal) Act 2018. The effect of the Act was to create "...a snapshot of EU Law as it applied in the UK on the 31st of December 2020" (Thorneloe 2021). The Retained EU Law (Revocation and Reform) Act 2023 was

[4] Section 29(2)(b) & Schedule 5 of the Scotland Act 1998.

[5] Section 94(3) and Schedule 5 of the Government of Wales Act 2006.

[6] Schedule 5, Head C8 of the Scotland Act 1998 and Schedule 5, Section C7 of the Government of Wales Act 2006.

[7] Transport lies within the remit of the North-South Ministerial Council.

[8] Schedule 3, Para 37 of the Northern Ireland Act 1998.

then passed. Coming into force on the 1st of January 2024, the Act operates so as to further diminish the scope of retained EU law in the UK. It provides that the supremacy of EU law is not part of UK law, revoking certain directly effective EU law rights and revoking UK courts' obligation to interpret domestic legislation in a way that is compatible with EU law. The UK's new relationship with the EU is governed by the Trade and Cooperation Agreement Between the EU and the UK. While the agreement calls for ongoing cooperation on product safety, product liability is not addressed (European Union et al. 2020, Article TBT 96).

1.6 International Law

The UK has been a member of the United Nations Economic Commission for Europe (UNECE) since 1947. It has been a party to the UNECE Agreement concerning the adoption of uniform technical prescriptions for wheeled vehicles and their parts since 1963. Having signed the Vienna Convention on Road Traffic in 1968, the UK ratified the treaty on the 28th of March 2018. Amendment 34bis to the Convention provides that regulation of the operation and use of vehicles with automated driving systems is a domestic matter, which only tends to reinforce the sense that divergent regulatory models cannot be avoided.

2 The Traditional Liability Model: SAE levels 0–2

2.1 Negligence

In contrast to the strict liability model for road traffic accidents employed in some other jurisdictions, Great Britain operates a fault-based system, which is backed by compulsory insurance.[9] The existing regulatory model continues to apply to vehicles that are not legally defined as an automated vehicle. This is likely to mean vehicles with automation features up to and including SAE Level 2.

In England and Wales, the branch of law governing civil wrongs that give rise to damage, injury or loss is tort. Specifically, in the case of road traffic accidents, the tort of negligence. Liability in tort "...arises from the breach of a duty primarily fixed by the law; such duty is towards persons generally..." (Winfield 1931, cited in Jones 2020, 1-03). The duty being, in broad terms, not to inflict damage carelessly (Riley-Smith and McCormick 2020, 5-05). Establishing liability in negligence requires: the existence of a relevant duty owed by the defendant to the claimant; that the defendant breached this duty; that the breach caused the defendant some

[9] As required by s 143 of the Road Traffic Act 1988.

form of legally actionable harm (*Rothwell v Chemical & Insulating Co* [2007]); and that the loss is not so unforeseeable as to be too remote (Jones 2020, at 7-04).

In Scotland, negligence claims fall to be dealt with under the law of delict. In contrast to tort, "...there is no exhaustive list of named delicts in the law of Scotland." (*Micosta S.A. v Shetlands Islands Council* (1986).) The law of delict operates on principle. Namely, *injuria* or *damnum injuria datum*, or "...breach of legal duty causing unjustifiable harm." (Walker 1981, p. 31; Hogg 2006; Thomson 2014). Consequently, "...If the conduct complained of appears to be wrongful, the law of Scotland will afford a remedy even if there has not been any previous instance of a remedy being given in similar circumstances." (*Micosta S.A.*)

2.2 A Driver's Duty of Care

Negligence is commonly understood to be the omission to do that which a reasonable man, guided by ordinary considerations that regulate the conduct of human affairs would do, or doing something which a prudent and reasonable man would not do (*Blyth v Birmingham Waterworks Co* (1856)). In *Donoghue v Stevenson*, the House of Lords defined the ambit of a person's duty of care by reference to what is now known as the neighbourhood principle. Namely, that a duty of care not to cause physical harm or damage is owed to persons sufficiently proximate to the duty holder that they ought reasonably to foresee that they will be directly affected by these acts or omissions (*Donoghue v Stevenson* (1932)). This duty extends to financial losses derived from physical harm or damage. However, the law does not typically allow recovery for purely economic loss absent physical damage or injury (Thomson 2014, 4.10). As such, whether they be drivers (Taylor 1986), pedestrians, cyclists or persons on horseback, road users owe a duty of care to other road users (Walker 1981, p.503; *Bourhill v Young* [1943]).

2.3 The Standard of Care

The standard of care owed by drivers to other road users is that of reasonable care (Walton et al. 2022, 11-199–200). In *Nettleship v Weston* (1971), the court held that this meant driving "...in as good a manner as a driver of skill, experience and care, who is sound in wind and limb, who makes no errors of judgment, has good eyesight and hearing, and is free from any infirmity..." This calls for "...care and skill of a high standard..." (*Nettleship*), as well as "...mutual respect and forbearance" (*Searle v Wallbank* [1947].) Reasonable care requires road users to drive avoiding "...excessive speed, keeping a good look out, observing traffic rules and signals..." (*Bourhill*). Driving needlessly slowly can amount to negligence (*Whitehead v Bruce* [2013]). Driving without due care and attention, or without reasonable consideration for other road users is an offence.[10] Activities secondary to driving should generally be minimised,[11] and some, such as using a mobile phone[12] and watching a television screen,[13] are currently proscribed under the Road Vehicles (Construction and Use) Regulations 1986;[14] breach of which is an offence.[15] Vehicle maintenance also falls within the scope of a vehicle user's duty of reasonable care (*Henderson v Henry E Jenkins & Sons* [1970]; *Rees v Saville* [1983]; Law Commissions 2018, 6.8-12). Under regulation 100 of the 1986 Regulations, vehicles must be maintained at all times in a condition such that they will not pose a danger to other road users.[16]

2.4 Breach of Duty

Whereas the standard of care is a question of law, breach of that duty in the context of road traffic accidents is typically a question of fact. The standard of reasonable care is judged objectively. Where a vehicle collides with another correctly proceeding vehicle, the former's collision is *prima facie* evidence of negligence. The case law in this area is extensive but for these purposes it is sufficient to note that this may include accidents caused by a driver: failing to allow enough room for an oncoming vehicle to pass (*Duncan v Wilson*, (1940)); cutting across the path of another vehicle (Department for Transport 2022b); as well as driving too close to the vehicle in front and failing to stop in time (*Thompson v Spedding* [1973]). Similarly, stopping suddenly in front of another vehicle may amount to negligence (*Brown and Lynn v Western S.M.T. Co*, (1945); Walton et al. 2022, 11-225-227). A breach of the terms of the Highway Code is also *prima facie* evidence of negligence (*Robertson v J Sidney Smith* 1999; Thomson 2014, 14.2).[17] Likewise, where a vehicle defect causes an accident, "...the starting point is that the insured person ...is liable for the accident" (Law Commissions 2018, 6.9.) A burst tyre is therefore "...*prima facie* indicative of fault" (*Barkway v South Wales Tpt. Co.* [1950]). Further, use of an unroadworthy vehicle is also an offence.[18] Similarly, a vehicle brake failure gives rise to a presumption of a breach of duty that requires positive evidence of reasonable care being exercised before the presumption can be dislodged (*Henderson.*)

---

[10] Section 3 of the Road Traffic Act 1988.

[11] Art 8.6 Vienna Convention on Road Traffic 1968.

[12] Regulations 110 of the Road Vehicles (Construction and Use) Regulations 1986.

[13] Regulation 109.

[14] Albeit, remote control parking via a hand-held device is now allowed, per regulation 110(5A).

[15] Section 41(d) Road Traffic Act 1988.

[16] Regulation 29 also requires steering gears be maintained in good and efficient working order.

[17] Section 38(7) of the Road Traffic Act 1988.

[18] Sections ss 40A, 41B & 42 of the Road Traffic Act 1988.

2.5 Causation

When pursuing a claim in negligence, the claimant requires to demonstrate a causal link between the breach of duty and the harm complained of. Traditionally, there have been two elements to causation. Namely, factual and legal causation. Factual causation requires a claimant to establish that *but for* the defender's breach of duty, the accident would not have occurred. In other words, that the breach was a necessary condition or *causa sine qua non* of the harm. Factual causation has therefore been said to operate so as to exclude irrelevant causes (Jones 2020, 2-09). However, the test 'may be relaxed where it would seem unjust to require the claimant to prove the impossible (Jones 2020, 2-100).

Meanwhile, "...legal causation tests whether something which qualifies as a factual cause is deemed significant enough to attract liability at law." (Hogg 2011). To establish legal causation, the claimant must show that the defendant's breach of duty was the *causa causans*—the real effective (*Global Process Systems Inc and another v Syarikat Takaful Malaysia Bhd* [2011], at [49]) or efficient cause (Stewart 2004, 9-20) of the loss, injury or damage. Specifically, that the breach was the proximate cause of the loss. A causal factor's mere proximity in 'point in time' will not suffice (*Global Process Systems Inc and another*, at [19]; *Leyland Shipping Co Ltd v Norwich Union Fire Insurance Society Ltd* [1918]).

The negligent intervening actions of a third party may break the chain of causation. Properly understood, the third party's actions are the proximate cause of the loss. This was the case in *Wright v Lodge* (1993), where the claimant's vehicle was hit by an articulated lorry. Despite the claimant's earlier negligence in leaving their broken-down vehicle on the road, the dangerous driving by the lorry driver was the effective cause of their injuries. However, where the subsequent actions of a third party are reasonably foreseeable, this is unlikely to break the chain of causation (*Rouse v Squires* [1973]; see also *Sayers v Harlow Urban District Council* [1958] and *Knightley v Johns* [1982]). This being so, drivers should maintain sufficient distance from the vehicle in front to pull up safely where the vehicle in front brakes suddenly (*Thompson*; see also *Stevenson v Rodbourne* LTL (5/11/98).[19] Running into the rear of a vehicle raises a considerable inference of negligence (Brodie 2021, A23-008); *Brown and Lynn v Western S.M.T.* (1945) at 332 and *Thomson v Brankin* (1968)). Where a vehicle shunts a vehicle in front into a third vehicle, the driver of the original collision is the real efficient cause of the subsequent collisions and therefore liable for this (although, see *Smith v Samuel Williams & Sons* (1972)).

The analogous law on causation in both jurisdictions is undoubtedly complex (Hogg 2011). In many instances, particularly where there is more than one causal factor, discriminating between causes that are sufficiently causally potent and those which are too remote comes down to questions of policy (Jones 2020, 2-100). This in turn is often largely informed by questions decided by common sense (See *Galoo Ltd v Bright Grahame Murray* [1994]). Consequently, where more than one party

[19]Rule 126 of the Highway Code.

contributes to an accident, the courts will frequently apportion liability between the parties according to their level of blameworthiness (See e.g., *Clift v Hawse, Bird and the MIB* [1999]; *Lindesay v Lamb & Tatner* [2007]; *Smith v Harris* [1939]). Given the challenges faced by claimants seeking to demonstrate causation described below, these policy and common-sense considerations can be expected to be live concerns for the foreseeable future.

2.6 Contributory Negligence

Where the contributory fault of the claimant is a causally material factor in the accident but not the proximate cause itself, s 1 of the Law Reform (Contributory Negligence) Act 1945 provides that the level of damages may be reduced by the court "...to such extent as the court thinks just and equitable having regard to the claimant's share in the responsibility for the damage." This has often involved consideration of both causal potency and blameworthiness of the claimant's contributory fault (Taylor 1986, p. 23). Findings of contributory negligence typically range between at least 10% (*Johnson v Tennant Bros Ltd* (1954) to 90%. A reduction of 100% is not possible in England and Wales (*Pitts v Hunt* [1990]). While there is authority for 100% reductions in Scotland (Stewart 2004, p. 313; *Robb v Salamis* (2003) and *Skipton Building Society v Lea Hough & Co* (2000)), it has nonetheless been suggested that the terms in which s 1 of the 1945 were framed apply only where there is shared liability (Thomson 2014, 6.13).

2.7 Limitation & Prescription

Claims for negligence in both jurisdictions are subject to statutory limitation periods. In England and Wales, the law is contained within the Limitation Act 1980. The Act provides that claimants have three years to raise personal injury claims, commencing from the date the injury occurred,[20] the age of majority[21] or, where the requisite knowledge of a ground of action postdates the occurrence of the injury, computation of the limitation period runs from the date of knowledge.[22] Claims in tort for property damage or financial loss must be raised within six years.[23] Limitation

[20] Section 11(4)(a) of the Limitation Act 1980.

[21] Section 28(6) provides that, in relation to s 11, time effectively does not run while a person is under a disability. Disability is defined at s 38(2) to include infants. Infants are not defined in the Act but, under s 1(1) of the Family Law Reform Act 1969, the age of majority is reached at 18. Section 1 (2) of the 1969 Act provides that, absent definition or indication of contrary intention, *inter alia* 'infant' is to be construed with reference to s 1(1).

[22] Section 11(4)(b) of the Limitation Act 1980.

[23] Section 2 of the Limitation Act 1980.

operates as a bar to the raising of proceedings. It does not extinguish the obligation to provide compensation. Notwithstanding the time bar contained in s 11, where it appears equitable to allow an injury claim to proceed out of time, a court has discretion under s 33 of the 1980 Act to override the time bar. Any period of time where a claimant does not pursue a claim as a result of fraud is not included in the computation of the limitation period/s.

In Scotland, the relevant legislation is the Prescription and Limitation (Scotland) Act 1973. Personal injury claims are also subject to a three-year limitation period.[24] As with its English and Welsh counterpart, depending on the circumstances, time runs from the date of harm,[25] the age of majority[26] or date of knowledge.[27] Claims for property damage and financial loss are subject to a five-year short negative prescriptive period.[28] The effect of which is to extinguish any duty to provide a remedy.

2.8 *Insurance*

Section 143 of the Road Traffic Act 1988 proscribes the use of a motor vehicle on a road or other public place without insurance (see Channon et al. 2019, p. 21). Doing so without suitable insurance is an offence. The requirements of insurance cover are set down in s 145, including, *inter alia*, cover for liability for third party death, personal injury or property damage arising out of or caused by the use of a vehicle. Mandatory insurance cover does not extend to the driver or their vehicle (Merkin 2021, 23-002; *Neto de Sousa v Portuguese State* (2017).[29] The degree to which s 143 adequately transposed the provisions of Article 3(1) of Council Directive (EEC) 72/166 has been called into question. The relevant CJEU jurisprudence (*Vnuk v Zavarovalnica Triglav* (2014)*; Rodrigues de Andrade v Proenca Salvador* (2017); *Torreiro v AIG Europe Ltd* (2017) suggested a broader construction of the phrase 'use of a vehicle' is called for than had traditionally been applied to s 143 (*R&S Pilling (t/a Phoenix Engineering) v UK Insurance Ltd* [2019]). However, following widespread unease with these cases (Vnuk in particular), the UK Government legislated to nullify retained case law insofar as it extended the domestic meaning of 'use of vehicles' in Article 3 of the 2009 Motor Insurance Directive beyond 'vehicles other than motor vehicles and locations other than on a road or other public place' (s1, Motor Vehicles (Compulsory Insurance) Act 2022). Motor vehicles are broadly defined at s 185 as a 'mechanically propelled vehicle intended or adapted for

[24] Section 17(2) of the Prescription and Limitation (Scotland) Act 1973.

[25] Section 17(2)(a).

[26] Section 17(3).

[27] Section 17(2)(b).

[28] Section 6(1).

[29] Section 145(4)(c).

use on roads.' Insurers must satisfy claims against their insured when the policy-holder is sued to judgment.[30] Claimants may also bring a claim against the insurer directly.[31]

2.9 Uninsured Drivers

Certain exclusions limiting the scope of motor policy cover are provided for at ss 148 and 151 of the 1988 Act. However, under Article 75 of the Motor Insurer Bureau's (MIB) Articles of Association, the insurer providing any insurance in respect of the vehicle must still pay any compensation due. Where the vehicle is uninsured or untraced, claimants will be able to obtain a remedy via the relevant MIB Agreements.[32] Under clause 3(1) of the Uninsured Drivers Agreement 2015, where a claimant obtains an unsatisfied judgment in respect of a relevant liability against any person in a court in Great Britain, the MIB will pay the claim or cause it to be paid (MIB 2015). A relevant liability is one for which there must be a suitable contract of insurance under Part Vi of the Road Traffic Act 1988 (MIB 2015, clause 1(4)).

2.10 SAE Level 2: Advanced Driver Assistance Systems (ADAS)

SAE Level 2 vehicles lie within the existing regulatory regime. Nonetheless, the addition of partial-automation features like ADAS warrants some specific consideration. There is currently no legal definition in GB law of SAE Level 2. Rather, it is effectively an industry standard. While the technical distinction between ADAS and ADS may become increasingly blurred,[33] the Law Commissions argued that the legal difference is profound (Law Commissions 2022, 3.6). In their third consultation paper, the Law Commissions suggested that classification of ADAS systems might indirectly fall within the scope of their proposed scheme for authorising automated driving systems (Law Commissions 2020, Figure 8.1, p. 126). Despite the Law Commissions' later change of heart on the point, the Automated Vehicles Bill potentially keeps some Level 2 vehicles within its scope where they contain

[30] Section 151.

[31] Under the European Community (Rights against Insurers) Regulations 2002 (SI 2002/3061).

[32] Specifically, the MIB's 'Uninsured Drivers Agreement' (2015) and 'Untraced Drivers Agreement' (2017). For a more detailed analysis of this issue, see Merkin and Hemsworth (2015) (ch. 7) and Channon et al. (2019) (ch. 3).

[33] There is evidence that the boundaries between SAE Levels may become blurred by increasing capacity to handle different dynamic driving tasks (DDTs). The SAE itself has considered the possibility of a Level 2+ classification (SAE 2020). See also The Autonocast (2021).

equipment that allows their motion to be controlled other than by an individual or to facilitate such control; or which is designed to incorporate or interact with software (Clause 91(1)). In any event, where ADS features fail to achieve authorisation for self-driving in Great Britain, such failure should not lead to their being treated as SAE Level 2 ADAS features by default (Law Commissions 2022, 5.54-8).

Under the Law Commissions' proposed scheme, civil liability was predicated upon the need for driver monitoring. Once a vehicle is classified as self-driving, the person in the driving seat—the 'User-in-Charge' (UiC)[34]— can "...relax and divert their attention, knowing that they are not responsible for anything that happens while the automated driving system (ADS) is engaged." (Law Commissions 2022, 3.4.) In contrast, vehicles with ADAS still require a "...human behind the wheel to monitor the vehicle and the driving environment, and to respond to events." (Law Commissions 2022, 3.3.) They therefore recommended 'a clear-cut distinction between ADAS and ADS, setting out when the user is no longer responsible for the dynamic driving task.' (Law Commissions 2022, 3.6.)

A corollary of this is that the duty of reasonable care subsists in respect of ADAS systems. The writer is unaware of any reported civil cases in the UK involving ADAS features. Nevertheless, in the event of an accident caused by a vehicle with Level 2 technology, the collision would likely be *prima facie* evidence of negligence on the part of the vehicle's driver (Channon et al. 2019, p. 34). Similarly, deliberate misuse of Level 2 automation technology, such as overriding driver monitoring devices or delegating complete control to the vehicle when it is inappropriate to do so, would clearly fall below the standard of reasonable care. Indeed, it would likely amount to an offence.[35]

Anecdotal video evidence suggests some Level 2 vehicles may attempt to turn onto or off the highway when it is not safe to do so (Loveday 2021). To take one example, Tesla's own terms make clear that unexpected manoeuvres can be expected and that drivers should keep their hands on the wheels at all times (Tesla 2022). This being so, it seems likely that drivers of Level 2 vehicles will be assumed to be on notice of the idiosyncrasies of their vehicle's automation features. Consequently, when Level 2 automation features are engaged, a driver's duty of reasonable care will likely be deemed to include engaged and proactive vehicle monitoring. While each case will turn on its facts, where a driver is aware that their vehicle may perform unpredictable manoeuvres, it is unlikely that defences periled on unexpected vehicle behaviour will succeed (*R v Spurge*, [1961]; *Hougham v Martins* (1964)). Nevertheless, insofar as the accident was caused by a Level 2 vehicle automation feature, a claimant may be able to pursue a claim against the producer of the vehicle directly under the Consumer Protection Act 1987. Where the

[34] Discussed below at para 3.4.

[35] Sections 2 (dangerous driving) and 3 (careless driving) of the Road Traffic Act 1988.

automation feature is partially responsible and the driver found liable, they may still have a right of relief against the producer.[36]

Technological naivety, ignorance or overreliance alone is unlikely to found a defence to a negligence claim (see BBC 2018). However, any standing presumption concerning drivers' awareness of the limits of their vehicle's automation capabilities *might* be capable of being overturned if a driver can show their misplaced expectations of the vehicle's capabilities were entirely reasonably formed. This might include reliance on inaccurate assurances about vehicle functionality by the vehicle's supplier (see **para 5.2.2–5.2.5**). This point has certainly been plead in several ongoing US cases (*Shawn Hudson v Tesla et al* (2018); *Kim Banner v Tesla Inc.* (2019); *Alexandro Filippini et al v Tesla Inc* (2021)). Although, this will be difficult to prove since driver knowledge will typically be assumed to derive from the vehicle's Driver Handbook (presumably replete with extensive caveats), as well as any supplementary warnings, instructions or advertising. Nonetheless, the Wettbewerbszentrale case concerning Tesla's 'Full Self-Driving' package (FSD) suggests that courts are at least sensitive to the issue of manufacturers overstating functionality (*Wettbewerbszentrale v Tesla* (2020).[37] Where liability attaches to a driver in these circumstances, they may still have a remedy for misrepresentation (see **para 4.2.3**.). Similarly, past performance may be a factor to which the courts have regard in circumstances where an over the air update (OTA) update has significantly and unexpectedly diminished or improved vehicle performance (O'Kane 2018; Lambert 2019). Nevertheless, courts are likely to be anxious to avoid drivers' duty of care being watered down by the novelty of new technology and so successful defences of this sort are likely to be vanishingly rare.

It is unlikely that any harm caused by a cyber-attack would give rise to a finding of negligence unless there is positive evidence of the driver failing to exercise reasonable care in relation to vehicle cyber security. Much will turn on the driver's awareness of a particular vulnerability, measures taken by the relevant manufacturer or software developer to notify drivers, recall the vehicles or provide some form of over-the-air (OTA) fix or patch to address the issue. Where a driver recklessly uses a vehicle known to be infected with a virus, it could, by analogy with the cases on automation, conceivably give rise to prosecution for dangerous or careless driving under the Road Traffic Act 1988 (*Watmore v Jenkins* [1962]). However, it would appear that any civil claim against the driver advanced on the basis of a breach of the Road Vehicles (Construction and Use) Regulations 1986 for failure to

---

[36] Under the Civil Liability (Contribution) Act 1978 in England and Wales, and s 3 of the Law Reform (Miscellaneous Provisions) (Scotland) Act 1940 in Scotland. See Section 4 of this chapter, which considers recoveries.

[37] Similarly, a court might also consider a successful claim under the Misrepresentation Act 1967 a material factor to which they may have regard when considering whether the user operated the automation features negligently. For more on this, see Dixon (2020); Lee and See (2004); Glassbrook et al. (2019), p. 106.

maintain the vehicle is unlikely to succeed (Brodie 2021, A23-004; *Reid v First Glasgow Ltd* (2003).[38]

3 Automated Driving Systems (ADS)

Vehicles with authorised Automated Driving System features (ADS) will be subject to a separate civil liability regime in Great Britain, which is primarily contained within the Automated and Electric Vehicles Act 2018 (AEVA). The Act is designed to reflect the existing scheme (Channon et al. 2019, p. 18), emulating the 'direct right of action against an insurer, which the 2002 regulations allow and extending it straight to its natural destination: the motor insurer (Glassbrook et al. 2019, p. 37). As the author notes elsewhere, the Act functions as a risk transfer mechanism, passing difficult questions of liability and causation to insurers to answer (Boon 2020; Davey 2017). However, in doing so, the Act conflates tortious and contractual principles, needlessly creating uncertainty and undermining many of the rules of reparation law that it seeks to uphold. Absent amendment, there must be doubts about the effective operation of any liability system based on the AEVA in its current form. Nonetheless, the Law Commissions concluded that the AEVA was 'good enough for now' (Law Commissions 2022, 13.23); albeit they recommend that the listing scheme provided for in s 1 is replaced with the new legislative authorisation scheme alluded to above. Although the listing scheme contained within s 1 of the 2018 Act is still in force, the Automated Vehicles Bill removes s 1 and replaces it with a new authorisation process (Schedule 2, Clause 5(2); Chapter 1). As such, while legislative changes are still pending, both schemes are discussed.

3.1 Defining Automated Vehicles: Section 1 and a New Automated Vehicles Act

While aware of the SAE Levels, Parliament deliberately chose not to employ them within the AEVA itself.[39] Instead, s 1 of the Act provides only a broad outline of the level of functionality necessary to bring vehicles within the ambit of the Act's regulatory purview and stops short of providing a substantive formal definition.[40]

[38] See also, *Balmer v Hayes* 1950 SC 477, which considered a similar claim pursued for breaches of the Road Traffic Act 1930 and The Motor Vehicles (Driving Licenses) Regulations 1947.

[39] During the passage of the Bill through Parliament, the UK Government confirmed "The Bill is intended to cover the use of highly automated and fully automated vehicles which are broadly equivalent to Level 4 and 5 as delineated by the Society of Automotive Engineers... the Bill does not cover conditionally automated (broadly equivalent to Level 3) vehicles" (Sugg 2018).

[40] This has led to a degree of uncertainty as to which automated vehicles would fall within the scope of the Act. Channon et al. (2019), p. 20; Glassbrook et al. (2019), pp. 80–81.

Instead, this is currently left to the unfettered discretion of the Secretary of State. Section 1(4) provides that an automated vehicle is one that appears on the Secretary of State's list, as required by s 1(1). As such, an automated vehicle is whatever the Secretary of State deems as such. Under s 1(1), the Secretary of State is required to prepare and keep up to date a list of vehicles that the Secretary of State feels are designed or adapted to be capable, in some circumstances, of safely and lawfully driving themselves. This is further supplemented by s 8(1)(a), which states: "*...a vehicle is "driving itself" if it is operating in a mode in which it is not being controlled, and does not need to be monitored, by an individual...*"

The Law Commissions noted that the AEVA's definition of a vehicle driving itself had been widely accepted and they concluded that it was broadly correct (Law Commissions 2022, 3.12). Flowing from this, they adopted the term 'self-driving' as a legal term of art, which is intended to have "...a precise meaning for legal rights and duties in Great Britain" (Law Commissions 2022, 3.2). Nonetheless, their longstanding concern that s 1 of the AEVA was underdeveloped and "...too simple..." remained (Law Commissions 2020, 8.73-8.882)].[41] Of particular note, the definition centred around 'two difficult terms' (Law Commissions 2022 (Summary), 2.2): 'safely' and 'monitoring', which are not defined under the AEVA.

3.2 *Authorisation of Self-Driving Vehicles*

The Law Commissions therefore recommended that future legislation should set down a demanding test that a vehicle with a self-driving feature must satisfy before being authorised for use. Namely, the vehicle should drive safely and legally, even where it is not being monitored (Law Commissions 2022, 3.67; 5.97.) This test sits at the heart of the Automated Vehicles Bill. Clause 3(1) provides that the Secretary of State may authorise a road vehicle for use as an automated vehicle if (a), in the opinion of the Secretary of State, the vehicle satisfies the self-driving test by reference to intended travel on a road (in at least some circumstances) and, where applicable, various initial authorisation requirements authorised under Clause 5 are met. Under Clause 1(2), a vehicle satisfies the self-driving test if (a) it is designed or adapted with the intention that a vehicle feature will allow it drive autonomously and (b) that feature will enable it to do so safely and legally. Under Clause 1(5), a vehicle travels autonomously if (a) it is controlled by the vehicle equipment rather than an individual and (b) neither the vehicle nor its surroundings are being monitored by an individual with a view to immediate intervention in the driving of the vehicle. By virtue of Clause 1(7), a vehicle travels (a) "safely" if it travels to an acceptably safe standard and (b) "legally" if it travels with an acceptably low risk of committing a

[41] The UK Government published a sample of what the Secretary of State's list might look like (CCAV 28 April 2021a, Annex C). However, the sample document appeared to only allow for limited information.

traffic infraction (see also Law Commissions 2022, 3.18). The Secretary of State must also prepare a statement of safety principles for the purpose of assessing whether a vehicle is capable of travelling autonomously and safely (Clause 2(1)). The principles must be framed with a view to securing that road safety in Great Britain will be better as a result of authorised automated vehicles on roads than it otherwise would be (Clause 2(2)). In assessing whether a vehicle is capable of travelling autonomously and safely, the Secretary of State must have particular regard to the statement of safety principles (Clause 1(8)). The Secretary of State is also able to vary, suspend or withdraw authorisation of an automated vehicle (Clauses 8 & 9). They may do so where (a) an authorisation requirement is/has not been met in respect of the automated vehicle; the vehicle has committed a traffic infraction; or (c) the vehicle no longer satisfies the self-driving test by reference to all authorised locations and circumstances (Clause 8(4)). Finally, the Secretary of State must keep a public register of automated vehicle authorisations (Clause 10(1)). Significantly, an automated vehicle authorisation takes effect when it is entered in the register (Clause 10(2)). The Bill also makes the Secretary of State the in-use automated vehicle regulator, with a wide range of powers. In addition to their authorisation role, the Secretary of State is also empowered to make regulations to make provision for the licensing of 'no-user-in-charge operators' (Part 1, Chapter 2; see para 3.16), include requirements for the provision of information from the bodies they regulate (Part 1, Chapter 3), and impose civil sanctions on regulated bodies (Part 1, Chapter 5). These sanctions include the issuing of compliance and redress notices, along with monetary penalties.

As with the AEVA, the proposed test for self-driving is anchored around vehicle safety and monitoring. However, the test for self-driving in the Bill is much more nuanced and developed than its s 1 predecessor. The requirement under the Bill that the vehicle be capable of travelling "legally" appears more onerous than the requirement for lawful use under s 1(b) of the AEVA. Similarly, regard must be had to the ODD of intended travel (Clause 1(3)). The Bill also requires the Secretary of State to have regard to the Statement of Safety Principles; the requirements of which are detailed at Clause 2. However, where their vehicle employs them, UiCs must still be receptive to Transition Demands. Namely, a demand "...communicated by equipment of a vehicle in which an authorised user-in-charge feature is engaged, that the user-in-charge assume control of the vehicle by the end of a period of time beginning with the communication of the demand (the "transition period")..." (Clause 7(2); see also United Nations Economic Commission for Europe (UNECE) Global Forum for Road Traffic Safety (2021), para 2(c)). While transition demands are felt to be compatible with self-driving, authorisation under the scheme is contingent upon the vehicle's transition demand being perceived by anyone (having regard in particular to users-in-charge with disabilities). Further, the period of transition must allow sufficient time for UiCs to prepare to assume, and assume, control of the vehicle. Vehicles should continue to drive autonomously, safely and legally during the transition period, and UiC's should be further warned once the transition period ends. Finally, there should be sufficient vehicular mitigation strategies to prevent crashes should the UiC fail to respond to the transition demand in time (Clause 7(3); see also Law Commissions 3.1-67).

3.3 Automated Lane Keeping Systems (ALKS)

There has historically been some uncertainty as to the level of autonomy necessary to reach the threshold for automated vehicle status under the AEVA. Questions as to the scope of the 2018 Act appear to have been settled with the advent of ALKS technology. Under UNECE Regulation 157,[42] ALKS is defined as a system that "...controls the lateral and longitudinal movement of the vehicle for extended periods without further driver command. ALKS is a system whereby the activated system is in primary control of the vehicle..." For this reason, vehicles with ALKS are commonly described as having SAE Level 3 functionality. Referred to as the 'traffic jam chauffeur,' ALKS is intended to operate in a specific and relatively narrow ODD. As of January 2023, the amended Regulation 157 limits its maximum operational speed to 130kph/80.78mph. It is intended for "...situations of heavy, slow-moving traffic on a Motorway (CCAV 2020, 2.2)." The Regulation also requires that "...the activated system shall not cause any collisions that are reasonably foreseeable and preventable... if a collision can be safely avoided without causing another one, it shall be avoided."[43] A consequence of this is that users may be able to engage in non-driving or secondary activities. Nonetheless, drivers must be ready for a transition demand. Given the potential for competing demands on a user's attention at this level of functionality, the period in which control is transferred between vehicle and user pose a significant risk. Regulation 157 requires that drivers have at least 10 seconds to take control before the vehicle requires to perform a minimum risk manoeuvre.[44] However, minimum risk manoeuvres may be initiated immediately in circumstances of severe system failure.[45]

ALKS has been considered in separate consultations by both CCAV (CCAV 2020) and the Law Commissions (Law Commissions 2020). CCAV and the Law Commissions provisionally reached the view that some vehicles with ALKS were capable of safe conditional automation and that they met the criteria for automated vehicles for the purposes of the AEVA.[46] CCAV's summary of the ALKS Consultation responses confirmed that: "...rather than list all vehicles fitted with ALKS by default, individual models will be listed [under s 1 of the AEVA] after they have

[42] UN Regulation 157 on uniform provisions concerning the approval of vehicles with regard to Automated Lane Keeping Systems ECE/TRANS/WP.29/2020/81 (25 June 2020).

[43] ALKS Regulation, para 5.1.1.

[44] ALKS Regulation, para 5.4.4.1.

[45] ALKS Regulation, para 5.4.4.1.1.

[46] However, the Law Commissions also note that some vehicles with ALKS technology may fall below the relevant threshold (Law Commissions 2020, 3.73).

received whole vehicle type approval (WVTA)" (CCAV 2021b, p. 41). In their Final Report, the Law Commissions suggested that, in addition to vehicle approval, "...an ALKS feature will only be lawfully used in self-driving mode once the vehicle has passed the second authorisation stage..." (Law Commissions 2022, 2.27). The Bill does not refer to ALKS specifically but it seems reasonably likely that some vehicles with ALKS technology will, in due course, satisfy the conditions necessary for authorisation in Part 1 of the Bill, thereby bringing them within the scope of the AEVA, and so this chapter proceeds on this basis. Helpfully, the Bill does not set out a minimum timescale for transition periods in Clause 7(3)(b). Rather they must simply be long enough for a user to prepare and assume control of the vehicle. Further, when Clause 7(3)(b) is read with Clause 1(3), it seems reasonable to assume that transition periods will not be a fixed global minimum window, but rather will vary according to the ODD and any other relevant contextual factors. This is on all fours with Law Commissions' suggestion that the minimum 10 second transition period required under UNECE Regulation 157 "...provides a floor for the acceptable transition time for that use case..." (Law Commissions 2022, 3.28). Noting this would likely be ODD-dependent, they stopped short of specifying how long any given transition period should be (Law Commissions 2022, 3.34).

4 The User-in-Charge (UiC)

One of the more significant developments to emerge from the work done by the Law Commissions is the creation of the 'User-in-Charge' (UiC). The Law Commissions' proposals were designed to address uncertainty around users' legal duties when an automated vehicle is driving itself. The Law Commissions recommended that a User-in-Charge should be 'defined as an individual who is in the vehicle and in position to operate the driving controls while an authorised ADS feature is engaged' (Law Commissions 2022, 8.9; 8.25).

Once an authorised ADS feature is engaged, 'The person inside the vehicle with access to its controls would become a "user-in-charge" rather than simply a "driver"' (Law Commissions 2022, 3.7). The Bill closely follows their proposals. Clause 4(3)(a) requires that the automation feature identified in the authorisation process must specify whether the mode of operation of the feature is "user-in-charge" or "no-user-in-charge". Likewise, under Clause 4(3)(b), the automation feature identified in the authorisation process must identify how that feature is engaged and disengaged. Clause 46 provides that "An individual is the "user-in-charge" of a vehicle if—(a) the vehicle is an authorised automated vehicle with an authorised user-in-charge feature, (b) that feature is engaged, and 20 (c) the individual is in, and in position to exercise control of, the vehicle, but is not controlling it" (see also Law Commissions 2022, 8.25; 8.18.) Clause 47 sets out a number of circumstances where a UiC is not liable for the manner of the vehicle's driving. However, that immunity is immediately qualified by a number of exceptions in Clause 48. Finally, Clause 49 lists various scenarios where the UiC is liable as a driver. However, in attempting to clearly delineate the legal obligations of driver and UiC, the focus of the Bill is on criminal, rather than civil, liability. Equally, since they do not feature in the AEVA, UiC status is arguably not a necessary condition for the liability to attach under the 2018 Act. Nevertheless, it will be relevant to the subject of any civil claim, and so a court would likely feel entitled to have regard to any offence committed or immunity afforded under the Automated Vehicles Act 2024 when considering the issue of negligence. There is no specific provision in the Bill around licensing of individual UiC's and so it appears that the licensing requirements under s 87 of the Road Traffic Act 1988 apply equally to UiC's (see Law Commissions 2022, 8.26-33). Use an automated vehicle while the user is intoxicated or asleep is proscribed under Clause 53, which inserts a new s 34B into the Road Traffic Act 1988 to make use of a road vehicle without someone exercising, or being in a position to exercise control of the vehicle, an offence (see also Law Commissions 2022, 8.42-50). Meanwhile, Clause 47(4)(b) provides that any immunity from criminal liability relating to the manner of the vehicle's driving does not include liability for the condition or qualifications of the driver. In a similar vein, a court is likely to have little difficulty in finding the user of an automated vehicle while intoxicated or asleep was negligent.

As noted above, the test for self-driving autonomous travel under Clause 1(5) means that a UiC has no monitoring obligations. They are therefore free to undertake secondary activities, such as checking their emails or using vehicle infotainment systems. The Law Commissions suggested that the prohibition of on-board screens under regulation 109 of the Road Traffic (Construction and Use) Regulations 1986 be removed on the proviso that the screen cuts out at the start of a transition demand. However, they argued that the existing restrictions on mobile phone use should apply equally to UiCs and that sleeping while being a UiC should also be an offence (Law Commissions 2022, 3.76-80). This proposal has already been given legislative effect, by way of regulation 3 of the Road Vehicles (Construction and Use) (Automated Vehicles) Order 2022, which amends regulation 109(1)(e) of the 1986 Regulations to allow the display of information on built-in apparatus in an automated vehicle. The vehicle must, however, be driving itself, be "driving itself" within the meaning of s 8(1)(a) of the AEVA. As s 8(1)(a) is currently drafted, this means the vehicle be operating itself in a mode where it is not being controlled, and does not need to be monitored, by an individual. However, the AEVA's definition of 'driving itself' is likely to change. By virtue of Schedule 2, Clause 5(6)(a) of the Bill, s 8(1)(a) will be amended so that a vehicle is driving itself if it is travelling while an authorised automation feature is engaged. Also, the vehicle must not be requesting that the driver take back control. As amended, regulation 109(1)(e)(iii) of 1986 Regulations refers to a driver and not a UiC, which suggests the prohibition would not be triggered until after the transition period has ended and the user has assumed the legal duties of a driver. As such, this would presumably mean it would not be an offence to look at an infotainment system during the transition period, which seems somewhat unsafe and might well satisfy the test for negligence (see **paras 3.37-39** and **3.46**). However, it is likely that any infotainment system will cut out at the pint a transition demand is made. Interestingly, while changes were made to the Highway Code whereby users 'do not need to monitor an automated vehicle if it is driving itself,' the update does not explicitly address transition demands (Department for Transport 01 July 2022a). However, where someone engages an unauthorised ADS feature, they will remain a driver with the accompanying legal responsibilities (Law Commissions 2022, 5.5; 5.40). In certain circumstances, the Law Commissions propose that the In-Use Regulator can recommend that authorisation of an ADS feature or the vehicle as a whole be withdrawn by the Authorisation Authority (Law Commissions 2022, 6.101-5). Absent vehicle recalls and/or satisfactory warnings, it is entirely conceivable that users might then unwittingly engage unauthorised ADS features.

Once handover of control is complete and the automated driving feature is disengaged, the conditions necessary for UiC status under Clause 46 are no longer met and the UiC becomes a driver again. It follows they will therefore owe other road users a duty of reasonable care again at this point (Law Commissions 2022, 8.122). This is presumably also the case where a UiC takes back control before the end of a transition period. Their legal duties as a *driver* would recommence at the point where they take control and not at the end of the transition period (see Law Commissions 2022, 8.132). Post-transition, 'Drivers' would be subject to the pre-AEVA scheme, regardless of the level of automation their vehicle is capable of. Similarly, UiCs will reacquire the legal responsibilities of a driver where they fail to respond to a transition demand (Clause 48(1); see also Law Commissions 2022, 8.134-7). Inappropriate engagement of an ADS feature is effectively criminalised under the Bill but only where that feature is engaged outwith its authorised locations and circumstances, and, where the UiC interferes with, or knows someone else has interfered with the vehicle equipment. Mistakenly activating an ADS feature when it is not appropriate to do so seems unlikely, as ADS features that operate outwith their ODD are unlikely to be authorised (See Law Commissions 2022, 8.79). Nevertheless, accidents occurring in these circumstances will likely lie outside the ambit of the AEVA and require claimants to pursue claims in negligence against the UiC instead (see **para 3.15–16** below).

The Law Commissions did not draw any distinction between UiCs and UiCs acting in a professional capacity for passenger service providers (Law Commissions 2022, 8.152). While the Bill does contain some provisions concerning passenger service permits in Part 5, the definition of a UiC does not distinguish between private use and passenger service provision. Consequently, the passenger service professional's legal responsibilities as a UiC or driver will also turn on whether the Clause 46 criteria are satisfied.

4.1 No-User-In-Charge Vehicles (NUiCs)

The Law Commissions also identified the need for a separate user classification. Namely, vehicles that do not require a UiC, which they called no-user-in-charge (NUIC) features.

The Law Commissions suggested NUiCs would most likely be employed by Highly Automated Road Passenger Services (HARPS), as well as automated delivery and goods transport services and some private users. More generally, remote operation would likely involve Fleet Operations (activities supporting ADS-equipped automated vehicles, including various non-dynamic driving tasks) and remote assistance (the remote provision of information or advice to an ADS-equipped vehicle) (Law Commissions 2020, 13.34-40.) These 'oversight duties' (Law Commissions 2022, 9.8) would be performed at some form of remote operations centre. Once again, the Bill follows this proposed model closely. Clause 12(2) of the Bill defines "no-user-in-charge journeys" as a journey by a vehicle with an authorised no-user-in-charge feature during which (at any point) (a) that feature is engaged and (b) there is no individual in the vehicle who is exercising control of it. Under Clause 12(1)(b), the Secretary of State may impose regulations on those persons in connection with NUiC journeys or the vehicles that undertake them. Clause 12(4) requires that, where the Secretary of State makes operator licensing conditions, they must do so in a way that, as far as the Secretary of State considers reasonably practicable, secure certain objectives. At Clause 12(5)(a), this includes a general responsibility on the part of a no-user-in-charge operator for the detection of, and response to, problems arising during a no-user-in-charge journey overseen by the operator. The prohibition under Clause 53 on the use, causing or permitting a person to use a mechanically propelled vehicle without someone exercising, or in a position to exercise, control also extends to NUiC journeys unless the NUiC feature is engaged at all times and the journey is overseen by a licensed NUiC Operator (see also (Law Commissions 2022, 9.72). Clause 82 of the Bill provides the appropriate national authority with a wide discretion when imposing conditions pursuant to the grant of an automated passenger services permit. Equally, Clause 13(4) allows the Secretary of State to make regulations that impose a requirement on individual NUiC Operators to comply with conditions attached to an individual license. The Bill does not specify specific requirements but, to obtain a license, NUiC operators would likely need to "...document its system of safety management, defining individual roles with clear responsibilities and competence requirements. This should cover individual staff, supervisors and senior managers, and include details of how safety concerns are to be escalated and addressed in the organisation..." (Law Commissions 2022, 9.102). NUiC Operators would also likely need to address vehicle oversight, incident management, systems expertise, data management, remote driving and mistake-driven learning (Law Commissions 2022, 9.110). Finally, along with ASDE's, NUiC Operators are referred to throughout the Bill as regulated bodies. As a regulated body, NUiC Operators may be subject to licensing requirements obliging them to disclose information to the Secretary of State and/or private businesses (including vehicle manufacturers and insurers) (Clause 14(2) & (3)). They may also be required to nominate an individual (Clause 15(2)) who would be responsible for ensuring that information is provided and that it is accurate. Failure to do so may constitute an offence of non-compliance under Clause 20. Further offences relating to information about the safety of automated vehicle operations are detailed in Clauses 24 and 25. Where a NUiC's nominated individual commits an offence under Clauses 20 or 24, that person would also be personally criminally liable (Clauses 26) The same is true of an NUiC Operator's senior managers (Clause 27). Where the Secretary of State is satisfied that a regulatory requirement is not or has not been met, they may issue a compliance notice to a regulated body (Clause 34(1)). Where this occurs, a compliance notice can be issued to them directing the NuiC Operator to take steps that the Secretary of State considers appropriate to secure, or make more likely, that a regulated body complies with the regulatory requirement in future (Clause 34(5)(a)).

While a NUiC Operator employee may not be driving the vehicle, they could still cause harm if they exercise poor judgement in the performance of their duties. Particularly if they are tired, badly equipped, poorly trained or responsible for too many vehicles (Law Commissions 2022, 9.19-22). It seems reasonable to infer from this that the standard of care owed by NUiC Operator employees will be at least equivalent to that owed by ordinary drivers.

4.2 Remote Driving

The Law Commissions viewed pure remote driving—i.e., real-time performance by a remote operator of some or all of the dynamic driving tasks—as falling outside the scope of their regulatory review. Accidents caused by a remote driver do not lie within the AEVA scheme discussed below. There is currently no separate regulatory model for remote driving and so the same rules appear to apply to remote operators as ordinary drivers. Namely, they must 'hold a license, be fit to drive, be in a position to have proper control of the vehicle with a full view of the road and traffic ahead' (Law Commissions 2022, 9.149; Law Commission 2023). The Law Commissions issued an Issues Paper on Remote Driving on 24 June 2022 (Law Commissions June 2022). They concluded that the current law is unsatisfactory and that remote driving is currently a legal grey area. The Law Commissions proposed that there should be a new statutory licensing scheme for companies deploying remote drivers beyond line-of-sight. They also suggest that remote driving abroad should be banned and that victims of road traffic collisions caused by remote driver should have access to some form of no-fault compensation scheme. They also recommended that the Road Vehicles (Construction and Use) Regulations 1986 be amended so that, absent a Vehicle Special Order in place, beyond line-of-sight driving without an in-vehicle safety should be prohibited (Law Commission (2023)). Their advice was published to Government in February 2023 and further legislative developments are still awaited at the time of writing.

4.3 Authorised Self-Driving Entities (ASDEs)

The Law Commissions also recommended the creation of Authorised Self-Driving Entities (ASDEs). The ASDE would be the vehicle manufacturer or software developer who puts an AV forward for authorisation as having self-driving features. It could be a vehicle manufacturer, or a software developer, or a partnership between the two. However, the ASDE should show that it was closely involved in assessing the safety of the vehicle. (Law Commissions 2022, 2.41.) The Bill provides for the creation of ASDE's at Clause 6. Under Clause 6(1), the Secretary of State must impose authorisation requirements designed to secure that, in respect of each automated vehicle authorisation, a person is designated as the "authorised self-driving entity" for the vehicle at all times. Remaining an ASDE may itself be subject to specific requirements under the authorisation (Clause 6(2)). Meanwhile, Clause 6(3) requires the Secretary of State impose authorisation requirements designed to secure, so far as the Secretary of State considers reasonably practicable, a number of safety-based objectives. These are set down in Clause 6(4):

> (a) that an authorised self-driving entity should have general responsibility for ensuring that an authorised automated vehicle continues to satisfy the self-driving test by virtue of its authorised automation features; and
> (b) that an authorised self-driving entity should be:
> (i) of good repute
> (ii) of good financial standing, and
> (iii) capable of competently discharging any authorisation requirements imposed on it for the purposes of paragraph (a).

It is important to note that the obligations under Clause 6(4)(a) is ongoing duty. The ASDE's duty to ensure a vehicle satisfies the self-driving test will subsist for as long as the vehicle authorisation itself lasts (see also Law Commissions 2022, 5.82; 5.93-97). ASDE's are also regulated bodies under the Bill. As such, their authorisation requirements may also include requirements concerning the collection and sharing of information (Clause 14(1) (see also Law Commissions 2022, 5.84). As with NUiC's, the disclosure contemplated is to the Secretary of State and/or private businesses (Clause 14(2) & (3). Authorisation may require an ASDE to nominate an individual (Clause 15(1)) who would be responsible for ensuring that information is provided and accurate. The various offences of non-compliance in Clause 20 also apply to ASDE's, as do the offences relating to information about the safety of automated vehicle operations in Clauses 24 and 25. Likewise, as with NUiC Operators, offences committed by an ASDE's nominated individuals or senior managers can also give rise to individual criminal liability (Clauses 26 & 27). As a regulated body, ASDE's are also subject to compliance notices issued under Clause 34(1). Where this occurs, a compliance notice can be issued requiring the ASDE to take actions that the Secretary of State considers appropriate to secure, or make more likely, that a regulated body complies with the regulatory requirement in future (Clause 34(5)(a)). Where an authorised automated vehicle commits a traffic infraction, the Secretary of State may issue a compliance notice under Clause 34(2) requiring the ASDE take actions that the Secretary of State considers appropriate in order to avoid, or make less likely, the future commission of similar traffic infractions by an authorised automated vehicle for which the ASDE is responsible (Clause 34(5)(b); see also Law Commissions 2022, 6.99; Law Commissions 2020, 10.24; 10.45).

Regulatory breaches by ASDEs are unlikely to be directly justiciable in the civil courts (*Reid v First Glasgow Ltd*). This is because any legislation would presumably be intended for the public at large, rather than a particular class of persons and the remedy available is likely to be a fine or civil penalty (*Morrison Sports Limited and others v Scottish Power* [2010]. Although, see *Roe v Sheffield City Council* [2004]). Somewhat surprisingly, the Law Commissions suggested that an ASDE may be able to satisfy the financial standing requirements in Clause 6(4)(b) by demonstrating that they have appropriate insurance in place (Law Commissions 2022, 5.89-91). The courts in both jurisdictions have held certain fines uninsurable on public policy grounds (*Safeway Stores Ltd v Twigger* (2010); *Les Laboratoires Servier v Apotex Inc* (2014); *Sainsbury's Supermarkets Ltd v Mastercard Inc and ors* (2016); *Geddes Ltd v Neil Johnson Health & Safety Services Ltd* (2017)). The Law Commissions sought to distinguish civil penalties imposed by an In-Use Regulator from fines imposed by the criminal courts (Law Commissions 2022, 6.74). Nevertheless, it is far from clear that ASDEs will be able to obtain insurance for regulatory sanctions arising under the proposed safety assurance scheme. Ultimately, much will turn on the nature of the breach.

4.4 Primary Liability: Section 2

Section 2 provides for the liability of insurers where an automated vehicle causes an accident while it is driving itself. Where these conditions are met, the effect of s 2 is to impose strict liability (Boon 2020; Glassbrook et al. 2019) on the CAV insurer or, in very limited circumstances where they are exempt from compulsory insurance, the vehicle owner. There is therefore no attempt to try to directly impose a standard of reasonable care on automated vehicle systems.[47] Save for a discreet class of exclusions relating to software updates provided for under s 4, insurers are prevented by s 2(6) from limiting cover for their s 2(1) liabilities in any way (Channon et al. 2019, pp. 29–31). Significantly, liability arising under s 2 does not affect any other person's liability in respect of the accident.[48] Therefore, any secondary tortious or contractual duty to compensate the person suffering damage owed by the other person subsists. As noted above, where control of the vehicle has been appropriately transferred to

[47] See written evidence by (Davey 2017) relating to the Automated and Electric Vehicles Bill. However, it is interesting to note that much of the discussion around safety levels in Chapter 5 of the Law Commissions' third consultation paper centres on performance levels that are analogous with concepts traditionally associated with tortious/delictual standards of care.

[48] S 2(7) of the Automated and Electric Vehicles Act 2018.

the driver, the driver will once again be subject to the rules of negligence. Where an accident is not caused, in whole or in part, by an automated vehicle driving itself, no liability arises under s 2 of the Act. If someone suffers damage in those circumstances, they would only be able to seek a remedy under the pre-AEVA regime. Claims under s 2 must be brought by claimants within 3 years of the accident.[49] Or, where the date of the injured person's knowledge of their right to claim against the insurer is later, and the claim relates to or includes damages for personal injury, 3 years from the date of the injured person's relevant date of knowledge.[50]

4.5 The Insurer

The insurer for the purposes of s 2(1) is the motor insurer providing cover to users of the vehicle.[51] The word 'insurer' is defined at s 8(2) as, when referring to an insured vehicle, 'meaning the insurer under that policy.' Under s 8(1)(b), "...a vehicle is "insured" if there is in force in relation to the use of the vehicle on a road or other public place in Great Britain a policy of insurance that satisfies the conditions in section 145 of the Road Traffic Act 1988." However, if passed, Schedule 2, Clause 5(6)(a) of the Automated Vehicles Bill will amend s 8(1)(b), so that "...a person is an "insured person" in relation to a vehicle, if there is in force in relation to that person's use of the vehicle on a road or other public place in Great Britain a policy of insurance that satisfies the conditions in section 145 of the Road Traffic Act 1988." Similarly, Clause 5(6)(b)(iii) amends the definition of insurer, so that ""insurer", in relation to an insured person, means the insurer under the policy in question."

Schedule 1 paragraph 19 of the AEVA amends s 145 of the 1988 Act, requiring that a policy conforming with s 145 must also provide for an insurer's obligations to their insured under s 2(1) of the AEVA. However, since a liability arising under s 2(1) wholly caused by the automated vehicle is not one which may be imposed on an insured (Boon 2020), no indemnity is required, as it is not a liability envisaged by s 145(3)(a) (see *Sahin v Harvard* [2017]). It is not clear that the drafters of the Act appreciated this point. Meanwhile, Schedule 1 Paragraph 19(3) amends s 145(4) of

[49] Section 11B(2)(a) of the Limitation Act 1980 in England & Wales; s 18ZA(2)(a) of the Prescription and Limitation Act 1973 in Scotland.

[50] s 11B(2)(b) of the Limitation Act 1980; s 18ZA(2)(b) of the Prescription and Limitation Act 1973.

[51] Significantly for CAV manufacturers wishing to offer insurance to their customers, s 145-(2) requires motor insurance policies be issued by an authorised insurer, which is defined at s 95(2) as being an insurer member of the Motor Insurers Bureau. By virtue of s 95(3), an insurer is (a) a person who has permission under Part 4A of the Financial Services and Markets Act 2000 to effect or carry out relevant contracts of insurance, or (b) an EEA firm of the kind mentioned in paragraph 5(d) of Schedule 3 to that Act, which has permission under paragraph 15 of that Schedule (as a result of qualifying for authorisation under paragraph 12 of that Schedule) to effect or carry out relevant contracts of insurance.

the Road Traffic Act, thereby removing any potential liability under s 2(1) for employer's liability insurers for accidents involving their employees.

3.23 Under the current version of the AEVA, difficulties could conceivably arise where the vehicle is subjected to a hack by an adverse actor. The AEVA's definition of an insurer is anchored to s 145 of the Road Traffic Act 1988. On its face, this would appear to rule out any liability arising under s 2(1) for cyber insurers. However, there appears to be a stateable argument that, where an accident occurs due to a cyber security breach allowing an unauthorised actor to take control, the vehicle cannot be said to be driving itself, as defined by s 8(1)(a) (see para 4.67). That being so, the test for causation will not be satisfied and so no insurer liability will arise under s 2(1). Such liability as there is will therefore fall to be determined by common law principles and therefore could give rise to a claim against a cyber policy. This would likely be an instance of 'silent cyber' cover (de Azevedo 2019, chs 2 & 3). The proposed amendment of s 8(1)(a) appears to address this potential lacuna. As noted above, the amended form of s 8(1)(a) will instead state that "...an authorised automated vehicle is "driving itself" if it is travelling while an authorised automation feature of the vehicle is engaged." However, 5(6)(a) would also introduce a new s 8(1A), through which, Clause 44(5) of the Automated Vehicles Bill (authorisation to determine when feature "engaged" or "disengaged") applies for the purposes of subsection (1)(a) as it applies for the purposes of Part 1 of the Bill. Under Clause 44(5), questions as to whether an authorised automation feature is engaged or disengaged are to be determined in accordance with what is specified or described in Clause 4(3). As such, it is submitted that, unless every authorisation by the Secretary of State explicitly states that a cyber security breach influencing the behaviour of an automated vehicle does not affect whether an authorised automation feature is engaged or disengaged for the purposes of Clause 4(3), there may still be room to argue that vehicle hacks constitute disengagement of the vehicle's authorised automation feature and thus accidents arising out of the hack lie outwith the AEVA.

For s 2 claimants, identifying the relevant insurer may not be straightforward. In part, this is because the law does not currently prescribe who the user of the vehicle is. As the author has noted elsewhere, where a qualifying accident occurs for the purposes of s 2 and there is more than one person suitably insured for use of the vehicle, there seems to be a reasonable argument that all the relevantly insured passengers' policies are engaged (Boon 2020). However, the Bill's proposed amendments to ss 8(1)(b) and 8(2) of the AEVA in the Bill appear to have removed this problem.

4.6 Uninsured Vehicles

Save for a narrow class of vehicles not requiring insurance under s 143 of the Road Traffic Act 1988, Section 2 of the AEVA does not apply to uninsured vehicles, as defined by s 8(1)(b). Since an accident caused by an automated vehicle does not fall within the scope of s 145(3)(a) of the 1988 Act, this cannot give rise to a 'relevant liability' for the purposes of the Uninsured Drivers Agreement 2015 and so there would be no requirement for the MIB to compensate a claimant suffering damage (Glassbrook et al. 2019, pp. 112–115). The claimant would be left to pursue a claim against the liable party themselves. The Law Commissions have therefore recommended that the Government takes steps to address this lacuna (Law Commissions 2022, 13.35-39). However, at there are still no agreements in place between the MIB and the Secretary of State for uninsured automated vehicles at the time of writing.

4.7 'Use' of an Automated Vehicle

At first blush, since the requirements for insurance under ss 143(1) and 145 apply equally to automated vehicles, it would appear that 'use' of an automated vehicle will be construed in the same way as use of an ordinary vehicle. However, it may be that, in time, new use cases and the emerging qualities of automation lead to a more expansive construction of the term (see Law Commissions 2022, Background Paper A, 1.20-26).

4.8 Damage

'Damage' is defined at s 2(3) as death, injury or damage to property other than the automated vehicle, goods carried on the vehicle for hire or reward, or property in the custody or under the control of the insured person or the person in control of the vehicle. Section 2(1)(c) provides that an insurer is liable to insured persons as well as other persons.

Another oddity of the AEVA is that the definition of 'damage' under the Act does not obviously appear to extend to financial losses, e.g., loss of earnings, medical treatment or ongoing care costs. Rather, the liability under s 2 is limited to the damage as defined by the Act. Namely, injury and/or property damage. This language is much more restrictive than s 145(3) of the Road Traffic Act 1988, which concerns *any liability* which may be incurred in respect of death or bodily injury to any person or damage to property. The effect of this would appear to be that financial losses are removed from the ambit of the AEVA, which would be bizarre. (However, it might explain the need for s 5(3).) There are reasonable grounds to assume Parliament intended financial losses be included within the scope of compensable losses available under s 2, since this accords with the existing practice in personal injury law and Parliament should be taken to have legislated against the background of common law principles (*Wisely v John Fulton (Plumbers) Ltd, Wadey v Surrey County Council* [2000]). Nonetheless, such a construction goes against the plain meaning of the text itself. Consequently, it may well be that the courts take the view that 'the word 'damage' does not mean damages' (*Birse Construction Ltd v*

Haiste [1996]). That being so, claimants would need to pursue separate parallel claims in contract, tort or statute for their economic losses, along with their injury and/or property damage claims under the AEVA.[52]

It is unclear at this time whether the phrase 'damage to property other than the automated vehicle' extends to third party software or data loss. It has been suggested that software erasure or corruption that impairs the usefulness of physical property may constitute property damage (Cannon et al. 2021, 20-061; *R v Whitely* [1993]). If such an expansive construction was allowed, there appears to be no reason why a claim should be limited to the third party's vehicle. Where a third party has a significant bank of connected physical assets affected by a qualifying CAV accident, the losses claimed might very quickly reach catastrophic levels.[53]

4.9 Causation

Liability under s 2(1) will only arise where the provision's two-limbed test for causation is satisfied. The first limb requires that the accident be caused by the automated vehicle. Under s 8(3)(b), 'caused by' includes accidents partly caused by an automated vehicle.[54] This is ostensibly a low bar. However, as Channon notes (Channon et al. 2019, p. 23), this is more onerous than the test contained within s 145 of the Road Traffic Act 1988. The 1988 Act concerns accidents 'caused by or arising out of the use of the vehicle' and so requires a weaker causal link to the use of the vehicle than that called for by the AEVA.[55]

The second required element is that the vehicle must be driving itself.[56] As previously noted, the AEVA's current definition of 'driving itself' is predicated on the vehicle operating outwith human control and without the need for human

[52] The duty of a compensator to pay recoverable benefits under the Social Security (Recovery of Benefits) Act 1997 subsists. However, where an employment contract contains a clause allowing the employer to recover wages paid to supplement an injured employee's statutory sick pay, employers would not be able to recover their losses under the AEVA either.

[53] Although, see Cannon et al. (2021, 20-108). Quantification will turn on the facts but may just be limited to diminution in value. See *Logical Computer Ltd v Euro Carparks Ltd* (2001) and *Ruxley Electronics v Forsyth* [1996].

[54] This provision is nonetheless broadly on all-fours with the approach to causation in the Consumer Protection Act 1987.

[55] The ambiguous terminology in the AEVA relating to causation is not without criticism. "The intention is for the insurer to then claim against whoever is "responsible" for the collision. The question of establishing who is "responsible", as well as questions of causation (see the terms "cause", "direct result of", "resulting from", "arising out of", which under conventional statutory interpretation do not mean the same things), are presumably left to the courts to decide on a case-by-case basis (or less charitably, for subsequent law review)." (Singapore Academy of Law 2020).

[56] As defined by s 8(1)(a). It is submitted that the test in s 2 qualifies the necessary conditions to establish causation in fact. Otherwise, the principles of causation outlined in Section 2.6–2.9 subsist.

supervision. In framing the causal test so that it is contingent upon the vehicle driving itself, the drafters of the AEVA arguably created a considerable hurdle for claimants to get over. Firstly, it would appear that these conditions must exist at the point in time when the accident is caused by the vehicle to satisfy the test for causation. It is worth recalling that by virtue of s 1(1)(a) of the AEVA, vehicles would only need to drive themselves safely in at least *some* circumstances or situations to qualify as an automated vehicle under the listing process. It follows from this that there might have been some circumstances where a vehicle on the list was unable to achieve these outcomes without the need for human monitoring and/or control. On those occasions, there seems to be a stateable argument that, where a vehicle needs to be monitored by an individual, it is not driving itself and therefore any accidents would need to be progressed under the pre-AEVA scheme.

Whether the vehicle needed to be monitored or controlled immediately prior to the accident is a question of fact. It will likely be judged by an objective test and may require expert evidence. Secondly, the burden of proving this falls to the claimant to discharge. This will be far from straightforward and seems at odds with the aims of the AEVA more generally.

The position for claimants will be simplified considerably by the proposed amendments contained within the Automated Vehicles Bill. If the Bill passes, the amended definition of 'driving itself' will change the second limb of the test for causation. All then that will be required is that an authorised automation feature is engaged when the accident is caused by an authorised automated vehicle.

Meanwhile, s 8(3)(a) provides that any reference to an accident should be construed to include two or more causally-related accidents.[57] This would appear to address multi-vehicle shunts. It has been suggested that s 8(3)(b) would give rise to a liability for insurers of vehicles that are "...not the primary cause of those accidents but just one of the partially-causative pieces in the concertina" (Glassbrook et al. 2019, p. 56). On one view, this is incorrect. Applying the first principles of causation, liability should attach to the insurer of the vehicle that is the real effective cause of each collision, just as it does with ordinary vehicles (**para 2.5**). It is noteworthy that the phrase 'caused or partly caused by' features in the Consumer Protection Act 1987 and no such issue concerning vehicles arises there (Fairgrieve and Goldberg 2020, 17.104). However, establishing the point is likely to require complex technical engineering evidence and it seems reasonable to assume insurers' approach to contributions from other CAV insurers will be driven by commercial pragmatism.

Existing authorities concerning erratic driving suggest that, where an automated vehicle pulls up suddenly and violently, or swerves violently, and an accident results, the vehicle's driving behaviour will be deemed to be the proximate cause of the accident (*O'Hara v Central Scottish Motor Traction Co Ltd* 1941; *Doonan v Scottish Motor Traction Co* 1950). Evasive manoeuvres that result in an accident are likely to satisfy the test for causation in s 2 but may, where the manoeuvre is

[57] Section 8(3)(a) of the Automated and Electric Vehicles Act 2018.

reasonable in all the circumstances, amount to a defence in recovery proceedings raised by a Section 5 Claimant.[58]

The ordinary meaning of the word accident in this context suggests some form of collision. However, 'accident' is not defined within the AEVA. Where the question has arisen in relation to the Road Traffic Acts, the courts have adopted a wide construction that also includes deliberately caused damage (Channon et al. 2019, p. 15; *Charlton v Fisher* [2002]; *Fenton v Thorley & Co* [1903]; *R v Morris* [1972]; *Chief Constable of the West Midlands Police v Billingham* [1979]; *Chief Constable of Staffordshire v Lees* [1981].

4.10 Primary Liability – Post-Transition Period Claims Against Drivers

Once control has been transferred from vehicle to driver, the vehicle is no longer driving itself. CAV accidents occurring immediately after transfer will lie outside the ambit of the AEVA and instead fall to be dealt with under the traditional systems of negligence in both jurisdictions. However, it is submitted that immediate post-transition liability of drivers is worthy of consideration in its own right. It is self-evident that any transfer of control will not take place in a vacuum. Consequently, where an accident occurs shortly after a completed transition request, a court considering the issue of an automated vehicle driver's negligence might find the circumstances of the case justify consideration of what a driver ought, or ought not, to have done immediately prior to transfer while they were still the UiC. The most obvious example would be a UiC carrying out some form of monitoring during the transition period and before taking back control. This seems conceptually analogous with checking one's mirrors before commencing a manoeuvre. Nonetheless, this kind of analysis seems at odds with the criteria for autonomous travel under the Bill's test for self-driving in Clause 1(5)(b), which requires a vehicle be capable of driving itself where "...neither the vehicle nor its surroundings are being monitored by an individual with a view to immediate intervention in the driving of the vehicle." Similarly, as noted above, it is hard to square this kind of approach with the Highway Code's guidance that UiC's do not need to monitor their vehicle while it is driving itself. As ever, the position is not straightforward. It is submitted that for vehicles with SAE Level 3 and even Level 4 functionality, monitoring of the vehicle and surrounding environment may well be necessary even where an authorised "user-in-charge" automation feature is engaged. As such, it is further submitted that a driver's duty of reasonable care does not necessarily commence at the end of a transition period. Rather, it may well begin before this. Namely, where the authorised "user-in-charge" automation feature is still engaged and the driver is still deemed to be a UiC.

[58] However, matters might be further complicated by questions of whether an ADS operating to prescribed safety levels should be caught off guard and respond in this way.

There are a number of reasons why this proposition may reasonably be stated. Perhaps most importantly, nothing in the AEVA or the Bill specifies when the driver's duty of care begins. This will therefore be a question of fact for the courts to determine. Further, the kind of monitoring contemplated by Clause 1(5)(b) is that which is necessary only with a view to immediate intervention in the driving of the vehicle. The word 'immediate' suggests a very narrow window of action. Likewise, "intervention" connotes a driver's unprompted taking back of control from the vehicle, which appears quite different from responding to a transmission request. Further, it is by no means guaranteed that when the driver takes control from the vehicle it will be safe for them to do so. For example, the minimum risk manoeuvre required under UNECE Regulation 157 and minimal risk conditions under SAE Level 4 may still lead to a vehicle coming to a stop in their lane, all while other traffic is still moving around them. Indeed, the Law Commissions noted that authorisation of an ADS feature as 'sufficiently safe without human intervention does not mean the elimination of all risk' (Law Commissions 2022, 3.43). Further, they acknowledged that where a UiC fails to respond to a transition demand in time, even where the vehicle is programmed to mitigate any risk, it may not be completely safe (Law Commissions 2022, 8.132). More generally, they argued that it was unrealistic to require that vehicles with ADS "...must always bring a vehicle to a safe, legal stop out of an active lane of traffic following a failed transition demand. With all-lane running on motorways, that would be to demand more of an ADS than of a human driver... Even the most sophisticated system will fail sometimes and may struggle to cope with the consequences of its own failure" (Law Commissions 2022, 3.41). Here it is worth recalling that the requirements necessary for autonomous travel under Clause 1(7) are that the vehicle travels acceptably safely and with an acceptably low risk of committing a traffic infraction. Therefore, notwithstanding the Law Commissions' view that vehicles should be safe enough to drive themselves without human monitoring, as a matter of general principle, it seems counterintuitive to argue that users have no need to monitor anything where their vehicle transition may place them at risk of serious harm.

More prosaically, the need for user monitoring is equally apparent when one considers driver's need to regain situational awareness in order to then safely take control of the vehicle. The Law Commissions suggest that, in order to regain situational awareness, users require to "...work out what is around them..." during any transition period. They also note that the necessary duration of any transition period will depend on a variety of factors, including the complexity of the environment, and whether it is day or night (Law Commissions 2020, 4.91). In their third consultation paper, the Law Commissions referenced the SAE's definition of monitoring. This included "...real-time human or machine sensing and processing of data..." concerning the 'driving environment, the vehicle performance for DDT performance-relevant system failures and the driving automation system performance' (Law Commissions 2020, 4.24). On this definition, regaining situational awareness appears to require some form of monitoring. Indeed, the Law Commissions indicate that, where users require to regain situational awareness quickly in an emergency, they would require to carry out prior monitoring of the environment

(Law Commissions 2018, 3.19). Consequently, it seems reasonable to conclude that drivers regaining situational awareness require to undertake some form of monitoring as a UiC and that this should take place during the transition period. This view finds some support in the drafting of Clause 7 on Transition Demands. Under Clause 7(3)(b), transition demands must, amongst other things, be long enough not just for the UiC to assume control of the vehicle but also to prepare to assume control. Failure to prepare to take control would likely lead to a finding that the Driver has failed to exercise reasonable care while they were a UiC.

The Bill's approach to driver immunity post-transfer is also instructive in this regard. Another recommendation of the Law Commissions, the immunity is concerned with offences unavoidably committed by a driver in the period immediately following a handover. The Law Commissions gave the example of a CAV mistakenly going the wrong way down a one-way street, which immediately gives rise to a traffic infraction at the point where the driver acquires responsibility for the vehicle. As one consultee noted, when faced with an immediate collision, drivers might try to take over without having had time to reengage with the driving task. "In this situation, their best efforts to avoid a crash may fall below the standard of a driver who is already fully engaged" (Law Commissions 2022, 8.127). Adopting the Law Commissions' recommendations on the point, the Bill provides that driver immunity for offences relating to the way in which a vehicle is driven is contingent upon the offence not resulting "...from the individual's conduct after ceasing to be the user-in-charge falling below the standard that could reasonably be expected of a careful and competent driver in the circumstances" (Clause 47(3)(b); see also Law Commissions 2022, 8.130). This provision is concerned with driving sufficiently dangerously to warrant criminal sanctions. Therefore, considerable care must be taken when attempting to extrapolate direct parallels in tort and delict. Nonetheless, the reference to a careful and competent driver – albeit, post-transfer - is consistent with the common law standard of reasonable care. Keeping a good lookout when driving is a constituent part of that duty. In Gorringe v Calderdale Metropolitan Council (2004), having concluded that drivers cannot rely upon local authorities to provide warnings of prospective dangers on the roads, Lord Rodger noted that drivers must take care for themselves. "By insisting that drivers always look out for dangers themselves and not rely on others, the common law supports the overall policy of promoting road safety" (Gorringe, [93]). Clearly, there are limitations to which Gorringe can be applied as a statement of general principle. Nonetheless, reading Clauses 7(3)(b) and 47(3)(b) of the Bill together, there seems to be scope for a line of authority to develop whereby drivers have a sliding scale of responsibility during and immediately after transition periods. Inevitably, liability is likely to be particularly fact-sensitive in each case. In any event, it would appear that drivers will need to do more than merely assert they were a UiC in order to resist civil claims that turn on purportedly negligent acts or omissions occurring during or immediately after a transition period.

While much of the discussion has focussed on whether 10 seconds is enough time to regain situational awareness (Law Commissions 2020, 4.90-99; Kinear et al, 2020; Stevens et al 2021), it is worth keeping in mind that nothing in the AEVA, the Automated Vehicles Bill or Regulation 157 prohibits a driver from attempting to take back control immediately after a transition request is issued. Faced with the prospect of a vehicle coming to a stop on a busy road, one can easily imagine drivers rushing to take back control before they regain full situational awareness. This may be due to concerns they will cause an accident or obstruct other traffic. Or simply because they are in a hurry and do not want to interrupt their journey. Decisions taken in the agony of the moment may cause accidents but might not be negligent (Brandon v Osborne, Garrett & Co [1924]; Tocci v Hankard (1966)). However, a user will clearly not be able to resist liability simply because his actions were designed to prevent an inconvenience (Adams v Lancashire and Yorkshire Rly (1869).

4.11 Contributory Negligence: Section 3

Section 3 of the AEVA contains a number of provisions concerning contributory negligence. Under s 3(1), liability arising under s 2 is subject to whatever reduction would apply under the Law Reform (Contributory Negligence) Act 1945 if the claim was brought against a person other than the insurer or owner. This is supplemented by s 6(3), which provides:

> For the purposes of section 3(1), the Law Reform (Contributory Negligence) Act 1945 and section 5 of the Fatal Accidents Act 1976 (contributory negligence) have effect as if the behaviour of the automated vehicle were the fault of the person made liable for the damage by section 2 of this Act.

The operation of s 3 is therefore contingent upon liability attaching under s 2. As a result, where an accident with an automated vehicle *not* driving itself is partly caused by a claimant's contributory fault, then issues of contributory negligence fall to be considered against the existing statutory framework of the 1945 Act. The same is true in accidents where, by virtue of the exclusions contained within s 4, no liability arises under s 2.

The awkward drafting of ss 3 and s 6(3) raises a number of issues. Unhelpfully, the thing reduced in s 3(1) is the 'amount of the insurer or owner's liability.' In framing matters this way, the Act conflates strict liability and negligence. This is especially conceptually awkward when one considers that the strict liability accruing under s 2 attaches notwithstanding the import of s 8(3)(b). As here, the term liability is often misused (Hogg 2015, p. 17). Properly understood, liability in tort is the tortfeasor's secondary obligation to compensate a party suffering harm caused by the tortfeasor's breach of the primary duty not to unreasonably cause such harm. The provision anchors the reduction of liability to the Law Reform (Contributory Negligence) Act 1945. However, under s 1 of the 1945 Act, it is the level of damages that is reduced, not the liability to pay them. Indeed, the computation of the reduction is more nuanced, and considers causal potency and blameworthiness (Taylor 1986, p. 23).

Section 6(3)'s difficult anthropomorphism requires an exercise in counterfactuals (Law Commissions 2018, 6.36). The language indicates no little uncertainty by the Act's drafters around how to frame the attribution of fault between human and machine. Glassbrook suggests that this uncertainty flows from a historic position whereby contributory negligence in "…road traffic accidents usually require a comparison of the faults of several [human] actors" (Glassbrook et al. 2019, p. 62). Further, the Act cannot avoid considerations of the blameworthiness of the automated vehicle as they have both contributed causatively to the accident. Such a comparison is therefore "…essential to the exercise of assessing contributory negligence…" (Glassbrook et al. 2019, p. 59). Glassbrook suggests that the Act makes the insurer vicariously liable for the conduct of the vehicle (Glassbrook et al. 2019, p. 66). Given the unwieldy terms in which its provisions are framed, the Act certainly invites this analysis. Nevertheless, it is submitted that this approach is incorrect and that nothing in S 6(3) alters the character or degree of liability arising under s 2. Not least because of the strict nature of the liability arising under s 2. Further, the phrase 'have effect' is qualified by '*as if*.' The purpose of s 6(3), therefore, is to enable the court to apply the equitable discretion arising under s 1 of the 1945 Act to awards of damages due as a result of liability imposed under s 2 of the AEVA; which, absent s 6(3), would otherwise lie outwith the scope of the 1945 Act.

Furthermore, as Glassbrook suggests, setting man against machine creates a false dichotomy (Glassbrook et al. 2019, pp. 62–68). However, it is instructive to note that the courts have previously addressed the issue of contributory fault in the context of employees using dangerous machinery without adopting this kind of approach (Hendy and Ford 2001, ch. 29). When undertaking assessments of contributory negligence under s 3(1), it is submitted that the courts can be expected to have regard to the destructive disparity between the claimant and the automated vehicle (*Eagle v Chambers* [2003]; *Jackson v Murray* [2015]; and Riley-Smith and McCormick 2020, 5-031), in particular the latter's causative potency, and adjust the balance of blameworthiness accordingly (*Jackson*).

Given the operation of ss 2 and 5, it is submitted that in accidents involving an automated vehicle and an ordinary vehicle, it is inappropriate to treat the parties as severally liable and apportion damages according to the respective levels of fault (Walker 1981, p. 361. See also Johnson 2021). Rather, the logical operation of the Act demands that proceedings are raised against the party liable under s 2, with such damages as are due to the claimant reduced under s 3 where it is appropriate to do so. The insurer's rights do not crystalise until after settlement of the s 2 claim and so presumably they would not have standing to counter claim under s 5 in an action founded upon s 2 of the AEVA. The insurer will therefore need to either raise separate proceedings or counter claim under the law of negligence for these losses. Meanwhile, the level of reduction in the s 2 claim would inversely inform the ratio of damages an insurer/owner is entitled to under their s 5 claim. In contrast, where a collision between two AVs is caused, or partly caused, by both the automated vehicles, each injured party would need to claim against the other's insurer under s 2, rather than s 5.

4.12 *Contributory Negligence of a User-in-Charge*

As previously noted, the AEVA specifically contemplates claims by automated vehicles' policy holders, as well as third parties.[59] As such, to the extent that a policyholder's own negligence is partially causative of the accident, their damages will be reduced in line with the principles expounded above. If the Automated Vehicles Bill is enacted, s 3(1) will nonetheless only apply to UiCs where an authorised vehicle is driving itself and therefore has an authorised automation feature of the vehicle engaged. This appears to open the door to considerations of whether the UiC has been negligent while an authorised automation feature is engaged; and, if so, it implies that, at least for the purposes of s 3(1), UiC's might still owe something akin to a duty of care while the vehicle is driving itself. Given the attempts made to insulate UiC's from criminal liability for accidents occurring while the vehicle is driving itself, it might seem counterintuitive to impose a duty of care upon them in a civil context. However, it is worth noting that the immunity against criminal liability under Clause 47 is not absolute. Likewise, the provisions of Clauses 47 & 48 suggest a variety of scenarios where a UiC's actions or inactions could contribute to an accident while the vehicle is driving itself. To take one example, where the vehicle comes to a stop in the middle of a busy road and the UiC exits the vehicle (see Clause 48(3)). Similarly, while the Highway Code makes clear that users do not need to monitor a vehicle driving itself, it nonetheless notes that a self-driving vehicle's ability to drive itself may be limited by its ODD. Drivers are therefore advised to follow the manufacturer's instructions about when and how to use the self-driving function safely (Department for Transport 01 July 2022). That being so, there seems to be a more than stateable argument that UiC's owe other road users a duty of care while an automated vehicle is driving itself; a breach of which will bring them within the ambit of s 3(1). Given the potential duties owed by drivers immediately prior to transition considered above, the existence of potentially co-extensive duties for UiC's seems apt to confuse and is therefore somewhat unfortunate.

As the Law Commissions and others note, there is ample evidence from human factors research that indicates that breaks in concentration can be anticipated even in careful drivers where Dynamic Driving Tasks (DDTs) have been automated. Nevertheless, it remains the case that the courts will be slow to forgive momentary lapses of concentration.[60] As a result, it is difficult to discern a bright line precisely marking out the parameters of the standard of care to be expected of a UiC. This in turn makes any assessment of a UiC's contributory fault challenging.

---

[59] Section 2(1)(c) of the Automated and Electric Vehicles Act 2018.

[60] Glassbrook argues that "...if Level 3 CAVs are to function safely on public roads, it seems that the courts would have no choice but to hold users to high standards. Accepting that Users in Charge will reasonably lose their focus and attention would, in effect, be an acceptance that Level 3 CAVs are not safe for general use." (Glassbrook et al. 2019, p. 16).

While their differences are self-evident, some general principles might usefully be drawn from other statutory strict liability schemes.[61] As Fairgrieve and Goldberg note, "…comparative fault is not easy to accommodate within a strict liability system since the defendant is, in principle, to be held accountable irrespective of fault" (Fairgrieve and Goldberg 2020, 17.132). The courts are likely to treat automated vehicles as something akin to a potentially lethal weapon[62] and so a high standard of care will be expected of UiCs. However, to borrow Lord Tucker's dictum on the Factories Acts, if too exacting a standard is imposed on UiCs, it will defeat the purpose of the AEVA (*Stavely Iron and Chemical Co ltd v Jones* [1956]). In not entirely dissimilar scenarios involving 'employees engaged in repetitive and monotonous work' (Hendy and Ford 2001, 29.28; *Lawrence v Syndic Manufacturing Co Ltd* [1952]), and where 'the dangers are obscured by repetition' (*General Cleaning Contractors Ltd v Christmas* [1953], per Lord Oaksey [189–190]), the courts have been reluctant to view 'the slackening of attention which naturally comes from constant repetition of the same operation' (*Caswell v Powell Duffryn Associated Collieries Ltd* [1940], per Lord Wright [176–178]; Hendy and Ford 2001, 29.16) as a form of contributory negligence. The case law in this area is likely to be particularly fact-sensitive. Nonetheless, the line of authorities concerning employer liability suggests that, in some circumstances at least, inadvertence and momentary lapses of concentration when operating dangerous machinery will receive at least a sympathetic ear. This seems even more likely in the context of a strict liability regime backed by compulsory insurance and where secondary activities are expressly permitted under the Highway Code.

Section 3(2) provides that, where the accident that it caused was wholly due to the person's negligence in allowing the vehicle to begin driving itself when it was not appropriate to do so, the insurer or owner of the automated vehicle is not liable under s 2. This is a complete defence. Much has been made of the fact that this amounts to a finding of 100% contributory negligence. As noted at **para 2.6**, findings of 100% contributory negligence have been criticised as conceptually misconceived, since the 1945 Act demands joint contributions to causation (*Jayes v IMI (Kynoch) Ltd* [1984]; Stewart 2004, p. 313). It is submitted that this analysis is equally true of s 3(2). Indeed, properly understood, where an accident is wholly caused by the claimant allowing the vehicle to drive when it is inappropriate to do so, it cannot then be said that the accident was caused by the vehicle driving itself (Hendy and Ford 2001, 29.06). This appears to hold true regardless of whether one applies the current definition of 'driving itself' under the AEVA or its prospective amendment under the Bill. Consequently, the test for causation under s 2 cannot be satisfied and so no liability should arise (*Boyle v Kodak Ltd* [1969]).

---

[61] Clearly, these circumstances are not directly on all fours with an employer/employee relationship, 'where Parliament has placed on the shoulders of the employer responsibility for compliance with the statutory duty' (Hendy and Ford 2001, 29.50).

[62] *Lunt v Khelifa* [2002], per Latham LJ at [20]: "…this court has consistently imposed on the drivers of cars a high burden to reflect the fact that a car is potentially a dangerous weapon…"

In any event, this provision appears to have very limited practical application. The phrase ‘begin driving itself’ makes clear that it only applies in situations where control is ceded by the user to the automated vehicle, and not the other way around. Furthermore, s 3(2) is contingent upon the accident being *wholly due* to the claimant’s negligence in allowing the vehicle to drive itself when it is inappropriate to do so. As Channon notes, this is a particularly onerous hurdle for insurers to get over (Channon et al. 2019, p. 31). The negligence envisioned here is of a very particular character too, since it relates specifically to the user *allowing* the vehicle to drive itself when it is inappropriate to do so. This provision appears to be clearly satisfied in circumstances where the UiC actively initiates the ADS. However, given its ordinary meaning, allow is more passive and suggests that a UiC’s failure to intervene where the vehicle itself initiates ADS may also give rise to a complete defence to a s 2 claim.[63] If so, this again tends to suggest some form of monitoring is required even where a vehicle is driving itself. What is ‘appropriate’ in all the circumstances for each case will clearly turn on its own facts. It is therefore noteworthy that when updates to the Highway Code were proposed, it was thought that it would be left to vehicle manufacturers to define “…when and how to use the self-driving function safely…” (Department for Transport 2022b). The Law Commissions acknowledged the risk that manufacturers “…might produce long and complex lists of when the ADS should not be engaged…” (Law Commissions 2022, 8.74). They therefore recommend that ADS features should not operate outwith their ODD (Law Commissions 2022, 8.75). This would further limit the application of s 3(2). Nevertheless, while the Bill allows the Secretary of State to impose requirements to this effect, nothing in the Bill seeks to proscribe such use and it is at least theoretically possible that the Secretary of State could authorise an authorised automation feature that is able to operate outside its ODD. In any event, parties seeking to resist liability under ss 2 and 5 will clearly argue that the courts should adopt an expansive construction of what is appropriate. There might be occasions where courts consider engaging an ADS system was entirely inappropriate. Nevertheless, they may well be anxious to avoid finding inappropriate engagement was the sole cause of the accident in all but the rarest of cases if the result of doing so prevents a claimant from obtaining a remedy under the AEVA.

Where the UiC’s inappropriate initiation of the ADS is the sole cause of an accident with a third party, the insurer’s defence under s 3(2) would appear to be something of a pyrrhic victory. The UiC’s actions would be *prima facie* evidence of negligence and so their motor insurance policy would presumably respond to cover any third party claim advanced outwith the AEVA scheme. As such, it seems unlikely that s 3(2) will be contentious unless the UiC is uninsured and MIB arrangements leave lacunae. Nonetheless, claimants may wish to join the UiC to proceedings, pleading breaches of their common law duty of care.

[63] See *Burgoyne v Phillips* [1983], which suggests that a failure to operate the controls may constitute driving, and (Law Commissions 2018, 2.65).

Conversely, the case law on automatism suggests that, where an automated vehicle develops a fault that renders it incapable of safely driving itself, if the UiC exercises some control, even if it is imperfect, and allows the vehicle to continue to drive itself, the UiC has arguably breached its duty of reasonable care.[64] In these circumstances, the vehicle might not be able to 'drive itself' for the purposes of s 8(1)(a) - as it is currently drafted - and, if so, liability is unlikely to attach under s 2 for the reasons outlined above. However, where the definition of 'driving itself' is only contingent upon an ADS feature being engaged, primary liability would likely attach to the insurer/owner under s 2.

4.13 Defences: Section 4

Section 4 concerns accidents resulting from unauthorised software or failure to update software. Section 4(1) allows insurers—but not the narrow class of owners addressed by s 2(2)—to entirely avoid any liability under s 2 to an insured person where an accident causing damage *occurs as a direct result* of either (a) the insured person making software alterations to the AV that are prohibited under the policy and/or (b) failure to instal safety-critical updates that the insured person knows, or ought reasonably to know were safety critical.

The exclusion in s 4 is predicated upon the acts or omissions of an 'insured person.' As per s 8(2), this currently refers to a person whose use of the vehicle is covered by the policy in question. This is narrower language than the phrase "...such person, persons or classes of persons as may be specified in the policy..." employed under s 145(3) of the Road Traffic Act 1988. It is therefore not certain that NUIC Operators would fall within the scope of this exclusion in its current form. However, if the courts are content to construe the word 'use' as extending to include an NUiC Operator's use of a vehicle, the proposed amendment to the definition of an insured person at s 8(1)(b) appears sufficiently wide as to capture NUiC Operations.

Curiously, s 4(1)(b) does not state who must be responsible for the failure to instal safety-critical updates for the policy exclusion to operate. Nor is it clear that s 4(6)(b) necessarily refers to safety-critical updates to the automated vehicle alone. As such, it is submitted that an overlooked safety critical software update by the vehicle's manufacturer,[65] software developer, and/or the vehicle's ASDE could fall within the scope of s 4(1)(b). Further, in certain circumstances, it might even

---

[64] Even if control retained is imperfect. See (*Roberts v Ramsbottom* [1980]; Law Commissions 2022, 8.138-151).

[65] See for example, Principle 6 of 'The key principles of vehicle cyber security for connected and automated vehicles' (CCAV et al. 2017). Likewise, the Department for Digital, Culture, Media & Sport's "Code of Practice for Consumer IoT Security" (Department for Digital, Culture, Media & Sport 2018) also envisages ongoing responsibilities regarding software updates for device manufacturers and IoT service providers. There is a more tenuous argument that, by virtue of s 4(6)(b), s

apply to omissions by a business providing software, cyber security or cloud services. It is not inconceivable that a CAV user may not be aware that a safety critical update making their vehicle unsafe to use has been overlooked. This might allow insurers to avoid liability in any circumstance where an accident is caused by a third party's failure to install a safety-critical update. Similar difficulties may also arise where a safety critical update is not fully downloaded due to interrupted or sub-standard Wi-Fi or 5G service. More generally, it remains to be seen how ASDEs' mooted regulatory duties, in addition to any related but separate obligations arising under service contracts, would overlap with and/or override policyholders' responsibilities in respect of vehicle software updates and cybersecurity. Whatever the effect, it seems unlikely to be straightforward.

Nonetheless, the point is potentially academic given the high bar this provision sets around causation. For the exclusion to bite, the software alteration or failure to install safety-critical software must be the proximate cause of the loss. This interpretation gains further support by virtue of the fact that Parliament employs different causal terminology.[66] Consequently, s 4(1) falls to be construed without reference to partial causation under s 8(3)(b). Establishing that these acts or omissions were the sole cause of the accident is likely to be extremely difficult and require complex technical evidence. Evidence that vehicle manufacturers and software developers are better placed to obtain than insurers. Consequently, it seems foreseeable that s 4(1) causation defences are more likely to be run in recovery actions pursued under s 5 than cases advanced under s 2(1).

Another matter rarely discussed is the insurer's right to prohibit certain software alterations. Depending on how broadly the policy terms are framed, this exclusion could conceivably be a significant issue. The basis for an insurer prohibiting a software alteration is currently left to the discretion of individual insurers. Likewise, the rather tautological definition of software updates and alterations given at s 4(6)(a) does little to shed any light on the point. Given its ordinary meaning, it could conceivably be argued that *any update* that alters any of the existing code is amenable to being proscribed under the policy by an insurer. This provision appears to indirectly allow insurers to police CAV system updates. It is not obvious that this is something insurers have the capacity or inclination to oversee. Similarly, given the potential tension between insurers, and CAV producers and service providers arising out of the operation of s 5 of the Act, it is not clear that such a role is entirely desirable. Since the insurer's liability under s 2(1) is capable of existing outwith the scope of the Motor Insurance Directive and s 145 of the Road Traffic Act 1988, it is submitted that insurers have considerable scope to use s 4(1) to limit their liability under s 2(1). Albeit, it may be that the effect of these prohibitions is tempered by s

4(1)(b) extends to cover updates on separate IoT devices, which are connected to the CAV and could foreseeably impact the vehicle's ability to drive itself safely.

[66] Glassbrook suggests that the language employed "...puts a more demanding burden upon the person (inevitably the insurer) seeking to rely upon the exclusion..." (Glassbrook et al. 2019, p. 76).

11(3) of the Insurance Act 2015, which provides exclusions can't be relied upon if: "…non-compliance with the term could not have increased the risk of the loss which actually occurred in the circumstances in which it occurred." It is therefore submitted that insurers' use of s 4 should be kept under careful scrutiny by the Secretary of State and addressed in any report produced pursuant to s 7(1)(b).[67]

Nevertheless, insurers will need to consider the wording of any exclusion carefully. As above, CAV manufacturers, software developers and service providers will presumably have a better understanding of the updates than the insurer, and so are potentially better placed to assess whether the prohibition applies. It is therefore not inconceivable that an insurer's decision to settle a s 2(1) claim may be subject to challenge in any subsequent recovery action on the basis that, on the facts, a prohibited alteration was made and therefore no liability arose under s 2(1).[68] It is not clear that an insurer would have any remedy if such an argument was successful, since their s 2(1) liability is not contingent on establishing liability under s 5.

Under s 4(6)(b), an update is safety-critical if it would be unsafe to use the vehicle without it.[69] Since the exclusions available through s 4 relate to liability arising under s 2(1), the use anticipated by s 4(1) would appear to be one which arises only when the vehicle is driving itself. If so, it follows that only those safety-critical updates that facilitate the vehicle driving itself safely fall within the scope of the exclusion. It is likely that courts will have regard to industry guidance in this area by the UNECE and other relevant bodies.[70] Nonetheless, as Channon notes, it is far from clear that a UiC would know whether any given update is safety critical, as defined by the Act (Channon et al. 2019, p. 29). Matters may become clearer as and when the Automated Vehicles Bill is enacted. Given their need to have particular regard to the Statement of Safety Principles, the Secretary of State would presumably be alive to the need to require ASDE's (and, where necessary, NUiC Operators) ensure automated vehicle users readily understand when and how to instal safety-critical software updates. Equally, it may be that ASDE's software and cyber security maintenance duties implied by Clause 6(4)(a) ultimately mean this exclusion is of little consequence (see also Law Commissions 2022, 9.50).

Should ADS-level functionality be provided as a subscription service and that service is not renewed, the question arises as to whether service loss that occurs while the vehicle is being driven falls within the scope of s 4(1)(b). It is submitted

[67] An interesting question arises as to whether insurers' regulatory and prudential obligations (such as ICOBS) apply where the exposure is the insurer's and not that of their insured.

[68] In this respect, the situation is not entirely dissimilar to some of the issues arising in the context of claims against reinsurance policies.

[69] As Matthew Channon notes, the term 'safety critical' is not without its problems. The term 'unsafe' in conventional motor insurance law tends to refer to mechanical defects affecting brakes and tyres, but not a vehicle driving at excessive speed. Not insignificantly for our purposes, the point appears to turn on the agency of the user (Channon et al. 2019, p. 27).

[70] What amounts to Safety critical software updates will also likely be informed by guidance issued by UNECE and other relevant bodies, e.g. the recently adopted UNECE Software Regulation (UN 2020) and Cybersecurity Regulation (UN 2020). See also, (Law Commissions 2020, 10.6-28).

that it does not and attempting to use automation features where the UiC is on notice of the impending rescission of service more properly falls within the ambit of s 3(2). The courts would then require to apply an objective test as to whether an ordinary CAV user ought reasonably to have known that failure to install the safety-critical update would compromise the safety of an AV when it is driving itself. Moreover, as Channon notes, insurers and manufacturers (as well as software developers and/or ASDEs) will need to tread with care to avoid direct or indirect discrimination where a protected characteristic affects a UiC's state of knowledge concerning a safety-critical update (Channon et al. 2019; see Law Commissions 2022, 10.36-8).

4.14 Software Time Bombs[71]

Similar concerns might arise where a software supplier incorporates a 'time bomb,' by which they are able to deactivate the software for a breach of the software's license terms (Cannon et al. 2021, 20-053-4). Where software deactivation adversely affects the safe operation of the vehicle and causes an accident, assuming the UiC is on actual or constructive notice that the software has been or is imminently likely to be deactivated, s 3(2) would appear to apply subject to the test outlined above.

'Holder of the policy' is not defined within the Act but the logic of s 4(1) indicates it is the insured keeper of the vehicle. By extension, an insured person who is not the holder of the policy can reasonably be assumed to be a named driver. Both the current definition of an insured person and the Bill's amended definition allow this construction. Whether the implicit assumptions around responsibility for software alterations and installing safety-critical updates are borne out remains to be seen.

4.15 Recoveries: Section 5

Section 5 provides insurers liable under s 2 with a right of recovery against the party otherwise liable to the claimant. The effect of s 5(1) is something akin to a statutory form of subrogation. It seems tolerably clear that Parliament intended the insurer or vehicle owner to 'stand in the shoes' of the claimant to recover their insured losses in much the same way that they do under traditional motor insurance. However, as the author has noted elsewhere, in traditional motor insurance, contract and tort operate in tandem. The law of tort imposes liability on the tortfeasor for their negligent driving. The tortfeasor then enjoys a right of indemnity under the contract of insurance, whereby the insurer effectively pays the injured party damages directly. In contrast, the liability referred to in s 5 of the AEVA is an awkward conflation of tort and contract that needlessly causes confusion and uncertainty. It is neither a form

[71] See also paragraph **4.28**.

of assignation, nor is it a species of joint and several liability (Boon 2020; Glassbrook et al. 2019, p. 89). Similarly, the language of s 5(1)(b) is jarringly imprecise and, as with s 3(1), it is regrettable that the drafters of the Act saw fit to refer to the 'amount of liability' in the context of a strict liability scheme.

The rights contained within s 5 only accrue upon settlement, and therefore 'full satisfaction' of the claim under s 2 (Boon 2020; *Tang Man Sit (Personal Representatives of) v. Capacious Investments Ltd* [1996], per Lord Nicholls of Birkenhead [522]. It would appear that the rights available to insurers and owners under s 5(1) do not crystalise until such time as *all* liability—including liability not imposed by s 2—has been settled. Again, the meaning of liability in this context is not wholly clear. On one view, there seems to be a stateable argument that this liability could extend to the insurer's contractual obligation to indemnify under the motor insurance policy, since the insurer cannot otherwise be liable in tort for the accident. However, it is equally true that a breach of duty by the owner could causally contribute to any losses suffered by a third party, supporting a different construction of the provision. As above, this uncertainty is regrettable. In any event, to avoid falling foul of the requirements of s 5(1), a belt and braces approach whereby insurers/owners ensure any settlement terms explicitly confirm that all liabilities have been discharged prior to pursuing claims under s 5 would appear sensible.

Section 5(2) prescribes the 3 scenarios in which settlement occurs for the purposes of s 5(1). Properly understood, an 'enforceable agreement' is effectively an extrajudicial settlement. While it is binding upon the parties to the agreement, it is not the product of judicial scrutiny. They are therefore *illiquid* debts. This is of no great consequence in England and Wales, and such agreements may competently be founded upon in litigation. However, this is not the case in Scotland, where the courts have declined to allow actions based on illiquid debts (*National Coal Board v Thomson* and *Comex Houlder Diving Ltd v Colne Fishing Co Ltd* 1987). Unfortunately, the AEVA's somewhat tin-eared approach to Scots law means that s 5(2)(c) is the exception to the well-established rule set out in *National Coal Board v Thomson* (see McNaughton 2003).

On its face, s 5(3) appears to be relatively benign. However, s 5(3) is arguably the most problematic provision within the AEVA. To have any meaningful effect, it requires some mechanism whereby parties compensated under s 2 are notified of improved settlements. Otherwise, s 5(3) would be redundant. Nevertheless, notwithstanding their liability to the party suffering damage, nothing in the Act requires a s 2 compensator to do anything to intimate to the compensated party details of the improved settlement. Similarly, the Act is silent on whether a s 2 compensator may keep the improved settlement in the event the original claimant cannot be traced. Likewise, the AEVA provides no limitation period for the s 5(3) liability. In England and Wales, this obligation would appear to be subject to a 6-year limitation period.[72] In Scotland, it is submitted that the improved settlement falls to be dealt with as a form of unjustified enrichment subject to a 5-year prescriptive period; expiration of

[72] S 9 of the Limitation Act 1980.

which extinguishes the obligation to compensate altogether (Boon 2020; *Dollar Land (Cumbernauld) Ltd v CIN Properties Ltd* 1998; *Cullen v Advocate General for Scotland* [2016]).

However, more controversial still is the implication within s 5(3) that settlements under s 2 might be improved upon by a s 2 compensator with their s 5 claim. The purpose of damages is *restitutio in integrum*, i.e. to put, as far as compensation allows, the claimant in the same position they were in but for the breach. Yet, in providing s 2 claimants a legal right to recover improved settlements, it follows that the AEVA specifically contemplates the improvement itself. Likewise, the requirement that the s 2 compensator pass on the difference indicates that the improved sum flows from the original cause of action. The logic of this is that any discharge of liability under s 2 by way of compensation is rendered uncertain. Further, that the original settlement can be topped up. This seems to run entirely against the principal of *res judicata* whereby actions previously decided by a competent court cannot be reopened for reconsideration (Jones 2020, ch. 7; Stewart 2004, 25-18; Walker 1981, p. 343).

When the implications of s 5(3) are appreciated, a series of difficult questions present themselves. What circumstance might warrant the reopening of a s 2 claim? Can a decision of an appeal court be reviewed by an inferior court via s 5(3)? What duty, if any, is a s 2 compensator under to seek an improved settlement where stateable grounds for an increase exist? Is a claimant entitled to a remedy where a s 2 compensator fails to pursue such an increase? The answers to these questions are far from clear and apt to confuse. Absent any discernible benefit, it is submitted that s 5(3) should be removed from the AEVA.

At s 6(5), the Act provides that a party with a claim under s 5 "...does not have a right to recover contribution from that person under the Civil Liability (Contribution) Act 1978 or under section 3 of the Law Reform (Miscellaneous Provisions) (Scotland) Act 1940." In so doing, the Act casually severs all ties with a significant body of settled case law in this area from both jurisdictions. In its place, Schedule 1 of the Act adds s 10A of the Limitation Act 1980 (the 1980 Act), and s 18ZC to the Prescription and Limitation (Scotland) Act 1973 (the 1973 Act) in respect of an insurer's s 5(1) recovery rights. In so doing, the two-year time bars are seemingly reinstated in both the jurisdictions' respective legislation on limitation and superficially the status quo subsists. However, this is not the case and once again the muddled juristic character of liability under the AEVA causes problems. It must be borne in mind that what the AEVA removes from Section 5 Claimants is a right, available under both the 1978 Act and the 1940 Act, to a contribution from another wrongdoer liable for the same damage. In return, the AEVA provides that any Other Person liable to the injured party in respect of the accident is under the same liability to the insurer or vehicle owner. This is not a like for like replacement. In effect, the AEVA simply extends the scope of an Other Person's pre-existing liability. This is not without consequence. Under the 1978 Act, liability in England and Wales "...to make contribution remains even if a person has ceased to be liable to the claimant by virtue of the expiry of a period of limitation, unless a period of limitation or prescription has extinguished the right on which the action against him was based"

(Jones 2020, 4.20). Likewise, in Scotland the same is true in respect of liability to contribute under the 1940 Act (Brodie 2021, A31-007). It is submitted that the position in respect of liability under s 5 of the AEVA is quite different, as it cannot extend the overall period for which an Other Person is liable to make some form of payment. Rather, the two-year limitation period on s 5 claims appears to be an additional time limit; one which is coextensive with any other limitation or prescriptive period governing an Other Person's underlying liability to the injured party. If this analysis is correct, there appears to be a stateable argument that, much like an assignment or assignation, an Other Person's liability to a Section 5 Claimant lasts for only as long as whichever of these limitation or prescriptive periods expire first. This would be a significant downgrade on the traditional statutory rights to contribution and is likely to catch unwary Section 5 Claimants off-guard.

Unfortunately, it also appears that the drafters of the AEVA did not appreciate the distinction between limitation and prescription in Scots law. The former is a matter of procedural law, rendering certain rights unenforceable but which, per s 19A of the 1973 Act, may be overridden by the courts where it is just and equitable to do so. The latter, in contrast, is a complete defence on the substantive merits, operating so as to extinguish completely any obligation to compensate (Welsh 2006, 2.114). There is no statutory discretion to allow prescribed claims to proceed out of time. Currently, claims advanced under s 3 of the Law Reform (Miscellaneous Provisions) (Scotland) Act 1940 are subject to a two-year *prescriptive* period, which is contained within s 8A of the 1973 Act. Section 8A makes clear that the obligation to make a contribution is extinguished upon the expiration of the two-year period. However, by contrast, the new time bar in s 18ZC of the 1973 Act has been placed beside the various species of personal injury in Scotland, which are subject to a limitation period[73] and the judicial override available under s 19A. While s 19A does not apply to s 18ZC, the language employed in s 18ZC is that of limitation, providing that a claim under s 5(1) of the Act 'may not be brought in Scotland after the expiration of the two-year time bar.' In effect, the obligation subsists but is no longer justiciable. Consequently, this may mean that the nature of any prescriptive period applicable to claims under s 5(3) discussed below is open to doubt. Moreover, it creates further confusion as to the character of s 5(3) claims in Scotland, since the 1973 Act now effectively treats s 5(1) claims—and, by extension, those under s 5(3)—as a type of personal injury, rather than an action for relief. Section 6(5) therefore arguably puts the law on time bar around s 5(1) recoveries in Scotland at large.

[73] These different forms of injury claim and their limitation period are set down in ss 17, 18, 18A or 18B of the 1973 Act.

5 Insurer Recoveries

Section Four of this chapter outlines the legal grounds by which parties liable under s 2 can pursue recovery of their losses from the Other Person responsible for the accident. Section Four consequently focusses primarily on the rights of insurers, as it is likely that they will be the party pursuing recoveries in the majority of cases. Nonetheless, the points discussed in this section apply equally to owners liable under s 2(2) unless otherwise stated. For ease of reference, they will collectively be referred to as Section 5 Claimants.

When exercising their recovery rights under the AEVA, Section 5 Claimants will potentially be able to advance claims on three footings. Namely, product liability claims under the Consumer Protection Act 1987, as well as claims in contract and negligence. These are dealt with in turn below. Much of the public discussion of the AEVA to date has been on claims by individuals, rather than commercial interests. However, given the likely cost of CAV ownership, anticipated changes to traditional models of vehicle ownership and possibility of work-based secondary activities, it seems reasonable to assume that commercial entities will also have a large stake in the CAV market. The detailed contractual analysis in this chapter is intended to reflect this.

5.1 Product Liability Claims: Consumer Protection Act 1987

The primary source of law in this area is found within the Consumer Protection Act 1987. The 1987 Act implemented the Product Liability Directive.[74] The Act was amended by the Product Safety and Metrology etc (Amendment etc.) (EU Exit) Regulations 2019. While the 1987 Act effectively remains on all fours with the Product Liability Directive, by substituting Member States for the UK, the Regulations effectively make the Act a wholly domestic piece of legislation.

5.1.1 Product Definition

'Products' are defined under s 1 of the 1987 Act as including goods and electricity. This includes a product within another product, whether by virtue of being a component part or raw materials or otherwise.[75] By virtue of s 45, the Act also applies to vehicles. Liability is imposed on the producer under ss 2(1) and 2(2)(a) of the act where harm is caused by a defective product. Liability is strict. However, if

[74] Directive 85/374/EEC, Official Journal L 210 of 7.8.1985, pp. 29–33.

[75] The Directive was amended so as to define a product under Article 2 as "...all movables even if incorporated into another movable or into an immovable. Dir 1999/34/EC, Official Journal L 141, 04/06/1999, pp. 20–21.

the defect is situated in something that does not meet the Act's definition of a product, no liability attaches under the Act. The complex ecosystem of hardware and software contained within automated vehicles, along with their reliance on external services and infrastructure, arguably create the greatest strain on the traditional concept of products as movables. The status of software for the purposes of the 1987 Act in particular has caused no little uncertainty (Fairgrieve and Goldberg 2020, 9.98-105). As McCormick and others note, there is a paucity of authority on this point. Such guidance as can be found arises instead in the context of the sale of goods. Software provided on a tangible medium contained within the CAV may well be deemed to be a product (Channon et al. 2019, 4.8-12; Glassbrook et al. 2019, p. 124; Fairgrieve and Goldberg 2020, 9.77). Consequently, where a defect in that software causes damage the CAV producer would be liable (Fairgrieve and Goldberg 2020, 9.84; Trib Cassino, 20-12-2007). Conversely, software not provided on a physical medium has not traditionally been seen as goods for the purposes of the Sale of Goods Act 1979 and thus, by analogy, is unlikely to be treated as a product (*St Albans City & District Council v International Computers Ltd* [1996]; *Your Response Ltd v Datateam Business Media Ltd* [2014]; *Computer Associates UK Ltd v The Software Incubator Ltd*; *Eurodynamics Systems Plc v General Automation Limited* (1988); *Saphena Computing Limited v Allied Collection Agencies Limited* (1995), *Horace Holman Group Limited v Sherwood International Group Limited* (2002); See also, Fairgrieve and Goldberg 2020, 9.98–9.105; Channon et al. 2019, 4.8–4.12; Glassbrook et al. 2019, pp. 123–126; Riley-Smith and McCormick 2020, 5-039-041). This 'subordination of the dominant characteristic of a complex product to its means of transmission' is, as Lord Penrose noted in *Beta Computers (Europe) Ltd v Adobe Systems (Europe) Ltd* (1996), rather odd and, it is submitted, somewhat unsatisfactory.

Superficially, the distinction appeared to have been resolved by the CJEU's decision in *Computer Associates (UK) Ltd v The Software Incubator Ltd* (2021). The Court held that software supplied electronically and not on a tangible medium could constitute goods for the purpose of Directive 86/653/EEC on the coordination of the laws of the Member States relating to self-employed commercial agents. However, in doing so, it noted that the Directive does not distinguish between tangible and intangible goods.[76] This is in contrast to the Product Liability Directive, which is concerned only with movables. Further, while the Court noted that the purchase of a computer program on DVD or CD-ROM was functionally similar to downloading the program from the internet, the analogy is an economic one anchored to the principle of equal treatment in respect of the distribution right under Art 4(2) of Directive 2009/24.[77] As such, while *Computer Associates (UK) Ltd* indicates that, at a conceptual level, software provided by download is capable of being treated as a form of goods, it is does not appear to be authority for the proposition that it should be treated as a product for the purposes of the 1987 Act.

[76] Para 37. The Court is referencing the Opinion of the Advocate General.

[77] Directive 2009/24/EC, Official Journal L111 of 5.5.2009, pp. 16–22.

That being so, it appears unlikely that the case will provide sufficient comfort to allow the courts to depart from the view that this is a problem for Parliament to resolve.

Under s 1(2), a product "...includes a product which is comprised in another product, whether by virtue of being a component part or raw material or otherwise." Consequently, a defective component within a CAV causing an accident will result in liability attaching jointly and severally to both the producer of the CAV and of the component.[78] This means s 2 insurers have a choice of parties to recover from (Glassbrook et al. 2019, p. 121). Where an automated vehicle component causes damage to the vehicle itself, no liability arises under s 2 of the AEVA since it is specifically excluded from the Act's definition of damage. Further, any direct claim made against the component producer under the Consumer Protection Act 1987 would also fail as "...the loss of or any damage to the whole or any part of any product which has been supplied with the product in question comprised in it..." is excluded under s 5(2) of the Act.[79] To the extent that a software update constitutes a new standalone product, this may allow a claim under the 1987 Act for damage to the vehicle (Glassbrook et al. 2019, p. 123).

5.1.2 Producers

Section 1(2) of the 1987 Act defines 'producers' as: (a) a product's manufacturer; (b) where a product is not manufactured but rather won or abstracted, the person who won or abstracted it; or (c) where it is processed, the processor. Section 2(2) extends the scope of Part 1 of the Act beyond producers to persons holding themselves out as the producer and any person who imports the product into the UK in the course of their business to supply to another. Fairgrieve and Goldberg categorise the latter as secondary producers, who, by virtue of s 2(3), can avoid liability under s 2 where they disclose the identity of the producer within a reasonable period of time (Fairgrieve and Goldberg 2020, 8.02).

5.1.3 Causation: Section 2

Section 2 provides that where damage is wholly or partly caused by a defective product, the producer, or person holding themselves out as the producer or the importer into the UK of the product in the course of their business to supply to another is liable for the damage. Liability is strict. The bar for causation appears, on its face, to be a low one, since the minimum required is that the defect partly caused

[78] S 2(5) of the Consumer Protection Act 1987.

[79] A claim may still lie in negligence though. See para 4.3.4.

the damage. While it is not necessary to show how a defect was caused,[80] claimants will still need to evidence the presence of a defect, damage and the causal relationship between defect damage.[81] It is submitted that this may well be more difficult than is often supposed. Not least because AV software may not be amenable to interrogation.

Where a hack causes an accident, there is at least a stateable argument that the actions of the hacker constitute a *novus actus interveniens*, which breaks the chain of causation (Riley-Smith and McCormick 2020, 5-022). However, this is likely to be time-limited and the longer the producer of the AV is on notice of a vulnerability capable of being exploited by hackers, the weaker any *novus actus* argument becomes. Nevertheless, where the vulnerability results from shortcomings in the vehicle design or manufacturing process, it is submitted that there are sound policy reasons why any subsequent security breach should not be held to break the chain of causation. In much the same way as CAV hacking, it has been suggested that a wholly aberrant and unpredictable output caused by changes resulting from self-learning code might constitute a *novus actus interveniens* (Riley-Smith and McCormick 2020, 5-023). Albeit, where the system is largely deterministic any defective operation may be traced back to the programmer (*B2C2 Ltd v Quoine Pte Ltd* [2019]. In practice, however, it seems likely the courts will be slow to allow producers to elide responsibility for a product capable of such dangerous independent iterations that they have knowingly put into the market.

5.1.4 Defective Product Definition: Section 3

Under s 3, a product is defective if the safety of the product is not such as persons generally are entitled to expect. In determining what safety levels persons generally are entitled to expect of a product, all the circumstances shall be taken into account, including: a product's marketing, get-up, instructions, and warnings; what reasonably might be expected to be done with a product; and the time when the product was supplied by a producer. Subsequent improvements in safety in and of themselves do not give rise to an inference of defectiveness. What expectations of safety users of an emergent and potentially dangerous technology should reasonably have is open to question. Whatever their expectations may be, perfection is too high a bar (Glassbrook et al. 2019, p. 127; Channon et al. 2019, p. 38). The question of how safe is safe enough is an objective one. Namely, what should users' legitimate expectations be? (Fairgrieve and Goldberg 2020, 10.57-67; *A v National Blood*

[80] See *Ide v ATB Sales Ltd* [2008]. The linked appeal of *Lexus Financial Service t/a Toyota F.S. UK Plc v Sandra Russell* [2008] is also relevant. In *Lexus*, the vehicle "...contained a complex and extensive electrical system..." and so there was "...much more to go wrong..." However, the case is particularly fact-sensitive and turned in large part on the witness evidence. The black box problem means that this kind of evidence is unlikely to be available for defective AI. See also Channon et al. (2019), p. 40.

[81] Article 4 of the Product Liability Directive 85/374/EEC.

Authority [2001]; *Wilkes v DePuy International Limited* [2016]). Since CAVs are unlikely to ever be entirely safe,[82] assessments of their safety will be 'inherently and necessarily a relative concept' (*Wilkes v DePuy International Limited*, per Hickinbottom J [13]. See also *Gee v DePuy International* [2018] and *A v National Blood Authority*). To the extent that users are aware of "...the impossibility of achieving complete safety due to code errors..." (Knight 2021, 7.140), it may be relevant to compare the software with other software on the market. In contrast, where standard CAV software is infected by a virus during the process of distributing the program onto a tangible medium within the CAV, the infected software would be treated as a non-standard product. The infected software would be assessed against the performance of its equivalent standard software prior to infection (Knight 2021, 7.137-9).

Section 3(2) invites the court to undertake a weighing exercise and it is therefore perhaps no surprise that courts have entertained a risk/benefit analysis of safety (see Fairgrieve and Goldberg 2020, 10.28–10.56; Riley-Smith and McCormick 2020, 5-043-4). Nonetheless, it is likely that, when assessing the levels of safety CAV users are entitled to expect, the courts will have regard as applicable to *inter alia* Secretary of State's Statement of Safety Principles and the terms of any automated vehicle authorisation, as well as additional regulatory requirements such as those set down in UNECE Regulation 157 (see **para 3.3**).

Given the risks posed by a serious system failure, it is submitted that users' reasonable expectations of safety can reasonably include mitigation strategies that operate when safety critical systems fail, along with robust and suitable redundancy measures. Likewise, there seems to be a stateable argument that an automated vehicle with an inadequate safety case may give rise to an inference that the vehicle falls below reasonable expectations of safety.

Any analysis of users' legitimate expectations should reflect the point in time at which the product was supplied. Given the levels of hyperbole in this space, this is no small thing. Indeed, as Milligan notes, against so much uncertainty, courts will likely attach greater significance to producers' communications when considering the level of safety to be expected of CAVs. It bears repeating that this is a new form of driving, which will be unfamiliar to even experienced drivers. That being so, since "good instructions and warnings can render safe products that come with potential dangers, whilst poor instructions and inadequate warnings can be reasons to find a product unsafe" (Riley-Smith and McCormick 2020, 5-045), any defences reliant upon CAV User Manuals, heavily caveated though they may be, are likely to come under particular scrutiny (see para. 4.34). Where any unsatisfactory training emanates from an ASDE, rather than the original producer, it is submitted that any failure by the producer to address issues of which it is on notice may well be sufficient to fix the producer with liability. The duty to warn extends to misuse that is predictable and not unreasonable (Fairgrieve and Goldberg 2020, 11.60; Channon et al. 2019,

[82]This would not necessarily render them defective. See *Bogle v McDonald's Restaurants Ltd* [2002].

4-17-18). Similarly, constructive knowledge of misuse is likely to be inferred where it is posted on social media sites. Particularly so where producers are tagged, retweet or reply to these posts. It seems readily foreseeable that a failure to warn against misuse, where it is specifically drawn to a producer's attention in this way, is likely to be a relevant factor to which a court may have regard when considering users' expectations of safety. McCormick raises the possibility that s 3 could extend to vehicle accessories on the basis that they might be deemed to be something 'done in relation to a product,' giving rise to a duty for producers to supervise the accessories market (Channon et al. 2019, 4-17-18). This could conceivably extend to software alterations using open-source code[83] and 3D Printing.

5.1.5 Defences: Section 4

Section 4 of the 1987 Act sets down a number of defences for producers of products. Under s 4(1)(a), producers can resist claims where product defects are due to conformity with any assimilated obligation. ('Assimilated obligations' are effectively those derived from retained EU law, which was relabelled by The Retained EU Law (Revocation and Reform) Act 2023 (Consequential Amendment) Regulations 2023.) There is very little authority on the point but it is clear that the defence is narrow in scope. For the defence to operate, the defect must flow directly from compliance with a specific regulation. As Fairgrieve and Goldberg note, it is difficult to envision a scenario where such a defence might realistically be employed (Fairgrieve and Goldberg 2020, 13.07).

Despite the considerable focus on whether CAV software is a product, there has been very little consideration of the liability implications arising from the means by which the software is supplied. Supply is defined in detail at s 46(1) of the Act. Even in the event that CAV software is ultimately deemed to be a product, Fairgrieve and Goldberg suggest that the "...downloading of information without transfer of physical matter..." (Fairgrieve and Goldberg 2020, 9.100) would not constitute *supply* for the purposes of s 46(1). That being so, "...the absence of supply would be a defence under s 4(1)(b)." (Fairgrieve and Goldberg 2020, 9.100-002). As such, if the courts were to adopt this analysis, it appears that where a defect transmitted by an OTA update causes an accident, the update producer will nonetheless be able to resist claims under the 1987 Act on the basis that there was no supply. Absent any remedy in tort/delict or contract, Section 5 Claimants will therefore not be able to recover their outlays. Given the degree to which CAV operating systems will rely on OTA updates, the significance to insurers of the s 4(1)(b) defence under the 1987 Act seems difficult to overstate. This will presumably inform insurers' attitudes to software alteration prohibitions.

Where a defect materialises after the CAV has been supplied, the producer will have a defence under s 4(1)(d). Identifying when the vehicle, as well as any hardware

[83] Particularly so given Tesla's open-source pledge (Tesla 2014; Duane Morris LLP 2018).

or tangible parts provided during repair, were supplied should be reasonably straightforward. The position relating to intangible elements is less easy. This could conceivably apply in situations where defects occur in 'self-learning' code post-supply. As Milligan notes, the availability of this defence turns on the degree of the supplier's control (Glassbrook et al. 2019, pp. 131–132). Establishing the extent to which a supplier does or should exercise control after supply is also likely to be challenging. Much will turn on whether the supplier has assumed some degree of responsibility for the ongoing operation of the software or system. Even where no such duties exist, the risk of anomalies outwith the Producer's control may well be treated as a risk warranting a warning from the producer. Again, the role of the ASDE in respect of software updates will need to be factored into any analysis of this point. Similarly, where a CAV AI system operates satisfactorily at the point of supply but, due to some pre-supply human error deteriorates thereafter (e.g. errors in the code, a defective data training set or insufficient operator training, etc.), deteriorates thereafter, where that system is supplied on something tangible, liability under s 3 will likely be established. Where the system is supplied OTA, the supplier may still have a liability in negligence or contract.

The development risks defence in s 4(1)(e) requires the producer to demonstrate that the state of scientific and technical knowledge was such that the defect was undiscoverable by *anyone* (*European Commission v United Kingdom* [1997]) at the point of supply (See *A v National Blood Authority* [2001]). Given the prevalence of external simulation testing in this space, it may be that data generated by testing companies will be relevant where this defence is run. Similarly, insofar as it is capable of disclosure, data collected pursuant to any Secretary of State's general monitoring duty, as set out in Clause 38 of the Bill, may be evidence of the state of scientific and technical knowledge (see also Law Commissions 2022, 6.36-7).

The 'state of the art' defence flows from the final lines of s 3(2). The effect of which being that the subsequent addition to the market of safer models does not, of itself, mean earlier iterations were unsafe (Fairgrieve and Goldberg 2020, 10.83). The pace of change in this space means that parties pursuing a recovery will need to be clear about the state of the art at the relevant time. This is likely to be a difficult task.[84]

5.1.6 Limitation Periods

Negligence actions consisting of, or including, claims for damages for personal injury in England & Wales under the 1987 Act must be brought within 3 years whichever is the later of (a) the date on which the cause of action accrued; or (b) the date of knowledge of the injured person or, in the case of loss of or damaged property, the date of knowledge of the plaintiff or (if earlier) of any person in

[84] However, it will be no defence to suggest that certain dynamic driving tasks are beyond the current state of the art. See Cannon et al. (2021, 20-075 and 20-091).

whom his cause of action was previously vested."[85] Similar periods apply in Scotland.[86] If the approach adopted in **para 3.19** is correct, this also applies to s 5 claims. The respective statutes for England & Wales, and Scotland contain a 10-year prescriptive longstop for claims.[87] Since the lifespan of many vehicles often exceeds this, this longstop is likely to bar many claims advanced under the 1987 Act.

5.2 Contract

Section 6(4) of the AEVA provides that the liability under s 2 is treated as a liability in delict or tort.[88] Further, under s 5(1), any other person liable to the injured party is under the same liability to the party liable under s 2. At first blush, this appears to suggest that the juristic character of the liability arising under s 5(1) is the same as that under s 2. However, s 6(4) makes no such qualification of s 5(1). The s 2 liability is borne of statute and defined by it. The Other Person's legal duties, and liability for breach thereof, arise independent of the AEVA. This is made explicit by s 2(7) of the Act, which provides that imposition of liability by s 2 does not affect any other person's liability in respect of the accident. Since the word 'liability' has historically been employed by turns to refer to both the primary obligation (the duty not to carelessly harm/performance of obligations) and secondary obligations (compensation for breach of the duty) in tort/delict and contract, any analysis of the word liability requires a degree of caution (Hogg 2015, p. 17). Standing such concerns, it presumably follows that, where any other person is liable in contract to the party suffering damage, the character of that liability subsists, unaffected by s 6(4). To construe this so that the liability becomes tortious/delictual could lead to the contracting party's rights under the contract being diminished. Likewise, such a construction would render s 2(7) redundant (Boon 2020; Glassbrook et al. 2019, p. 90).

Following the logic of s 5(1), if an automated vehicle causing an accident while driving itself gives rise to a third-party contractual liability, this liability appears to be capable of being transferred to a Section 5 Claimant. If this analysis is correct, the parameters of this contractual liability will be informed and defined by the terms of the contract; and, for reasons outlined below, the character of the damage sustained. This will include any relevant implied terms imported into the contract from statute.

[85] S 11A(4) of the Limitation Act 1980.

[86] S 22B of the Prescription and Limitation Act (Scotland) Act 1973.

[87] Respectively, s 11A(3) of the Limitation Act 1980 and s 22A of the Prescription and Limitation (Scotland) Act 1973.

[88] For the purposes of any enactment conferring jurisdiction on a court with respect to any matter.

5.2.1 Basis for Contractual Liability

While there are some not insignificant differences, for the purposes of this chapter contract law in England and Wales, and Scotland can be considered to be broadly analogous.[89] Contracts are formed where an offer is made and accepted.[90] Both jurisdictions require that one party assumes some form of obligation, along with sufficient certainty and an intention to be legally bound. The law of England and Wales requires that that party does so in return for consideration or under seal.[91] However, consideration is not required under the Scots law of contract and so a unilateral promise can give rise to enforceable obligations even where nothing is provided in return.[92] The terms giving rise to contractual obligations may be express or implied. Contractual exclusions in both consumer and commercial contracts purporting to exclude a contracting party's liability for personal injury or death are unenforceable.[93] Once a breach of a relevant contractual term is established, liability is usually strict. There are various remedies available for breach of contract. For our purposes, however, the primary remedy is usually damages (See Beale 2021, Ch. 29; McBryde 2007, Ch. 22). Absent express terms governing the point, damages for breach of contract are generally intended to compensate for damage, loss or injury suffered through the breach. The purpose of damages being to put the claimant, as far as money can do so, in the same position they would have been had the contract been performed (*Robinson v Harman* (1848); *Morris-Garner v One Step (Support) Ltd* [2018]; *Govan Rope and Sail Co Ltd v Weir & Co* (1897)) Meanwhile, contributory negligence is generally not available in relation to breaches of contract (Cannon et al. 2021, 20-128 & 20-130; *Forsikringsaktieselskapet Vesta v Butcher* [1989]; *Logical Computer Ltd v Euro Carparks Ltd*. Finally, contract claims are subject to a 6-year limitation period in England and Wales.[94] In Scotland, a prescriptive limitation period of 5 years applies to contract claims.[95]

[89] Where relevant, material differences are highlighted. Nonetheless, for a more detailed account of the law in either jurisdiction, readers are advised to consult the relevant primary texts. Namely, *Chitty on Contracts*, *The Law of Contract in Scotland*, as well as *Product Liability*.

[90] This includes 'click-wrap' contracts, typically provided on the supplier's standard terms, where acceptance is achieved through clicking on a button acknowledging the supplier's terms and thereafter another confirming that the terms are accepted. On this, see Davey in Edwards (ed) (2018), pp. 340–341; Smith (2020, 6.136-7), and on the differences between *click-wrap*, *browse-wrap* and *web-wrap*, (Smith 2020, 10-161-171).

[91] The UK left the EU before the Digital Content Directive (Directive (EU) 2019/770 EC, Official Journal L136 22.5.2019) was transposed into UK law. Consequently, where data is provided as consideration for digital services, those services will remain outwith the scope of the Consumer Rights Act 2015, per s 33 of the Act.

[92] This nonetheless requires evidence of an intention to be legally bound.

[93] See s 2 of the Unfair Contract Terms Act for England and Wales, and s 16 for Scotland. See also Part 2 and Schedule 2 of the Consumer Rights Act 2015.

[94] Section 5 of the Limitation Act 1980. The same period applies in respect of actions for sums recoverable under statute by virtue of s 9 of the Act.

[95] Section 6 of the Prescription and Limitation (Scotland) Act 1973.

As noted above, there are several accident scenarios where a Section 5 Claimant will be unable to obtain a remedy under the Consumer Protection Act 1987.[96] However, in these circumstances it may be that the law of contract allows a Section 5 Claimant to recover their losses from the *Other Person*. Likewise, unlike tort and delict, pure economic losses are usually recoverable in contract. Nonetheless, even where a contractual remedy exists for pure economic loss, they may not be recoverable under s 5(1). The extension of any Other Person's liability under s 5 is contingent on liability arising under s 2, which is unlikely to extend to pure economic loss (see **para 2.2**). As such, absent any s 2 liability, a claimant suffering legally actionable pure economic loss will likely only be able to recover their losses directly from the contracting party and then only if the contract allows for it.

5.2.2 Express Terms: Goods and Services

To the extent that any express terms in a CAV contract address description, sample, quality or fitness for a purpose, they are unlikely to prove controversial unless they seek to limit or exclude the seller's liability (Bridge 2021, 26-164.). Where a CAV or related services do not conform to any warranty provided,[97] this may well amount to an actionable breach justifying *inter alia* an award of damages. Notwithstanding the criteria by which 'automated vehicles' are listed or authorised, experience has shown it is entirely foreseeable that false or erroneous representations about the vehicle's capabilities may be made by the seller (Law Commissions 2022, 7.13-15). This in turn could materially alter how the vehicle is operated, which may in turn cause an accident. Clearly, where an ADS feature is not engaged in most instances these sorts of accidents can safely be categorised as driver error, regardless of whether it was made in good faith or not. That being so, the vehicle would not have been driving itself for the purposes of s 2 and so any claim would lie outwith the scope of the AEVA. Where an insurer advances a s 3(2) defence in circumstances where inappropriate operation of an ADS feature allegedly arose through manufacturer misstatement about functionality, it is submitted that the burden of proving the ADS feature was inappropriately engaged will fall to the insurers to discharge. However, it may nonetheless seem a far more attractive option than settling the claim and seeking recovery of its outlays by way of a misrepresentation claim against the relevant supplier. Establishing whether these inaccurate descriptions are actionable by Section 5 Claimants will be difficult (See *Oscar Chess Ltd v Williams* [1957]).[98]

[96] Most notably, where the defective 'thing' causing the accident was not a product; the property damaged by the defective product was not intended for private use, occupation or consumption; or where the defective product was supplied more than ten years before the accident.

[97] In Scotland, "Warranty… which has a special meaning in England, Wales and Northern Ireland in respect of sale of goods, is not a term of art…" (McBryde 2007), pp. 20–93.

[98] The position in Scotland concerning innocent misstatement and negligent misrepresentation is less clear, and has not been considered by the courts. The English cases are to be approached with caution, since they are informed by "…devices to avoid the rule that damages were unobtainable for

Much turns on whether the statement was one of opinion or one of fact (Fairgrieve and Goldberg 2020, 3.02-04; *Hopkins v Tanqueray* (1845), and therefore whether it is incorporated as a term into the contract. Context is a material consideration in any analysis of the point but it appears true of both jurisdictions that matters turn on whether there is: "... evidence of an intention by one or both parties that there should be contractual liability in respect of the accuracy of the statement" (*Heilbut, Symons & Co v Buckleton* [1913]; Beale 2021, 15-003). Even where no warranty is given, knowledge of its intended use may mean a statement about the state of the vehicle is deemed to be a term incorporated into the contract of sale. Similarly, attempts to limit or exclude liability for errors of description can be overridden by verbally expressed warranties (*Couchman v Hill* [1947]). This may be significant where vehicle sales staff give assurances around performance or capability that exceed the levels described in any written materials incorporated into the contract of sale.[99] The expertise of the person making the statement is also relevant. An inexperienced hobbyist in a private sale may be afforded greater latitude than an experienced professional in a showroom making the same statement (Fairgrieve and Goldberg 2020, 3.04).

5.2.3 Misrepresentation

Depending on whether it was innocent, fraudulent or negligent, a misrepresentation that induces a person to enter into a contract that causes loss may give rise to a variety of remedies in England and Wales (*Derry v Peek* (1889); *Hedley Byrne & Co Ltd v Heller & Partners Ltd* [1964]; *Esso Petroleum Co Ltd v Mardon* [1976]. See Beale 2021, 9-001-133; Fairgrieve and Goldberg 2020, 3-04).[100] In particular, damages may be available in respect of non-fraudulent misrepresentation under the Misrepresentation Act 1967, unless reasonable grounds can be shown for believing the misrepresented facts were true.[101] There seems to be a stateable argument that a case could be made out under the 1967 Act where misrepresentation about an automated vehicle's capabilities leads to an accident. While decided in a different jurisdiction on different points of law, the recent Wettbewerbszentrale case against Tesla over the latter's use of the terms 'Autopilot' and 'Full Self-Driving' nonetheless indicate that a misrepresentation claim might receive a sympathetic judicial response. However, a claim for fraudulent misrepresentation is likely to be

innocent misrepresentation..." and the doctrine of collateral warranty (McBryde 2007, 5-48-53; *Esso Petroleum v Marden* [1976]).

[99] As alleged in the civil case brought by Shawn Hudson against Tesla.

[100] The 1967 Act does not apply to Scotland. Under Scots law, s 10 of the Law Reform (Miscellaneous Provision) (Scotland) Act 1985 provides that a claim for damages is available where a contract is entered into as a result of a negligent misstatement – *Hamilton v Allied Domecq* 2001 G.W.D. 13-517. See also McBryde (2007, 15-01-94).

[101] S 2(1).

an extremely difficult hurdle to get over.[102] Conversely, where the statement is a vague and impressionistic 'puff' typical of salespersons' patter, it has no legal effect (Fairgrieve and Goldberg 2020, 3.02). Such was the concern around misrepresentation of vehicle functionality, the Law Commissions recommended criminalising the use of terms "self-drive", "self-driving", "drive itself", "driverless" and "automated vehicle" by commercial entities in this space to describe technology that has not been appropriately authorised as self-driving under the authorisation scheme (Law Commissions 2022, 7.37). While specific terms are not referred to in the Bill, Clause 78 would enable the Secretary of State to criminalise the use of certain terms. Likewise, Clause 79 sets out a number of further offences relating to communications likely to confuse end-users of vehicles as to whether the vehicle is an authorised automated vehicle capable of travelling safely, autonomously and legally on the roads or other public places in Great Britain.

5.2.4 Commercial Contracts for the Supply of Goods: Implied Terms

Implied terms may be imported into a contract from statute,[103] at common law (*Marks and Spencer Pl v BNP Paribas Securities Services Trust Co (Jersey) Ltd* [2015]) or by mercantile custom (*Luxor (Eastbourne) v Cooper* [1941]). In addition to any relevant express terms, goods sold to consumers and commercial entities are subject to various terms implied by statute. In the context of non-consumer sales of automated vehicles, the main legislation governing the sale of goods is the Sale of Goods Act 1979. Subject to the provisions of the Unfair Contract Terms Act 1977, implied terms can be excluded or varied by express agreement, dealings or usage between the parties.[104]

5.2.5 Description, Quality and Fitness for Purpose

Under s 13(1) of the Sale of Goods Act 1979, in contracts for the sale of goods by description there is an implied term that those goods will conform with that description.[105] Purchasers will likely rely on descriptions of CAV software or cyber security measures installed on an automated vehicle, since they are unlikely to be able to ascertain its quality even with inspection (Fairgrieve and Goldberg 2020, 4.27-8). Section 14(2) provides that there is an implied term that goods sold in the course of

[102] Nevertheless, it is interesting to note that in the US case of *Alexandro Filippini v Tesla Inc*, the plaintiff specifically pleads *inter alia* fraudulent misrepresentation.

[103] For our purposes in particular, under the Sale of Goods Act 1979, the Supply of Goods and Services Act 1982 and the Consumer Rights Act 2015.

[104] Section 55(1).

[105] Section 13(1A) provides that the implied term is a condition in England and Wales, and Northern Ireland. The distinction between conditions and warranties—not a feature of Scots law—affects whether recission of the contract is permissible.

business[106] are of satisfactory quality. Satisfactory quality is an objective standard, which takes account of its description, price and all other relevant circumstances.[107] Furthermore, under s 14(2B) the quality of an automated vehicle will be informed by its state and condition, as well as but not limited to, its fitness for all the purposes for which goods of that kind are commonly supplied; appearance and finish; freedom from minor defects; safety; and durability. However, s 14(2C) states that any implied term under s 14(2) does not extend to any matter making the quality of the goods unsatisfactory where it is drawn to the buyer's attention or it the goods are examined, where examination would reveal the issue, prior to the contract being made. Equivalent terms were included in the Supply of Goods and Services Act 1982[108] and the Consumer Rights Act 2015.[109]

5.2.6 Safety

A serious defect in the vehicle that causes an accident, such as a battery catching fire during vehicle use, is likely to lead to a finding that the vehicle was not of merchantable quality (See *Bernstein v Pamson Motors (Golden Green) Ltd* [1987]; *Ward v MGM Marine Ltd* [2012]). While there are no authorities directly on the point, there seems to be a stateable argument that assessments of the state and/or condition of the vehicle should include consideration of whether the software is up to date when sold. This is likely to be a particularly fact-sensitive area. The presence of a virus within the vehicle operating system at the point of sale would, by analogy, appear to constitute a breach (Fairgrieve and Goldberg 2020, 4.70; *Chaproniere v Mason* (1905)). Where such defects cause damage, the contractual standard for implied terms relating to safety is likely to be assessed by a standard aligned with that in tort (Fairgrieve and Goldberg 2020, 4.70). Failure to meet minimum standards of design is likely to give rise to a breach (*Bernstein*). Likewise, industry guidance about best practice regarding vehicle design and cybersecurity are likely to be just as relevant here. The Secretary of State's Statement of Safety Principles, as well as the terms and conditions of the automated vehicle's authorisation will likely also be material factors for these purposes. For example, where industry guidance warns against it (see Department for Digital, Culture, Media and Sport 2018), the use of default passwords might arguably amount to a breach where cybersecurity and consequently vehicle safety is compromised. Conversely, high risk aspects of CAVs like the handover process, which are effectively an inherent feature of

[106] As Fairgrieve and Goldberg note, the concept of business sales in this context becomes more challenging where the sale is carried out as an ancillary commercial activity by organisations like charities or as a hobby by an individual. In such circumstances, an otherwise stateable claim might not succeed (Fairgrieve and Goldberg 2020, 4.44-46). See *Smith v Lazarus* (1981).

[107] Section 14(2A).

[108] Sections 3 & 4 in England and Wales; s 11 in Scotland.

[109] Sections 9–18 of the Consumer Rights Act 2015.

automated driving, are unlikely to amount to a breach under s 14 (Glassbrook et al. 2019, p. 135). Similarly, it may be possible for a seller to qualify the scope of terms implied under s 14 where they specifically draw a buyer's attention to defects (*Hurley v Dyke* [1979]).

5.2.7 Durability

Given the number and complexity of interactions between an automated vehicle's constituent parts, a greater degree of fragility can be expected. However, this alone will not afford producers greater latitude in the discharge of contractual obligations. Indeed, where parts are performing safety critical functions, courts are likely to apply a high standard in this regard. A perhaps more interesting question is how a court might approach claims for an unsatisfactory level of durability where there is evidence of planned obsolescence. It is not clear that s 14 would necessarily capture this. Particularly where a dangerous deterioration in performance was the result of a software update affecting, for example, battery life.[110] It may be the better argument in those circumstances was that designed obsolescence amounts to a repudiatory breach and an interference with the buyer's quiet possession of the goods, which would be implied either under s 12(2)(b) or at common law (Cannon et al. 2021, 20-053-4). A similar analysis applies where the supplier employs a 'time bomb' for a breach of the license terms (Cannon et al. 2021, 20-053-4; *Rubicon Computer Systems Ltd v United Paints Ltd* [1999]).

5.2.8 Standard CAV Software

As alluded to above, the question of whether CAV software constitutes goods and thus subject to implied terms of satisfactory quality and fitness for purpose under the Sale of Goods Act 1979 is not straightforward.[111] Consequently, absent legislative reform, claims by Section 5 Claimants for accidents caused by defective software under the 1979 Act are likely to be fiercely contested. Much turns on the manner in which the software is installed in the vehicle. There appears be a general consensus that software supplied on a tangible form, i.e. a disk, would be goods under the 1979 Act and thus capable of supporting a claim under the Act by a Section 5 Claimant. Conversely, software supplied by download would not appear to be goods for the purposes of the Act (Glassbrook et al. 2019; *Hervey and Lavy 2020*; *Channon* et al. 2019; *Fairgrieve and Goldberg* 2020*;* Smith 2020; Cannon et al. 2021. See *St Albans City & District Council*; *Your Response Ltd*; *Computer Associates UK Ltd*

[110] As alleged by the plaintiff in *Rasmussen v Tesla* (Case No. 5:19-cv-04596-BLF).

[111] In this context, the author is referring to standard software, as opposed to tailored standard software or bespoke software. These are likely to be produced in the context of a contract for the supply of services or goods and services, which are considered below.

v The Software Incubator Ltd [2018]). However, other decisions indicate the point is far from settled (*Thunder Air Ltd v Hilmar Hilmarsson* [2008]; *Southwark LBC v IBM UK Ltd* [2011]).[112]

In *Computer Associates UK Ltd v The Software Incubator Ltd*, the Court of Appeal concluded that a download was not a sale of goods for the purposes of the Commercial Agents (Council Directive) Regulations 1993. On appeal to the UK Supreme Court, a reference was made to the CJEU. In reaching the view that software can be goods for the purpose of the Regulations, the Court held that, "…it must be stated that software can be classified as 'goods' irrespective of whether it is supplied on a tangible medium or, as in the present case, by electronic download" (*The Software Incubator Ltd v Computer Associates (UK) Ltd* (2021). It would appear that the granting of a perpetual license *may* be a necessary condition for the supply of software to constitute the sale of goods (See also Cannon et al. 2021, 20-013; *Eurodynamic Systems Plc;* and *Southwark LBC*). However, following the decision in *St Albans City & District Council*, much of the debate on this point is likely of academic interest only. Courts are "…likely to imply contractual terms akin to the statutory implied terms of satisfactory quality and fitness for purpose in all, or virtually all, cases of software supply" (Cannon et al. 2021, 20-012). Consequently, while some bugs may reasonably be expected in tailored standard software or bespoke software (*Saphena Computing Ltd v Allied Collection Agencies Ltd* [1995]), their presence in standard CAV software may amount to a breach of ss 14(2B)(c) and (d) where the 1979 Act applies, or at common law in the alternative (Glassbrook et al. 2019, p. 135; Cannon et al. 2021, 20-084.).

5.2.9 Artificial Intelligence (AI)[113] Appliances

Supply of a CAV AI appliance comprised of hardware, software and data is likely to be treated as goods for the purposes of the 1979 Act (Lavy and Munro 2020, 6-006). Nonetheless, the same uncertainties arise in relation to the application of the 1979 Act to AI systems supplied by way of download.

While assessments of software quality and fitness for purpose will likely turn on questions of defect number or severity, adherence to accepted standards with regards user interfaces and use of resources, AI is harder to assess than traditional software (Lavy and Munro 2020, 6-006). That a CAV AI system does not appear to work well does not automatically mean it falls short of implied terms of quality or fitness for purpose. Equally, that the AI system seems to work well does not necessarily mean it "…is in fact accurate, robust, secure and reliable" (Lavy and Munro 2020, 6-007).

---

[112] In *Thunder Air Ltd*, software on hard disk was not goods for the purposes of the Torts (Interference with Goods) Act 1977. Subject to the contract, the software in *Southwark LBC* might be goods.

[113] Artificial Intelligence is a term not without misunderstanding. This chapter gratefully adopts the analysis of John Zerilli and Adrian Weller (Zerilli and Weller 2020).

Sub-optimal performance may be the result of a poor customer specification or flawed third party data sets. However, where the AI system adversely affects other aspects of a CAV operating system, the supplier may well be in breach for any failure to provide warnings or instructions for use (Fairgrieve and Goldberg 2020, 4.62). For these reasons, Lavy and Munro suggest that considerations of satisfactory quality may also require to take account of how the AI system has been built and "...the extent to which the assumptions, models and principles governing its design have been validated" (Lavy and Munro 2020, 6-008). Consequently, rather than relying upon standard terms, one can expect to see commercial entities increasingly adding carefully defined express terms on CAV AI standards (Lavy and Munro 2020, 6-009).

5.2.10 Consumer Contracts

Contracts for the sale of goods to consumers[114] are primarily governed by the Consumer Rights Act 2015. As a tangible moveable item, an automated vehicle would constitute goods for the purposes of the Act.[115] The provisions relating to satisfactory content and fitness for purpose in the 1979 Act are effectively replicated for consumer contracts for the sale of goods in Chapter 2 and for digital content in Chapter 3 of the 2015 Act,[116] and so much of the earlier analysis is equally relevant here. Where the digital content is supplied for free, Chapter 3 does not apply unless it is supplied with goods or services purchased by the consumer,[117] or it is not generally available on this basis.[118] Where a contract is for goods, digital content and/or services, it is a mixed contract, and the relevant chapters continue to apply to each element.[119] The 2015 Act thereby avoids many of the problems relating to software under the Sale of Goods Act 1979. Where supplied under a consumer contract, it is submitted that the qualities and capabilities of an AI program and its associated data would constitute digital content under the 2015 Act. Where a CAV is supplied with digital content, this will be treated as a contract for the supply of goods. Any defect in the digital content therefore has the effect of rendering the goods disconform to the contract of supply too (s 16(1) of the Consumer Rights Act 2015). It applies equally to on-premises,[120] and distance or off-premises

[114] Someone acting for purposes wholly or mainly outside their trade, business, craft or profession, per s 2(3).

[115] Section 2(8) of the Consumer Rights Act 2015.

[116] Digital content is defined as data produced and supplied in digital form, per s 2(8).

[117] Section 33(2)(a).

[118] Section 33(2)(b). However, where digital content is supplied without payment (e.g., in exchange for personal data), the related contracts and notices may still be subject to Part 2 of the 2015 Act, as well as the Consumer Protection from Unfair Trading Regulations 2008.

[119] Section 1(4).

[120] Schedule 1.

contracts.[121] Under the 2015 Act, terms purporting to exclude or limit liability arising out of injury or death, as well as under those terms governing quality, fitness for purpose and expectations, are unenforceable.[122]

5.2.11 Contracts for Services, and Goods and Services

One area developing rapidly is the provision of CAV-related services.[123] In addition to the likely extensive suite of customer agreements created at the point of supply (Knight 2021, 6.007), CAV users may then be expected to enter into any number of contracts with third party entities for CAV-related services thereafter. These might include anything from repair and maintenance (including CAV software and cyber-security), infotainment, real-time diagnostics, CAV training, to remote operators (See Law Commissions 2019, 5.43-46). Consequently, the provision of these services *ex post* sale may well create a growing pool of commercial entities who could potentially face professional liability claims by Section 5 Claimants.[124]

5.2.12 CAV Software Services: Professional Liability[125]

The preceding paragraphs on software concerned the supply of standard mass-produced software supplied on a tangible form, i.e. the vehicle, and which falls to be treated as a form of goods. In contrast, the supply of commercial CAV-software by OTA update is arguably better understood as a service (Cannon et al. 2021, 20-002-06). In time, it may be that commercial enterprises have use cases that call for bespoke or tailored standard application or system software. Similarly, notwithstanding the requirements for authorisation of an ADS feature, Tesla's roll out of FSD Beta to small groups of pre-approved consumer testers hints at a potential market for "bespoke" CAV software for private individuals. Also, vehicle adaptions for people with disabilities might include bespoke or tailored software modifications designed to meet the individual's specific needs. More generally, the writing of bespoke or modification of standard CAV software, as well as any associated CAV software services such as implementation, testing and staff training would all potentially be dealt with as contracts for professional IT services (Cannon et al. 2021, 20-002). To the extent that they arise from any breach of warranty and/or duty

[121] Schedule 2.

[122] Sections 31, 47, 57 and 65.

[123] For example, the Law Commissions note that many maintenance issues like sensor alignment, software updates and cyber security will require specialist understanding and technical skills (Law Commissions 2020, 13.99).

[124] This may be true even where the profession in question 'does not have a recognised qualification, universal code of conduct, regulatory framework or other guidelines governing the appointment or work of its professionals' (Cannon et al. 2021, 20-001).

[125] Professionals' related duties in tort and delict are discussed below.

of reasonable skill and care in the performance of those services, claims against service providers will fall within the realm of professional liability (Cannon et al. 2021, 20-001).

As with all Professionals, the scope of an IT professional's duty of care to their clients will be informed by the express and implied terms of their contractual retainer. Professionals performing these services will be expected to exercise reasonable skill and care (Cannon et al. 2021, 20-044). They will be judged against the ordinary skill and care typical of their profession. This is true regardless of whether the service is provided to a business[126] or a consumer.[127] However, the scope of the professional's duties will be more onerous where their clients are inexperienced or unsophisticated (Cannon et al. 2021, 2-032); *Carradine Properties Ltd v DJ Freeman & Co* [1999]). Moreover, it is open to a professional service provider to contract on more onerous terms (Cannon et al. 2021, 2-039-46). In any event, the nature of the risk posed by new CAV technology means that professional service providers will require to adopt a cautious approach. Whether it be an IT professional performing system maintenance on a vehicle or an external consultant advising on cloud-based services for a commercial NUIC Operator, professionals should be circumspect about what can be achieved. Where a course of action is not available due to a deficiency in the state of the art, it arguably should not be attempted at all (Cannon et al. 2021, 20-075).

Where the contract is a mixed contract for the supply of goods and services, Section 5 Claimants will need to identify whether the requirement to exercise reasonable skill and care, or more onerous terms concerning fitness for purposes and satisfactory quality, apply.[128] Where the subject matter of the contract is tailored standard software, the distinction may not always be clear (Cannon et al. 2021, 20-015; *St Albans City & District Council*). Similarly, what might ostensibly look like one contract could in fact be several, with each element subject to different standards (*Gretton & Starkey (t/a Open Systems Design) v British Millerain Co Ltd*; Cannon et al. 2021, 20-021). Moreover, the approach to standard terms for each contract may vary over the course of negotiations, which could make it difficult to establish whether a contract is subject to statutory tests of reasonableness (*Salvage Assoc v CAP Financial Services* (1995)).

Where the CAV software is provided under a contract of service, it may be that the courts will allow a greater degree of latitude with regards the presence of bugs.[129] However, while minor issues may be tolerated, where their aggregated effect is sufficiently serious to fundamentally affect the proper operation of a safety-critical

---

[126] Section 13 of the Supply of Goods and Services Act 1982.

[127] Section 49 of the Consumer Rights Act 2015.

[128] Sections 3, 4, 11, and 11C, 11D and 11L of the Supply of Goods and Services Act 1985.

[129] The authors of *Jackson & Powell on Professional Liability* suggest that "…the expression "bug free" has to be seen as meaning 'acceptably free from bugs…'" (Cannon et al. 2021, 20-037). Indeed, it is widely accepted that "…software can never be error-free" (Knight 2021, 6.401). See also *Stephenson Blake (Holdings) Ltd v Streets Heaver Ltd* [2001].

CAV system, a breach will likely be established (Cannon et al. 2021, 20-080. See also *GB Gas Holdings v Accenture (UK) Ltd* [2010]). Contracts for the supply of software services usually include an express term requiring that the supplier provide debugging/testing services for a fixed period after supply (Cannon et al. 2021, 20-028); absent same, the courts will usually allow them a reasonable period of time (*Astea (UK) Ltd v Timegroup Ltd* [2005]). Consequently, it has been suggested that any obligation that bespoke software be defect free applies only after reasonable testing, refinement and adjustment (Cannon et al. 2021, 20-086). Indeed, failure to allow testing and modification may actually be deemed a repudiatory breach of the contract (Cannon et al. 2021, 20-031). A corollary of this is that the supply of CAV software in "BETA" form may not, in and of itself, constitute a breach of the contract. Similarly, if a software provider is taking reasonable steps to address a particular bug, where that bug causes an accident it still may not amount to an actionable breach by which a Section 5 Claimant can recover its losses (Cannon et al. 2021, 20-097). Nevertheless, even allowing for some debugging, there appears to be a stateable argument that prototype or BETA software that cannot perform the function expected of it at the point of supply is not fit for purpose (*St Albans City & District Council*). Particularly where these functions are safety-critical. Likewise, terms stating that the software is supplied 'as is' or in Beta format may be treated as unfair (Reed and Edgar 2021, 3.2).

Where an employer is aware that additional training for the safe operation of the CAV by their staff is necessary but fails to provide it, they will likely be liable where their omission causes an accident. However, where an IT consultant has recommended a CAV system that, unbeknownst to the client, is outwith the reasonable capabilities of its staff, the consultant may be liable for their failure to warn of the need for extra training (*SAM Business Systems Ltd v Hedley & Co (TCC)* (2002), [83]). Conversely, a Section 5 claim may fail if the client has failed to communicate the purpose of its equipment to a consultant (*Micron Computer Systems Ltd v Wang (UK) Ltd* (1990)).

Under Recital 19 of the Consumer Rights Directive,[130] contracts for digital content not supplied on a tangible medium should be classified as neither sales contracts nor service contracts. However, where digital content supplied to a consumer damages a device or other digital content belonging to the consumer, and the damage would not have occurred but for the failure of the trader to exercise reasonable skill and care,[131] the Consumer Rights Act 2015 provides that the trader is liable *inter alia* to pay compensation.[132] The terms of supply are likely to be governed by a 'normal' contract and so the standard of care owed will presumably be such ordinary skill and care as should be expected from another trader in the same

[130] Directive 2011/83/EU, Official Journal L304 of 22.11.2011.

[131] Section 46(1). Where digital content is supplied subject to the right of the trader or third party to modify the digital content, these updates are also subject to the same terms concerning quality, fitness for purpose and description, per s 40(1).

[132] Section 46(2).

profession **para**. Liability under s 5(1) is contingent upon liability for *damage* under s 2. Since the definition of damage excludes damage to 'the automated vehicle,' a s 2 compensator would not owe a duty to compensate for damage caused by digital content to the vehicle and so no liability arises under s 5(1). Where a claimant suffers damage to their vehicle as a result of defective digital content negligently supplied, they will need to claim against the supplier directly instead of relying upon the AEVA; potentially by way of a claim under s 46(1) of the Consumer Rights Act 2015 for damage to a device or other digital content.

5.2.13 AI Services: Professional Liability

Insofar as CAV services using and relating to AI are provided by IT Professionals, the same legal analysis applies to any consideration of their professional liability. Assessments of whether the duty of reasonable care and skill has been discharged will be fact sensitive and potentially more extensive where the subject of the contract concerns AI. Lavy and Munro emphasise the need for effective AI governance, and identify several factors courts will need to consider when assessing what constitutes reasonable skill and care. These include the nature of the AI service (in the context of CAV's this may well be safety-critical); whether its role was fundamental or peripheral; whether process verification and governance—i.e., detecting and addressing aberrant system behaviour—were within the service provider's remit; the potential impact of errant or unexpected behaviour and the steps taken to mitigate them; the technical state of the art; and state of AI regulation (Lavy and Munro 2020, 6-011-12). Given the potential risk to life posed by automated vehicles, AI-related service providers can expect to be held to a high standard. Absent express exclusions limiting the scope of their retainer, courts may well find extensive and ongoing implied duties in the contract. Indeed, express exclusions or limitations may well be deemed unfair and thus unenforceable.

Where providing a service that utilises AI, the duty of reasonable skill and care extends beyond "...the underlying service being supplied... to the selection, deployment and validation of AI processes, as well as ongoing training" (Lavy and Munro 2020, 6-014). Lavy and Munro characterise them as 'non-trivial' activities requiring significant resources and expertise ((Lavy and Munro 2020, 6-014)). Their conclusion that many businesses operate AI systems without adequate oversight suggests this may be a fertile ground for Section 5 Claimants.

Where a service provider utilising AI or providing AI-related services can demonstrate it has taken satisfactory steps to discharge its duty of reasonable skill and care, and the AI's behaviour was unforeseeable and the accident unavoidable, it may well be able to resist claims by a Section 5 Claimant (Ebers 2021; Tanna and Turner 2019).[133] However, if it has contracted to perform to a specification, it will be

[133] However, see para 3.1.4.

strictly liable (Lavy and Munro, 6-015). Even then, assessing whether an AI system is performing sufficiently well is likely to be a challenging task for the courts.[134]

5.3 Negligence

Where no remedy exists under the Consumer Protection Act 1987 or in contract, Section 5 Claimants may still be able to pursue a claim in negligence. The scope of a manufacturer's duty of reasonable care expounded in *Donoghue v Stevenson* was soon extended in *Grant v Australian Knitting Mills Ltd* (1936) (Fairgrieve and Goldberg 2020, 14.01-04), and the law governing producers' negligence has developed significantly in complexity and scope in the intervening period. In addition to manufacturers, the term producer can be understood to include "...wholesalers, retailers, assemblers, repairers and those who hire products out to domestic or business user" (Fairgrieve and Goldberg 2020, 14.07).

It is uncontroversial that a manufacturer owes a duty of care in relation to the production of vehicles. For example, failure to remedy known wheel bearing faults (*Walton v British Leyland UK Ltd* (1978)), supplying a reconditioned car with insecure wheel fastenings (*Herschtal v Stewart and Arden Ltd* [1940]) and supplying tyres knowingly produced by a defective manufacturing process (*Carroll v Fearon* [1998]) have all given rise to a finding of negligence. This could include CAV software contained within the vehicle at the point of supply, and so "...there is there is no reason in principle to treat defective code differently from a wrongly designed electronic circuit or a defective camshaft." (Knight 2021, 7.19). This duty extends to all classes of person likely to suffer injury, damage or loss if the vehicle is defective. The standard of care owed for real time safety-critical software will likely be higher than other applications where the risk of harm is more remote (Knight 2021, 7.86-7). If a novel question arises as to whether a duty of care is owed, the courts will consider whether it is just and reasonable to impose a duty in the circumstances (*Caparo Industries Plc v Dickman* [1990]).[135] Given the risk posed by vehicles, the standard of care expected of vehicle producers is likely to be high (*Lunt v Khelifa*; Fairgrieve and Goldberg 2020, 14.10; Riley-Smith and McCormick 2020, 5-009.). Nonetheless, there may be no liability in relation to inherently dangerous aspects of CAV use where the user is fully aware of the risk of harm relating to use of the vehicle (*McTear v Imperial Tobacco Ltd* [2005]).

Any failure by a vehicle manufacturer to issue suitable warnings and instructions about their product may amount to negligence (Fairgrieve and Goldberg 2020,

[134] Lavy and Munro highlight AI's complexity, opacity, openness (debugging, further work on delivery, training on buyer's data, etc), autonomy, unpredictability, data reliance and vulnerability (Lavy and Munro 2020, 6-019).

[135] This includes Scotland, where *Caparo*'s "...tripartite test has now been transplanted into Scots law to determine the existence of a duty of care in novel situations not governed by existing precedent" (Thomson 2014, 3.8).

14.78-123). Likewise, failure to warn of defects discovered after supply may amount to a breach of duty (*E Hobbs (Farms) Ltd v Baxenden Chemical Co Ltd* [1992]; *United Central Bakeries Ltd v Spooner Industries Ltd* [2013]. See also *Hutton v General Motors of Canada Ltd* 2010 and *Comstock v General Motors Corporation* (1961)). This extends to any failure to initiate a recall campaign where there is evidence of a defect (*Walton v British Leyland UK Ltd* (1978)). It is not clear how the courts will view fixes carried out OTA instead of a product recall. Much will turn on whether an OTA update is sufficient, and whether a full recall was necessary to identify and address the issue. Conversely, a suitable warning by a supplier may be sufficient to discharge their duty of care (*Hurley v Dyke*; Glassbrook et al. 2019, pp. 141–146). The duty to warn extends to concealed dangers or unpredictable behaviour (Fairgrieve and Goldberg 2020, 14.91).[136] Evidencing that any failure to warn the injured party was the proximate cause of the loss will presumably prove a challenging evidential threshold to meet for Section 5 Claimants.

Manufacturers of components will likely face strict liability for their defective products (*Caroll v Fearon* (1998); Fairgrieve and Goldberg 2020, 14.36–14.37). Where the component produces flawed data that is fed directly into the vehicle ADS, the situation is conceptually no different from a CAV defect and so liability will attach on the basis of *Donoghue v Stevenson* principles (Knight 2021, at 7.34). However, where the flawed data is relied upon by a driver when operating the vehicle or, it is submitted, a UiC during a transition period, a duty of care arises by virtue of that reliance (*Hedley Byrne & Co Ltd v Heller & Partners Ltd* [1964]; See Knight 2021, 7.33 and Cannon et al. 2021, 20-057). Where the flawed data is "...directed at a specific proposed transaction or activity...," there will likely be sufficient proximity to warrant the imposition of a duty of care (Knight 2021, 7.35). Less clear is whether a duty of care arises in respect of flawed safety-critical data emanating from another vehicle. Where the producer ought reasonably to know that vehicle data transmitted to third party vehicles will be relied upon by the vehicle and/or vehicle user—for example, prompting the vehicle to make unnecessary emergency manoeuvres—this too appears to be data directed to a specific proposed activity and might create sufficient proximity for a duty of care to arise. However, such an argument would very likely turn on its own particular facts. Where a duty is held to exist, the risk of injury, damage or loss to road users is entirely foreseeable and so any duty of care owed will be onerous.

A manufacturer is also under a duty to inspect and test not only parts they have made, but also the components made by sub-manufacturers (See *Taylor v Rover Co Ltd* [1986]; Fairgrieve and Goldberg 2020, 14.38–14.40; Riley-Smith and McCormick 2020, 5-027). Given that the performance of CAV software and hardware components may vary depending on the other components they are partnered with, it would appear that this duty may well extend not just to testing parts individually but

[136]The General Product Safety Regulations 2005 also require producers to provide warnings about risks that may not be immediately obvious. 4 General Product Safety Regulations 2005 (SI 2005 no 1803), reg 7.

also in combination.[137] Indeed, it may well be that specialist integration services were utilised in the manufacturing process. More generally, to the extent they are not mandated, manufacturers may also owe a duty to install safety features where their inclusion could have prevented or reduced the level of damage ((Fairgrieve and Goldberg 2020, 14.63-72). Moreover, once industry best practice is discernible, failure to incorporate particular features, e.g. LIDAR or RADAR to the vehicle sensor suite may itself be evidence of negligence.

Suppliers of products are under a duty to take reasonable care when carrying out commercial duties (*Kubach v Hollands* [1937]; *Watson v (1) Buckley (2) Osborne, Garrett & Co Ltd and (3) Wyrovoys Products* [1940]; See also Riley-Smith and McCormick 2020, 5-027-28.). CAVs suppliers will not typically be under a duty to test the vehicles prior to resale where they are obtained from a reputable source (*Taylor v Root Co Ltd* [1966]). However, where the source is less reputable, there may be a duty to inspect (*Clarke v Army and Navy Cooperative Society Ltd* [1903]). Particularly since other hardware or software suppliers' recommendations may reflect vested interests (*Stephenson Blake (Holdings) Ltd*). It is submitted that, subject to relevant and specific warranties given by the producer, any duty to inspect may include testing the vehicle's software and cybersecurity measures (*Faisal v Younis* (t/a Safa Superstore) [2018]). This would appear to be particularly true of second-hand vehicles, where the eminence and safety of software, particularly open source code, may not always be clear.

Vehicle repairers have traditionally owed a duty of reasonable care (Fairgrieve and Goldberg 2020, 15.23-24). Software servicing is likely to become an increasingly common feature of vehicle maintenance. For obvious safety and proprietary reasons, the matter is likely to be closely controlled by CAV producers and subject to strict disclaimers. Nonetheless, where a suitably qualified repairer instals a software update from a nominated supplier that contains defective code but can demonstrate that they exercised reasonable skill and care in doing so, the repairer may avoid liability in negligence (See *G.H. Myers & Co v Brent Cross Service Co.* [1934]). However, the repairer may still be liable for negligently installing, specifying, or omitting to install a particular software update (Law Commissions 2018, 7.106; See also *Stennett v Hancock* [1939]). While the risk of serious harm is entirely foreseeable, whether there is sufficient proximity between organisations carrying out real world and simulation testing to road users will likely turn on the scope of their client's instructions. Where the terms of their retainer are sufficiently narrow, they may well be able to resist any claim against them (*Harrison v Technical Sign Co Ltd* [2013]). Particularly when further in-house testing by the manufacturer is presumably likely. Conversely, where the system under review is safety-critical and duty to inspect for defects is broadly defined, the risk of harm from any negligent failure to identify a defect is likely to create sufficient proximity to persons likely to be affected for a duty of care to arise. Insofar as the testing service gives rise to a

[137] Albeit the component manufacturer may also require to provide suitable warnings too (Fairgrieve and Goldberg 2020, 14.77).

measure of control and responsibility for a hazardous situation occurring, liability may well attach (*Perrett*).

5.3.1 Standard Software & AI Appliances

Where the software (including AI applications) is supplied with the vehicle and constitutes goods, the principles in *Donoghue v Stevenson* apply. As such, the supplier will likely be liable for any injury or property damage caused by their negligence (Cannon et al. 2021, 20-057; Riley-Smith and McCormick 2020). This would also appear to be true where the software is supplied by way of an OTA update (Cannon et al. 2021, 20-058). Where the AI involves machine learning or neural nets, the AI application may behave unpredictably. Consequently, the duty of reasonable care is likely to be more extensive than that typically required of an ordinary product. This may well include proper due diligence around the selection of the AI system's data sets, along with thorough and ongoing testing of the system outputs (Riley-Smith and McCormick 2020, 5-009-011). Riley-Smith and McCormick suggest such testing may well include: (a) outcome monitoring against a benchmark; (b) outcome monitoring against a non-machine learning model; (c) black Box testing; (d) explainability tools; (e) validation of engineer features; (f) data quality validation (Riley-Smith and McCormick 2020, 5-011).

As above, damage to the vehicle caused by the software will not give rise to a s 2 liability. However, it may still be open to the party suffering damage to pursue the software supplier directly under ordinary common law principles. Further, where a bug in the software causes damage to other software or files contained within the CAV, it has been suggested that, rather than being pure economic loss, these other files might instead be 'other property' whose negligent damage warrants compensation (Cannon et al. 2021, 20-061-3; *D&F Estates Ltd v Church Commissioners* (1988)).

Where an update results in the transfer of a virus, the supplier may, by analogy with dangerous animals, be liable in trespass in England and Wales for dissemination of the virus (Knight 2021, 7.175-9). While the risk of harm is by now self-evident, it is debatable whether a virus would be treated by the Scottish courts as having the necessary *ferae naturae* to warrant strict liability. However, on ordinary delictual principles of *culpa*, where the virus is disseminated as a result of negligence then liability will likely attach (Thomson 2014, pp. 192–193).

5.3.2 Professional Negligence: IT Professionals

In addition to their contractual obligations, CAV IT professionals are also likely to owe a concurrent duty of reasonable care to their clients not to cause physical harm or damage. This tortious duty flows from their client's reliance on their work and so CAV IT professionals will be deemed to have assumed responsibility towards the client (Cannon et al. 2021, 2-121; Keating and Wright 2020, 7-002). As the risk of personal injury and property damage is an entirely foreseeable outcome of negligence by a CAV software producer, the courts will almost certainly hold that a duty of care was owed not just to their clients but to other road users (*Donoghue v Stevenson*; Cannon et al. 2021, 20-056-7). Subject to more onerous express terms, discharge of the duty of reasonable care owed in tort or delict is likely to be on all fours with the contractual standard of reasonable skill and care. Whereas service providers are globally responsible for performance of obligations arising under contract, obligations in tort may be discharged by relying upon performance by another party (e.g. another supplier, sub-contractor or consultant) (Cannon et al. 2021, 9-051). Consequently, where reasonable care has been exercised in the selection of the other party, professionals may be able to resist claims in tort founded on a sub-contractor's negligence.

CAV IT Professionals may also be under a concurrent duty of care to prevent economic loss from their negligence, unless precluded by the contract (*Henderson v Merrett Syndicates Ltd* [1995]). As economic loss resulting from the negligently performed services is foreseeable, along with the client's reliance upon their performance of the services, it would appear fair, just and reasonable to impose a duty on CAV IT Professionals in respect of their client's economic losses (*Caparo Industries Plc*; Cannon et al. 2021, 20-045; Glassbrook et al. 2019, p. 137.). In contrast, third party economic losses would be recoverable only in limited circumstances. Namely, where there has reasonably been reliance upon a misleading statement or advice by the CAV professional (*Hedley Byrne & Co Ltd*), or where the professional such as a supplier assumes responsibility for the task (*Henderson).*[138] As such losses fall outwith the ambit of the AEVA, claims would need to be pursued by injured parties directly.

The standard of care expected will be that of an IT Professional of ordinary skill and experience exercising reasonable skill and care (Cannon et al. 2021, 20-045; Keating and Wright 2020, at 7-002. *Bolam v Friern Hospital Management Committee* [1957]; *Hunter v Hanley* [1955]).[139] Where the provision of services is governed by a 'normal' contract not calling for professional skill or judgement, the courts are unlikely to find a duty of care exists in relation to the client's economic losses (*Robinson v PE Jones (Contractors) Ltd* [2011]). Conversely, where a tradesperson working under a normal contract fails to exercise reasonable skill and care, defences premised on the more onerous *Bolam* or *Hunter* test of professional care are unlikely to succeed (*Morrisons Associated Companies Ltd v James Rome & Son* (1962)). Where the professional advising on the choice of software is not the supplier, they should be able to resist any claim resulting from a defect provided they exercised due care in its selection (Cannon et al. 2021, 20-060).

[138] Where the supply is for standard software, no such duty is likely to arise unless the supplier assumes responsibility for advising an unsophisticated client. (See the Canadian case of *Jonas & Erickson Software v Fitz-Wright Co Ltd* 1990, extracted from Cannon et al. 2021, 20-046).

[139] In certain cases, it falls to the courts to decide what the standard of care is, per *Montgomery v Lanarkshire Health Board* [2015].

5.3.3 Highways and Roads Authorities

CAV accidents caused by the failure to maintain or repair highways may give rise to a liability for the relevant authority responsible for maintaining it. In England and Wales, the duty to maintain highways arises under s 41(1) of the Highways Act 1980.[140] Where a Highway Authority fails to maintain the highway, a special defence is provided for under s 58 where the Highway Authority can demonstrate that it has exercised such care as was reasonable in all the circumstances.[141] While this maintenance duty extends to repairing potholes (*AC v Devon County Council* [2013]), it does not oblige the Highway Authority to ensure the road is in a perfect condition at all times (*Goodes v East Sussex County Council* [2000]). In Scotland, s 1(1) of the Roads (Scotland) Act 1984 provides that Local Authorities are responsible for maintaining public roads within their area.[142] However, there is no equivalent to the s 58 special defence available under the 1984 Act. In both jurisdictions, the limited resources available to relevant authorities means a hierarchical, priority-driven approach is taken to discharging their duties and, as a matter of policy, informs their standard of care. While there were significant differences historically, the two jurisdictions are now broadly aligned on the issue of public authority liability (*Mitchell v Glasgow City Council* (2009)). Consequently, while the authorities of one jurisdiction may not be binding on the courts of another, they are often likely to be highly persuasive.

Accidents resulting from sensor displacement or damage which is caused by a CAV hitting a pothole are likely to pose particular challenges. CAV's can reasonably be expected to be designed to withstand the physical demands of ordinary use. However, given the sensitivity of the onboard sensor suite, where a CAV hits a pothole with sufficient force to damage, say, the wheel rim or crack a windscreen, there appears to be a stateable argument that the user should anticipate damage elsewhere in the vehicle. It is likely that vehicle instruction manuals will contain warnings to that effect. In these circumstances, any unreasonable delay in having the vehicle inspected by a suitably qualified repairer may be sufficient to break the chain of causation.

Highway Authorities in England & Wales are also under a duty to ensure, so far as is reasonably practicable, that safe passage along a highway is not endangered by snow or ice.[143] Section 34 of the Roads (Scotland) Act 1984 imposes a duty on Scottish Roads Authorities to take such reasonable steps as they think are necessary to prevent snow and ice endangering the safe passage of pedestrians and vehicles on public roads. Questions about whether the CAV driving along icy roads would still be in its ODD aside, any unreasonable failure to grit the roads may be actionable.

[140] Per s 329(1), this includes the duty to repair.

[141] This defence is subject to various factors to which the court must have regard under s 58(2).

[142] Scottish Ministers are responsible for trunk roads.

[143] S 41(1A) of the Highways Act 1980.

Where a known hazard poses a risk to road users driving with due skill and care, a breach of duty is likely to be established (*Russell v West Sussex County Council* [2010] EWCA). Similarly, where a failure to provide warning signs renders a road feature manifestly dangerous to road users, liability is likely to attach (*McKnight v Clydeside Busses* 1999). Nonetheless, the failure to provide information concerning obvious risks, whether by way of street furniture or painted signs (*MacDonald v Aberdeenshire Council* [2013]), is not thought to give rise to any liability for failure to maintain the highway (*Gorringe v Calderdale Metropolitan Borough Council* [2004]). Where there is an obvious risk, drivers exercising due skill and care and, by implication, CAVs, will be expected to approach such road features carefully (*MacDonald v Aberdeenshire Council,* [85]). There appears to be no sound reason why this should not also apply to street infrastructure using smart technology where the duty to maintain a good look out subsists (*Gorringe v Calderdale Metropolitan Borough Council,* [93]). However, the relevant authority may be liable for accidents caused by it installing misleading signs (Reid 2016, p. 6) or introducing a hazard (*Yetkin v Mahmood* [2011]; *Gorringe v Calderdale Metropolitan Borough Council.*). It is submitted that, where a relevant road authority becomes aware that its connected infrastructure is causing or poses a foreseeable risk of harm to CAV users (e.g. through the spread of viruses or facilitating a cyber security breach) but unreasonably fails to abate the risk, a breach of duty is likely to be established. Similarly, where the street furniture itself poses a risk of injury, a failure to repair it may give rise to a liability in negligence (*Shine v Tower Hamlets London Borough Council* (2006)).[144]

Under s 39(3) of the Road Traffic Act 1988, relevant authorities in both jurisdictions are required to carry out studies into accidents arising from the use of vehicles. However, in *Gorringe v Calderdale Metropolitan Borough Council* the House of Lords held that this particular kind of public law 'target duty' was incapable of giving rise to a duty of care.[145] Nonetheless, *Gorringe* did not close the door entirely in this respect. In *Poole Borough Council v GN* (2019), having considered both *Goringe* and *Stovin v Wise,* the UK Supreme Court noted that where a public authority creates a source of danger or assumes a responsibility to protect a claimant from harm, they can owe a common law duty of care (Lord Reed, [85]). This concept of assumption of responsibility is resonant of the principles of professional negligence (Yarrow 2020; *Hedley Byrne & Co Ltd v Heller & Partners Ltd* [1964]). However, the latitude historically afforded to professionals when assessing their standard of care means that establishing a breach of duty for professional negligence (*Bolam v Friern Hospital Management Committee; Hunter v Hanley*) regarding

[144]It is also noteworthy that in the US case brought by the family of Walter Huang against Tesla, the plaintiffs also sued State of California on the basis that they were allegedly responsible for the concrete median with which Mr Huang's vehicle collided (*Sz Huang et al v. Tesla Inc. dba Tesla Motors, inc. et al.*).

[145]Nonetheless, it may be that such studies inform reports by the Secretary of State on the effectiveness of the AEVA, required by s 7(1).

street or road design will be an onerous task for Section 5 Claimants.[146] Since roads/highway authorities are unlikely to have sufficient resources or the technical expertise needed to develop the necessary infrastructure for CAV's by themselves, it reasonable to assume that much of the work will be undertaken by private companies as part of Public/Private Partnerships.[147] That being so, the roll out of CAV-related road design and infrastructure can be expected to lead to further developments in the law of professional negligence.[148]

5.3.4 Cyber Security

Given the complex and interconnected ecosystem that CAVs inhabit, the digital and physical attack surface for automated vehicles[149]—and the resulting risk of harm posed by cyber security breaches—is significant. Similarly, the sheer number of connected stakeholders in this space means the range of potential attack vectors[150] is extensive (ENISA 2019, pp. 18–26). The attack surface is further extended by vehicle-to-vehicle and vehicle-to-infrastructure interfaces (ENISA 2019, p. 5). For Section 5 Claimants, identifying that an accident was caused by a cyber security breach and the attack vector involved will be all but impossible without expert technical evidence. Similarly, in light of the novel ways in which an unauthorised user can exploit a vulnerability, accidents arising from cyber security breaches are likely to raise complex questions about causation, as well as remoteness and scope of duty. To take one example, that a hacker was only able to access a CAV system through a vulnerability in an external software update provided by a Cloud Service Provider, allowing them to target the central processing unit (CPU) controlling the vehicle's acceleration, does not mean that the vehicle manufacturer's failure to design out single points of failure should be beyond causal enquiry (Law Commissions 2022, 9.26).[151]

Suppliers of CAV-related products or services to businesses can be expected to seek to limit their contractual liability for security breaches to the fullest extent permissible under the applicable law. Much then turns on whether the other contracting party is able to negotiate (or renegotiate) more favourable terms, along

[146] This approach was adopted unsuccessfully in *Lara Lovell (née Geraghty) v Leeds City Council* [2009], [25].

[147] Albeit, it has been suggested that the introduction of automated vehicles on the roads in Great Britain is contingent upon their being able to cope with existing infrastructure (Law Commissions 2022, 3.44).

[148] In Scotland, the issue has turned on traditional common law principles of negligence (Brodie 2021, A22-011).

[149] The collective points of vulnerability that an unauthorised user can exploit, from hardware, software, applications, servers and websites, to end user devices like laptops, mobile phones and hard drives.

[150] The method by which a cyber criminal can exploit a vulnerability.

[151] Namely, a system failure that can stop the entire system working. See Koopman (2014), p. 34.

with any relevant statutory restrictions concerning clauses excluding or limiting liability. However, it is not inconceivable that in a market where a reputation for robust cyber security is a valuable commodity, products or services that offer customers very limited comfort around cyber security might start to become less palatable (Knight 2021, 6-068; Michells and Walden 2021, 3.4). In the case of standard business-to-business or business-to-consumer terms, the question is whether or not any terms limiting the supplier's liability are unfair. Where the scope of a consumer contract extends to cyber security matters, provision of the service will be subject to an express or implied term requiring reasonable skill and care in the performance of providing the service. Further, insofar as they are part of the supply chain of a regulated operator of essential services (OES), CSPs may also be indirectly regulated by the Network and Information Systems Regulations 2018 (Michells and Walden 2021, ch. 11). Where a vulnerability is contained within digital content provided to a consumer with the vehicle, this arguably render the whole vehicle disconform to terms implied into the contract under s 16(1) of the Consumer Rights Act 2015.

As above, the risk of harm posed by a cyber security breach is entirely foreseeable. Consequently, suppliers of CAV-related products or services may well owe a tortious/delictual duty of reasonable care to their customers and third parties at risk of injury or damage by cyber security breaches resulting from the supplier's negligence. For now, the UK Government has avoided legislating on cyber security matters, preferring instead to issue guidance (Channon et al. 2019, p. 50). These and other domestic guidelines are likely to inform the tortious/delictual and contractual standard of care expected in relation to cyber security (Knight 2021, 6-228. See also, Riley-Smith and McCormick 2020, 5-015.). Of particular note are BSi PAS 1885 (PAS 1885: 2018), the UK Government's *Key principles of vehicle cyber security for connected and automated vehicles*, (CCAV et al. 2017) and BS 10754-1 (BS 10754-1:2018). The *Key principles* emphasise the need for, *inter alia*: board level ownership of cyber security and security-by-design (Principle 1); product aftercare and incident response to ensure lifetime security (Principle 3); a defence-in-depth approach to system design to avoid single point failures (Principle 5); software lifetime management (Principle 6); and that systems should be resilient to attacks (Principle 8) (CCAV et al. 2017; See also Law Commissions, 13.22). Elsewhere, as Riley-Smith and McCormick note, guidance in other high-risk fields suggests that manufacturers and suppliers of CAV-related goods and services will also need to anticipate reasonably foreseeable misuse and potentially undertake post-sale cyber security surveillance.[152]

Domestic guidance will likely be supplemented by guidelines and publications issued by international bodies like UNECE (UN Regulation No 155), ENISA (ENISA 2019) and the SAE (see Channon et al. 2019, ch. 5). Indeed, industry

---

[152] They discuss the EU's Medical Device Coordination Group's *Guidance on cybersecurity for medical devices*. See (Riley-Smith and McCormick, 5-015). This is also a requirement of UN Regulation 157.

cyber-security schemes are anticipated under the GDPR and the Directive on the Security of Network and Information Systems (NIS Directive), and likely to be seen as evidence of 'the adequacy of their technical and organizational measures' (Hon et al. 2021, 5.5.1). Likewise, it may be the supplier has provided a warranty that its security policies reflect good industry practice (Knight 2021, 6.357). In the same way, where a manufacturer or software developer owes a common law duty of care, any failure in their capacity as ASDE or NUIC operator to discharge their regulatory duties in relation to cyber security could be *prima facie* evidence of negligence. Nonetheless, demonstrating that cyber security breaches resulted from negligence and/or failure to adhere to industry practice will likely be particularly difficult for Section 5 Claimants. Not least because the waters may be muddied by issues such as pre-contractual pen testing by third party vendors or, where the contract allows for it, the efficacy of cyber security audits during the performance of the contract (Hon et al. 2021, 5.5.2; Knight 2021, 6.104-113. On ALKS system auditing, see para 3.5.5 and Annex 4 Para 7 of UN Regulation 157).

Where a CAV security weakness allows an adverse actor to take control of the vehicle and who, in the exercise of that control, causes an accident, there appears to be some doubt whether an accident would be covered by a motor insurance policy. As Channon notes, as s 8(1)(a) is currently defined, it would be difficult to argue that the vehicle is driving itself (Channon et al. 2019, 5.22.). By anchoring the definition of "driving itself" to the engagement of an authorised automation feature, the Bill potentially removes that hurdle. Nevertheless, there appears to be a stateable argument that such interference breaks the chain of causation, so no s 2 liability would arise (Riley-Smith and McCormick 2020, 5-022). Similarly, absent any negligence by the insured, it is not clear how such an accident would trigger the duty to indemnify under a traditional motor insurance policy. Moreover, to take one example, where there is a risk of criminal sanctions[153] and potentially no insurance cover available, it is difficult to see how a refusal to pay a ransom leading to a catastrophic Denial-of-Service Attack is actionable. As above, this would likely raise challenging questions about causation, as well as scope and breach of duty. Consequently, there is some force to Channon and Marson's argument that a specific insurance regime is necessary for CAV cyber security breaches (Channon and Marson 2021; Glassbrook et al. 2019, ch. 8).

5.3.5 Soft Law: Industry Guidance and Standards

Given the pace of change in this space, it has been felt necessary to produce a number of non-binding or 'soft law' advisory instruments to help define and inform

[153] Depending on the circumstances, payment of a ransom could amount to an offence under s 328 of the Proceeds of Crime Act 2002 (money laundering); s 15(3) of the Terrorism Act 2000 (financing of terrorism); or s 12(1) of the Terrorist-Asset Freezing, etc. Act 2010 (making money available, directly or indirectly, to designated persons).

best practice. These include CCAV's CCAV's *Code of Practice: Automated Vehicle Trialling* and a number of Publicly Available Specifications (PAS).[154] While not directly justiciable in and of themselves, they are nonetheless material factors to which the courts may have regard when considering whether a party has discharged its duty of care.[155] More generally, failure to conform to voluntary standards that are widely accepted as establishing an industry norm will likely be evidence of defectiveness, as it is at odds with users' reasonable expectations of safety (Fairgrieve and Goldberg 2020, 10.101, 10-98-102; Fairgrieve et al. 2013). In this, they are likely to hold a similar status to the Highway Code.[156] Indeed, CCAV's *Code of Practice: Automated Vehicle Trialling* states that: "...Failure to follow the Code may be relevant to liability in any legal proceedings. Similarly, compliance with the expectations set by the Code does not guarantee immunity from liability in such circumstances." Moreover, the suite of BSi Publicly Available Specifications addressed below are the result of extensive consultation by an interdisciplinary pool of industry experts. To the extent that they are relevant to matters at large in proceedings, it is submitted that the courts are likely to find them persuasive evidence of best practice (see Riley-Smith and McCormick 2020, 5-013).

6 Conclusions

In recent years, the UK has sought to position itself at the forefront of automated vehicle research and development (HM Government 2017). To this end, there has been significant private and public investment in this space in the UK (Zenzic 2020). Eight linked, world-leading test centres for automated vehicles have been established. Testing of automated vehicles in public spaces has been sanctioned from Oxford to Edinburgh.

---

[154]BSi PAS 11281:2018 Connected automotive ecosystems. Impact on security of safety. Code of practice, 31 December 2018; PAS 1881:2020 Assuring the Safety of Automated Vehicle Trials and Testing. 29 February 2020. PAS 1881 specifies the minimum requirements for safety cases for automated vehicle trials and development testing in the UK to demonstrate activities can be undertaken safely. See also (Law Commissions 2020, 7.63–7.98); PAS 1880:2020 Guidelines for Developing and Assessing Control Systems for Automated Vehicles, 30 April 2020; Draft PAS 1884: 2021 Safety operators in automated vehicle testing and trialling – Guide; and PAS 1882:2021 Data collection and management for automated vehicle trials for the purpose of incident investigation, 29 March 2021.

[155]They may also be a relevant consideration in criminal proceedings concerning statutory offences under the Road Traffic Act 1988, along with regulatory proceedings advanced under the Corporate Manslaughter and Corporate Homicide Act 2007 and the Health and Safety at Work etc. Act 1974.

[156]Under s 28(7) of the Road Traffic Act 1988, whereby a person's failure to observe a provision of the Highway Code may be relied upon by any party to proceedings as tending to establish or negative any liability which is in question in those proceedings. See also *Wakeling v McDonagh* [2007] and *Goad v Butcher* [2011].

However, the UK's ability to compete in this space is heavily dependent on ensuring that any regulatory regime for automated vehicles is fit for purpose. The AEVA may have been 'good enough for now' but there is still work to do. While the Automated Vehicles Bill addresses some of the issues in the 2018 Act, a number of problems remain. There is more legislative work to be done if the AEVA is going to be a success but there is seemingly little appetite to revisit the Act at this time. Given the pace of change in this space, the UK runs the risk that standing still on automated vehicle legislative reform is tantamount to going backwards.

Cases

A v National Blood Authority [2001] 3 All ER 289

AC v Devon County Council [2013] EWCA Civ 418

Adams v Lancashire and Yorkshire Rly (1869) LR 4 CP 739

Airbus SAS v Generali Italia SpA (formerly INA Assistalia SpA) [2019] EWCA Civ 805

Alexandro Filippini et al v Tesla Inc, (20CV01141), Santa Barbara, California. Court papers available at: <https://www.plainsite.org/dockets/download.html?id=298295760&z=2b2f7966> Accessed 09 January 2022.

Astea (UK) Ltd v Timegroup Ltd [2005] EWHC 725.

AXA v Lord Advocate [2011] UKSC 46.

Balmer v Hayes 1950 SC 477

Barkway v South Wales Tpt. Co. [1950] A.C. 185

Bernstein v Pamson Motors (Golden Green) Ltd [1987] 2 All ER 220

Beta Computers (Europe) Ltd v Adobe Systems (Europe) Ltd 1996 SLT 604.

Binnie v Rederu Theodoro BV 1993 SC 71.

Birse Construction Ltd v Haiste [1996]

Blyth v Birmingham Waterworks Co (1856) 11 Ex 781

Bogle v McDonald's Restaurants Ltd [2002] EWHC 490 (QB).

Bolam v Friern Hospital Management Committee [1957] 1 WLR 582

Bourhill v Young [1943] AC 92

Boyle v Kodak Ltd [1969] 2 All ER 439.

Brandon v Osborne, Garrett & Co [1924] 1 KB 548

Brown and Lynn v Western S.M.T. Co, 1945 S.C. 31

Burgoyne v Phillips [1983] RTR 49

Caparo Industries Plc v Dickman [1990] 2 A.C. 605.

Carroll v Fearon [1998] PIQR P 416, The Times, 26 January 1998.

Case C-410/19 21 September 2021.

Caswell v Powell Duffryn Associated Collieries Ltd [1940] AC 152

Chaproniere v Mason (1905) 21 TLR 633.

Charlton v Fisher [2002] QB 578

Chief Constable of Staffordshire v Lees [1981] RTR 506, QBD.

Chief Constable of the West Midlands Police v Billingham [1979] 2 All ER 182

Clarke v Army and Navy Cooperative Society Ltd [1903] 1 K.B. 155; [1902] 11 WLUK 96.

Clift v Hawse, Bird and the MIB [1999] All ER (D) 1315

Comex Houlder Diving Ltd v Colne Fishing Co Ltd 1987 SC (HL) 85.

Computer Associates UK Ltd v The Software Incubator Ltd [2018] EWCA Civ 518.

Comstock v General Motors Corporation 99 NW 2d 627, 78 ALR 2d 449 (Mich, 1961).

Couchman v Hill [1947] KB 554

Cullen v Advocate General for Scotland [2016] CSOH 170

D&F Estates Ltd v Church Commissioners (1988) 41 B.L.R. 1.

Derry v Peek (1889) 14 App Cas 337, HL

Dollar Land (Cumbernauld) Ltd v CIN Properties Ltd 1998 S.C. (H.L) 90

Donoghue v Stevenson (1932) 1932 S.C. (HL) 31

Doonan v Scottish Motor Traction Co 1950 SC 136.

Duncan v Wilson, 1940 S.C. 221

E Hobbs (Farms) Ltd v Baxenden Chemical Co Ltd [1992] 1 Lloyd's Rep 55

Eagle v Chambers [2003] EWCA Civ 1107

Esso Petroleum Co Ltd v Mardon [1976] 2 QB 801.

Eurodynamics Systems Plc v General Automation Limited (unreported, 6 September 1988)

European Commission v United Kingdom [1997] All ER (EC) 481.

Faisal v Younis (t/a Safa Superstore) [2018] EWHC 1111(QB).

Fenton v Thorley & Co [1903] AC 443, HL

Forsikringsaktieselskapet Vesta v Butcher [1989] A.C. 852.

Fortress Value Recovery Fund I LLC & ors v Blue Skye Special Opportunities Fund LP (A Firm) & 18 ors [2013] EWCA Civ 367

Galoo Ltd v Bright Grahame Murray [1994] 1 W.L.R. 1360

GB Gas Holdings v Accenture (UK) Ltd [2010] EWCA Civ 912.

Geddes Ltd v Neil Johnson Health & Safety Services Ltd (2017)

Gee v DePuy International [2018] EWHC 1208 (QB)

General Cleaning Contractors Ltd v Christmas [1953] AC 180

Global Process Systems Inc and another v Syarikat Takaful Malaysia Bhd [2011] UKSC 5

Goad v Butcher [2011] EWCA Civ 158.

Goodes v East Sussex County Council [2000] 1 W.L.R.

Gorringe v Calderdale Metropolitan Borough Council [2004] 1 W.L.R. 1057

Govan Rope and Sail Co Ltd v Weir & Co (1897) 24 R. 368

Grant v Australian Knitting Mills Ltd [1936] AC 85

Gretton & Starkey (t/a Open Systems Design) v British Millerain Co Ltd [1998] 7 WLUK 570

Hamilton v Allied Domecq 2001 G.W.D.

Harrison v Hill 1932 JC 13

Harrison v Technical Sign Co Ltd [2013] EWCA Civ 1569

Hedley Byrne & Co Ltd v Heller & Partners Ltd [1964] A.C. 465.

Heilbut, Symons & Co v Buckleton [1913] A.C. 30, 50.

Henderson v Henry E Jenkins & Sons [1970] AC 282
Henderson v Merrett Syndicates Ltd [1995] 2 A.C. 145.
Herschtal v Stewart and Arden Ltd [1940] 1 KB 155.
Hopkins v Tanqueray (1845) 15 CB 130
Horace Holman Group Limited v Sherwood International Group Limited (2002) 146 SJLB 25
Human Factors 50 DOI: 10.1518/hfes.46.1.50_30392.
Hunter v Hanley [1955] SLT 213
Hurley v Dyke (1979) R.T.R. 265
Hutton v General Motors of Canada Ltd 2010 ABQB 606
Ide v ATB Sales Ltd [2008] EWCA Civ 424
Jackson v Murray [2015] UKSC 5
Jayes v IMI (Kynoch) Ltd [1984] LS Gaz R 3180, CA
Johnson v Tennant Bros Ltd, (CA) Unreported, 19 November 1954
Jonas & Erickson Software v Fitz-Wright Co Ltd (unreported) 5 October 1990, Supreme Court of British Columbia
Kim Banner v Tesla Inc., (50-2019-CA-009962-XXXX-MB) Palm Beach County, Florida
Knightley v Johns [1982] 1 W.L.R. 349
Kubach v Hollands [1937] 3 All ER 907
Lara Lovell (née Geraghty) v Leeds City Council [2009] EWHC 1145 (QB)
Lawrence v Syndic Manufacturing Co Ltd [1952] 1 All ER 935.
Les Laboratoires Servier v Apotex Inc (2014)
Lewis v The Laird Line Limited and Others 1925 S.L.T. 316.
Lexus Financial Service t/a Toyota F.S. UK Plc v Sandra Russell [2008] EWCA Civ 424
Leyland Shipping Co Ltd v Norwich Union Fire Insurance Society Ltd [1918] AC 350
Liesbosch Dredger v S.S. Edison [1933] A.C. 449
Lindesay v Lamb & Tatner [2007] EWHC 2948 (QB)
Logical Computer Ltd v Euro Carparks Ltd unreported 19 June 2001 QB
Lunt v Khelifa [2002] EWCA Civ 801
Luxor (Eastbourne) v Cooper [1941] A.C. 108
MacDonald v Aberdeenshire Council [2013] CSIH 83
Marks and Spencer Pl v BNP Paribas Securities Services Trust Co (Jersey) Ltd [2015] UKSC 72.
McKnight v Clydeside Busses 1999 S.L.T. 1167
McTear v Imperial Tobacco Ltd [2005] 2 SC 1
Micosta S.A. v Shetlands Islands Council 1986 S.L.T. 193, per Lord Ross at [198]
Micron Computer Systems Ltd v Wang (UK) Ltd (Unreported) 9 May 1990 QBD
Mitchell v Glasgow City Council 2009 S.C. (H.L.)
Montgomery v Lanarkshire Health Board [2015] UKSC 11.
Morris-Garner v One Step (Support) Ltd [2018] UKSC 20
Morrisons Associated Companies Ltd v James Rome & Son (1962) S.L.T. (Notes) 75
National Coal Board v Thomson 1959 SC 353

Neto de Sousa v Portuguese State (C-506/16) EU:C:2017:642.
Nettleship v Weston (1971) 2 QB 691
O'Donnell v Murdoch McKenzie & Co Ltd 1967 SC (HL) 63
O'Hara v Central Scottish Motor Traction Co Ltd 1941 SC 363
Oscar Chess Ltd v Williams [1957] 1 All ER 325.
Parker v South Eastern Railway (1877) 2 CPD 416
Perrett v Collins [1998] Lloyd's Rep.255
Pitts v Hunt [1990] 3 All ER 344
Poole Borough Council v GN [2019] UKSC 25
R v Morris [1972] 1 All ER 384
R v Spurge, [1961] 2 QB 205; *Hougham v Martins* (1964) 108 Sol Jo 138, Div Ct.
R v Whitely [1993] F.S.R.168 CA
R&S Pilling (t/a Phoenix Engineering) v UK Insurance Ltd [2019] UKSC 16.
Rahman v Arearose Ltd [2001] Q.B.351
Randall v Newson (1876) 45 LJQB 364
Rasmussen v Tesla (Case No. 5:19-cv-04596-BLF), California Northern District Court. Court papers available at: <https://www.plainsite.org/dockets/download.html?id=283443796&z=b61c9562> Accessed 09 January 2022
Re Williams (1887) 36 Ch. D. 573, 578.
Rees v Darlington Memorial Hospital NHS Trust [2003] UKHL 52
Rees v Saville [1983] RTR 332 CA
Reid v First Glasgow Ltd, 2003, Rep. L.R. 66
Richardson, Spence & Co v Rowntree [1894] A.C. 217
Robb v Salamis 2003 G.W.D. 33-949
Roberts v Ramsbottom [1980] 1 All ER 7
Robertson v J Sidney Smith 1999 GWD 4-214
Robinson v Harman (1848) 1 Ex. 850
Rodrigues de Andrade v Proenca Salvador (C-514/16) EU:C:2017:908
Rothwell v Chemical & Insulating Co [2007] UKHL 39
Rouse v Squires [1973] Q.B.889
Rubicon Computer Systems Ltd v United Paints Ltd [1999] 11 WLUK 408.
Russell v West Sussex County Council [2010] EWCA Civ 71.
Ruxley Electronics v Forsyth [1996] 1 A.C. 344.
Safeway Stores Ltd v Twigger (2010)
Sahin v Harvard [2017] 1 WLR 1853
Sainsbury's Supermarkets Ltd v Mastercard Inc and ors (2016)
Salvage Assoc v CAP Financial Services [1995] F.S.R.654
SAM Business Systems Ltd v Hedley & Co (TCC), unreported 19 December 2002
Saphena Computing Limited v Allied Collection Agencies Limited (1995) FSR 616
Saphena Computing Ltd v Allied Collection Agencies Ltd [1995] F.S.R. 616
Sayers v Harlow Urban District Council [1958] 2 All ER 342
Scott v London and St Katherine Docks Co (1865) 3 Hurl. & C. 596.
Searle v Wallbank [1947] A.C. 341
Shawn Hudson v Tesla et al (2018-CA-011812-O), 9th Judicial Circuit, Orange County, Florida

Shine v Tower Hamlets London Borough Council (2006) EWCA Civ 852.
Skipton Building Society v Lea Hough & Co [2000] P.N.L.R. 545.
Smith v Harris [1939] 3 All ER 960
Smith v Lazarus (unreported, 23 June 1981, CA).
Smith v Samuel Williams & Sons (1972) Times, 3 March
Southwark LBC v IBM UK Ltd [2011] EWHC 549 (TCC).
St Albans City & District Council v International Computers Ltd [1996] 4 All ER 481
St Albans DC v International Computers Ltd [1996] 4 All E.R. 481
Stavely Iron and Chemical Co ltd v Jones [1956] AC 627
Stennett v Hancock [1939] 2 All E.R. 578.
Stephenson Blake (Holdings) Ltd v Streets Heaver Ltd [2001] Lloyd's Reg.P.N.44.
Stevenson v Rodbourne LTL 5/11/98
Stovin v Wise [1996] A.C. 923
Sz Huang et al v. Tesla Inc. dba Tesla Motors, inc. et al, Superior Court of California, County of Santa Clara, Case No. 19CV346663.
Tang Man Sit (Personal Representatives of) v. Capacious Investments Ltd [1996] A.C. 514
Taylor v Root Co Ltd [1966] 2 All ER 181.
The Salvage Association v CAP Financial Services [1995] FSR 654.
The Software Incubator Ltd v Computer Associates (UK) Ltd (Case C-410/19) Judgment of the CJEU 4th Chamber 21 September 2021
Thompson v Spedding [1973] R.T.R. 312
Thomson v Brankin 1968 S.L.T. (Sh. Ct) 2
Thunder Air Ltd v Hilmar Hilmarsson [2008] EWHC 355 (Ch)
Tocci v Hankard (1966) 110 Sol Jo 835, CA.
Torreiro v AIG Europe Ltd (Case C-334/16).
Toyota motor Corporation (2013) WL5763178 (Texas).
United Central Bakeries Ltd v Spooner Industries Ltd [2013] CSOH 150.
Vnuk v Zavarovalnica Triglav [2014] CJEU C-162/13
Wakeling v McDonagh [2007] EWHC 1201 (QB)
Walton v British Leyland UK Ltd (1978) (Willis J) (unreported).
Ward v MGM Marine Ltd [2012] EWHC 4093 (QB)
Watmore v Jenkins [1962] 2 QB 572
Watson v (1) Buckley (2) Osborne, Garrett & Co Ltd and (3) Wyrovoys Products [1940] 1 All ER 174.
West Tankers Inc v Ras Riunione Adriatica Di Sicurta SpA and Generali Assicurazioni Generali SpA [2005] EWHC 454 (Comm)
Wettbewerbszentrale v Tesla (judgment of 14.07.2020, Az. 33 O 14041/19.
Whitehead v Bruce [2013] R.T.R. 25, CA
Wilkes v DePuy International Limited [2016] EWHC 3096 (QB)
Wisely v John Fulton (Plumbers) Ltd, Wadey v Surrey County Council [2000] 2 All ER 545)
Wisniewski v Manchester Central Health Authority [1998] PIQR 324
Wright v Lodge [1993] 4 All E.R. 299

Yetkin v Mahmood [2011] Q.B. 827; *Gorringe v Calderdale Metropolitan Borough Council.*
Your Response Ltd v Datateam Business Media Ltd [2014] EWCA Civ 281

References

Books

Beale HG (ed) (2021) Chitty on contracts. Sweet & Maxwell, London
Bridge MG (ed) (2021) Benjamin's sale of goods. Sweet & Maxwell, London
Brodie D (ed) (2021) Reparation: liability for delict. Sweet & Maxwell, London
Cannon M, Evans HL, Stewart R (gen eds) (2021) Jackson & Powell on professional liability. Sweet & Maxwell, London
Channon M, McCormick L, Noussia K (2019) The law and autonomous vehicles. Informa Law, Oxford
Davey F (2018) Making contracts online: old wine in smart bottles? In: Edwards L (ed) Law policy and the internet. Hart Publishing, Oxford
de Azevedo C (2019) Cyber risks insurance. Law and practice. Sweet & Maxwell, London
Fairgrieve D, Goldberg R (2020) Product liability. Oxford University Press
Glassbrook A, Northey E, Milligan S (2019) A practical guide to the law of driverless cars. Law Brief Publishing, Somerset
Hendy J, Ford M (eds) (2001) Munkman on employer's liability. Butterworths Tolley, London
Hogg M (2006) Obligations. Avizandum Publishing Ltd, Edinburgh
Hon WK, Millard C, Singh J (2021) Control, security and risk in the cloud. In: Cloud computing law. Oxford University Press, Oxford
Jones M (ed) (2020) Clerk & Lindsell on Torts. Sweet & Maxwell, London
Keating R, Wright L (2020) AI and professional liability. In: Hervey M, Lavy M (eds) The law of artificial intelligence. Sweet & Maxwell, London
Knight A (ed) (2021) Encyclopaedia of information technology law. Sweet & Maxwell, London
Lavy M, Munro I (2020) Liability for economic harm. In: Hervey M, Lavy M (eds) The law of artificial intelligence. Sweet & Maxwell, London
McBryde WW (2007) The law of contract in Scotland. W. Green & Son, Edinburgh
McKnight v Clydeside Busses 1999 S.L.T. 1167
Merkin RM (ed) (2021) Colinvaux's law of insurance. Sweet & Maxwell, London
Merkin R, Hemsworth M (2015) The law of motor insurance, 2nd edn. Sweet & Maxwell, 2015, Chapter 1
Michells JD, Walden I (2021) Cybersecurity, cloud and critical infrastructure. In: Cloud computing law. Oxford University Press, Oxford
Reed C, Edgar L (2021) Cloud computing law
Rees v Darlington Memorial Hospital NHS Trust [2003] UKHL 52
Rees v Saville [1983] RTR 332 CA
Riley-Smith T, McCormick L (2020) Liability for physical damage. In: Hervey M, Lavy M (eds) The law of artificial intelligence. Sweet & Maxwell, London
Smith G (2020) Internet law and regulation. Sweet & Maxwell, London
Stewart WJ (2004) Delict. W. Green & Son, Edinburgh
Taylor JA (ed) (1986) Bingham's motor cases. Butterworth & Co (Publishers) Ltd, London
Thomson J (2014) Delictual liability. Bloomsbury Professional, West Sussex
Walker DM (1981) The law of delict in Scotland. W. Green & Son Ltd, X

Walton CT, Armitage M, Cooper R, Hyde R, Kramer P, Todd S (eds) (2022) Charlesworth & Percy on negligence. Sweet & Maxwell, London

Welsh T (ed) (2006) Macphail's Sheriff Court Practice. W. Green & Son, Edinburgh

Zerilli J, Weller A (2020) The technology. In: Hervey M, Lavy M (eds) The law and artificial intelligence. Sweet & Maxwell, London

Articles, Reports and Other Sources

Bank of England and the Financial Conduct Authority (October 2019) Joint report on Machine Learning in UK Financial Services. Available at: <https://www.fca.org.uk/publication/research/research-note-on-machine-learning-in-uk-financial-services.pdf>. Accessed 09 Jan 2022

Baroness Sugg (13 March 2018) Letter to Baroness Randerson. Available at: http://data.parliament.uk/DepositedPapers/files/DEP2018-0264/Baroness_Sugg_-_Baroness_Randerson_AEV_Bill_2nd_reading.pdf

BBC News (28 April 2018) Tesla driver banned for M1 autopilot seat-switch. Available at: <Tesla driver banned for M1 autopilot seat-switch - BBC News>. Accessed 09 Jan 2022

Boon F (2020) Two bites of a peculiar cherry? *Res judicata*, time bar and illiquid debts: insurer recoveries under the Automated and Electric Vehicles Act 2018. Br Insur Law Assoc J, No. 133

BSI (2022) CAV resources. Standards, guidelines, research and viewpoints. Available at: <CAV resources - Connected and automated vehicles | BSI (bsigroup.com)>. Accessed 09 Jan 2022

BSi 10754-1:2018, Information technology. Systems trustworthiness - governance and management specification. Available at: <https://shop.bsigroup.com/products/information-technology-systems-trustworthiness-governance-and-management-specification/standard>. Accessed 09 Jan 2022

BSI Connected and automated vehicles – Vocabulary BSI Flex 1890 v3.0:2020-10

BSi PAS 11281:2018 Connected automotive ecosystems. Impact on security of safety. Code of practice, 31 December 2018. Available at: <https://shop.bsigroup.com/products/connected-automotive-ecosystems-impact-of-security-on-safety-code-of-practice/standard>. Accessed 09 Jan 2022

Burgoyne v Phillips [1983] RTR 49

Centre for Connected and Autonomous Vehicles (July 2016) Pathway to Driverless Cars: Proposals to support advanced driver assistance systems and automated vehicle technologies, p 8. Available at: <https://assets.publishing.service.gov.uk/government/uploads/system/uploads/attachment_data/file/536365/driverless-cars-proposals-for-adas-and_avts.pdf>. Accessed 09 Jan 2022

Centre for Connected and Autonomous Vehicles (August 2020) Safe Use of Automated Lane Keeping System (ALKS). Call for Evidence. Available at <Safe Use of Automated Lane Keeping System (ALKS): Call for evidence (publishing.service.gov.uk)>. Accessed 09 Jan 2022

Centre for Connected and Autonomous Vehicles (28 April 2021a) Annex C: sample list of Automated Vehicles (AVs) for use in Great Britain. Available at: <https://assets.publishing.service.gov.uk/government/uploads/system/uploads/attachment_data/file/979495/annex-c-sample-list-of-automated-vehicles-avs-for-use-in-great-britain.pdf>. Accessed 09 Jan 2022

Centre for Connected and Autonomous Vehicles (April 2021b) Safe Use of Automated Lane Keeping System (ALKS): Summary of Responses and Next Steps. Available at: <https://assets.publishing.service.gov.uk/government/uploads/system/uploads/attachment_data/file/980742/safe-use-of-automated-lane-keeping-system-alks-summary-of-responses-and-next-steps.pdf>. Accessed 09 Jan 2022

Centre for Connected and Autonomous Vehicles and Department for Transport (28 April 2022) Self-driving vehicles listed for use in Great Britain. Available at: <https://www.gov.uk/guidance/self-driving-vehicles-listed-for-use-in-great-britain>

Centre for Connected and Autonomous Vehicles, Department for Business, Energy and Industrial Strategy, and the Department for Transport (06 February 2019) Code of Practice: Automated Vehicle Trialling. Available at: <https://www.gov.uk/government/publications/trialling-automated-vehicle-technologies-in-public/code-of-practice-automated-vehicle-trialling>. Accessed 09 Jan 2022

Centre for Connected and Autonomous Vehicles, The Centre for the Protection of National Infrastructure and the Department for Transport (06 August 2017) The key principles of vehicle cyber security for connected and automated vehicles. Available at: <https://www.gov.uk/government/publications/principles-of-cyber-security-for-connected-and-automated-vehicles/the-key-principles-of-vehicle-cyber-security-for-connected-and-automated-vehicles>. Accessed 09 Jan 2022

Centre for Connected and Autonomous Vehicles, The Centre for the Protection of National Infrastructure and the Department for Transport (06 August 2017) The key principles of vehicle cyber security for connected and automated vehicles. Available at: <The key principles of vehicle cyber security for connected and automated vehicles - GOV.UK (www.gov.uk)>. Accessed 09 Jan 2022

Channon M, Marson J (2021) The liability for cybersecurity breaches of connected and autonomous vehicles. Computer Law and Security Review, vol 43

Connected and Autonomous Vehicles. Position Paper, by Society of Motor Manufacturers and Traders, p 5, February 2017. https://www.smmt.co.uk/wp-content/uploads/sites/2/SMMT-CAV-position-paper-final.pdf

Davey J (October 2017) Automated and Electric Vehicles Bill. UK Parliament: Written answers, 01 November, Session 2017-19. Available at: <https://publications.parliament.uk/pa/cm201719/cmpublic/automated/memo/aevb04.htm>. Accessed 09 Jan 2022

Department for Digital, Culture, Media & Sport (October 2018) Code of Practice for Consumer IoT Security. Available at: <https://assets.publishing.service.gov.uk/government/uploads/system/uploads/attachment_data/file/971440/Code_of_Practice_for_Consumer_IoT_Security_October_2018_V2.pdf>. Accessed 09 Jan 2022

Department for Transport (February 2015) The Pathway to Driverless Cars: a detailed review of regulations for automated vehicle technologies. Available at: <https://assets.publishing.service.gov.uk/government/uploads/system/uploads/attachment_data/file/401565/pathway-driverless cars-main.pdf>. Accessed 09 Jan 2022

Department for Transport (28 April 2021) Rules on safe use of automated vehicles on GB Roads. Available at: <https://www.gov.uk/government/consultations/safe-use-rules-for-automated-vehicles-av/rules-on-safe-use-of-automated-vehicles-on-gb-roads>. Accessed 09 Jan 2022

Department for Transport (25 April 2022a) Consultation outcome: rules on the safe use of automated vehicles: summary of responses and government response. Available at: <https://www.gov.uk/government/consultations/safe-use-rules-for-automated-vehicles-av/outcome/rules-on-the-safe-use-of-automated-vehicles-summary-of-responses-and-government-response#draft-amendment-to-the-highway-code%2D%2Da-new-section-for-self-driving-vehicles>. Accessed 13 June 2022

Department for Transport (2022b) The Highway Code. Available at: <The Highway Code - Guidance - GOV.UK (www.gov.uk)>. Accessed 09 Jan 2022

Directive (EU) 2016/1148 of the European Parliament and of the Council of 6 July 2016 concerning measures for a high common level of security of network and information systems across the Union (2016) OJ L 194

Directive 1999/34/EC of the European Parliament and of the Council of 10 May 1999 amending Council Dir 85/374/EEC on the approximation of laws, regulations and administrative provisions of the Member States concerning liability for defective products [1999] OJ L141/20

Directive 2009/24/EC of the European Parliament and of the Council of 23 April 2009 on the legal protection of computer programs [2009] OJ L111/22

Directive 85/374/EEC of 25 July 1985 on the approximation of the laws, regulations and administrative provisions of the Member States concerning liability for defective products (the Product Liability Directive) [1985] OJ L210/29

Directive 86/653/EEC of 18 December 1986 on the coordination of the laws of the Member States relating to self-employed commercial agents [1986] OJ L382/17

Dixon L (May 2020) Autonowashing: The Greenwashing of Vehicle Automation. Transportation Research Interdisciplinary Perspectives, vol 5

Draft PAS 1884: 2021 Safety operators in automated vehicle testing and trialling – Guide. Available at: <https://standardsdevelopment.bsigroup.com/projects/2020-02212#/section>. Accessed 09 Jan 2022

Duane Morris LLP (04 December 2018) A Closer Look at Tesla's Open-Source Patent Pledge. Available at: <https://www.lexology.com/library/detail.aspx?g=ca6c332f-2cc5-401b-b80d-36473d0754c7>. Accessed 09 Jan 2022

Ebers M (April 2021) Liability for Artificial Intelligence and EU Consumer Law. J Intellect Prop Inf Technol E-Commerce Law, Issue 35

EC Expert Group on Liability and New Technologies – New Technologies Formation (21 November 2019) Liability for Artificial Intelligence and other emerging technologies. Available at: <https://op.europa.eu/en/publication-detail/-/publication/1c5e30be-1197-11ea-8c1f-01aa75ed71a1/language-en>. Accessed 09 Jan 2022

ENISA (2019) Good practices for security of Smart Cars. Available at: <https://www.enisa.europa.eu/publications/smart-cars/@@download/fullReport>. Accessed 09 Jan 2022

EU Medical Device Coordination Group (December 2019) Guidance on cybersecurity for medical devices. Available at: <https://ec.europa.eu/health/sites/default/files/md_sector/docs/md_cybersecurity_en.pdf>. Accessed 09 Jan 2022

European Union, European Atomic Energy Community and the United Kingdom (2020) Trade and Cooperation Agreement Between The European Union and The European Atomic Energy Community, of the One Part, and The United Kingdom of Great Britain and Northern Ireland, of The Other Part, Official Journal L444/14

Fairgrieve D, Howells G, Pilgerstorfer M (2013) The Product Liability Directive: time to get soft. J Eur Tort Law 4(1):1–33. https://doi.org/10.1515/jetl-2013-0001

HM Government (2017) Industrial Strategy: Building a Britain Fit for the Future. Available at: <https://assets.publishing.service.gov.uk/government/uploads/system/uploads/attachment_data/file/664563/industrial-strategy-white-paper-web-ready-version.pdf>. Accessed 09 Jan 2022

Hogg M (2011) Developing causal doctrine. In: Goldberg R (ed) Perspectives in causation. Hart Publishing, Oxford

Hogg M (2015) Saying what we mean: fundamental structural language in contract law. In: DiMatteo LA, Hogg M (eds) Comparative contract law: British and American perspectives. Oxford University Press, Oxford

Johnson A (2021) Contribution claims against third parties under the Automated and Electric Vehicles Act 2018. Available at: https://aevlaw.com/blog/contribution-claims-against-third-parties-under-the-automated-and-electric-vehicles-act-2018/. Accessed 09 Jan 2022

Kinnear N et al, "Safe performance of other activities in conditionally automated vehicles" (TRL, December 2020), published as Appendix A to Department for Transport, Safe Use of Automated Lane Keeping System (ALKS): Summary of Reponses and Next Steps (April 2021). Available at: <https://www.gov.uk/government/consultations/safe-use-of-automated-lane-keeping-system-on-gbmotorways-call-for-evidence>. Accessed 09 Jan 2022

Koopman P (2014) A case study of Toyota unintended acceleration and software safety. Available at: https://users.ece.cmu.edu/~koopman/pubs/koopman14_toyota_ua_slides.pdf. Accessed 09 Jan 2022

Lambert F (14 November 2019) Tesla Model 3 Performance's 0-60 mph acceleration dips below 3 seconds after software update. Available at: <https://electrek.co/2019/11/14/tesla-model-3-performance-0-60-mph-acceleration-dips-below-3-sec-software-update/>. Accessed 09 Jan 2022

Law Commission of England & Wales, and the Scottish Law Commission (08 November 2018) Automated vehicles – a joint preliminary consultation paper. Available at: https://s3-eu-west-2.amazonaws.com/lawcom-prod-storage-11jsxou24uy7q/uploads/2018/11/6.5066_LC_AV-Consultation-Paper-5-November_061118_WEB-1.pdf. Accessed 09 Jan 2022

Law Commission of England & Wales (November 2021) Smart Legal Contracts. Advice to Government. Available at: <https://s3-eu-west-2.amazonaws.com/lawcom-prod-storage-11jsxou24uy7q/uploads/2021/11/Smart-legal-contracts-accessible.pdf>. Accessed 09 Jan 2022

Law Commission of England & Wales, and the Scottish Law Commission (25 January 2022) Automated vehicles: joint report background paper A. Available at: <https://s3-eu-west-2.amazonaws.com/lawcom-prod-storage-11jsxou24uy7q/uploads/2022/01/Background-papers-24-01-22.pdf>. Accessed 01 Mar 2022

Law Commission of England & Wales, and the Scottish Law Commission (18 December 2020) Automated Vehicles: Consultation Paper 3 - a regulatory framework for automated vehicles. Available at: <https://s3-eu-west-2.amazonaws.com/lawcom-prod-storage-11jsxou24uy7q/uploads/2021/01/AV-CP3.pdf>. Accessed 09 Jan 2022

Law Commission of England & Wales, and the Scottish Law Commission (25 January 2022) Automated vehicles: joint report. Available at <https://s3-eu-west-2.amazonaws.com/lawcom-prod-storage-11jsxou24uy7q/uploads/2022/01/Automated-vehicles-joint-report-cvr-03-02-22.pdf>. Accessed 01 Mar 2022

Law Commission of England and Wales (1996) *Privity Of Contract: Contracts For The Benefit Of Third Parties*, July 1996. Available at: <https://s3-eu-west-2.amazonaws.com/lawcom-prod-storage-11jsxou24uy7q/uploads/2021/11/Smart-legal-contracts-accessible.pdf>. Accessed 09 Jan 2022

Law Commission of England & Wales (2023) Remote driving: advice to government. Available at: <Automated Vehicles Consultation Paper 2 on Passenger Services and Public Transport> Accessed 01 Jan 2024

Law Commission of England and Wales, and the Scottish Law Commission (16 October 2019) Automated Vehicles: Consultation Paper 2 on Passenger Services and Public Transport: a joint Consultation Paper. Available at: <https://s3-eu-west-2.amazonaws.com/lawcom-prod-storage-11jsxou24uy7q/uploads/2019/10/Automated-Vehicles-Consultation-Paper-final.pdf>. Accessed 09 Jan 2022

Lee J, See K (2004) Trust in automation: designing for appropriate reliance. Hum Factors 46(1):50. https://doi.org/10.1518/hfes.46.1.50_30392

Loveday S (19 March 2021) Anxiety-inducing drone footage of tesla FSD beta unprotected lefts. Available at: <Anxiety-Inducing Drone Footage Of Tesla FSD Beta Unprotected Lefts (insideevs.com)>. Accessed 09 Jan 2022

Mayer Brown (16 September 2021) Tech Talks Podcast, Cloud + Strategic Partnership: The Evolution of Cloud Services. Available at: <https://urldefense.com/v3/__http://cld3097web.audiovideoweb.com/il80web20029/2021-QTR4/Mailing-210923-CHI-PODCAST-TT-Cloud.mp3__;!!FVajH6CFHhmYMuQ!tE2kLm8v4rOn3WHIp6z-RmjEzTez016S0QoJ7YJMEBiEhRUukzc0qvCtaZl3Nlrw2zC4$>. Accessed 09 Jan 2022

McNaughton D (2003) When is a liability not a liability? Third party procedure: traps for the unwary. Scots Law Times

Ministry of Justice (01 October 2020) Civil Procedure Rules, Part 31: Practice Direction 31B. Available at: <https://www.justice.gov.uk/courts/procedure-rules/civil/rules/part31/pd_part31b>. Accessed 09 Jan 2022

Motor Insurers Bureau (2015) Uninsured Drivers Agreement 2015. Available at: <https://www.mib.org.uk/media/166917/2015-uninsured-drivers-agreement-england-scotland-wales.pdf>. Accessed 09 Jan 2022

O'Kane S (02 June 2018) Tesla can change so much with over-the-air updates that it's messing with some owners' heads. Available at: <https://www.theverge.com/2018/6/2/17413732/tesla-over-the-air-software-updates-brakes>. Accessed 09 Jan 2022

Office for Product Safety & Standards (2021) UK Product Safety Review. Call for Evidence. Available at: https://assets.publishing.service.gov.uk/government/uploads/system/uploads/attachment_data/file/1035916/uk-product-safety-review-call-for-evidence2.pdf

Parasuraman R, Manzey D (2010) Complacency and bias in human use of automation: an attentional integration. Hum Factors 53(3):381. https://doi.org/10.1177/0018720810376055

PAS 1880:2020 Guidelines for Developing and Assessing Control Systems for Automated Vehicles, 30 April 2020. Available at: https://www.bsigroup.com/globalassets/localfiles/en-gb/cav/pas-1880%2D%2D-guidelines-for-developing-and-assessing-control-systems-for-automated-vehicles.pdf. Accessed 09 Jan 2022

PAS 1881:2020 Assuring the Safety of Automated Vehicle Trials and Testing. Available at: <https://shop.bsigroup.com/products/assuring-the-safety-of-automated-vehicle-trials-and-test ing-specification/standard>. Accessed 09 Jan 2022

PAS 1882:2021 Data collection and management for automated vehicle trials for the purpose of incident investigation, 29 March 2021. Available at: <https://standardsdevelopment.bsigroup.com/projects/2019-03951#/section> 09 January 2022

PAS 1885: 2018. The fundamental principles of automotive cyber security – specification. Available at: <https://shop.bsigroup.com/products/the-fundamental-principles-of-automotive-cyber-security-specification/standard>. Accessed 09 Jan 2022

Regulation 2019/2144 on type-approval requirements as regards general safety and the protection of vehicle occupants and vulnerable road users (Revised General Safety Regulation 2019/2144)

Reid E (2016) Different and yet the same? Delictual liability of roads authorities in Scotland and in England. Juridical Rev 2016(1):6

Singapore Academy of Law (September 2020) Report on the Attribution of Civil Liability for Accidents Involving Autonomous Cars. Available at: <https://www.sal.org.sg/sites/default/files/SAL-LawReform-Pdf/2020-09/2020_Report%20on%20the%20Attribution%20of%20Civil%20Liability%20for%20Accidents%20Involving%20Autonomous%20Cars.pdf>. Accessed 09 Jan 2022

Skipton Building Society v Lea Hough & Co [2000] P.N.L.R. 545

Society for Automotive Engineers International (SAE) (2020) 'Level 2+': making automated driving profitable, mainstream. Available at: <https://www.sae.org/news/2020/12/rise-of-sae-level-2>. Accessed 09 Jan 2022

Society for Automotive Engineers International (SAE) (April 2021) J3016 Taxonomy and Definitions for Terms Related SAE Taxonomy J3016 to Driving Automation Systems for On-Road Motor Vehicles (SAE Taxonomy J3016)

Stevens A, Knight I, Reed N (03 February 2021) The benefits and risks of introducing ALKS. Available at: <https://urldefense.com/v3/__https://www.iten.global/alks-benefits-and-risks/__;!!FVajH6CFHhmYMuQ!oi8GQfs_t1Zucr9wRl6JNw_QDby8IJgPRB9pJsDalQC1-SjA1vpoz2RflUdBTRF_7xnH$>. Accessed 09 Jan 2022

Tanna M, Turner J (27 September 2019) AI-powered investments: who (if anyone) is liable when it goes wrong? Tyndaris v VWM. <http://disputeresolutionblog.practicallaw.com/ai-powered-investments-who-if-anyone-is-liable-when-it-goes-wrong-tyndaris-v-vwm/>. Accessed 09 Jan 2022

Tesla (12 June 2014) Patent Pledge. Available at: <https://www.tesla.com/about/legal#patent-pledge>. Accessed 09 Jan 2022

Tesla (2022) Autopilot and Full Self-Driving Capability. Available at: <https://www.tesla.com/support/autopilot>. Accessed 09 Jan 2022

Thatcham Research and the Association of British Insurers (September 2019) Defining Safe Automated Driving: Insurer Requirements for Highway Automation. Available at: <https://www.abi.org.uk/globalassets/files/publications/public/motor/2019/defining-safe-automation-technical-document-aug-2019.pdf>

The Autonocast (27 May 2021) 'Aine Denari of ZF on the Rise of ADAS,' #191. Available at: <#191: Aine Denari of ZF on the rise of ADAS — Autonocast>. Accessed 09 Jan 2022

Thorneloe D (05 January 2021) Retained EU law in the UK after Brexit. Available at: <https://www.pinsentmasons.com/out-law/guides/retained-eu-law-uk-after-brexit>. Accessed 09 Jan 2022

Transport Research and Innovation Monitoring and Information System Digest (2020) Leveraging blockchain for open Mobility-as-a-Service Ecosystems. Issue 14, May. Available at <https://trimis.ec.europa.eu/sites/default/files/documents/article_3_-_leveraging_blockchain_.pdf>. Accessed 09 Jan 2022

Transport Scotland (December 2019) A CAV Roadmap for Scotland. <https://www.transport.gov.scot/transport-network/roads/connected-and-autonomous-vehicles-cav/a-cav-roadmap-for-scotland/>. Accessed 09 Jan 2022 (Roadmap for Scotland)

UK Government (October 2022) EM on the repeal of EU directive 85/374 (13134/22, COM(2022) 495). Available at: <EM on the repeal of EU directive 85/374 (13134/22, COM(2022)495) - GOV.UK (www.gov.uk)> (06 February 2023)

UK Government (2022) Queen's Speech 2022: background briefing notes. Available at: <https://assets.publishing.service.gov.uk/government/uploads/system/uploads/attachment_data/file/1074113/Lobby_Pack_10_May_2022.pdf>. Accessed 13 June 2022

UK Government Press Release (21 February 2021) Government announces plan to scrap EU law, ensuring British drivers avoid £50 a year insurance hike. Available at: <https://www.gov.uk/government/news/government-announces-plan-to-scrap-eu-law-ensuring-british-drivers-avoid-50-a-year-insurance-hike>. Accessed 09 Jan 2022

UK Government Press Release (25 May 2016) Driverless cars are the future, not science fiction. Available at: <https://www.gov.uk/government/speeches/driverless-cars-are-the-future-not-science-fiction>. Accessed 09 Jan 2022

UK Government Press Release (04 September 2019) New system to ensure safety of self-driving vehicles ahead of their sale. Available at: <https://www.gov.uk/government/news/new-system-to-ensure-safety-of-self-driving-vehicles-ahead-of-their-sale>. Accessed 09 Jan 2022

UK Government Press Release (28 April 2021) Government paves the way for self-driving vehicles on UK roads. Available at: <https://www.gov.uk/government/news/government-paves-the-way-for-self-driving-vehicles-on-uk-roads>. Accessed 09 Jan 2022

UN Regulation 157 on uniform provision concerning the approval of vehicles with regards to Automated lane Keeping System ECE/TRANS/WP.29/2020/81 (25 June 2020) (ALKS Regulation)

UN Regulation 157 on uniform provisions concerning the approval of vehicles with regard to Automated Lane Keeping Systems ECE/TRANS/WP.29/2020/81 (25 June 2020)

UN Regulation No 155 – Uniform provisions concerning the approval of vehicles with regards to cybersecurity and cybersecurity management system UN Regulation on uniform provisions concerning the approval of vehicles with regard to cyber security and of cybersecurity management systems, ECE/TRANS/WP.29/2020/79 (Cybersecurity Regulation). Available at: <https://unece.org/fileadmin/DAM/trans/doc/2020/wp29grva/ECE-TRANS-WP29-2020-079-Revised.pdf>. Accessed 09 Jan 2022

UN Regulation on uniform provisions concerning the approval of vehicles with regard to software update processes and of software update management systems ECE/TRANS/WP.29/2020/80 (Software Update Regulation). https://unece.org/fileadmin/DAM/trans/doc/2020/wp29/ECE-TRANS-WP29-2020-080e.pdf

UN Regulation on uniform provisions concerning the approval of vehicles with regard to cyber security and of cybersecurity management systems, ECE/TRANS/WP.29/2020/79 (Cybersecurity Regulation): <https://www.unece.org/fileadmin/DAM/trans/doc/2020/wp29/ECETRANS-WP29-2020-079e.pdf>. Accessed 09 Jan 2022

UNECE (2020) Amendment proposal to Article 34 in the 1968 Convention on Road Traffic ECE/TRANS/WP.1/2020/1/Rev.1, section I. <https://unece.org/fileadmin/DAM/trans/doc/2020/wp1/ECETRANS-WP1-2020-1-_Rev1e_.pdf>. Accessed 09 Jan 2022

United Nations Economic Commission for Europe (UNECE) Global Forum for Road Traffic Safety (WP1), proposed resolution on safety considerations for activities other than driving undertaken by the driver when the automated driving system is exercising dynamic control (July 2021) ECE/TRANS/WP.1/2021/2

Yarrow T (November 2020) Road design claims: non feasance, misfeasance and the *Bolam* test. 1 Chancery Lane Personal Injury Briefing. Available at: https://1chancerylane.com/wp-content/uploads/2020/11/Highways-Briefing.pdf

Zeeb K, Buchner A, Schrauf M (2016) Is take-over time all matters? The impact of visual-cognitive load on driver take-over quality after conditionally automated driving. Accid Anal Prev 92:230. https://doi.org/10.1016/j.aap.2016.04.002

Zenzic (2020) UK Connected and Automated Mobility Roadmap to 2030. https://zenzic.io/content/uploads/2020/10/Zenzic_Roadmap_Report_v3.pdf

Book Project on Self-driving Cars

Questionnaire

1. **What is the legal system you discussed in your chapter?**

   ____

   ____

2. **In your legal system, there is:**

   □ a specific legal regulation of civil liability for semi-autonomous and/or autonomous vehicles
   □ the amendment of an existing legislation to be applied also to semi-autonomous and/or autonomous vehicles for civil liability or testing.
   □ the legislation has never been amended to be applied to semi-autonomous vehicles and/or autonomous vehicles as well

3. **If one or more of the previous statements apply in relation to your legal system, please indicate the title and/or reference of the new law (or amendment of the existing law) and attach the legislative text currently in force.**

   ____

   ____

   Attachment

4. **In the event of an accident caused by a conventional vehicle (up to SAE level 2), who is usually held responsible in your legal system?**

   □ As a rule, the driver and/or the owner/keeper/equivalent subjects, except in cases of a defective product in which the manufacturer is liable
   □ As a rule, the manufacturer, except in cases of improper driving by the driver or incorrect maintenance of the owner/keeper/equivalent subjects).

H. Steege et al. (eds.), *Autonomous Vehicles and Civil Liability in a Global Perspective*, Data Science, Machine Intelligence, and Law 3,
https://doi.org/10.1007/978-3-031-41992-8

5. **In the event of an accident caused by a conditionally and highly automated vehicle (SAE levels 3 and 4), who is usually held responsible in your legal system?**

   □ As a rule, the driver-operator and/or the owner/keeper/equivalent subjects, except in cases of a defective product in which the manufacturer is liable
   □ As a rule, the manufacturer, except in cases of improper driving/operation by the driver-operator or incorrect maintenance of the owner/keeper/equivalent subjects.

6. **In the event of an accident caused by a fully automated vehicle (SAE level 5), who is usually held responsible in your legal system?**

   □ As a rule, the vehicle user and/or the owner/keeper/equivalent subjects, except in cases of defective product in which the manufacturer is liable
   □ As a rule, the manufacturer, except in cases of improper driving/operation/use by the vehicle user or incorrect maintenance of the owner/keeper/equivalent subjects).

The manufacturer's authorised representative in the EU is Springer Nature Customer Service Centre GmbH, Europaplatz 3, 69115 Heidelberg, Germany. If you have any concerns regarding our products, please contact ProductSafety@springernature.com

Printed and bound by CPI Group (UK) Ltd, Croydon, CR0 4YY
30/09/2026
02225314-0018